Visual and Computational
Plasma Physics

Visual and Computational Plasma Physics

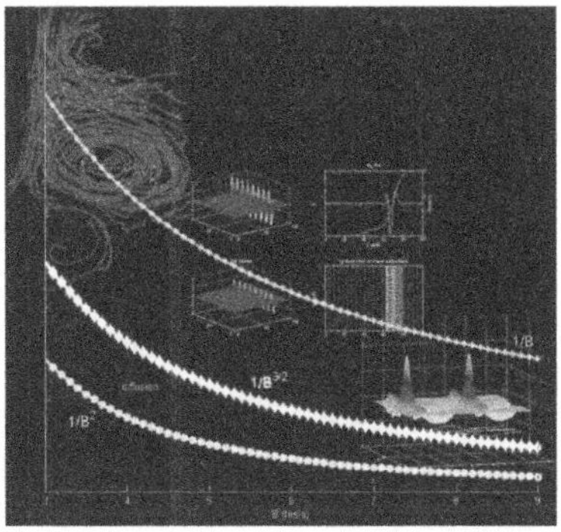

James J Y Hsu
National Cheng Kung University, Taiwan

World Scientific

NEW JERSEY • LONDON • SINGAPORE • BEIJING • SHANGHAI • HONG KONG • TAIPEI • CHENNAI

Published by

World Scientific Publishing Co. Pte. Ltd.
5 Toh Tuck Link, Singapore 596224
USA office: 27 Warren Street, Suite 401-402, Hackensack, NJ 07601
UK office: 57 Shelton Street, Covent Garden, London WC2H 9HE

Library of Congress Cataloging-in-Publication Data
Hsu, Jang-Yu, author.
Visual and computational plasma physics / James J.Y. Hsu (National Cheng Kung University, Taiwan).
pages cm
Includes bibliographical references and index.
ISBN 978-9814619516 (alk. paper)
1. Plasma (Ionized gases) 2. Mathematical physics. I. Title.
QC718.H75 2014
530.4'40151--dc23
2014031061

British Library Cataloguing-in-Publication Data
A catalogue record for this book is available from the British Library.

In-house Editor: Rhaimie Wahap

Typeset by Stallion Press
Email: enquiries@stallionpress.com

Preface

This book is derived from the courses of Plasma Physics and Computational Physics I taught at the National Cheng Kung University since 2006. Programs in this book are mainly written in MATLAB to demonstrate the numerical algorithms, the analytical approaches, and the physical principles. The MATLAB basics and some Object Oriented Programming concepts are introduced to prepare students for writing their own codes.

We start with single particle, single fluid, and single wave, then the kinetic theory, the transport, the magnetohydrodynamics, and the nonlinear physics. The book emphasizes on the numerical algorithm and the analytical asymptology to tackle problems in plasma physics, and to demonstrate the underlying physics principles by graphical visualization. Students are introduced to the multiple time and multiple space scales as they learn the basic plasma phenomena, and are requested to solve the problems with either MATLAB or C++.

This book aims its audience at the senior and graduate level. The emphasis of this book is to teach students to solve problems from the features and characteristics of the problem itself, and not from a presumed methodology or a predefined tool. It tries to avoid the students from falling into the mind frame of what the old saying said best, "If you are a hammer, everything else is a nail." *The rightful problem solving mentality is let the problem reveal where the solution might be, and study the clues to find the answers.* Therefore, start from the asymptotic analysis once the problem is translated into a mathematical equation, and get all the hints possible even if a numerical solution is inevitable.

There is homework along the teaching material in each Chapter. The students are urged to work out the solutions on their own. It is important to understand that *the most important learning is not knowing the solution, but knowing how to figure out the solution.*

許正餘
James J. Y. Hsu
July 2014

Contents

Chapter One

A Challenging Physics Discipline

"The most exciting phrase to hear in science, the one that heralds new discoveries, is not 'Eureka!' (I found it!), but 'That's funny...'"

Isaac Asimov (1920–1992)

Among the physics disciplines, plasma physics is younger than relativity and quantum mechanics. Since the 1950s, evidence of solar energy being derived from nuclear fusion attracted great talents to explore plasma physics in search for an inexhaustible energy source. Thus far, plasma physics has achieved certain maturity. But challenges are great and many, and will remain so for a long time to come. Yet, the fruits of labor can be sweet and rewarding, since the advancement in plasma physics will have great impact on important areas such as controlled fusion, astrophysics, space and earth physics, nanobio and industrial applications, and not the least, the computational physics.

Plasma is a common term to describe an ionized gas of ions and electrons. It is usually referred to as the fourth state of matter. It was recognized that the earth has a plasma roof, viz., the ionosphere, which is a layer of partially ionized gas in the upper atmosphere. It protects us from the cosmic rays and solar winds, and reflects radio waves that can be received across continents. The earth's inner iron core in the crystalline phase is surrounded by the plasma molten. Its turbulence is believed to maintain through the **dynamo effect** the earth's magnetic field, the staple for the very existence of the ionosphere. The precursor of **VHF** (**very high frequency**) seismic-related emissions before an earthquake could be a signature of the stress induced **surface plasma waves** or **electron cyclotron waves**. These down-to-earth examples illustrate the importance of matters in the plasma state. In fact, we live in a universe with 99% of the material in the plasma state; e.g., the nebula — the cluster in the star formation, the solar flare, the interplanetary space, and the earth's magnetosphere.

Plasma, after the Greek word $\pi\lambda\alpha\sigma\mu\alpha$ which means "moldable substance" or "jelly", is traditionally a medical term to describe the clear liquid in the blood, which carries various corpuscles. Langmuir coined the term, while working at General Electric on extending the life in light bulb of tungsten filament, to describe an ionized gas that carries electrons and ions, similar to how the blood plasma carries red and white corpuscles. As a result of these studies, the "**Langmuir probe**" was conceived and the "**Langmuir waves**" were discovered. Later, plasma was found to be present in various places, that leads the scientific research to where man has never been before.

Hannes Alfvén, who won the 1970 Nobel for his discovery of the **Alfvén wave,** developed the theory of ***magnetohydrodyamics*** (**MHD**) in the 1940s, in which plasma is treated as a conducting fluid. MHD theory has been widely employed to study solar physics and fusion plasma. On the other hand, **James Van Allen**'s discovery in 1958 of the radiation belt surrounding the earth, helped open up the field of **space plasma physics.** The field of ***laser plasma physics*** was advanced after the development of high power lasers in the 1960s, and the **inertial confinement** for fusion plasmas was pursued in the 1970s. Plasma physics for particle accelerator was actively pursued since the 1980s to reduce the size and the cost. **Plasmonics** has become an active field in nanobio studies at the dawn of this century with applications to biosensor, photonics, solar energy devices, telecommunication, and near-field instrumentation. The hydrogen bomb test in 1952 generated great interest in the ***controlled thermonuclear fusion*** deemed as capable of providing the unlimited energies. Fusion physicists were concerned with understanding how thermonuclear plasma could be confined by the magnetic field, and investigated the plasma instabilities that might defeat this purpose. The effort for fusion in the fifites has grown into a great international undertaking, notably the construction of **ITER** (originally an acronym of **International Thermonuclear Experimental Reactor** and Latin for "the road"), scheduled to be completed in 2020. It is a **tokamak** machine, characterized by the externally imposed toroidal magnetic field of azimuthal symmetry and the plasma current generated poloidal magnetic field to achieve stable MHD equilibrium. The term tokamak is an acronym made from the Russian words and refers to a toroidal chamber with magnetic coils. It was invented in the 1950s by Soviet physicists **Igor Yevgenyevich Tamm** and **Andrei Sakharov**. In the recent fusion experiment in **Joint European Torus (JET)**, it managed to reach $Q = 0.7$. The energy gain factor Q is the ratio of the fusion power released to the input power required to maintain the plasma. A commercially viable fusion reactor will require an energy amplification factor at least 20 and beyond. The other promising magnetic fusion scheme is the **stellarator** featuring the helically symmetric magnetic field with a magnetic well to confine the plasma. It has no plasma current and presumably has the advantage of sustaining the fusion plasma in a steady state. There are other magnetic confinement machines such as the Malmberg and Penning trap, which holds a **nonneutral plasma** for a long time, and has applications to storage and measurement of subatomic particles. The **National Ignition Facility** (**NIF**) in USA declared the success of laser fusion in achieving the $Q = 1$ ignition condition in 2013.

Industrial applications of plasma technology made great progress in this century. Most notably are the plasma TV and the man-made diamond. The laboratory synthesis by the microwave plasma yielded the nearly perfect yellow diamonds.

Yet, many promising opportunities for plasma applications in the industrial sector remain as such, perhaps owing to the fact that the full potential is still hindered by the lack of perfect understanding and precise knowledge of plasma physics. For examples, the production of **carbon nanotubes** (**CNTs**) by the **plasma vapor chemical deposition** (**PVCD**) is unable to control the chirality or the radius of the produced CNTs, and the commercialization of some **plasma torches** needs longer life span to make them practical. Moreover, some other applications require an interdisciplinary approach. The microwave therapy, which attaches nanoparticles to cancer cells to kill tumors by the RF power, is one good example. Basic plasma physics research will benefit applications of plasma technologies such as the **plasma rocket** and the plasmonics, and our understanding of the earth's environment and **space weather** such as the aurora of polar lights, the sprite of large-scale electrical discharges, the geomagnetic storm, the sun spots, and the solar flare. The curiosity is a human desire to explore the universe and beyond.

1.1 Fundamentals in Plasma Physics

As physics is an approximation in describing nature, and as nature can be better appreciated when it is made simple but not simpler, plasma phenomena may be described through the behavior of a **particle**, such as a test particle, a **wave**, such as the electrostatic wave, or a **fluid**, such as in magneto hydrodynamics. Plasma physics is very rich in its wave phenomena, and is a superset of hydrodynamics. Plasma physics may also require the relativistic or even the quantum description, depending on where and how the problems need be tackled. Plasma physics has the discrete particle effects on top of its collective many-body phenomena. Therefore, kinetic and statistical descriptions are indispensable to further the understanding.

It cannot be over emphasized that plasma physics is basically a nonlinear many-body system. The nonlinear many-body phenomena often surprise us in both the classical and quantum systems as beyond our imagination. As such, it cannot be said that plasma physics is fully understood. The difficulty in the nonlinear many-body problem is often the lack of the **self-consistency** in describing a phenomenon. In this regard, computer simulation can serve as a great tool to unravel the underlying physics. The purpose of this book is to emphasize on understanding the physics principles by visualization and solving problems with the use of computer algorithm. There are however advantages of an analytical approach. The physical law would have the clarity and the insightfulness from the compact mathematical solutions however primitive. We thus emphasize on the asymptology as a means to extend to regimes where the exact closed formula in terms of known functions is lacking.

Plasma physics covers an enormously wide range of space and time scales with parameters varying by many orders of magnitude. It is therefore important to keep the key parameters at the finger tips so that the relevance of some physical effects can be ruled in or ruled out by simple comparison. Furthermore, by examining the length and time scales of an event, it may reveal unmistakable signatures of the underlying physics. Thus, it is instructive to take a look at few key parameters. As is the custom in the field we will be using the CGS Gaussian system of units.

1.1.1 *Key Parameters*

Plasma frequency: The Langmuir wave has the characteristic timescale, viz., the inverse of the electron plasma frequency. It is an intrinsic natural frequency in the plasma and is given by $\omega_{pe} \equiv \sqrt{4\pi n_e q^2/m_e}$. Here, n_e is the electron density, $e = -q < 0$ is the electron charge, and m_e is the electron mass. Similarly, one may define the ion plasma frequency as $\omega_{pi} \equiv \sqrt{4\pi n_i (Zq)^2/m_i}$, where n_i is the ion density, Zq is the charge of the ion of atomic number Z, and m_i is the ion mass.

Gyrofrequency: A charged particle gyrates in a magnetic field B at a frequency $\Omega_\sigma \equiv q_\sigma B/m_\sigma c$, where c is the speed of light, and the subscript σ refers to the particle species with q_σ its charge and m_σ its mass. This frequency is an important measure of the time scale to determine whether the magnetic effect is important.

Thermal velocity: By assuming that plasma is at a thermal temperature T, the particles would have a thermal velocity given by $v_{th} \equiv \sqrt{k_B T/m}$, where k_B is the Boltzmann constant.

Alfvén velocity: An Alfvén wave is a relatively low-frequency mass oscillation travelling at the speed of $V_A = B/\sqrt{4\pi n_i m_i}$ under the tensile force of the magnetic field line. There are two types of Alfvén waves. The one propagating along the field line with its wave magnetic component perpendicular to the equilibrium magnetic field is the Shear Alfvén wave. The other with wave magnetic component along the equilibrium magnetic field is the compressional Alfvén wave, which is more energetic and needs more free energy to excite.

Homework 1.1: Show that the compressional Alfvén wave with wave magnetic component along the equilibrium magnetic field is more energetic and therefore more difficult to excite than the shear Alfvén wave with perpendicular component.

Gyroradius: A particle gyrating in a magnetic field has an orbital at the gyroradius or Larmor radius, $\rho \equiv v_{th}/\Omega$. The averaged value over the finite Larmor radius (FLR) is often the net electromagnetic field effect on the particles.

Debye length: The electrons are light and mobile. They can screen the Coulomb force of the ions. Beyond a Debye length $\lambda_D \equiv \sqrt{k_B T/4\pi n_0 q^2} = v_{th}/\omega_p$ of an ion, an electron will not experience its $1/r$ Coulomb potential, and vice versa. Note that the Debye length is the same for electrons and ions at the same density and temperature despite the mass disparity.

Plasma skin depth: The strength of electromagnetic radiation can be severely attenuated by electrons in a metallic or plasma surface since the electrons are highly mobile; the field is limited in a narrow range, termed as plasma skin depth and given by $\lambda_s \equiv c/\omega_p$.

Plasma parameter: Typical plasma has a huge number of particles in a Debye sphere, $N_D \approx n\lambda_D^3 \gg 1$. Therefore, its inverse $\varepsilon_p \equiv (n\lambda_D^3)^{-1} \propto n^{1/2}/T^{3/2}$ tends to be a very small parameter. Many effects related to the plasma discreteness are measured by this plasma parameter. The fluctuating electric field energy $|\delta E|^2 \sim \varepsilon_p n k_B T$ becomes vanishingly small in the very limit of $\varepsilon_p \to 0$, and the plasma is considered to be collisionless in the sense that the binary collisions are insignificant, and the collective effects dominate. A dimensionless parameter important in the collisional effect is $\Lambda \equiv \lambda_D/r_{\min} = 4\pi n\lambda_D^3 = 4\pi N_D \gg 1$, the ratio of the Debye length to the closest distance of like particles $r_{\min} \equiv e^2/k_B T$.

Plasma pressure: The plasma pressure can be written as the sum of the thermal energy density for both the electrons and the ions $P \equiv n_i k_B T_i + n_e k_B T_e$.

Magnetic pressure: The magnetic pressure $H_M \equiv B^2/8\pi$ is an important measure of the magnetic energy associated with a plasma situation. For fusion device, this reflects the important investment on plasma confinement. For space physics, this may determine whether the solar wind will intrude upon earth's electromagnetic environment. For example in the solar corona, the self generated magnetic field is the cause of solar flares.

Plasma beta: The ratio of the plasma pressure to the magnetic pressure $\beta \equiv P/(B^2/8\pi)$ is termed as the plasma beta. It is crucial to a viable fusion device since it measures the amount of magnetic energy needed to hold the plasma pressure energy. High β also implies that the plasma thermal energy is enough to modify the magnetic field.

1.1.2 *Single Particle*

As often is the case, a single entity can foretell the governing principle and the physical characteristics to enable an insight into a difficult situation. While this is a far cry from the self-consistent theory, it can be the first step in the right direction.

A test particle or a test wave can probe into a complex system to give important hints. These techniques have wider applications in both the theoretical pursuit and the experimental investigation.

Consider the particle dynamics in a strong magnetic field. The length scale can be characterized by its gyroradius $\rho \equiv V_T/\Omega$. A diffusive process due to collisions would have the step size no other than $\Delta x = \rho$. The characteristic time is $\Delta t = 1/\nu_c$. The classical diffusion coefficient may then be estimated to be $D = \Delta x^2/\Delta t = \nu_c V_T^2/\Omega^2$ with the $1/B^2$ scaling, very optimistic for magnetic fusion confinement.

```
function TestParticle(NuC)                       % Test particle pitch angle
dt=2*pi/100;                                     scattering
rand('state',sum(100*clock));                    % a gyro orbit is completed in
%initialize a random sequence                    100 steps
%by a seed number
% tied to the clock.
vp=1; x=0; y=-1; theta=0;
%start the same initially
for it=1:2000
     theta=theta+dt*(1+NuC*(rand(1)-0.5));
     vx=vp.*cos(theta);
     vy=vp.*sin(theta);
     x=x+vx*dt; y=y+vy*dt;
     X(it)=x; Y(it)=y;
     end;
plot(X,Y,'g-'); xlabel('X'); ylabel('Y');
title('Particle gyro orbit with pitch angle scattering')
axis equal;
```

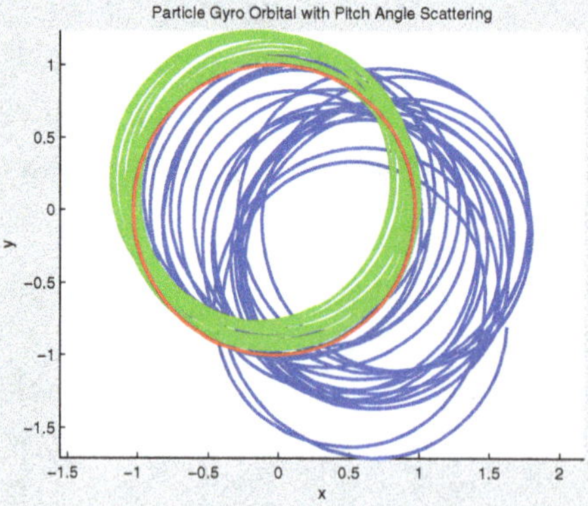

The following MATLAB program calculates the pitch angle scattering effect on the particle motion by the Monte Carlo method with use of the random number generator. The pitch angle scattering is energy conserving. The code has time normalized to the gyrofrequency Ω and the velocity normalized to the thermal velocity. Setting $\nu_c = 0$ the particle has a gyroradius of unity (red trajectory), $\nu_c = 0.1$ the particle motion is diffusive (green trajectory), and $\nu_c = 1$ drastic diffusive process occurs (blue trajectory).

Homework 1.2: Run TestParticle.m with an input of ν_c at 0, 0.01, 0.1, 1, 10, and 100. Make observation of the particle dynamics.

1.1.3 *Single Wave*

In the early days of tokamak research, the Russian scientists achieved remarkable plasma density around $10^{13}/\text{cm}^3$ and the electron temperature around 1keV. Yet, no confirmation from the experimental measurement was realized until the British sent a scientific team to set up the microwave interferometry. The electromagnetic wave could not shine through when the wave frequency ω_0 is lower than the electron plasma frequency ω_{pe}. So long as $\omega_0 > \omega_{pe}$ the wave would be detected on the other side of the plasma. This test wave experiment, capable of measuring the plasma density, can be understood from the wave dispersion relation, $\omega^2 = c^2k^2 + \omega_{pe}^2$ for the electromagnetic wave propagating in the plasma. Without the plasma it is reduced to the familiar $\omega^2 = c^2k^2$ that has the nondispersive wave number k in the vacuum. Rewrite the dispersion relation in terms of space and time variables, we end with the propagation equation,

$$\frac{\partial^2 E}{\partial t^2} - c^2\frac{\partial^2 E}{\partial x^2} + \omega_{pe}^2(x)E = 0. \tag{1.1a}$$

Define the dimensionless variables of time $\tau \equiv \omega_0 t$ and space $\chi \equiv x\omega_0/c$. By assuming the following density profile $\rho \equiv \omega_{pe}^2(x)/\omega_0^2 = \aleph \exp(-\alpha\chi^2)$, Eq. (1.1a) becomes

$$\frac{\partial^2 E}{\partial \tau^2} - \frac{\partial^2 E}{\partial \chi^2} + \aleph e^{-\alpha\chi^2} E = 0. \tag{1.1b}$$

Equation (1.1b) is implemented in the accompanying MATLAB program.

The value $\aleph$ represents the peak density of the plasma, and the frequency of the EM wave is set to unity. When $\aleph = 1.5$, the incident wave launched from the left hand side (LHS) is reflected almost in its entirety, except the high frequency components of the wave packet to go through the plasma. When $\aleph = 0.5$, the incident wave would penetrate to the other side of the plasma with the exception of some low frequency components. The program EMPW produces the animation that demonstrates the physics as described.

Homework 1.3: Assume $\alpha = 0$ in Eq. (1.1) and the solution as $E \sim \exp(ik\chi - i\tau)$ in the Fourier form, show that if $\aleph > 1$, k is imaginary and the wave is evanescent.

```
function EMPW(N)                                          % type EMPW(1.5) or
global z k0 ro dz dt Nz Nt Eend Estart tau N;             EMPW(0.5) to run.
clc;
figure(1);
init;
E=0*z; W=E;
for it=0:Nt
    [E,W]=advance(E,W,it);
    if(mod(it,10)==0)
        plotPW(z,E,ro,tau); getframe; end;
end;
clear all;

function [E,W]=advance(E,W,it)                            %advance the wave
global z k0 ro dz dt Nz Nt Eend Estart tau N;
d=dt/dz/dz;   tau=it*dt;
if(tau<50)
    E(1)=getX(k0,dz,tau+100); end;
eN=E(Nz)*(1-dz/dt)+Eend*dz/dt;
eS=E(1)*(1-dz/dt)+Estart*dz/dt;
W=W-E.*ro*dt+d*(-2*E+[E(2:Nz),eN]+ [eS,E(1:Nz-1)]);
Eend=E(Nz); Estart=E(1);
E=E+dt*W;

function X=getX(k,dz,t)                                   %launch a wave packet
A=0.5; omega0=1; k0=k; dOmega=0.05; dk=0.05;
X=A^2*sin((omega0+2*dOmega)*t-(k0+2*dk)*(-dz));
X=X+A*sin((omega0+dOmega)*t-(k0+dk)*(-dz));
X=X+sin(omega0*t-k0*(-dz));
X=X+A*sin((omega0-dOmega)*t-(k0-dk)*(-dz));
X=X+A^2*sin((omega0-2*dOmega)*t-(k0-2*dk)*(-dz));

function plotPW(z,E,ro,it)                                %plot the wave
L=max(z);                                                 propoation
plot(z,real(E),'r-', z,ro,'k-');
s=sprintf('Plasma Wave Propagation iT=%7.1f',it);
title(s);
axis([-L L -2 2]);

function init   %initilize the parameters                  % omegap=2e11,
global z k0 ro dz dt Nz Nt Eend Estart tau N;             lambda=0.15,
alpha=0.001; k0=1; dz=0.25; dt=0.1;                       k0=2pi/lambda=40;
Nz=1001; Nz2=(Nz-1)/2;
z=(-Nz2:1:Nz2)*dz;
ro=N*exp(-alpha*z.^2);
Eend=0; Estart=0; Nt=5000; tau=0;
plot(z,ro);
title('density profile');
xlabel('x'); ylabel('ro');
```

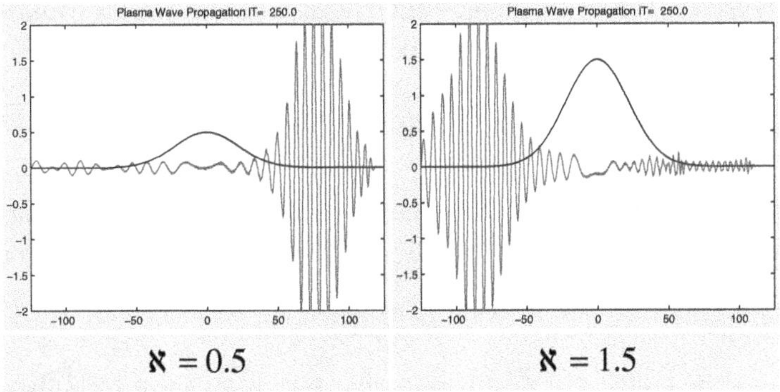

1.1.4 *Single Fluid*

Like a single wave, the fluid description can reveal the plasma collective behavior. Plasma has the distinct property of being highly conducting due to the high mobility of electrons. As a result, it tends to remain charge neutral locally, and is suitable for the single fluid model, as in the context of the magnetohydrodynamics (MHD) in plasma physics. The MHD is a superset of hydrodynamics, and as such, it has many open questions to be answered. The fluid mechanics treats the lower velocity moments in the continuity, the momentum and the energy equations. The flow velocity $\vec{v}(\vec{x}, t)$ varying in both space and time, has the time derivative

$$\frac{d\vec{v}(\vec{x},t)}{dt} = \frac{\partial \vec{v}(\vec{x},t)}{\partial t} + \frac{\partial \vec{v}(\vec{x},t)}{\partial \vec{x}} \cdot \frac{\partial \vec{x}}{\partial t} = \frac{\partial \vec{v}}{\partial t} + \vec{v} \cdot \frac{\partial \vec{v}}{\partial \vec{x}}.$$

Here $d/dt = \partial/\partial t + \vec{v} \cdot \nabla$ following the flow, is often termed as the material derivative and is the sum of local and convective derivatives. The convective term of the flow velocity $\vec{v} \cdot \nabla \vec{v}$ is often the source of numerical difficulty, since it can cause the steepening effect, among other things. It is very much responsible for the shock wave formation and wave breaking. It may be rewritten as $\vec{v} \cdot \nabla \vec{v} = \frac{1}{2} \nabla v^2 - \vec{v} \times (\nabla \times \vec{v})$. The first term on the right hand side is the velocity pressure (head) that is important in aviation for airplane lifting. The second term can be expressed as $\vec{\Omega} \times \vec{v}$, where $\vec{\Omega} \equiv \nabla \times \vec{v}$ is the **vorticity**. It has some analogy to the $\vec{J} \times \vec{B}$ force, well known for the current pinch that was one of the important mechanisms in fusion research by applying current to hold the plasma together since current filaments flowing in the same direction will attract each other. The $\vec{\Omega} \times \vec{v}$ force is expansive and can cause vortexes of the same polarity in its circulation direction to merge, thus allowing the small whirlwinds to cumulate into humongous hurricanes and typhoons. It also has the capability of self-organizing the small-scale turbulent flow to form large-scale regular flow pattern. Fluid mechanics often deals with the **Navier Stokes equation** that includes the pressure p and the viscosity μ,

$$\frac{\partial \vec{v}}{\partial t} + \vec{v} \cdot \frac{\partial \vec{v}}{\partial \vec{x}} = -\frac{1}{\rho}\nabla p + \mu \nabla^2 \vec{v}. \tag{1.2}$$

The change of the mass density at the local position is due to the net mass flux $\rho\vec{v}$, and is governed by the continuity equation, which states the conservation of mass,

$$\frac{\partial \rho}{\partial t} + \nabla \cdot (\rho \vec{v}) = 0. \tag{1.3}$$

Together with an expression for the pressure, they are complete in describing the hydrodynamics. Since

$$\nabla \cdot (\rho \vec{v}) = \vec{v} \cdot \frac{\partial \rho}{\partial \vec{x}} + \rho \nabla \cdot \vec{v},$$

Eq. (1.3) may be rewritten as

$$\frac{d\rho}{dt} + \rho \nabla \cdot \vec{v} = 0, \tag{1.3'}$$

It is clear that if $\nabla \cdot \vec{v} = 0$, the fluid will have no density variation along the flow, and is termed as the **incompressible flow**.

To gain some physical insight into the fluid dynamics, let us examine a nontrivial solution of the **Taylor-Green vortex** for the incompressible flow with constant density in space. The pressure is assumed to balance the velocity head, namely,

$$\vec{v} \cdot \frac{\partial \vec{v}}{\partial \vec{x}} = -\frac{1}{\rho}\nabla p. \tag{1.4}$$

Hence in a square box with four sides of length 2π, one gets from Eq. (1.2)

$$\frac{\partial \vec{v}}{\partial t} = \mu \nabla^2 \vec{v}. \tag{1.5}$$

Equation (1.5) can be solved through separation of variables to give the following:

$$\begin{pmatrix} v_x \\ v_y \end{pmatrix} = \begin{pmatrix} \sin mx \cos my \\ -\cos mx \sin my \end{pmatrix} F(t), \quad F(t) = e^{-2m^2 \mu t}. \tag{1.6}$$

The vortex flow is $\vec{\Omega} = \nabla \times \vec{v} = 2m \sin mx \sin my F(t)\vec{e}_z$, shown below for $m = 1$ in the quiver plot. The pressure is found from Eq. (1.4) to be

$$p = \frac{1}{4}\rho(\cos 2mx + \cos 2my)F^2(t),$$

presented in the contour plot. It is also clear from Eq. (1.6) that higher m modes will decay faster, and the neighboring vortices have the opposite vorticity.

The vortex flow is an important physical entity in fluid mechanics as well as plasma physics. Consider the electron flow that is mainly driven by the electromagnetic field so that $m\partial\vec{v}/\partial t = q\vec{E}$. Since

$$\nabla \times \vec{E} = -\partial\vec{B}/\partial t/c = -\partial\nabla \times \vec{A}/\partial t/c,$$

where $\vec{A}$ is the vector potential, we may neglect the electrostatic field and write $m\partial\vec{v}/\partial t = -q\partial\vec{A}/\partial t/c$. This shows that the canonical momentum $\vec{p} \equiv m\vec{v}+q\vec{A}/c$ is a constant of the motion. Thus,

$$\nabla \times \vec{p} = \nabla \times (m\vec{v} + q\vec{A}/c) = m\vec{\Omega} + q\vec{B}/c.$$

This may result in the spontaneous generation of the vortex flow and the magnetic field, and it is not difficult to recognize that the magnetic loop formation in the solar flare will accompany with the vorticity.

It is not yet proven that in three dimensions solutions always exist for Eqs. (1.2) and (1.3), or that if they do, they are singularity free. These are called the Navier–Stokes existence and smoothness problems, considered to be one of the seven most important open problems in mathematics and offered by the Clay Mathematics Institute a prize of one million US dollars for a solution or a counter-example.

```
function TaylorGreenVortex(m)
% plot the vortexes
n=300;
dx=2*pi/n;
x=0:dx:2*pi;
y=x; n=n+1;
X=repmat(x,n,1);
Y=repmat(y',1,n);
U=sin(m*X).*cos(m*Y);
V=-cos(m*X).*sin(m*Y);
P=cos(2*m*X)+cos(2*m*Y);
figure(1);
d=10;
quiver(X(1:d:n,1:d:n),Y(1:d:n,1:d:n),
%make the quiver plot
U(1:d:n,1:d:n),V(1:d:n,1:d:n));
axis([0,2*pi,0,2*pi]);
xlabel('X'); ylabel('Y');
title('Quiver Plot of Taylor-Green
Vortex');
figure(2);
contour(X,Y,P);
%make the contour plot
axis([0,2*pi,0,2*pi]);
xlabel('X'); ylabel('Y');
title('Contour Plot of Pressure Profile');
```

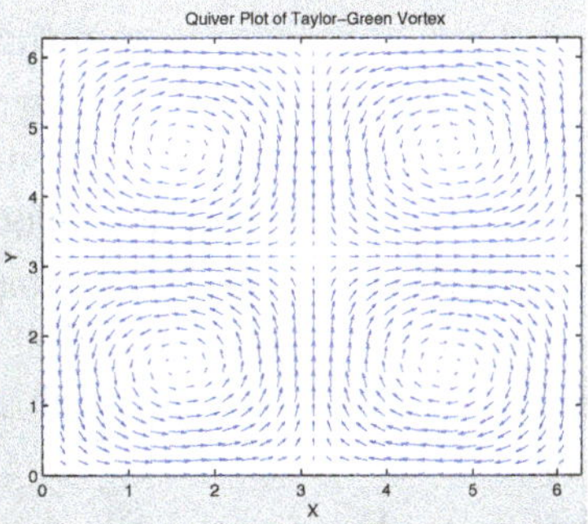

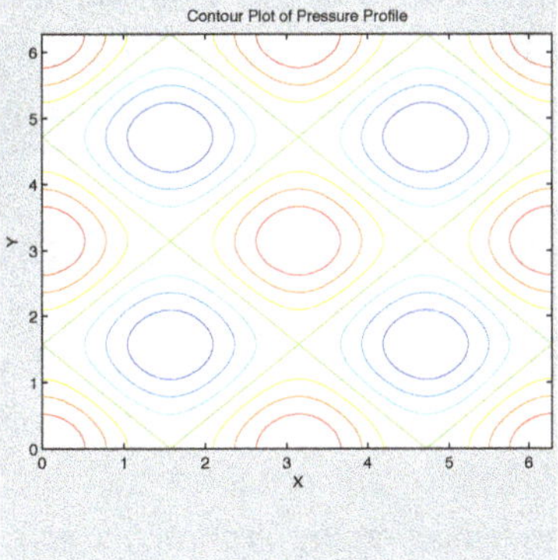

1.1.5 *Kinetic and Statistical Approach*

Beyond the fluid description, the kinetic theory considers the phase space distribution function, and it may include higher order correlations. It is mathematically rigorous and physically elaborate. It may, for example, derive the fluid equations with transport effects. While the kinetic approach can be mathematically tedious and challenging, the statistical concepts often lead to global understanding free from the detailed mechanical constraints that can be too stringent and too many. One good example is the H-theorem that predicts the entropy is ever increasing (an irreversible process), derived by Boltzmann from the binary collision process which, by contrast, is a reversible process with the conservation of energy and momentum. The Boltzmann theorem was not widely accepted until Poincaré pointed out that a system of enormous large number of particles indeed returns to the initial phase point, but only after a **Poincaré cycle** of astronomical time scale. By truncating away the higher order correlation, Boltzmann pushed the Poincaré cycle to infinity, and allowed the irreversible process to prevail as the real world experience. The nonlinear evolution of a complex system can vary tremendously, time and again, by minute deviations in the initial condition or the boundary values. Thus, *the* ***causality*** *(cause and effect) through a detailed mechanical description while accurate in its own right can be often misleading as it could be statistically insignificant.*

As a practical example, we here calculate the diffusion coefficient $D = \langle(\Delta x)^2/\Delta t\rangle$ in the following program. The diffusion coefficient, as described in the test particle picture, needs an ensemble average, denoted by the angular bracket, to obtain the statistically meaningful result. The ensemble average is taken over particles of a normal distribution of the particle velocity with uniform angular distribution in velocity. The plot shows that the radial expansion $\langle(\Delta x)^2\rangle^{1/2} \propto \sqrt{\Delta t}$. The code finds the diffusion coefficient, and can be used to study the dependency of the diffusion on the magnetic field and the collisional frequency.

Homework 1.4: Fix the collision frequency ν_c but vary the magnetic field OMEGA from 1 to 10 in the TPD program. Find out the scaling law of the diffusion coefficient as function of the magnetic field.

1.2 Fundamentals in Theoretical Description

To describe a physical phenomenon, we want to define the relevant variables and relate them in mathematical formulation. Before the equations are formulated, whether all the relevant variables are included can be a serious issue in the scientific pursue. The hand waving arguments and cartoon physical pictures may well start the mind mapping for searching the underlying mechanism, followed by the back

```
function TPD(NuC,OMEGA)                              % Test Particle Diffusion
close all; clc;                                      % type(0.05,1) to run
N=2000; Nr=100; Nt=50000;
% NuC collisional frequency, OMEGA
% Cyclotron frequency
% It shows dR~sqrt(t).
dt=2*pi/(Nr+7);
rand('state',sum(100*clock));
vp=sqrt(2*randn(N,1));
x=randn(N,1)*0.01;
y=randn(N,1)*0.01;
theta=rand(N,1)*2*pi;
T=1:Nt/Nr;
R=T*0; j=0; E=R;
X=T*0; Y=X;
tic;
for it=1:Nt
    theta=theta+OMEGA*dt+NuC*(rand(N,1)-0.5)*dt;
    vx=vp.*cos(theta); vy=vp.*sin(theta);
    x=x+vx*dt; y=y+vy*dt;
    X(it)=x(1); Y(it)=y(1);
    if(mod(it,Nr)==0)
        j=j+1;
        R(j)=sum(sqrt(x.^2+y.^2))/N;
        E(j)=sum(vp.^2)/N;
        X(j)=x(1); Y(j)=y(1);
        end;
    end;
toc; datestr(now),
y=sqrt(T).*R(j)./sqrt(T(j));
plot(T,R,'g*',T,y,'r-');    %D=max(R)^2/T,
D=sum(sum(R.^2,1),2)/Nt/dt/N/2, NuC,
EnergyConservation=abs(E(Nt/N)/E(1)-1),
xlabel('T'); ylabel('dR');
title('Diffusion Process as Function of Time');
```

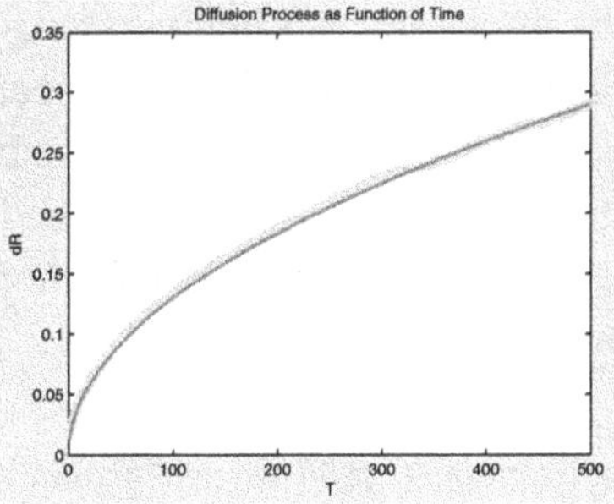

of the envelop calculation that could help reveal the relevance of the conjecture. Then a formal theory, substantiated and error pruned by the numerical solutions and asymptotic limits, may help establish the correct answer. Finally, the theoretical prediction should be checked against the experimental observations and further refined as the understanding progresses. Thus, the ***N***umerical computation, ***E***xperimental

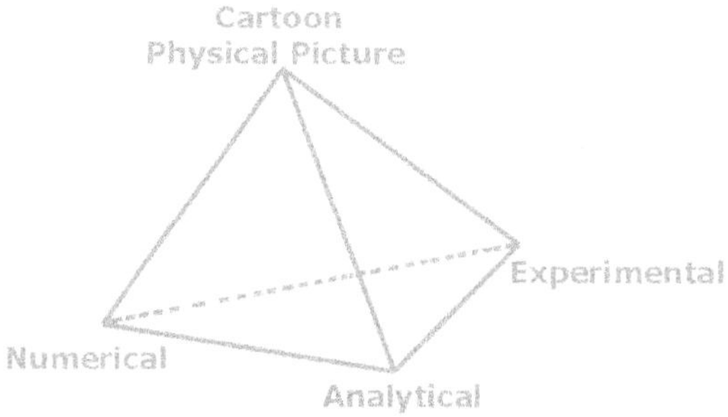

Fig. 1.1. NEAP view of the theoretical investigation.

observation, ***A***nalytical calculation, and the ***P***hysical picture persuasion, properly abbreviated as ***NEAP***, depicted in the drawing in a tetrahedron, represents elements needed to elucidate the physics principles. This will not only help ensure a correct conclusion, but also in itself a practical process to scientific discovery. Scientific computing, especially modeling and simulation, can be an important part of the knowledge creation. Here we introduce a few general concepts useful in starting a theoretical investigation.

1.2.1 *Dimensionless Variables*

The quantity of a physical entity as measured by its absolute magnitude is often insufficient to judge its physical significance until it is compared with other key parameters in its physical environment. The dimensionless quantities are often more revealing than otherwise. The governing equations normalized to the proper space and time scales not only ready the computer programming, but also foretell when and where each term can be important, especially in studying the nonlinear physics when many mechanisms and energy channels are competing. The magnitude in its dimensionless form will enable us to compare the strength of individual terms, even if they have different units and are not obviously related, so to spot the dominant mechanisms and relevant effects. Take a resonance phenomenon, an important topic in physics, as described by

$$\frac{d^2\chi}{d\tau^2} + \chi = p\sin(\chi - \tau + \Delta\tau), \tag{1.7}$$

in the dimensionless form. Here τ is the normalized time variable, χ is the normalized displacement, p represents the normalized amplitude of the external force, and Δ represents a frequency mismatch. Without the χ term in the sine function on the right hand side of Eq. (1.7), it is a linear equation with the particular (inhomogeneous) solution $\chi_p = p\sin(1-\Delta)\tau/(2\Delta - \Delta^2)$. As $\Delta \to 0$, $\chi_p \propto 1/\Delta$

can be unbounded. This shows that a linear theory often has its limitations in describing a physical phenomenon as no physical quantity can be infinitely large. This is the very reason why the nonlinear equations need be addressed in plasma physics or otherwise when and where the linear theory breaks down. The displacement dependency in the sine function causes the frequency mismatch or the phase variance to curtail the strong resonance permitted by the linear theory. On the other hand, a frequency mismatch in a resonance phenomenon can be compensated by the external field amplitude, as demonstrated by the accompanying MATLAB program.

The trajectory is sampled by the stroboscopic method, which takes one data point per wave period and shows the bounded trajectories in the phase space. The larger amplitude is able to compensate for the frequency mismatch and results in larger excursion in the phase space. While the two quantities are of quite different origin, their strength can be measured and compared in the dimensionless form as in this case. Therefore, when $p \sim O(\Delta)$, the resonance effect becomes more pronounced.

```
function FM(p) %Frequency
Mismatch
dT=2*pi/200;
Nt=3000000;
delta=0.2;
rand('state',sum(100*clock));
v=0.1*rand(1); x=0.1*rand(1);
x=x+v*dT/2;
E0=v^2/2;
% E0 initial energy, E average
energy
X=0*(1:200:Nt+1);
V=X; icnt=1;
X(1)=x; V(1)=v;
for it=2:Nt
v=v-x*dT+
dT*p*sin(x-it*dT*(1-delta));
x=x+v*dT;
if(mod(it,200)==0)
        icnt=icnt+1;
        X(icnt)=x;
        V(icnt)=v;
        end;
end;
plot(X,V,'g.'); axis equal;
s=sprintf('%3.2f p=%3.2f',delta,p);
title(strcat('Frequency Mismatch
\Delta=',s));
xlabel('x'); ylabel('v_x');
E=sum(V.^2)/length(V);
dE=abs(E-E0)/E0,
%the energy amplification factor
```

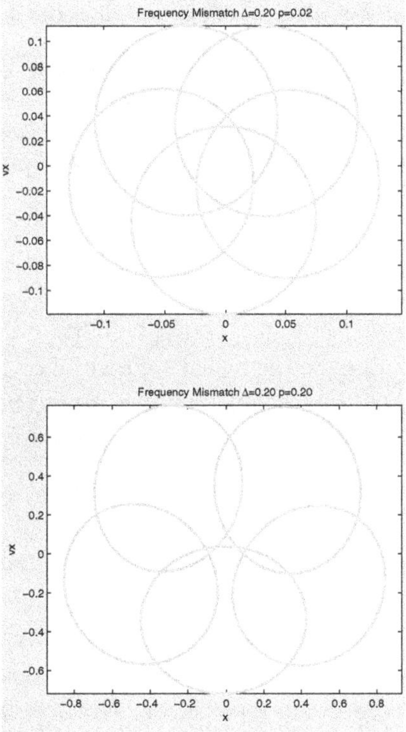

Homework 1.5: Modify the MATLAB code of FrequencyMismatch by setting delta to 0 and 0.5, respectively, and compute for $p = 0.02$ and $p = 0.2$. Explain what you get.

Homework 1.6: Given the electric field amplitude in a laser experiment at 30 GV/ cm and 1 μm wavelength, determine the value of the parameter $p \equiv eEk/m\omega^2$.

1.2.2 *Order of Magnitude*

The order of magnitude analysis is an important process to sort out physically related effects. The mere order of magnitude estimate could point out the underlying principles to rule in or rule out a conjecture. We are often quite satisfied with order of magnitude descriptions to acquire the insight into a scientific inquiry. For example, as quoted from the book of Biology by Campbell, Reece and Mitchell,

> *"Earth formed about 4.5 billion years ago, and life probably began only a few hundred million years later. "*
> *"Biologists have identified and named about 1.5 million species, including over 260,000 plants, almost 50,000 vertebrates and more than 750,000 insects."*
> *"In the continuum of life spanning over 3.5 billion years, humans and apes have shared ancestry for all but the last few million years."*

It would be meaningless and even erroneous to provide more than two digits of accuracy as cited above. Yet the magnitude in the proper unit is of great significance and defines the exact science of the imprecision nature of the verification methodology. Thus, the concept of the order of magnitude provides a way of communicating the imprecise nature of the scientific facts.

Suppose a test particle is under the thermal pressure, the electric field, and the gravitational force as given in the following equation,

$$m\frac{dv}{dt} = -\frac{dk_BT}{dx} + eE - mg. \tag{1.8}$$

The importance of each term can be measured by translating the quantities in terms of the unit eV/cm. They are $m_pg = 1.0e^{-9}$ eV/cm, and $k_BT = 8.6e^{-5}$ T(Kelvin)eV. Here, m_p is the proton mass. We immediately recognize that given an electric field on the order of 1 eV/cm, and a temperature gradient length of 1 cm, the temperature needs be around ten thousand Kelvin for the thermal pressure to be important, and unless the ion mass is around 10^9 times the proton mass, the gravitational force will not be significant. On the other hand, a low temperature plasma at a temperature around 10^4 K, the plasma could result in an electric potential around

1 eV to allow chemical reaction to occur. This is where low temperature plasma finds good application as in the PVCD since the electrons have enough energy to make ionization and chemical bonding among the atoms. The **plasma sheath** of narrow **boundary layer** can make the electric potential gradient stronger, and lower the workable temperature to 10^3K in practical PVCD applications. The term, "low temperature plasma", is referred to this temperature range, which is high compared to the room temperature, but low in contrast to the fusion condition. The PVCD is capable of producing carbon nanotubes or diamonds, for example.

> **Homework 1.7:** Determine the gravitational pulls on the moon by the sun and the earth. Find out which of the two has stronger pull.

Proper unit can make the underlying physics more transparent. The visible light is often referred to by the frequency range from 400 to 790 THz or by the wavelength from 3900 to 7500 Å. To make sense out of photon material interaction, it is however easier to expressed in terms of the electron volts: 1.8 to 3.1 eV. Similarly, a pressure at 1 megabar = 100 GPa is equivalent to 0.625 eV/Å^3. To have 1 eV effect at the atomic scale, 150 GPa is needed, and 2 eV 300 GPa. It is not surprising that these are the pressures the experiments would begin to alter the material properties, and it is where the plasma molten resides in the earth outer core.

> **Homework 1.8:** The earth has a radius $R = 6350$ km. Its inner core of radius 1220 km is under a pressure around 350 giga Pascal (GPa). Assume the pressure is given by $p(z) = p_0 + g\int_0^z \rho(r)dr$, where z is the depth into the earth. Determine the depth from the earth surface where the material would become ionized plasma. Note that 100 GPa = 0.625 eV/Å^3.

1.2.3 *Asymptotic Limits*

Taking a variable to its large and small limits can help reveal its physical significance as demonstrated in Eq. (1.7) regarding the resonance when $\Delta \to 0$. It may serve to uncover, in time, the error in the analytical or numerical approaches, and provide the physical insight for the better solution to be discovered, even if a numerical approach is inevitable. One of the important asymptotic expansions is the **Stirling's formula** $\log n! \approx n \log n - n + O(\log n)$ often applied in statistical mechanics as the number of particles $n \gg 1$. This can be derived from the **Euler-Maclaurin formula** to give $\ln N! = \sum_1^N \ln n \approx \int \ln x dx + \frac{1}{2}\ln N + O(1) \approx N\ln N - N + \frac{1}{2}\ln N + O(1)$. Another good example of the asymptotic limit is the **Gamma function**, which gives

$$\lim_{z\to -n} \frac{1}{\Gamma(z)} \to 0, \tag{1.9}$$

when z approaches a negative integer. This formula is needed to show the quantization condition in hydrogen atom. The Gamma Function can be defined by

$$\Gamma(z) = \int_0^\infty e^{-t} t^{z-1} dt. \tag{1.10}$$

There is a singularity near $t = 0$ if the real part of z is negative. But, **aysmptology** allows us to manage even negative z by making good use of the asymptotic limit. Since the singularity is near $t = 0$,

$$\Gamma(z) = \int_0^\infty e^{-t} t^{z-1} dt = \int_0^1 e^{-t} t^{z-1} dt + \int_1^\infty e^{-t} t^{z-1} dt, \tag{1.11}$$

the first term dominates. Expanding e^{-t} in serial expansion

$$\Gamma(z) = \sum_{m=0}^{\infty} \frac{(-1)^m}{m!(m+z)} + \int_1^\infty e^{-t} t^{z-1} dt \tag{1.12}$$

If z is near $-n$ with a small deviation, viz., $z = -n + \varepsilon$, we have

$$\Gamma(z) \approx \left(\frac{1}{\varepsilon}\right) \frac{(-1)^n}{n!} \gg 1. \tag{1.13}$$

Therefore, $1/\Gamma(z) \approx (-1)^n \varepsilon n! \to 0$, as $\varepsilon \to 0$, i.e, $z \to -n$. Thus, Eq. (1.9) is proved. The MATLAB function invGamma evaluates $1/\Gamma(z)$ and plot the same that shows the characteristics at different n values.

```
function invGamma
z=-4.05:0.005:5;
g=z*0;
f=1./gamma(z);
plot(z,f,'g.-',z,g,'r-');
xlabel('z','fontsize',16);
ylabel('f','fontsize',16);
title('Inverse of Gamma
Function','fontsize',16);
grid on;
```

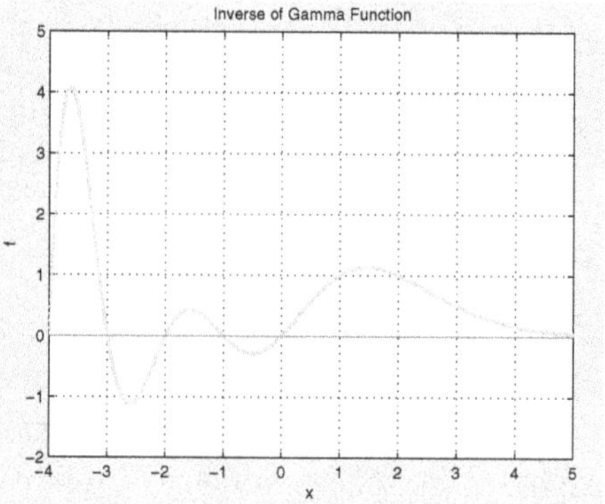

Homework 1.9: Find the asymptotic limit of $\mathbf{log}(e^{2x} + e^x)$ for $x \gg 1$.

Homework 1.10: Prove that $N^N \gg N! \gg C^N \gg N^C \gg CN$ for $N \gg C \gg 1$.

1.2.4 *Multiple Time and Multiple Space Scales*

The importance of multiple time scale perhaps can be vividly explained by the following description of relevant time to specific event:

> *"To realize the value of one year, ask the student who has failed a class.*
> *To realize the value of one month, ask the mother who has given birth to a premature baby.*
> *To realize the value of one week, ask the editor of a weekly.*
> *To realize the value of one day, ask a daily wage laborer.*
> *To realize the value of one hour, ask the man waiting for his girlfriend.*
> *To realize the value of one minute, ask the person who has just missed his flight.*
> *To realize the value of one second, ask the person who has survived an accident.*
> *To realize the value of one millisecond, ask the person who has just won silver in the Olympics.*
> *To realize the value of one nanosecond, ask a hardware engineer.*
> *And if you still don't realize the value of time, you must be a government employee."*
>
> *Author unknown*

Plasma physics has a wide range of time scales. If we are interested in the phenomenon in the electron time scale, quite often we may treat the ions as the immobile background. If we are interested in the phenomenon of ion time scale, we may treat the electrons by the dynamic equilibrium so that $\delta n/n_0 \approx e\delta\varphi/k_BT$, which simply states that the electrons quickly reach the thermal equilibrium and the density perturbation redistribute accordingly.

Boundary layer phenomena are well recognized in fluid mechanics. Plasma physics has a wide range of space scales, and entails the simplification by the expansion in the space scales. For example, we may take finite gyroradius expansion if the length scale of interest is longer than the gyroradius, or simply make the guiding center approximation. Mode conversion is one other important example where a boundary layer exists for two waves of different physical properties such as a longitudinal wave and a transverse wave to have met the conversion conditions: $\omega_1 = \omega_2$ and $k_1 = k_2$, corresponding to the energy and momentum conservation, respectively. While the two waves may exist in the similar frequency range, the wave numbers can be quite off until the density gradient compensates in a narrow layer for the wave number mismatch, namely, $k_1 = k_2 + k_L$, where k_L is the equivalent wave number (momentum) imparted by the density gradient.

Homework 1.11: Describe the electron by the fluid equation $nm_e dv/dt = -dn\, k_BT/dx - neE$ (cf. Eq. (1.3)). Show that $\delta n/n_0 \approx e\delta\varphi/k_BT$ is reached in equilibrium, where the electron density $n = n_0 + \delta n$ and $\delta n \ll n_0$. The electric field is electrostatic and expressed by its potential as $E = -d\delta\varphi/dx$.

1.2.5 *Causality*

One simple but important physical law is the principle of causality: the cause and its effect, and yet it is most easily denied, misinterpreted, or ignored. For examples, before the industrial revolution, even the emperor had at most a six-horsed wagon. There used to have only a handful of emperors or empresses in the world, but nowadays, there are billions driving cars at 150 to 250 horsepower and beyond, and US leads the world at 12 KW per capita energy consumption. Our wishful thinking is that this will not cause global warming with gigantic hurricane, arctic oscillation and sporadic weather extremes. We continue to pump oil from the deep sea, and believe there is no reason that it will cause the tectonic plates to become more unstable, or result in more frequent and stronger earthquakes despite the data from the last century suggested otherwise. We continue to deplete the fish population in the ocean and expect the seafood will continue to be plentiful for the future generations to enjoy. Our actions are inconsequential to the inconveniences or even catastrophes on earth, and yet the new normal is pretended as mysterious, perhaps due to some invisible hand. In scientific studies, any effect has a cause, and every action produces consequence, perhaps more dramatically illustrated by the 'butterfly effect' — a sensitive dependency on a small deviation in the initial conditions to result in large differences in the later stage. The earth is finite in everything. It is also perfectly vacuum sealed, except the black body radiation. We could be slowly 'cooking' ourselves to death without recognizing it. While the plasma physics is exploring to tame the fusion energy, we cannot afford not to think whether smart energy is the right approach than unlimited energy. The UN advice: reduce, reuse, recycle, will have to be our permanent life style to give the earth a chance.

Homework 1.12: Write a short essay on the smart energy policy. What options are available to replace the unlimited energy infrastructures currently fashioned in the developed countries? Explain also how to reach the vision of 2 KW society.

Further Reading

Keeping a copy of Naval Research Lab's Plasma Formulary is useful for the course learning and beyond. Visit: http://wwwppd.nrl.navy.mil/nrlformulary/.

Irving Langmuir's coining of "plasma" was described by Harold M. Mott-Smith in a letter to *Nature*, Vol. 233, p. 219, 17 September 1971.

A solar flare is a thunderous explosion that occurs in the solar corona and chromosphere within the atmosphere of the sun. The incredible energy level of a solar flare is equivalent to tens of millions of atomic bombs exploding at the same time! The powerful energy commonly associated with solar flares can take as long as several days to build up, but only minutes to release. During the occurrence of a solar flare, plasma is heated to tens of millions degrees Kelvin, while electrons, protons and heavier ions are accelerated to near the speed of light. For further reading, visit: http://en.wikipedia.org/wiki/Solar_flare. Last accessed May 2014.

Check into the web site of Clay Mathematics Institute for the millennium problems: http://www.claymath.org/millennium/. Last accessed May 2014.

The Ph.D thesis by Gary James Weisel on "Containing plasma physics: A disciplinary history, 1950–1980", University of Florida, has studied the first 30 years of American fusion research in depth.

Cold fusion of **low temperature nuclear effect** as a viable fusion scheme has not been accepted by the main stream plasma physicists. Watch CBS 60 minutes on Cold fusion: "More than Junk Science", http://www.cbsnews.com/video/watch/?id=4967330n&tag=contentMain;contentBody and investigate the pros and cons of this path of research.

The conservation of the canonical momentum $\vec{p} \equiv m\vec{v} - q\vec{A}/c$ results in the spontaneous generation of the dc magnetic field and the dc vortex flow is discussed in Chapter 11.

Nature has plenty of plasma phenomena that may reveal important signatures of events such as the earthquake, the lightening, and even the oil reserve. The hydrocarbons could have formed deep within the earth, where the plasma molten due to the pressure of the earth core bounds water and carbon.

Theoretical effort is cost-effective. With the improved understanding, mistakes can be avoided and new ideas and new inventions could follow. Capitalism or the doctrine of free enterprise has the underlying assumption of unlimited resources to work with. The earth is, however, finite in everything, and an unrealistic assumption could lead to disasters. The weather extremes cannot be uncorrelated with the continually increased energy consumption and heat waste. For more information on the 2 KW society, first introduced in 1998 by the Swiss Federal Institute of Technology in Zürich, check into http://en.wikipedia.org/wiki/2000-watt_society. Last accessed May 2014.

Homework Hints

Homework 1.1: Show that the compressional Alfvén wave with wave magnetic component along the equilibrium magnetic field is more energetic and therefore more difficult to excite than the shear Alfvén wave with perpendicular component.

Given $\vec{B} = \vec{B}_0 + \delta\vec{b}$, where $B_0 \gg \delta b$, we have the total magnetic energy

$$\varepsilon_m = \frac{1}{8\pi}|\vec{B}|^2 = \frac{1}{8\pi}[|\vec{B}_0|^2 + 2\vec{B}_0 \cdot \delta\vec{b} + |\delta\vec{b}|^2],$$

and the perturbed magnetic energy

$$\delta\varepsilon_m = \frac{1}{8\pi}[2\vec{B}_0 \cdot \delta\vec{b} + |\delta\vec{b}|^2].$$

For the shear Alfvén wave, since $\vec{B}_0 \cdot \delta\vec{b} = 0$, we have its local energy $\delta\varepsilon_s = |\delta\vec{b}|^2/8\pi$. For the compressional Alfvén wave, since $\vec{B}_0 \cdot \delta\vec{b} \neq 0$, we have its local energy $\delta\varepsilon_c = [2\vec{B}_0 \cdot \delta\vec{b}]/8\pi$. Therefore, $|\delta\varepsilon_c| \gg |\delta\varepsilon_s|$. Thus, it needs a greater amount of energy to cause the disturbance of a compressional Alfvén wave.

Homework 1.3: Assume $\alpha = 0$ in Eq. (1.1) and the solution as $E \sim \exp(ik\chi - i\tau)$ in the Fourier form, show that if $\aleph > 1$, k is imaginary and the wave is evanescent.

$k^2 = 1 - \aleph$.

Homework 1.5: Modify the MATLAB code of Frequency Mismatch by setting delta to 0 and 0.5, respectively, and compute for $p = 0.02$ and $p = 0.2$. Explain what you get.

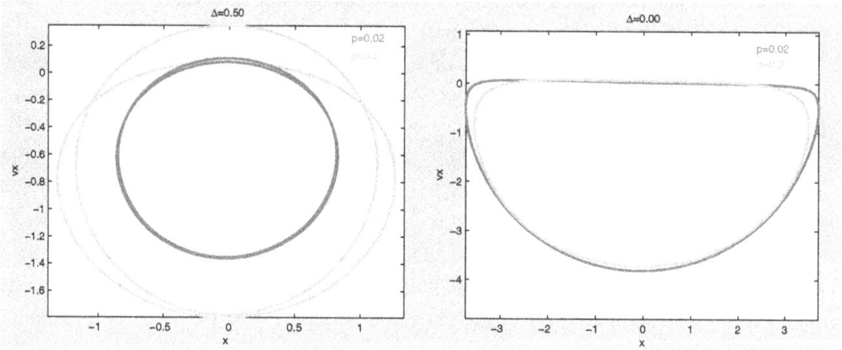

Homework 1.7: Determine the gravitational pulls on the moon by the sun and the earth. Find out which of the two has stronger pull.

The sun has a mass 3.3e^5 times that of the earth, while the moon has only 1/81 of the earth. The sun is 1.5e^8 kilometers (1 AU) away from the earth, and the moon is 4e^5 km away. Thus, the sun has a stronger pull on earth than the moon.

Homework 1.9: Find the asymptotic limit of $\mathbf{log}(e^{2x} + e^x)$ for $x \gg 1$.

2x.

Homework 1.11: Describe the electron by the fluid equation $nm_e dv/dt = -dn\, k_B T/dx - neE$ (cf. Eq. (1.3)). Show that $\delta n/n_0 \approx e\delta\varphi/k_B T$ is reached in equilibrium, where the electron density $n = n_0 + \delta n$ and $\delta n \ll n_0$. The electric field is electrostatic and expressed by its potential as $E = -d\delta\varphi/dx$.

$k_B T d\delta n/dx = n_0 e\delta\varphi/dx$ by letting $d/dt \to 0$.

Chapter Two

Tools for Analysis

"Prediction is difficult, especially about the future."

Yogi Berra

A theoretical investigation may get initiated by following the hints from the experimental observations, physical considerations or both. To arrive at a conclusion and to ensure its correctness, however, a theoretical undertaking nowadays may require both numerical and analytical computations. The compact analytical formulae have the advantages of clarity, insight, and convenience, while the numerical analysis provides the ultimate solution and the vivid graphical visualization. MATLAB offers the symbolic computation and reduces the overhead in programming and consequently, the time to get the answer, and the effort to visualize the result. This is useful at the beginning of a scientific endeavor when the problem may not be well defined, and quick answers on a few conjectures are desired. After the problem gets mathematically formulated and well defined, and an efficient code is geared toward production phase or even commercial application, C++ or its variants of internet ready programming languages such as Java, PhP, or Python is the recommended choice. Here, we start by introducing MATLAB for use in computer programming.

2.1 MATLAB

The name MATLAB is derived from MATrix LABoratory. It is a matrix based language, and provides great convenience for programming the mathematical formula in terms of vectors and matrixes. For beginners, it is easy to execute in the command window the interactive mode. By default, any output is immediately printed in the window. MATLAB is conveniently an interpreter, therefore easy to debug. It is a high level language with many mathematical and graphical facilities to present a solution as requested. It is a powerful tool simple enough for nonprogrammer to go beyond the spread sheet and yet it is sophisticated enough for serious programmers to implement codes to solve problems by the **finite element method** (**FEM**) or in the **finite difference time domain** (**FDTD**).

A script file in the extension '.m' will be a better way to run a program of substantial coding effort. MATLAB is convenient in its function overload and operator overload as in C++. The strengths of MATLAB include algorithms, versatile functions, matrix manipulation and programming tools. While it was not designed for symbolic computation, it makes up for this weakness by allowing the direct link to Maple, and has many functions available to execute a mathematical manipulations such as integration (INT), differentiation (DIFF), limit (LIMIT), Taylor expansion (TAYLOR), summation (SYMSUM), factorization (FACTOR), polynomial roots (ROOTS), and solution of algebraic equations (SOLVE), etc. The parenthesized letters are the function names that will be spelled out in upper case in this Chapter. When applied to MATLAB, it is understood that only lower case will be recognized.

The output of each command line is automatic unless a semi-colon (;) ends the line. This is by far the smartest design in all available languages. C coding would require for example a 'PRINTF' function to output a text line, say "printf("Hello, World!" \n);" C++ may use "COUT<<" as in "cout<<"Hello, World!"<<endl;", and JAVA coding would have to import the applet and graphics to draw a string at the position follows the string: "g.drawString ("Hello, World!", 50,25), while FORTRAN needs "WRITE" function as in "write("Hello, World!"). A line can be commented out by adding % in the front. There is no notation for commenting out a section like C++'s '/*' and '*/' pair, but you may comment out the whole section by Control R and repeal that by Control T.

MATLAB's ability to create vector and matrix variables, and evaluate numerically with the built in functions for matrix inversion (INV), Fast Fourier Transform (FFT), eigenvalue and eigenfunction solver (EIG), singular value decomposition (SVD), cubic spline data interpolation (SPLINE), etc., allows the programmer to get the result with the least number of lines of coding. It also provides for the sparse matrix the declaration (SPARSE) to save memory, and the corresponding functions such as the eigenvalue solver (EIGS). The ease in getting the graphical output by plot (PLOT), contour plot(CONTOUR), 3d plot (PLOT3), 3d colored surface plot (SURF), is rather impressive and rewarding. There are also tools for data analysis such as the curve fitting (CFTOOL).

2.1.1 *Symbolic Analysis*

To get a sense of how the programming in MATLAB works, we shall start with a few examples of symbolic analysis. At the MATLAB prompt, we may define the variables first by "SYMS". Sometimes we need to define the types of the variable, either "REAL", "INTEGER", "COMPLEX", or "POSITIVE". These can be placed following the sequence of the variables. Let us examine the following:

syms x; limit(besselj(1,x)/x,x,0) ans = ½	%Define a symbolic variable x; %Obtain the formulae: $\lim_{x\to 0}\frac{J_1(x)}{x}=\frac{1}{2}$
syms x; taylor(besseli(0,x),x,4) ans = 1+x^2/4	%Taylor expand $I_0(x)=1+x^2/4+\ldots$ to four terms, or the third order in x.
syms x; taylor(besseli(0,x),x,5) ans=1+x^2/4+x^4/64	%Taylor expand $I_0(x)=1+x^2/4+x^4/64+\ldots$ to five terms, or the fourth order in x.
c = tan(pi) c = -1.2246e-016	%Evaluate $\tan(\pi)$. The answer is returned in c and is given by the limit of precision, eps=2.2204e-016, which is considered to be zero.
syms f z; f=factor(z^4-1) f=(z-1)*(z+1)*(z^2+1)	%Factorize z^4-1.
syms x; z=diff(besselk(0,x)) z=-besselk(1,x)	% $\frac{dK_0(x)}{dx}=-K_1(x)$
syms x lambda positive; int(exp(-lambda*x),x,0,inf) ans=1	%Find $\int_0^\infty dx e^{-\lambda x}$ and the answer is $1/\lambda$. %The POSITIVE declaration is necessary for λ.
roots([1, 0, -1]) ans = -1 1	%Find roots of the polynomial x^2-1. %The answer is $x=\pm 1$.
syms x A; A=solve('3*x^2-4*x+1=0') A= [1/3] [1]	%Find roots of a polynomial $3x^2-4x+1=0$. %The answer is $x=1/3$ and $x=1$.
syms n integer; symsum(-1^n/n^4,1,inf) ans=-pi^4/90	% $\sum_{n=1}^{\infty}(-1)^n/n^4=-\pi^4/90$.
syms x A B; A=[1 x; x 1]; B=inv(A) B= [1/(1+x^2), -x/(1+x^2)] [x/(1+x^2),1/(1+x^2)] I=eye(2);	%Find the inversion of a matrix by inv. % Define a 2x2 unit matrix

Note that if the variable is not spelled out, then "ANS" is the default variable, which could be used for the next evaluation. It is a temporary holder of the output and will be replaced in the subsequent operations. There are other functions handy for the symbolic operations. The "SIMPLE" and "SIMPLIFY" would reduce the algebraic expression to a more manageable form, and "PRETTY" would format the equation to a typesetting style.

```
B=A\I                                  % Find the inversion of a matrix by A\I.
B=
[1/(1+x^2), -x/(1+x^2)]
[x/(1+x^2),1/(1+x^2)]
B*A
ans =                                  % Verify B*A gives a unit matrix
[ 1/(1+x^2)+x^2/(1+x^2),
0]
[ 0,
1/(1+x^2)+x^2/(1+x^2)]                 % Indeed it is true, but you need to simplify the
simplify(ans)                          algebra.
ans =
[ 1, 0]
[ 0, 1]
```

Homework 2.1: Diffusion Equation: Verify that both

$$P(x,t) = \frac{1}{\sqrt{4\pi Dt}} e^{-x^2/4Dt} \quad \text{and} \quad P(x,t) = \frac{x/t}{\sqrt{4\pi Dt}} e^{-x^2/4Dt}$$

are solutions to the diffusion equation

$$\frac{\partial P(x,t)}{\partial t} = D \frac{\partial^2 P(x,t)}{\partial x^2}.$$

Homework 2.2: Show that **a.** $\sigma \equiv \int d\tau \int d\tau' \frac{e^{-r-r'}}{|\vec{r}-\vec{r}'|} = \frac{5}{4}$

b. $\eta \equiv \int d\tau \int d\tau' \frac{e^{-r^2-r'^2}}{|\vec{r}-\vec{r}'|} = \frac{\sqrt{2\pi}}{16}$

where $\int d\tau \equiv \int_0^\infty r^2 dr \int_{-1}^{1} \frac{d\mu}{2} \int_0^{2\pi} \frac{d\phi}{2\pi}$, and $\mu \equiv \cos\theta$.

Homework 2.3: Hydrogen atoms: A hydrogen atom has a proton of measure zero size and an electron of charge density $\rho_e(\vec{r}) = -qe^{-2r/r_b}/(\pi r_b^3)$ centered at the proton, where r_b is the Bohr radius.. Work out and plot the interaction potential for two hydrogen atoms as a function of their distance. Do these two hydrogen atoms attract or repel each other?

2.1.2 *Vector Calculation*

The matrix operation in MATLAB is both convenient and powerful for scientific and engineering analysis. The ease in creating the vector and matrix variables with the built-in functions could allow a beginner to start some serious ccomputations. There are these following matrix constructions: "EYE" for identity matrix, "RAND" for random number matrix, "ZEROS" and "ONES" for matrices with elements of zeros and ones. MATLAB has the operator overloaded with convenient notation. When two matrixes are multiplied together, a '*' represents the normal matrix multiplication, while '.*' represents the element-element multiplication. The same principle applies to division and power. The exponential, logarithmic, transcendental and other more elaborate functions are by default operating on element by element basis.

A =[1 2 3 4 5 6 7 8 9 10] A = 1 2 3 4 5 6 7 8 9 10 B = [1:10] B = 1 2 3 4 5 6 7 8 9 10 C = 1:10	%"A", "B", and "C" are all vectors. They are equal. Note the use of the colon ":" operator. "m:n" counts from m to n by a default addition of unity.
C = 1 2 3 4 5 6 7 8 9 10 D = 0:2:10 D = 0 2 4 6 8 10 A(3) ans = 3 A(2:7) ans = 2 3 4 5 6 7 A(:) 1 2 3 4 5 6 7 8 9 10	%More uses of the colon. Note that you can use it to get slices of a vector or get the whole thing. Equivalent to a wild card in choosing elements.
E = [1; 2; 3]; E=E' E = 1 2 3	%A single quote " ' " computes the transpose of a matrix, or in this case, switches between row and column vectors.
E * E' ans = 14 E .* E ans = 1 4 9	%* is a matrix multiplication, and requires the correct dimensions. " .* " is entry-by-entry multiplication.
G = [1 2 3; 4 5 6; 7 8 9] G = 1 2 3 4 5 6 7 8 9	%Entering a matrix.
G^2 ans = 30 36 42 66 81 96 102 126 150	%This multiplies the matrix by itself.
G .^ 2 ans = 1 4 9 16 25 36 49 64 81	%The second squares each entry in the matrix.

rand('state',sum(100*clock));	% resets the seed of random number generator to a different state each time
R=rand(2) R = 0.9115 0.3773 0.9243 0.7640 R=rand(2,3) R = 0.0359 0.5924 0.2779 0.6580 0.8660 0.5482	% New implementation may require the following coding: RandStream.setDefaultStream(RandStream ('mt19937ar','seed',sum(100*clock))); %Generates a matrix with entries randomly distributed between 0 and 1
N=randn(5) N=-0.0056 1.8185 -0.0380 0.9136 0.9326 -1.1437 0.0256 -0.2912 0.1606 0.1108 0.4290 0.0730 0.6506 -1.1748 0.4845 -0.6803 -0.1913 -0.7063 0.2482 0.7983 -0.1576 -0.0147 1.7010 -0.0367 0.4278	%Generates a 5x5 random numbers in normal distribution.
N=zeros(2) N = 0 0 0 0	%Generates a 2x2 null matrix.
U=ones(2) U = 1 1 1 1	%Generates a unitary matrix.

Always type "help FUNCTION" at the command prompt to get the helpful hint, where FUNCTION is the particular function name you are unfamiliar with. For example, 'help rand' would suggest to add this at the command prompt: "rand ('state', sum(100*clock));", or "RandStream.setDefaultStream (RandStream ('mt19937ar','seed',sum(100*clock)));" the random number generation may be reset to a different state every time. This is particularly useful when you are doing **Monte Carlo** or **Simulated Annealing** computations. There are vector operators such as the inner product "**DOT**", the cross product "CROSS", the determinant 'DET', "GRADIENT", "DIVERGENCE", "CURL", the discrete Laplacian "DEL2", the inversion "INV" and the eigenvalues and eigenvectors of a matrix "EIG". It is most convenient to find the solution to a matrix equation AX = B by the simple operation 'X = A\B'.

I=eye(2) I = 1 0 0 1	%Identity matrix I.
W=[I, I, I] W = 1 0 1 0 1 0 0 1 0 1 0 1	%This shows the convenience of symbolic operation.
V=[I; I; I] V=W' V = 1 0 0 1 1 0 0 1 1 0 0 1	%The two expressions give the same result.
W=[1,4,7;2,5,8;3,6,9], V=reshape(W,1,9), W= 1 4 7 2 5 8 3 6 9 V= 1 2 3 4 5 6 7 8 9	%Sometimes, one needs to rearrange the dimension of a matrix. It is important to notice that the memory arrangement is starting from the column vectors. Therefore, the first column is loaded and then the second, and so on. RESHAPE returns the same number of elements but rearrange them according to the specified dimension. The example converts matrix to a vector.
U=reshape(V,3,3), U= 1 4 7 2 5 8 3 6 9	%Convert the vector to the matrix. It follows the same rule that the column comes first.

Homework 2.4: Given M = [2 1;1 0]; and C = (1.8 < M&M < 2.1)*0.5 + (1.1 > M&M > 0)*0.25. What is C?

Homework 2.5: Find the eigenvalues of matrix

$$X = \begin{vmatrix} 1 & 2 \\ 2 & 1 \end{vmatrix},$$

and verify the result.

2.1.3 *Graphical Presentation*

A picture is worth a thousand words. Sometimes to see is to understand since visualization may eliminate some blind spots. The simplest way to plot is using 'PLOT' as in "x = 0.01:0.1:10; plot(x,exp(-x));". This will produce an x-y plot of the function y = exp(-x) from $x = 0$ to $x = 10$. Here are a few examples:

```
function plotMaxwell                    %Plot the Maxwellian distribution
v=-5:0.025:5;                           %The data points are marked with *,
f=exp(-v.^2)/sqrt(pi);                  connected with a solid line and in green color.
plot(v,f,'*-g')                         %A text of description is added.
axis([-5 5 0 0.601]);
title('Maxwellian Distribution');
xlabel('v');
ylabel('f');
text(1,.3,'f=exp(-v^2)/sqrt(\pi)')
```

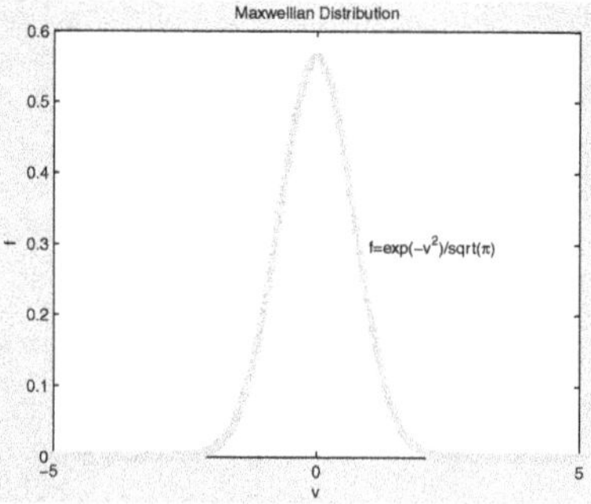


```
function PlotDatawErrorBar                      %Plot Data with Error Bar
err=[0,0.014,0.028,0.042,0.056,0.069,...        %This also demonstrates how to HOLD ON
0.083,0.097,0.111,0.125,0.111,0.097,...         to a graph for additional plots.
0.083,0.069,0.056,0.042,0.028,0.014,0];         %The red diamonds are marked on the second
eta=[0,0.210,0.395,0.556,0.691,0.803,...        data curve.
0.889,0.951,0.988,1.000,0.988,0.951,...
0.889,0.803,0.691,0.556,0.395,0.210,0];
eta2=[0,0.105,0.198,0.278,0.346,0.401,...
0.445,0.475,0.494,0.500,0.494,0.475,...
0.445,0.401,0.346,0.278,0.198,0.105,0];
L=eta.*err; U=eta.*err;
theta=0:5:90;
errorbar(theta,eta,L,U,'*-g')
axis([0 90 0 1.25]);
title('Efficiency');
xlabel('Theta Angles');
ylabel('eta');
hold on;
L=eta2.*err; U=eta2.*err;
errorbar(theta,eta2,L,U,'d-r')
```

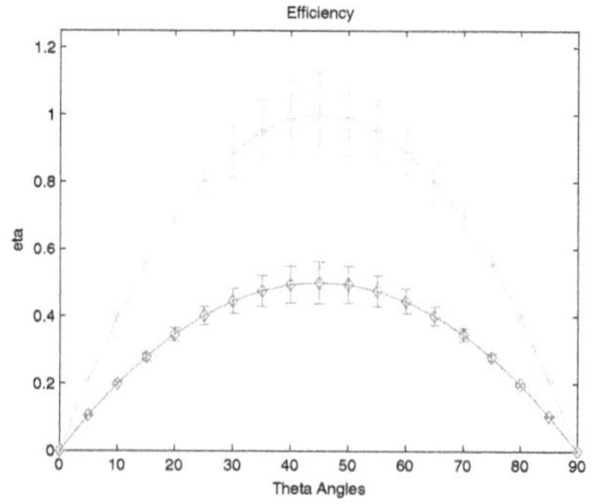

function ESPP
d=0.1; R=5;
N=fix(2*R/0.1+1);
x=-R:d:R;
y=x;
X=repmat(x,N,1); Y=repmat(y',1,N);
PHI=2./sqrt(X.^2+(Y-2).^2+eps);
PHI=PHI+2./sqrt(X.^2+(Y+2).^2+eps);
PHI=PHI-1./sqrt((X-1).^2+Y.^2+eps)-1./sqrt((X+1).^2+Y.^2+eps);
value=[-1.5,-1,-0.75,-0.5,-0.35,-0.25,0,0.25,0.35,0.5,0.75,1,1.5];
contour(PHI,value);
	%Electrostatic potential plot
	%Evaluate the phi value due to two negative charges placed along the x-axis, and two positive charges on the y-axis
	%Replicate the x,y coordinates into the square matrixes
	%Create the phi values in (x,y) plane
	
	% Request a contour plot

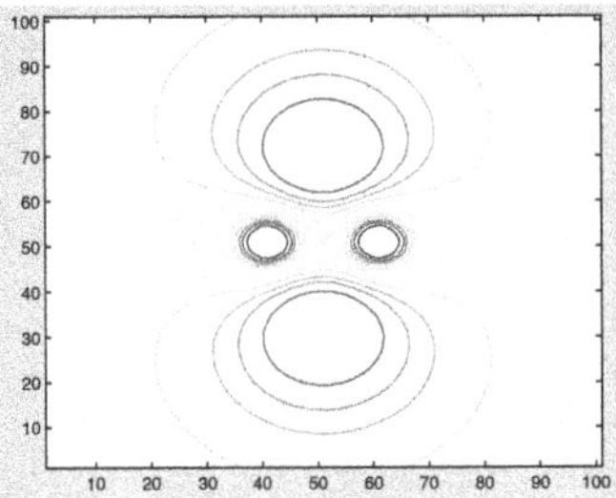


```
function gyroorbit
dt=0.025;
vz=rand(1),
vp=rand(1),
theta=0;
z=0;
figure; hold on;
title('Gyro Orbit');
xlabel('X');
ylabel('Y');
zlabel('Z');
for i=1:1000
   z=z+vz*dt;
   theta=theta+dt;
   x=cos(theta);
   y=sin(theta);
   plot3(x,y,z,'r-*');
   if(mod(i,10)==0)
        j=fix(i/10);
        M(j)=getframe;
        view(-2*j,30);
        end;
   end;
```

% A movie clip on gyrating orbit that also has the longitudinal motion.

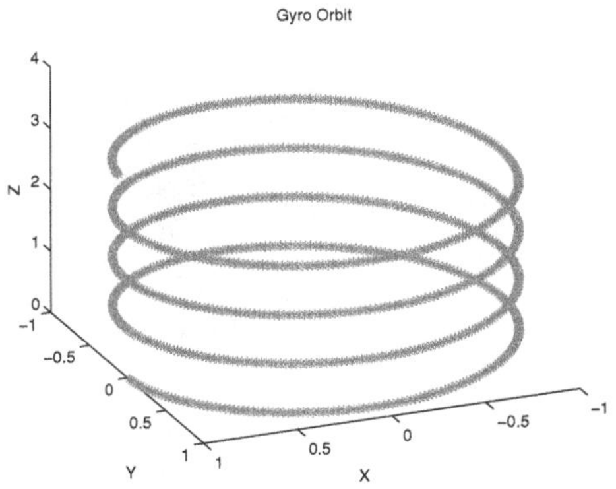

The last two examples plot three dimensional graphics, which may be rotated inside MATLAB runtime interface platform. It may also be recorded as a movie file with use of MOVIE2AVI for visualization although at a lower resolution. The command to play the movie is MOVIE.

```
function bifurcation(LAMBDA)   %LAMBDA=1.5;
x=-1:0.01:1; y=x;
N=length(x);
X=repmat(x',1,N); Y=repmat(y,N,1);
r=sqrt(X.^2+Y.^2);
y01=2*log(4/LAMBDA*(1+sqrt(1-LAMBDA/2)));
y02=2*log(4/LAMBDA*(1-sqrt(1-LAMBDA/2)));
alpha=LAMBDA*exp(y01)/8;
J=1./(1+alpha*r.^2).^2;
figure;
surf(X,Y,J);
xlabel('X'); ylabel('Y'); zlabel('J');
title('Bifurcated States - Peaked Current Profile');
%shading flat;
alpha=LAMBDA*exp(y02)/8;
J=1./(1+alpha*r.^2).^2;
figure;
surfc(X,Y,J);
xlabel('X'); ylabel('Y'); zlabel('J');
title('Bifurcated States - Flat Current Profile');
```

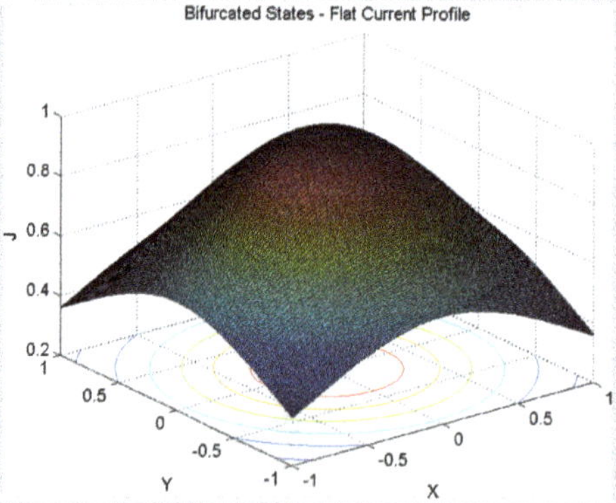

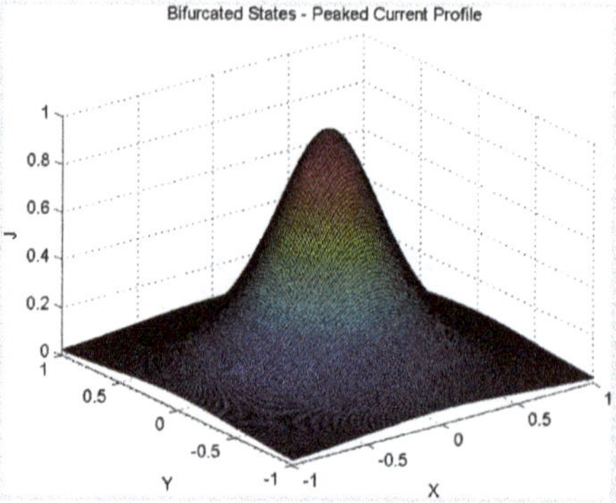

% Calculate the helical flux function given by

$$\Psi(r,\theta)=\tfrac{1}{2}B_0\varepsilon(r/a)^2-\frac{I_0}{2\pi}\ln\left(\frac{1-2(r/a)^2\cos(\theta-\delta/2)+(r/a)^4}{1-2(r/a)^2\cos(\theta+\delta/2)+(r/a)^4}\right)$$

```
function HelicalB
B0=1;
I0=0.2;
delta=0.5;
x=-0.9:0.025:0.9;
y=x;
N=length(x);
X=repmat(x',1,N);
Y=repmat(y,N,1);
r=sqrt(X.^2+Y.^2);
theta=atan(X./Y);
F=(1-2*r.^2*cos(theta-
delta/2)+r.^4);
F=F./(1-
2*r^2*cos(theta+delta/2)+r.^4);
PSI=B0*r.^2/2-I0*log(F);
surfc(X,Y,PSI)
xlabel('X');
ylabel('Y');
zlabel('B');
```

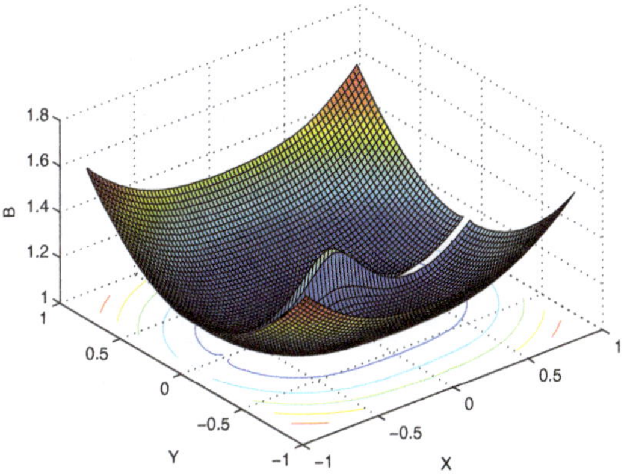

Homework 2.6: Run the following program and explain the meaning of each line of coding.

```
function plotFFT
for k=1:50
    plot(fft(eye(k+5)));
    axis equal;
    M(k)=getframe;
end;
movie(M,2);
```

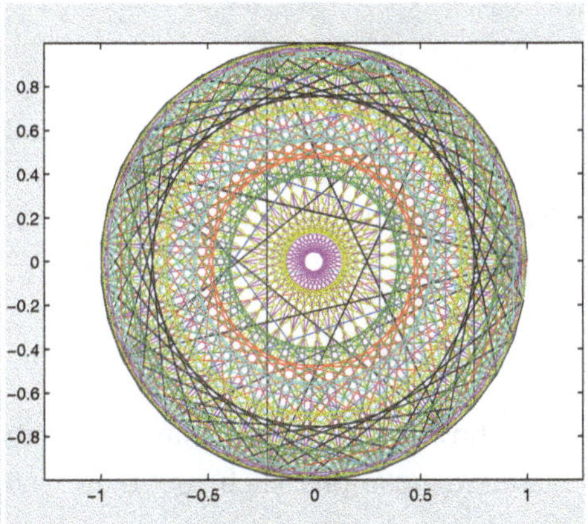

Homework 2.7: A student designed the following program. The density profile as shown looks nice. However, its density gradient is unacceptable. Plot the density gradient and show the unacceptable spikes. Modify the code to make the density gradient continuous.

```
function DensityProfile
N=1500;  Lz=100; dz=Lz/N;
z=(1:N)*dz;
nH=5;
nL1=0;
nL2=0.5;
z1=5.5*Lz/10;
z2=7*Lz/10;
zc1=4*Lz/10;
zc2=9*Lz/10;
Lc1=5; Lc2=7;
n0=tanh((z-zc2)/Lc2)*(nH-nL1)/2+(nH+nL1)/2;
n0(1:fix(z1/dz))=tanh((z(1:fix(z1/dz))-zc1)/Lc1)*(nL2-nL1)/2+(nL2+nL1)/2;
n0(fix(z1/dz+1:z2/dz))=nL2*ones(1,fix(z2/dz-z1/dz));
n0(fix(z2/dz)+1:N)=tanh((z(fix(z2/dz)+1:N)-zc2)/Lc2)*(nH-nL2)/2+(nH+nL2)/2;
plot(z,n0,'g-');
```

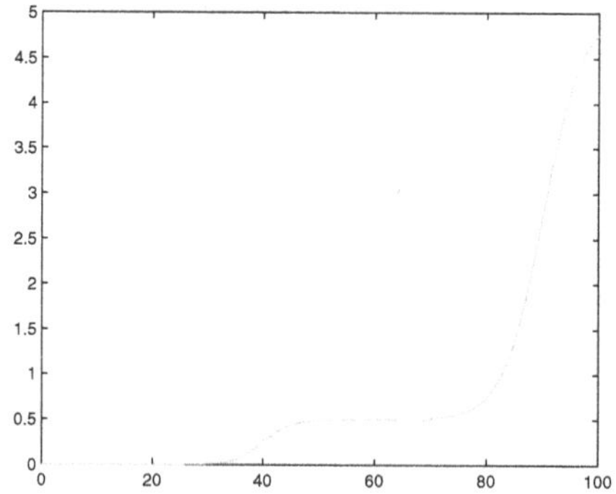

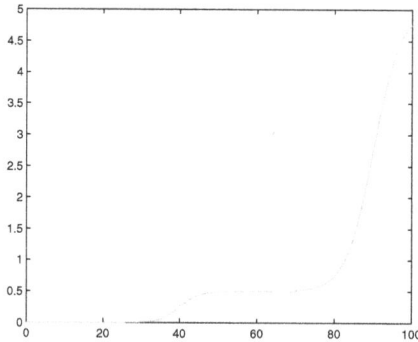

Homework 2.8: **Electric Field Line Tracing:** A charge q is placed at a distance d from a conducting wall of infinite dimension. Draw the electric field lines.

2.1.4 *Program Control*

This section shows how to utilize functions to ease the program control. MATLAB adopts fairly standard notations such as "FOR" or "WHILE" loop. The vector operation in MATLAB often eliminates "FOR" loop, speeds up the calculation, and makes the coding more understandable, as demonstrated in the function below.

function A=mcPI(n)
p=rand(2,n);
r=sqrt(sum(p.*p,1));
hits=(r<=1);
A=4*sum(hits)/n;
	% This function finds π by the Monte Carlo method. It compute the probability of hits inside a circle of unitary radius.% type mcPI(20000000)
	ans=
	3.1419

A structured program will at least adopt the functional approach. The following has a main program solving the **Fibonacci problem** of walking up a stair with one or two steps each time to find out how many ways to walk up a star of N steps. It applies the **recursive formula** that can be very powerful to instruct the computer to solve a problem that would otherwise be difficult to program as demonstrated in the Tower of Hanoi in Chapter One. This is a **divide-and-conquer** strategy that often reduces the computational effort.

function Stairs
N=input('What is the total number of stairs? ');
The_Number_of_Ways=Fibonacci(N)
%--
function W=Fibonacci(n)
if(n<=2) W=n; end;
if(n>2)
W=Fibonacci(n-1)+Fibonacci(n-2);
end;
	% Define a function to count the number of ways to walk up N stairs.
	% With use of the recursive formula, the problem is easily programmed.
	% There is two ways to walk up two stairs. Then walk to n steps is simply the sum of the number of ways to arrive at n-1 and n-2 stairs.

Functions accept arguments, and they remove the redundancy of coding repeated operations. It also helps debugging effort. The main program is best written in a format consisting of pseudo codes, which simply spell out the intent and define the flow chart of work as requested, but with the functions to perform the real tasks so to complete a job. An interaction session is particularly useful for quick checking on ideas or data analysis. But for continuous and more productive effort, it is always as a practical matter to save the instructions into an m-extension file for later use. MATLAB statements can be prepared with any editor, and stored in a file, referred to as a script, or an "m-file", since they must have names of the form xxx.m to easily rerun. MATLAB may be run in the "batch mode." At the MATLAB prompt type:

```
Job=batch('runBatch.m');
```

This allows runBatch.m to run in the background thread, and other front running job can still proceed.

Homework 2.9: Convert for-loop to matrix operation: Rewrite the following program to eliminate the for-loops in favor of the matrix computation. Create the matrix H from random numbers. Compare the computing speed for n=10, 100, 1000, and 3000.

```
function K=getK(H,n)
for i=1:n
  for j=1:n
    K(i,j)=(i-1)*H(i,j);
  end;
end;
```

Homework 2.10: You may utilize the function rand(m,n,p,...) to create a matrix/vector of random numbers within (0,1). This is useful for Monte Carlo approach to simulate many kinds of events such as Q-line, traffic jam, particle simulation, etc., and form a theoretical basis by statistical approach. Given needles of unit length and a piece of paper horizontally lined at unit spacing, you are to randomly toss these needles onto the paper. Find out the probability of a needle crossing a line by both the analytical method and MATLAB simulation, and compare the result.

2.1.5 *Utilities*

Besides the usual mathematical functions, many other utility functions are available such as the easy 3d plot (EZSURF, EZPLOT3), eigenvalue solver (EIG, EIGS), Fast Fourier Transform (FFT), Delaunay triangulation (DELAUNAY), and many more. File handling routines are standard with FSEEK, FSCAN, FWRITE, FOPEN, FREAD, FREWIND, FEOF, FCLOSE, FGETS etc., to manage the file handle. There are many built in mathematical functions such as Bessel functions, BESSEL(H,I,K,J,Y), elliptic fnctions, ELLIP(J,KE), error functions (ERF), cosine integral function (COSINT) and sine integral function (SININT). A quiver plot (QUIVER), which displays velocity vectors as arrows with components at specified coordinates, is also very useful. It is always advantageous to use sparse (SPARSE) if the matrix is not dense, which would be compressed for storage saving and handled with speedy operations.

The next example demonstrates the eigenvalue solver by solving a model equation for the Alfven eigenmode in the toroidal geometry:

$$\frac{\omega^2}{\omega_A^2}\Psi = -\frac{1}{q^2}\frac{\partial^2}{\partial\theta^2}\Psi + n^2(1 - 2\varepsilon\cos\theta)\Psi. \tag{2.1}$$

```
function MATHIEU
close all; clear all; clc;
N=1000; epsilon=0.2; q=2; n=1;
dTheta=2*pi/N,
theta=-pi+dTheta/2:dTheta:pi-dTheta/2;
dt2=(dTheta*q)^2;
E=2/dt2+n^2*(1-epsilon*cos(theta));
A=diag(E)-diag(ones(1,N-1),1)/dt2-
diag(ones(1,N-1),-1)/dt2;
A(1,N)=-1/dt2;
A(N,1)=-1/dt2;
```

% Find the eigenvalues with use of the eigenvalue solver for sparse matrix. The option structure stops verbose.

% lambda =
2.0191 1.3883 1.1988 0.8053

% f =
0.8974 ω / ω_A
is the lowest frequency.

% The mode resides mainly in the bad curvature region.

```
options=struct('disp',0,'tol',1.0e-5,'maxit',900);
[v,lambda]=eigs(A,4,'SM',options);
lambda=diag(real(lambda))',
[lambda,idx]=sort(lambda,2,'ascend');
f=sqrt(lambda(1)),
plot(theta,-v(:,idx(1)),'g*');
title('Alfven Eigenmode');
axis tight;
xlabel('\theta');
figure;
plot(theta,-v(:,idx(2:4)));
axis tight;
```

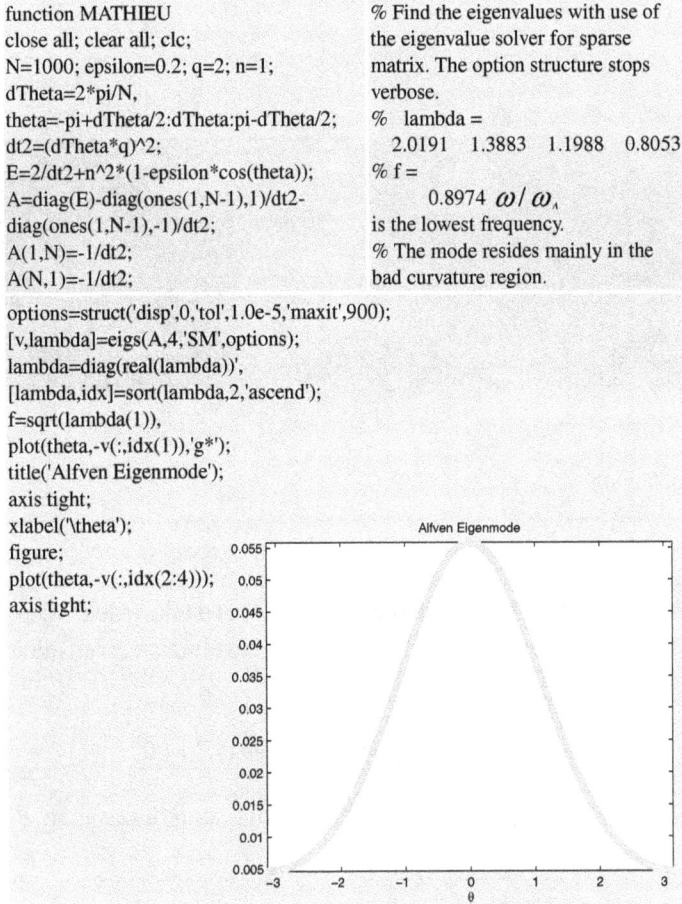

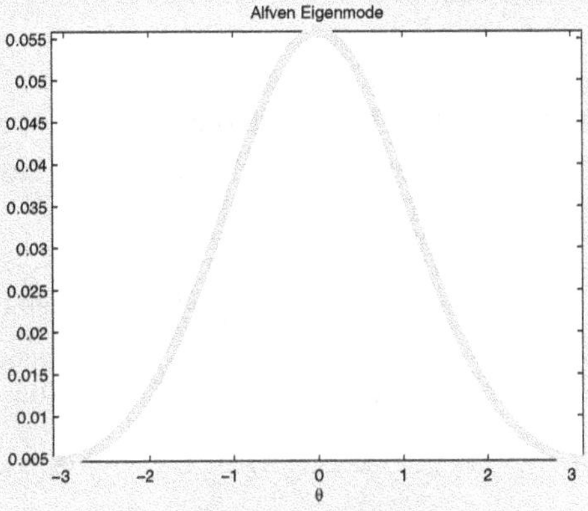

There are additional utility suites for specific purposes. The differentiation matrix suite allows the spectral methods to solve partial differential equations. The calling of C or C++ programs from MATLAB allows the convenient interface with source codes written in a difference language. Other utilities such as optimization, data analysis, data acquisition, signal processing, mechanical design and computational biology are currently available. SIMULINK uses an icon-driven interface for the construction of a block diagram representation for multidomain simulation and model-based design for dynamical systems. It provides an interactive graphical environment and a customizable set of block libraries, readily extendable to specialized applications. The following potential flow passing around an infinitely long cylinder that is sticking out of the paper is solved by the PDETOOL in MATLAB. The arrows show the quiver plot of the velocity $\vec{v} = \nabla\phi$, and the color map indicates the values of the potential ϕ.

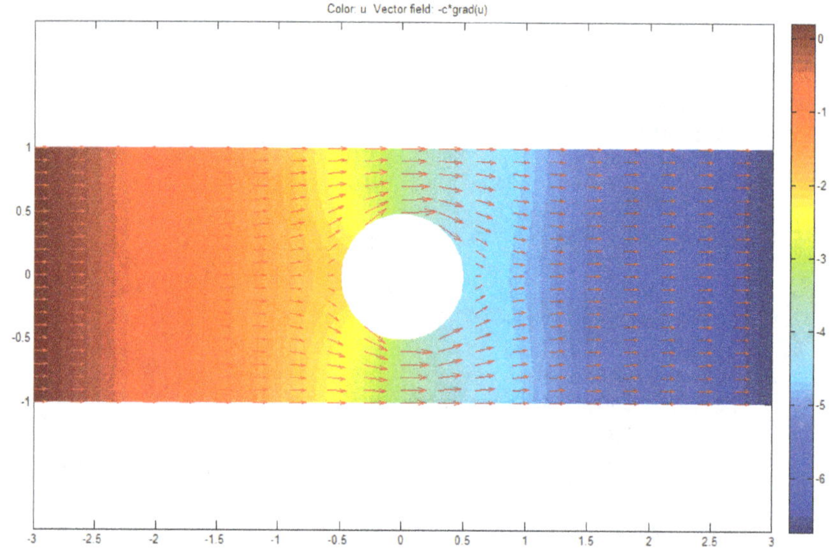

Homework 2.11: Electric potential in a Conical Hole (Jackson p.104). The Laplace equation for the electric potential can be cast into an eigenvalue Legendre equation,

$$\frac{d}{d\mu}(1-\mu^2)\frac{d}{d\mu}P + \nu(\nu+1)P = 0$$

Solve in MATLAB this equation given the boundary conditions that: $P(\mu = \beta < 1) = 0$.

Homework 2.12: King Arthur's Round Table:
Long time ago, King Arthur was getting old and wanted his beautiful Princess who had grown up to find a husband. The princess told the king that she wanted someone who was good at math. The king asked how to do that. The Princess suggested to put the knights around a table with the seats marked from 1 to N. The first number would be kept alive, the next one killed, and going around the table until only the last person survived. The Princess smiled and said she would marry the person who knew where to sit, and of course, she would not really kill anyone. The first few survivors are given as:

N	1 2 3 4 5
S	1 1 3 1 3

(a) Write down the survivor number s for the given N, where N ranges from 1 to 17.

(b) What is the number after 17 that would give s = 1 as the survivor?
(c) Figure out the condition when the survivor is the number 1 seat.
(d) Figure out a formula so that s = F(N).
(e) Write a simple function in Matlab to solve the problem.

Homework 2.13: Laplace Transform:
Solve in MATLAB the following equation $d^2x/dt^2 + \lambda dx/dt + x = 0$given the initial values: $x(t = 0) = 1$ and $x'(t = 0) = 0$, making use of the Laplace transform.

Homework 2.14: Laplace Equation:
Consider the problem of the electrostatic potential in two dimensions. Divide the space into 201 grid points in x and y directions. The outermost box has a potential of 100. The inner boundary box bounded by the four lines at $x = \pm 10$, $y = \pm 10$ has the potential at 0. Solve the Laplace equation by iteration. Take the potential at any point other than the boundary equal to the average of its nearest four neighbors. Make a contour plot of the equipotential surfaces at the potential values of (1, 20, 40, 60, 80, 99). Set up an error checking routine to ensure the desired accuracy.

2.2 Object Oriented Programming

C and Fortran languages are popular in scientific computing. The modern computer languages such as Java, Perl, PhP, and Python, etc., have features derived from the first language of **Object Oriented Programming** (**OOP**), namely, the C++. It is important for any programmer to understand the basic concepts of OOP, such as **data abstraction**, **encapsulation**, **inheritance**, **modularity** and **polymorphism**, and the basics in computer architecture such as memory allocation, **message passing interface** and pointer. These concepts are helpful in both developing a numerical solution and communicating with coworkers. The intricacy of many programming languages represents important evolution of the programming techniques such as case sensitive, type sensitive and content sensitive, operator and function overload. Many concepts are commonly adopted in modern programming languages, and are routinely used in MATLAB programming as well. Good programming sense may be developed from criteria to evaluate software, while subjective and qualitative, consisting of the following: robustness, usability, scalability, portability, versatility.

They are important for developing a better code. A word of caution however is that before indulging oneself with heavy computing, always use brain power before computer power, and always use one CPU before multi-CPU.

OOP emphasizing on modularity, is easier to develop and to be understood later on. OOP has a collection of collaborating objects, as opposed to the traditional instructions or procedures. Moreover, an object may own the methods and data that are classified carefully into the private, the protected, and the public type. This is the encapsulation that allows inheritance selection, as well as the modularity. Each object is like a toolbox, containing many gadgets to handle a specific type of mission. Each object is also capable of sending/receiving messages to/from other objects. This is an idea first conceived in the Smalltalk at Xerox PARC by **Alan Kay** and associates in the early 1970s.

2.2.1 *A Quick Tour of C++*

Shorthand symbols: The C language utilizes extensively the shorthand notations. Almost every symbol on the keyboard has a place in the C language. This simplifies scientific writing greatly and makes the coding more readable. Long, hyphenated or underscored terms often result in more confusion than clarity; these were taught in the old days as a product from the era of ambiguous function names as in the IMSL library. The convention of "capitalize the first letter of a word" offers good clarity, such as MonteCarloDistribution, or minX. The simplicity in variable names could make a code more readable especially in a large code development.

Preprocessing: In the header files, C++ defines the constants, synonymous operations, common objects, and fetches the standard libraries of working data, procedures and objects. The declarations in the header files provide a very good programming style in forward looking, and make C++ a very structured and portable language. Additional comments enclosed by /* and */ improve the code readability and documentation of coding purpose. For single line statement, the // would do. The ease to comment out a section of coding also makes the debugging and diagnostics by elimination or substitution practical.

Types: C++ is a type sensitive language. As such, it checks on the syntax and programming errors by keeping track of the type consistency. It may, however, sometimes cause trouble for the programmer. For example, a numerical calculation with the single/double precision of floating numbers may result in a compiler error when mixed with integers in a formula. Modifying a number 4 to 4.0 to make the compiler happy and get fatal error eliminated can be quite an experience. The stringent rule on type comes with another ingenious design of the function overload that allows a procedure to perform same tasks or otherwise for different types, and eases the utilization of the library functions, default or customized alike. The C++ language also allows bit operation, direct memory address accessing (with pointer type), and execution of functional modules by giving the function address pointer. It is rather versatile to have the capability as the low level assembly and the convenience as a high level language.

Expression: Some expressions in C++ language are special and unique, and provide the convenience and power. For example, $i = i + 1$ can be written as $i++$; that can be conveniently included in do loops for index increment. The notation of ++i and i++ are not the same and $y = (\text{++i}) * x$ would not equal to $y = (i\text{++}) * x$. The former expression is equivalent to $i = i + 1$; $y = i * x$; while the latter is to $y = i * x$; $i = i + 1$; resulting in different values in y. As defined here, the function pointer: int f(void), (*pf)(void); and the address pointer: int a[10], *p; conveniently passes the data/code segment to other objects/procedures/ functions. The expression $x+ = y$ means: add y to x, that is excellent in its shorthand writing. This convention applies to almost all the operations such as -,*,/,^,|,&,etc. The logical expression: z? x:y evaluates the formula z and assign x if true, or y otherwise.

Library: There are many libraries in C language that can be fetched by including the header files and readily applicable. The stdio.h has the file and string handling, the stdlib.h has the memory allocation and free functions, the math.h is indispensible for numerical work, and time.h allows you to monitor your CPU effort to find the bottleneck of computation.

Key Words: Many keywords are important as they allow the best program construct to be effectively designed. A summary of important key words are included the following:

class: to define an object that contains properties (data) and methods (functions)
delete: release the memory after new has fetched
friend: a friend class can access all members of the class in question; a friend function needs not be defined in the class in question
inline: simple replacement just like MACRO
new: to fetch an instance of a class, or a memory block of any data type
operator: to define an operator overload

protected: only derived class can share
public: shared with all other classes, be it a data or a function
template: applying data abstraction to function overload or operator overload
this: a pointer to the specific object for which the fucntion is being invoked.
virtual: virtual function is a generic function the abstract base class can't define and is to be implemented in the derived class

Many concepts in C++ are useful for good programming. *Data abstraction* allows function overload and operator overload, and generalizes class as a data type. Object allows encapsulation, inheritance, and polymorphism. The virtual functions do different jobs for different derived classes with dynamical binding. The construct of class include constructor, destructor (annihlator), data (private, protected, public), function (member, virtual, private, protected, public, friend). On top of that the MPIC would allow the distributed computing to run on multi-CPUs.

2.3 Asymptology

For complicated problems, exact mathematical solutions in terms of known functions are often unattainable, but important physics may readily be revealed by taking the serial or asymptotic expansions. The behavior of the solutions at large or small parameter limits may help us figure out correct answers even if a numerical solution is inevitable and may serve to find out timely the errors in analytical or numerical approaches. Moreover, it may offer the physical insight for the better solution to be discovered. The following are some of the benefits in using **asymptology**:

(1) It can be the poor man's life saver if no other solution is likely.
(2) It provides the error checking and the needed clarity even when analytical or numerical solutions are possible.
(3) It guarantees consistency in series and asymptotic expansions.
(4) It finds the solution within the domain of validity and the condition for the theory to incur serious error or break down in its entirety.
(5) It facilitates the powerful analytical tool especially for problems of multiple time and multiple space scales. When the numerical method fails to deliver the accuracy, the asymptology may come to rescue especially in the boundary layer problem.

The ordering and the expansion are emphasized in Asymptology no matter whether the solutions are convergent or not. As long as the equations are not ill-posed, the overall physical solutions will be found by, for example, the matching procedure. It provides good roots finding, summation, and integration methods, and the capability to handle the multiple scales in time or space.

2.3.1 *Ordering*

The order of magnitude analysis is an important process for sorting out physically related effects. Scientists and physicists in particular, are often concerned about the order of magnitude of a property, which would provide the insight for the relevance of a particular effect as measured by the said quantity. As demonstrated in Chapter One regarding the pressure effect we recall that 1 Pa $= 6.25e^{12}$ eV/cm^3. This relates the pressure to the energy content in a unit volume. If we further make the following substitution 100 GPa $=$ 0.625 eV/Å^3 we may readily conclude that this order of magnitude in pressure begins to alter the atomic structure.

The orders of magnitude of physical parameters are generally distinguished by the power of 10-based numbers. For example, the wavelengths of γ-rays are of the order of 10^{-10} cm, x-rays of the order of 1 pm to 10 nm, visible lights 10^3 Å, FM wavelengths 1 meter, AM wavelengths 1 km.

There are two types of order symbols: the big O and the small o. The expression $f(x) = O(g(x))$ as $x \to x_0$ means $f(x)/g(x)$ has a finite bound as $x \to x_0$, while $f(x) = o(g(x))$ means that $f(x)/g(x) \to 0$ as $x \to x_0$. The generalized notion of the order of magnitude can best be denoted by the ordering parameter O. We may refer to a quantity as order of unity by O(1), and order of ε by $O(\varepsilon)$. When $\varepsilon \ll 1$, $O(\varepsilon)$ implies a smaller quantity when compared with the regular properties, and entails the serial expansion. On the other hand, $O(1/\varepsilon)$ implies a larger quantity.

To elaborate on the significance of the O-notation, recall that the limit function gives a precise value if it does converge, for example,

$\lim_{x\to 0} \frac{J_1(x)}{x} \to \frac{1}{2}$	syms x; limit(besselj(1,x)/x,x,0) ans=1/2	%This has less information than the second expression.
$J_1(x)/x = 1/2 + O(x^2)$	taylor(besselj(1,x)/x,x,3) ans = /2 -x^2/16	%This provides an easy error estimate.

The first expression is less informative than the second expression, $J_1(x)/x = 1/2 + O(x^2)$ for $x \ll 1$, that provides an easy error estimate for nonzero x. For example, if $x = 0.1$, the second expression tells us that there is an error of the order of 1%. It turns out that the numerical solution is 0.4994, showing that if it is approximated by 0.5, the error is 0.1%.

The o-symbol asserts the bounded nature of the investigated formula by the perhaps better known function so that $F(x) = o(G(x))$ for $x \to x_0$ means $\lim_{x\to 0} F(x)/G(x) \to 0$. For example, $I_0(x) = 1 + o(x)$, for $x \to 0$, since $\lim_{x\to 0} I_0(x) \to 1 + x^2/4 + \cdots$. However, fewer cases are likely for the o-symbol to be as enlightening as the O-symbol.

The following formula has a familiar asymptotic limit for $\Lambda \gg 1$

$$\left[1 - \frac{x}{\Lambda}\right]^{-\Lambda} \xrightarrow{\Lambda\to\infty} e^x.$$

How do we prove this? It is natural to take logarithm of the formula so that

$$\log\left\{\left[1-\frac{x}{\Lambda}\right]^{-\Lambda}\right\} = -\frac{1}{\varepsilon}\log(1-\varepsilon x) \xrightarrow{\varepsilon\to 0} x + O(\varepsilon),$$

where $\varepsilon \equiv 1/\Lambda$. Therefore,

$$\left[1-\frac{x}{\Lambda}\right]^{-\Lambda} = e^x + O\left(\frac{1}{\Lambda}\right). \tag{2.2}$$

We may also utilize the summation for $\Lambda \gg 1$ as in the following to prove the same:

$$\begin{aligned} y = \left(1-\frac{x}{\Lambda}\right)^{-\Lambda} &= 1 + (-\Lambda)\left(-\frac{x}{\Lambda}\right) + \frac{(-\Lambda)(-\Lambda-1)}{1\cdot 2}\left(-\frac{x}{\Lambda}\right)^2 \\ &+ \frac{(-\Lambda)(-\Lambda-1)(-\Lambda-2)}{1\cdot 2\cdot 3}\left(-\frac{x}{\Lambda}\right)^3 + \cdots \\ &= 1 + x + \frac{1}{1\cdot 2}x^2 + \frac{1}{1\cdot 2\cdot 3}x^3 + \cdots + O\left(\frac{1}{\Lambda}\right) \approx e^x + O\left(\frac{1}{\Lambda}\right). \end{aligned}$$

This involves the summation technique we will discuss later.

Homework 2.15: With the use of the MATLAB function **LIMIT** to find the following integral: $\int_1^x (1+\frac{1}{t})^t dt$ **for** $x \gg 1$ to leading order.

2.3.2 *Roots and Singular Solutions*

Simple polynomials can be solved for their roots in MATLAB by utilizing ROOTS and SOLVE. Consider the following:

```
syms x e I A;
I=e*x^3+x^2-6*x+8;
A=solve(I)
e=0.01;
roots([e, 1, -6, 8])
ans =
        -105.7455
          3.7019
          2.0436
```

% Find roots of a polynomial
$\varepsilon x^3 + x^2 - 6x + 8 = 0.$
% Although analytical solutions are available, the answer is cumbersome.
A=...
% Take $\varepsilon = 0.01$, and find the roots.
% The numerical solutions are: -105.7455, 3.7019, 2.0436, respectively.

Recast the equation into $\varepsilon x^3 + (x-3)^2 = 1$, the two roots solved by the regular perturbation treatment would be to lowest: $x_0 = 3 \pm 1$, and to next order, $x_1 = -\varepsilon x_0^3/2/(x_0-3)$, so that $x = 2+4\varepsilon+O(\varepsilon^2)$ and $x = 4-32\varepsilon+O(\varepsilon^2)$, in agreement with the numerical solutions. There is however a singular solution that has to be solved by the singular perturbation treatment. Taking $x = \varepsilon^{-1}x_{-1} + x_0 + \varepsilon x_1 + \cdots$ so that

$$O(\varepsilon^{-2})\colon \varepsilon x_{-1}^3 + x_{-1}^2 = 0 \Rightarrow x_{-1} = -1/\varepsilon,$$
$$O(\varepsilon^{-1})\colon 3\varepsilon x_{-1}^2 x_0 + 2x_{-1}x_0 - 6x_{-1} = 0$$
$$\Rightarrow x_0 = 6/(2 + 3\varepsilon x_{-1}) = -6,$$

we find $x = -1/\varepsilon - 6 + O(\varepsilon)$. The agreement with the numerical solutions is obvious, and the solutions are more manageable having both the accuracy and the clarity.

Given another example,

$$xe^x = t, \tag{2.3}$$

we are to solve for x. Considering small x limit, equivalent to small t regime, we have then $x + x^2 \approx t$, so that $x \approx t - t^2 + O(t^3)$. On the other hand, if we are to look for a large x solution, which is the same as the large t limit, we take a logarithm on both sides of Eq. (2.3) to give $x + \log x = \log t$, so that we have

$$x \approx \log t - \log(\log t) + O(\log(\log(t))).$$

Our next example is this equation

$$\tan(\Delta/\gamma) = \gamma\lambda, \tag{2.4}$$

which happens to be the dispersion relation for the parametric instability of plasmon in the nanolayer of near field optics, where Δ is related to the layer thickness, and γ is the growth rate. We are to solve for the weak instability limit, viz., $\gamma \ll 1$. Assuming that $\gamma\lambda \to 0$, we find $\sin(\Delta/\gamma) \to 0$. This translates to the solution $\gamma \approx \Delta/n\pi$, where n is an integer. For $\Delta/\pi < 1$, it is clear that the assumption of $\gamma \ll 1$ can be satisfied if n is nonzero. There is a singular situation when $n = 0$, however, that gives an unbounded γ. This solution can be attempted by assuming that $\Delta/\gamma \ll 1$ instead, so that Eq. (2.4) gives $\Delta/\gamma \approx \gamma\lambda$ and $\gamma \approx (\Delta/\lambda)^{1/2}$. It gives a larger γ than the other solutions for as long as $(\Delta\lambda)^{1/2} < \pi$, consistent with the assumption $\Delta/\gamma \ll 1$ or $(\Delta\lambda)^{1/2} \ll 1$.

Homework 2.16: Find roots of $\sqrt{1+x} = 1 - \sqrt{1+4x}$.

Homework 2.17: Find the roots of $\cos x = x \sin x$ to leading order.

Homework 2.18: Find the solutions to Eq. (2.4) up to the first order.

2.3.3 *Summation and Integral*

Summations are often encountered for evaluating physical properties. For example, the Madelung's constants evaluate the interaction energy of the ionic crystals. In one dimension

$$M_1 = 2\sum_{k=1}^{\infty}\frac{(-1)^k}{k} = -2\ln 2 \approx -1.3862943611989, \tag{2.5}$$

in two dimension

$$M_2 = \sum_{k,l=-\infty}^{\infty}{}' \frac{(-1)^{k+l}}{\sqrt{k^2+l^2}} = 2M_1 + 4\sum_{k=1}^{\infty}\sum_{l=1}^{\infty}\frac{(-1)^{k+l}}{\sqrt{k^2+l^2}} \approx -1.6155426267, \tag{2.6}$$

and in three dimension,

$$M_3 = \sum_{k,l,m=-\infty}^{\infty}{}' \frac{(-1)^{k+l+m}}{\sqrt{k^2+l^2+m^2}} = 3M_1 + 8\sum_{k=1}^{\infty}\sum_{l=1}^{\infty}\sum_{m=1}^{\infty}\frac{(-1)^{k+l+m}}{\sqrt{k^2+l^2+m^2}}$$
$$\approx -1.747564594633182. \tag{2.7}$$

where the prime on summation excludes the point where all indexes are zero. The Madelung's constants indicate that limited space does make a difference. M_1 may be evaluated with use of the generating function $\ln(1 + x)$, but the other two has no exact form. The question is how to evaluate these summations without an exact formulation.

The major contribution from a summation may come from a relatively small number of terms either at the beginning or at the end or somewhere in the middle. The alternating series generally have neighboring terms cancel each other and can converge relatively quickly. Consider a general alternating series as given by

$$S_{2n}^{2N+2n+1} = \sum_{i=2n}^{2N+2n+1}(-1)^i f(i) = \sum_{i=0}^{N} f(2n+2i) - \sum_{i=0}^{N} f(2n+2i+1)$$
$$= \sum_{i=0}^{N}[f(2n+2i) - f(2n+2i+1)]$$

where i could be even or odd, and $f(i)$ is always positive. The two series sums in the third expression are generally close to each other. The last expression would make each bracketed term small, which may be carried out by utilizing the following identity

$$\int_a^b (b-x)F''(x)dx = (b-x)F'(x)|_a^b + \int_a^b dx F'(x)$$

$$= -(b-a)F'(a) + F(b) - F(a). \tag{2.8}$$

By taking $a = 2n+2i$ and $b = 2n+2i+1$ and $b = 2n+2i+2$, the two expressions in Eq. (2.8) result in

$$f(2n+2i+1) - f(2n+2i) = f'(2n+2i) + \int_{2n+2i}^{2n+2i+1} (2n+2i+1-x)f''(x)dx$$

$$\frac{1}{2}\int_{2n+2i}^{2n+2i+2} f'(x)dx = f'(2n+2i) + \frac{1}{2}\int_{2n+2i}^{2n+2i+2} (2n+2i+2-x)f''(x)dx.$$

Subtracting the first equation from the second, we arrive at

$$f(2n+2i) - f(2n+2i+1) + \frac{1}{2}\int_{2n+2i}^{2n+2i+2} f'(x)dx$$

$$= \frac{1}{2}\int_{2n+2i}^{2n+2i+2} (1 - |2n+2i+1-x|)f''(x)dx.$$

Therefore,

$$\left| f(2n+2i) - f(2n+2i+1) + \frac{1}{2}\int_{2n+2i}^{2n+2i+2} f'(x)dx \right|$$

$$\leq \frac{1}{2}\int_{2n+2i}^{2n+2i+2} |f''(x)|dx.$$

For $f(x) \to 0$ as $x \to \infty$, we find

$$\left| S_i^\infty - \frac{1}{2}f(i) \right| \leq \frac{1}{2}\int_i^\infty |f''(x)|dx.$$

Therefore,

$$M_1 = 2\sum_{k=1}^{n-1}\frac{(-1)^k}{k} + S_n^\infty \approx 2\sum_{k=1}^{n-1}\frac{(-1)^k}{k} + \frac{(-1)^n}{n} + O\left(\frac{1}{n^2}\right). \tag{2.9}$$

Thus, we may sum up the needed number of terms to ensure the desired accuracy.

syms i;
M1=eval(symsum((-1)^i/i,1,inf)*2)
N=100
n=1:N;
m1=(-1).^n./n;
m1=sum(m1)*2+(-1)^(N+1)/(N+1)
	% symbolic summation at the proper limit
	M1=
	-1.38629436111989
	% vector summation
	% the error is in the fifth digit after the decimal.
	m1=
	-1.38624534871940

Extending the 1D result to 2D would give

$$M_2 \approx 2M_1 + 4\sum_{k=1}^{n-1}\sum_{l=1}^{n-1}\frac{(-1)^{k+l}}{\sqrt{k^2+l^2}} + 4\sum_{k=1}^{n-1}\frac{(-1)^{k+n}}{\sqrt{k^2+n^2}} + O\left(\frac{1}{n}\right). \tag{2.10}$$

Since the error would rise to O(1) for 3D case, to improve upon the accuracy, the error in 1D must be reduced to O(1/n^3). It is clear that if the interaction force is not ionic but dipole or quadruple, the convergence is faster and the error will be more acceptable. The following sums up the interaction energy of the dipoles aligned in the same direction.

```
syms k integer;
symsum(1
/k^3,1,inf)
ans=zeta(3)
```

$$S_N = \sum_{k=1}^{N} k^{-3},$$

$$S_N = S_\infty - \sum_{k=N+1}^{\infty} k^{-3} = S_\infty - O(N^{-2})$$

% Find the summation of $\sum_{n=1}^{\infty} 1/n^3$.

The value $S_N = S_\infty - \sum_{k=N+1}^{\infty} k^{-3} = S_\infty - O(N^{-2})$, where the last term is approximated by an integral for evaluation as $\int_{N+1}^{\infty} dk/k^3 \approx 3/(N+1)^2$. This may serve good purpose for checking on the numerical result if you write a code to evaluate S_N. Note that $S = \zeta(3) \approx 1.202$, where ζ is the Riemann Zeta function. Similarly, the summation $L_N = \sum_{k=1}^{N} k^{-1}$ for $N \gg 1$, may be evaluated by the following:

$$L_N = \ln N + \sum_{k=1}^{N} k^{-1} - \int_1^N \frac{dz}{z} = \ln N + \gamma + \sum_{k=N+1}^{\infty} k^{-1} - \int_N^{\infty} \frac{dz}{z} \tag{2.11}$$

where we have defined the Euler's constant

$$\gamma = \lim_{N\to\infty}\left[\sum_{k=1}^{N} k^{-1} - \int_1^N \frac{dz}{z}\right] \approx 0.5772157.$$

The remaining part is

$$l_N = \lim_{N\to\infty}\left[\sum_{k=N+1}^{\infty} k^{-1} - \int_N^{\infty} \frac{dz}{z}\right],$$

that vanishes as $N \to \infty$, so we expect $l_N = O(N^{-1})$. Here is a simple program in MATLAB to evaluate the Euler's constant.

```
N=20000;
M=1:N;
M=1./M;
S=sum(M,2)-log(N)
I=limit(symsum(1/n,n,1,N)-
int(1/n,n,1,N),N,inf)
```

```
% vector summation.
S=
  -0.5772
% symbolic summation at the
proper limit
I=
  eulergamma
```

These summations can easily be found from MATLAB

$$\sum_{k=1}^{N} k = \frac{1}{2}N(N+1), \quad \sum_{k=1}^{N} k^2 = \frac{1}{6}N(N+1)(2N+1),$$

$$\sum_{k=1}^{N} k^3 = \frac{1}{4}N^2(N+1)^2, \ldots$$

syms k N S p positive;	% declare positive variables
p=input('type the power ');	% get the power value
S=simplify(symsum(k^p,1,N))	% symbolic summation at the proper limit

The general formula may be expressed in terms of the Bernoulli polynomials,

$$\sum_{k=M}^{N} k^{\alpha} = \frac{1}{\alpha+1}\{B_{\alpha+1}(N+1) - B_{\alpha+1}(M)\}. \tag{2.12}$$

The first few Bernoulli polynomials are listed as below:

$$B_0(x) = 1, \quad B_1(x) = x - \frac{1}{2}, \quad B_2(x) = x^2 - x + \frac{1}{6},$$

$$B_3(x) = x^3 - \frac{3}{2}x^2 + \frac{1}{2}x, \quad B_4(x) = x^4 - 2x^3 + x^2 - \frac{1}{30}, \ldots.$$

They satisfy the following properties:

$$\int_0^1 B_n(x)dx = 0, \quad n > 1, \quad B_n(1-x) = (-1)^n B_n(x), \; B_n'(x) = nB_{n-1}(x). \tag{2.13}$$

A rather sophisticated but useful approximation of sums by integration through the Euler–Maclaurin formula is given below,

$$\sum_{n=1}^{N} f(n) = \int_1^N f(x)dx + C + \frac{1}{2}f(N) + \sum_{n=1}^{m} \frac{b_{2n}}{(2n)!} f^{(2n-1)}(N)$$

$$- \int_1^N f^{(2m)}(x) \frac{B_{2m}(x-[x])}{(2m)!} dx. \tag{2.14}$$

Here $[x]$ is the largest integer of x, and C is defined as

$$C = \frac{1}{2}f(1) - \sum_{n=1}^{m} b_{2n} \frac{f^{(2n-1)}(1)}{(2n)!}.$$

The last term of Eq. (2.13) can be estimated to be bounded by the following:

$$|B_{2m}(x-[x])| \le |b_{2m}| = 2(2m)!(2\pi)^{-2m} \sum_{k=1}^{\infty} k^{-2m}.$$

Let us use Eq. (2.13) to evaluate the Stirling's approximation. Note that

$$\ln N! = \sum_{n=1}^{N} \ln n \approx \int_1^N \ln x dx + \frac{1}{2}\ln N + O(1)$$
$$\approx (x\ln x - x)|_1^N + \frac{1}{2}\ln N + O(1)$$
$$\approx N\ln N - N + \frac{1}{2}\ln N + O(1) \tag{2.15}$$

Therefore, $N! \approx CN^{N+1/2}e^{-N}$. To determine the constant, it is easier to utilize the integral representation of $N!$, $N! = \int_0^\infty e^{-t}t^N dt$. The integrand is peaked at $t = N$, we may therefore expand $t = N + z$. Since

$$\ln(e^{-t}t^N) = -t + N\ln t = -N - z + N\left(\ln N + \frac{z}{N} - \frac{1}{2}\frac{z^2}{N^2} + \cdots\right)$$
$$\approx N\ln N - N - \frac{1}{2}\frac{z^2}{N},$$

we have

$$N! = \int_0^\infty e^{-t}t^N dt \approx \int_0^\infty e^{-N}N^N e^{-\frac{1}{2}\frac{z^2}{N}}dz = \sqrt{2\pi N}N^N e^{-N}.$$

This last example naturally leads us to the asymptotic expansions in integral. It often happens that an integral is dominated by the contribution from the neighborhood of a peak, and the approximation in that neighborhood would well account for the asymptotic expansion to the leading order. This is the spirit of the Laplace Method for integrals. The following Exponential Integral may demonstrate this effect,

$$E_i(\varepsilon) = \int_\varepsilon^\infty e^{-t}t^{-1}dt \tag{2.16}$$

when $\varepsilon \ll 1$. Since it is singular at $\varepsilon = 0$, the major contribution comes from the region near ε, the integration by parts would then give

$$E_i(\varepsilon) = -e^{-\varepsilon}(1 + 2\varepsilon)\ln\varepsilon + O(\varepsilon). \tag{2.17}$$

This formula may readily be applicable to find the following integral

$$\Xi = \lim_{\varepsilon\to 0}\int_\varepsilon^\infty (e^{-\lambda t} - e^{-\gamma t})t^{-1}dt = \lim_{\varepsilon\to 0}\{E_i(\lambda\varepsilon) - E_i(\gamma\varepsilon)\} = \ln(\gamma/\lambda), \tag{2.18}$$

which is exact by applying Eq. (2.17).

A similar function is the Gamma Function which is defined by

$$\Gamma(z) = \int_0^\infty e^{-t}t^{z-1}dt. \tag{2.19}$$

In order to have a convergent integral the real part of z has to be positive. The following two identities are of great importance: $\Gamma(n) = (n-1)!$ and $z\Gamma(z) = \Gamma(z+1)$, where n is a positive integer. As shown in Section 1.2.3, we have

$$\Gamma(z) = \int_0^\infty e^{-t}t^{z-1}dt = \int_0^1 e^{-t}t^{z-1}dt + \int_1^\infty e^{-t}t^{z-1}dt$$

$$= \sum_{m=0}^{\infty} \frac{(-1)^m}{m!(m+z)} + \int_1^\infty e^{-t}t^{z-1}dt. \tag{2.20}$$

Since the singularity is near $t = 0$, the first term dominates. If z is near n with a small deviation, viz., $z = -n + \varepsilon$, we have

$$\Gamma(z) \approx \left(\frac{1}{\varepsilon}\right)\frac{(-1)^n}{n!}, \tag{2.21}$$

Therefore, the following conclusion is reached,

$$\lim_{z\to -n} \frac{1}{\Gamma(z)} \to 0, \tag{2.22}$$

which is needed to show the quantization condition in hydrogen atoms. Another great property of Gamma function for very large $z \gg 1$ (cf. Eq. (2.15)),

$$\Gamma(z) \approx \sqrt{\frac{2\pi}{z}}\left(\frac{z}{e}\right)^z\left[1 + \frac{1}{12z} + \cdots\right], \tag{2.23}$$

that readily gives the Stirling's formula.

The last example we will discuss is an integral similar to what we used for the Madelung's constant

$$S(x) = \int_0^x \sin t \frac{dt}{t}. \tag{2.24}$$

First observe that

$$\lim_{x\to\infty} S(x) = \int_0^\infty \sin t \frac{dt}{t} = O(1)$$

since there is no free parameter, and the integral is convergent. The constant value will be shown to be $S_\infty = \pi/2$. We may rewrite the integral as follows:

$$S(x) = S_\infty - \int_x^\infty \sin t \frac{dt}{t} = S_\infty - \frac{\cos x}{x} - \frac{\sin x}{x^2} + 2\int_x^\infty \sin t \frac{dt}{t^3}$$

$$= S_\infty - \frac{\cos x}{x} - \frac{\sin x}{x^2} + O\left(\frac{1}{x^3}\right). \tag{2.25}$$

This gives an asymptotic expansion rather useful for large x. To prove $S_\infty = \pi/2$, recall that

$$I = \int_{-\infty}^\infty e^{it}\frac{dt}{t} = 2\pi i.$$

Since there is a pole at the origin on the real axis, the contour integration can be found by taking the residue for encircling the pole, while the integrand vanishes on the upper half of the complex plane. This would then give $S_\infty = \pi/2$.

The Laplace method for integrals may be stated formally as in the following:

$$\int_{-a}^{b} e^{tf(x)}dx \sim (2\pi)^{\frac{1}{2}} e^{tf(0)}/\sqrt{[-tf''(0)]} \quad \text{for } t \gg 1 \tag{2.26}$$

by assuming the peak occurs at the origin. There are limitations and variations to this theorem, for example, if the maximum happens to be at the integration boundary or other peaks are equally important. The general concept however, is well described by Eq. (2.26). The Laplace method could be the basis for other approximate methods such as the Saddle Point Method.

> **Homework 2.19:** Compare the sizes of these terms: $N^N, N!, N^C, C^N, CN$ for $N \gg C \gg 1$.

> **Homework 2.20:** Prove that
>
> $$I_n \equiv \int_0^{\pi} x^n \sin x dx \approx n^{-2}\pi^{n+2}$$
>
> for $n \gg 1$.

2.3.4 *Examples in Asymptology*

The following example is to find the asymptotic limits of a coefficient in the relativistic theory for the electrostatic plasma wave:

$$A(\mu) \equiv \frac{\mu}{K_2(\mu)} \int_1^{\infty} \frac{(\gamma^2 - 1)^{3/2}}{\gamma} e^{-\mu\gamma} d\gamma. \tag{2.27}$$

The thermal plasma is governed by the Gibb's distribution $f_0 = N\exp(-\mu\gamma)$, where $N = n_0\mu/4\pi m_0^3 c^3 K_2(\mu)$ is the normalization constant for the distribution function, $\mu = mc^2/k_B T$ is the inverse temperature normalized to the electron rest energy, and $\gamma = \sqrt{1 + p^2/m^2c^2} = 1/\sqrt{1 - v^2/c^2}$. We first examine the asymptotic limit for the modified Bessel function $K_2(\mu)$, which might be expressed as

$$K_2(\mu) = \frac{\sqrt{\pi}\,\mu^2}{4\Gamma(\frac{5}{2})} \int_1^{\infty} e^{-\mu\xi}(\xi^2 - 1)^{\frac{3}{2}} d\xi \xrightarrow{\mu \ll 1} 2/\mu^2,$$

where $\Gamma(\frac{5}{2}) = 3\sqrt{\pi}/4$. As $\mu \ll 1$, the main contribution to the integral comes from $\xi \gg 1$. Thus, we have applied to the leading order, $\int_0^{\infty} e^{-\mu\xi}\xi^3 d\xi = 6/\mu^4$. Similarly, to the leading order,

$$A(\mu) \equiv \frac{\mu}{K_2(\mu)} \int_1^{\infty} \frac{(\gamma^2 - 1)^{3/2}}{\gamma} e^{-\mu\gamma} d\gamma \rightarrow \frac{\mu^3}{2} \int_1^{\infty} \gamma^2 e^{-\mu\gamma} d\gamma = 1. \tag{2.28}$$

As $\mu \to 0$, to ensure the numerical accuracy of evaluating $A(\mu)$, the γ limit for direct numerical integration has to extend to a very large number, which is often impractical if not impossible. This asymptotic expansion alerts the programming effort to adopt the variable length accordingly to ensure speed and accuracy as demonstrated in the following MATLAB code.

```
function getA
muStart=0.01; muEnd=100; dmu=0.1;
N=fix((muEnd-muStart)/0.1)+1;
mu=dmu*(1:N)+muStart;
for i=1:N
  k(i)=getK(mu(i));
end;
MaxPoints=1/muStart*40,
K=k; K(1)=1;
semilogx(mu, k,'-g*',mu,K,'r-');
xlabel('\mu'); ylabel('A');
title('A as function of MU');

function k=getK(mu)
N=max(1/mu,1); ; qEnd=40*N;
k2=besselk(2,mu); C=mu/k2;
dq=0.004/N;
k=0;
coeff=4;
for q=1+dq:dq:1+qEnd
  if(coeff==4) coeff=2; else coeff=4; end;
  k=k+coeff*(q^2-1)^(3/2)/q*exp(-mu*q);
end;
k=k+((qEnd+dq)^2-1)^(3/2)/(1+qEnd+dq)*exp(-mu*(1+qEnd+dq));
k=dq*C*k/3;
```

%The maximum number of data points has to increase rapidly for mu<<1 in order to ensure accuracy.

%This line offers the variable length control.

%Simpson integration rule (cf Chapter VII).

Homework 2.21: Find the asymptotic leading order term to Eq. (2.27) for $\mu \gg 1$.

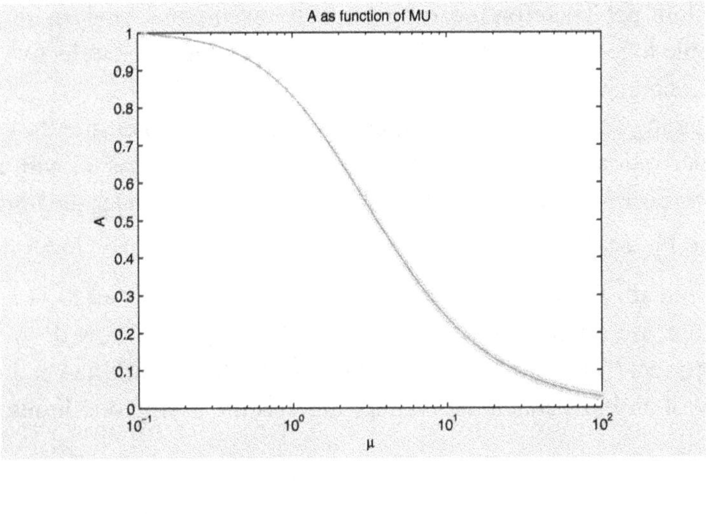

The next example is the diffusion coefficient

$$D = \mathrm{Im}\left(\frac{q}{m}\right)^2 \frac{1}{\Omega} \int d^3\vec{k} \sum_{n,n',l} S_{nn'l} \frac{k_x^2}{k^2} \frac{1}{k_\perp^2 D + i(n'+l)\Omega} \frac{1}{k_\perp^2 D + i(n-l-1)\Omega} \tag{2.29}$$

for a strongly magnetized plasma given the flute like thermal fluctuations of electric fields characterized by $\Omega \gg k_\perp v_T \gg k_\perp^2 D \gg k_{||} v_T$ as in a two-dimensional approximation.

The $\vec{\vec{S}}_{nn'l}$ is the electric field autocorrelation given by

$$\vec{\vec{S}}_{nn'l} = \frac{2n_0 q^2 \vec{k}\vec{k}}{\pi k^4 \varepsilon_n \varepsilon_{n'}^*} < J_n^2\left(\frac{k_\perp v_{0\perp}}{\Omega}\right) J_{n'}^2\left(\frac{k_\perp v_{0\perp}}{\Omega}\right) > I_l(\lambda) e^{-\lambda}, \tag{2.30}$$

where $\lambda \equiv k_\perp^2 v_T^2 / 2\Omega^2$and there are three integer indices n, n',l to be summed up from $-\infty$ to ∞ in Eq. (2.29). The dominant solution has to come from the lowest integers since the electric field decays as $(B^{-2})^{n,n',l}$. Thus, the choice is clear, allowing $n = n' = l = 0$ would avoid B^{-2} decay in favor of a $B^{-1}$from the last factor in Eq. (2.29). This quantization rule leads to

$$D = \left(\frac{q}{m}\right)^2 \frac{1}{\Omega} \int d^3\vec{k} S_{000} \frac{k_x^2}{k^2} \frac{1}{k_\perp^2 D} \frac{1}{\Omega} = \frac{c}{B} \sqrt{\int d^3\vec{k} S_{000} \frac{k_x^2}{k^2} \frac{1}{k_\perp^2}}. \tag{2.31}$$

This gives the Bohm diffusion scaling law $D \propto 1/B$, a very unfavorable scaling law for magnetic confinement.

It is clear that asymptology helps solve problems of great complexity by offering a first glimpse of the truth. The order of magnitude and the limits at large or small value often allow a conjecture to be quickly checked before any serious effort is taken. The rightful problem solving mentality is to let the problem reveal where the solution might be. Therefore, starting from the asymptotic analysis to get all the hints possible together with the MATLAB programming offers effective means for a scientific enquiry.

Asymptology can serve a very important function as to ensure the correctness of a new derivation. Take for an example, the relativistic plasma with the Gibb's distribution $f_0 = N\exp(-\mu\gamma)$, where $N = \mu/4\pi m_0^3 c^3 K_2(\mu)$ is the normalization constant, and $\gamma = \sqrt{1 + p^2/m_0^2 c^2}$. When a flow is included, the Gibb's distribution is generalized to $f_0 = N\exp(-\mu\gamma + \vec{\alpha} \cdot \vec{p}/m_0 c)$ with $\alpha = \mu\sqrt{1-\lambda^2}$. After some algebra, we arrive at$N = \mu\lambda^2/4\pi m_0^3 c^3 K_2(\mu\lambda)$. Now how do we think the new N is correct? The two N's are identical when flow is nullified at $\lambda \to 1$. The probability of an independent derivation, that has the asymptotic limits consistent with the known results but is still incorrect, can be extremely small. This raises the confidence interval greatly. It is in analogy to the **CRC (cyclic redundancy check)** in communication theory by the expected **checksum** to ensure the transmitted digital signals are error free at an extremely high probability.

> **Homework 2.22:** Physics is an approximation to describe nature; Knowing how to make good approximation can speed up the scientific discovery. Consider the multiplication of 8.5*9.5. Use single digit multiplication to find the answer and estimate its error.

Further Reading

There are a few good books on Asymptology. Some of them could be more mathematically oriented. Nevertheless, to get a complete understanding of the subject, students are suggested to read the books by Murray (1974), De Brujin (1961), Kruskal (1969), or Estrada and Kanwal (2002).

In the web site of Mathworks, additional documentations on MATLAB may be useful for further learning: http://www.mathworks.com/help/matlab/index.html. Last accessed May 2014.

Homework Hints

> **Homework 2.1:** Verify that both $P(x,t) = \frac{1}{\sqrt{4\pi Dt}} e^{-x^2/4Dt}$ and $P(x,t) = \frac{x/t}{\sqrt{4\pi Dt}} e^{-x^2/4Dt}$ are solutions to the diffusion equation $\partial P(x,t)/\partial t = D\partial^2 P(x,t)/\partial x^2$.

```
function Diffusion
syms D x t;
P=exp(-x^2/4/D/t)/sqrt(4*pi*D*t);
Q=x/t*P;
dPt=diff(P,t);
dPx2=diff(P,x,2);
eq1=simplify(dPt-D*dPx2),
dQt=diff(Q,t);
dQx2=diff(Q,x,2);
eq2=simplify(dQt-D*dQx2),
```

```
eq1 =0

eq2 =0
```

> **Homework 2.3:** **Hydrogen atoms:** A hydrogen atom has a proton of measure zero size and an electron of charge density $\rho_e(\vec{r}) = -qe^{-2r/r_b}/(\pi r_b^3)$ centered at the proton, where r_b is the Bohr radius.. Work out and plot the interaction potential for two hydrogen atoms as a function of their distance. Do these two hydrogen atoms attract or repel each other?

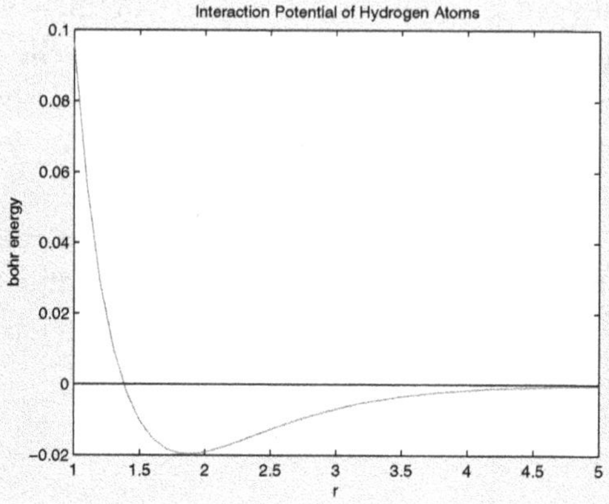


```
function HH
clear all; close all; clc;
syms d r rp re MU positive;
ROe=-exp(-2*rp)/pi;
Q=int(ROe*rp^2*4*pi,rp,0,inf),
Efield=int(ROe*rp^2*4*pi,rp,0,r)/r^2;
Vep=int( Efield,r,r,inf);
Eep=simplify(Vep),
Ee=subs(Eep,r,sqrt(re^2+d^2-2*re*d*MU));
Ee=simplify(Ee);Vee=int(-Ee,MU,-1,1)/2;
Vee=simplify(Vee);
Vee=simple(Vee);
Eee=int(Vee*re^2*exp(-re*2)*4,re,0,d)+ int(Vee*re^2*exp(-
re*2)*4,d,inf),Eee=simplify(Eee);
Eee=simple(Eee),
pretty(Eee),
Vep=subs(Vep,r,d);
E=1/d+2*Vep+Eee;
E=simplify(E),
pretty(E),
Energy=[]; x=[];
for d=1:0.1:5
  V=eval(E);
  Energy=[Energy,V];
  x=[x,d];
end;
Eeeinf=eval(Eee),
plot(x,Energy,'r-',x,x*0,'k-');
 title('Interaction Potential of Hydrogen Atoms');
xlabel('r'); ylabel('bohr energy');
```

```
Eep =1/exp(2*r) + (1/exp(2*r) - 1)/r
Eee =(9*exp(4*d) + 9)/(16*exp(6*d)) - d^2/(6*exp(2*d)) - (93*exp(4*d) +
27)/(48*exp(6*d)) - (2*exp(4*d) - 2)/(16*d*exp(6*d)) - (42*exp(4*d) - 48*exp(6*d) +
6)/(48*d*exp(6*d)) + (d*(32*exp(4*d) + 8))/(16*exp(6*d)) - (d*(132*exp(4*d) +
24))/(48*exp(6*d))

Eee =-(33*d - 24*exp(2*d) + 18*d^2 + 4*d^3 + 24)/(24*d*exp(2*d))
E =(- 4*d^3 - 18*d^2 + 15*d + 24)/(24*d*exp(2*d))
```

Homework 2.5: Find the eigenvalues of matrix

$$X = \begin{vmatrix} 1 & 2 \\ 2 & 1 \end{vmatrix},$$

and verify the result.

```
function getEIG
X=[1 2;2 1];
[V,I]=eig(X),
V1=V(:,1); V2=V(:,2);
lambda(1)=I(1,1);
lambda(2)=I(2,2);
eq1=X*V1 -lambda(1)*V1,
eq2= X*V2-lambda(2)*V2,
```

```
I =

  -1  0
   0  3
```

Homework 2.7: A student designed the following program. The density profile as shown looks nice. However, its density gradient is unacceptable. Plot the density gradient and show the unacceptable spikes. Modify the code to make the density gradient continuous.

```
function DensityProfile
N=1500; Lz=100; dz=Lz/N;
z=(1:N)*dz;
nH=5;
nL1=0;
nL2=0.5;
z1=5.5*Lz/10;
z2=7*Lz/10;
zc1=4*Lz/10;
zc2=9*Lz/10;
Lc1=5; Lc2=7;
n0=tanh((z-zc2)/Lc2)*(nH-nL1)/2+(nH+nL1)/2;
n0(1:fix(z1/dz))=tanh((z(1:fix(z1/dz))-zc1)/Lc1)*(nL2-nL1)/2+(nL2+nL1)/2;
n0(fix(z1/dz+1:z2/dz))=nL2*ones(1,fix(z2/dz-z1/dz));
n0(fix(z2/dz)+1:N)=tanh((z(fix(z2/dz)+1:N)-zc2)/Lc2)*(nH-nL2)/2+(nH+nL2)/2;
dn=diff(n0);
dn=[0,dn];
plotyy(z,n0,z,dn);
dn=dn.*(z<54|z>70.5);
n0=cumsum(dn);
plotyy(z,n0,z,dn);
```

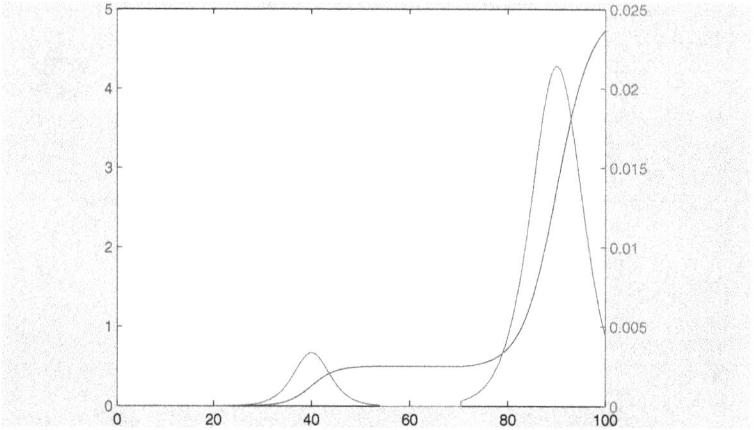

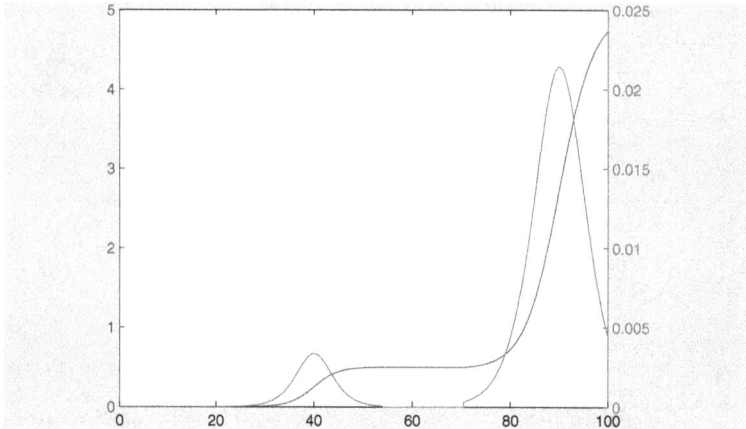

Homework 2.9: Convert for-loop to matrix operation: Rewrite the following program to eliminate the for-loops in favor of the matrix computation. Create the matrix H from random numbers. Compare the computing speed for n=10, 100, 1000, and 3000.

```
function getK(n)
H=rand(n);
tic;
m=(0:n-1)';
M=repmat(m,1,n);
K1=H.*M;
toc;
tic;
K=zeros(n);
for i=1:n
   for j=1:n
      K(i,j)=K(i,j)+(i-1)*H(i,j);
   end;
end;
toc;
D=norm(K1-K),
```

```
% Matrix operation runs faster and its coding can often be
easier to follow.
>> getK(10)
Elapsed time is 0.000152 seconds.
Elapsed time is 0.000015 seconds.
>> getK(100)
Elapsed time is 0.000214 seconds.
Elapsed time is 0.000378 seconds.
>> getK(1000)
Elapsed time is 0.014385 seconds.
Elapsed time is 0.058500 seconds.
>> getK(3000)
Elapsed time is 0.123809 seconds.
Elapsed time is 0.582366 seconds.
% The speed difference is substantial for large n, while at
small n, the overhead by matrix operation in fact makes the
run slower.
%n>3000 may run into virtual memory, and the time
measure can be erroneous.
```

Homework 2.11: Electric potential in a Conical Hole (Jackson p.104). The Laplace equation for the electric potential can be cast into an eigenvalue Legendre equation,

$$\frac{d}{d\mu}(1-\mu^2)\frac{d}{d\mu}P + \nu(\nu+1)P = 0$$

Solve in MATLAB this equation given the boundary condition that: $P(\mu = \beta < 1) = 0$.

Express the first term in the following

$$\frac{d}{d\mu}(1-\mu^2)\frac{d}{d\mu}P = (1-\mu^2)\frac{d^2}{d\mu^2}P - 2\mu\frac{d}{d\mu}P$$

and make the finite difference expansion.

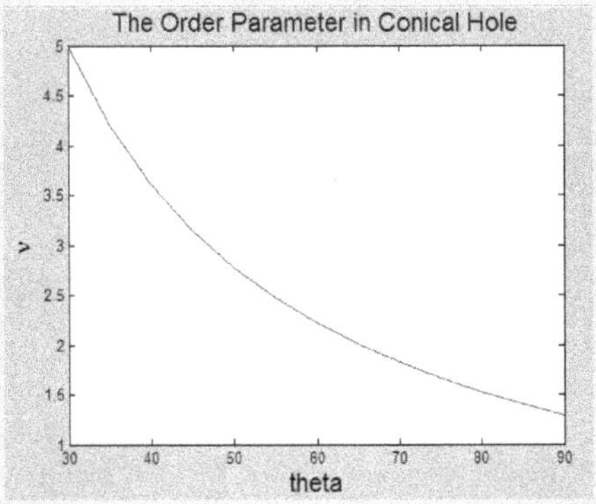

```
function ConicalFunction
N=3000;
nu=[];
for theta=30:5:90
    beta=cos(theta*pi/180);
    dx=(1-beta)/N;
    x=beta:dx:1;
    N=N+1;
    d=-2*(1-x.^2)/dx/dx;
    M=diag(d)-diag(d(1:N-1),1)/2-diag(d(2:N),-1)/2;
    f=x/dx;
    M=M-diag(f(1:N-1),1)+diag(f(2:N),-1);
    M(1,1)=1;
    M(1,2)=0;
    M(2,1)=0;
    [V,e]=eigs(M,5,'SM');
    e=diag(e),
    lambda(:,1)=-0.5+0.5*sqrt(1-4*e);
    lambda(:,2)=-0.5-0.5*sqrt(1-4*e);
    lambda,
    nu=[nu,lambda(2,1)];
end;
plot(30:5:90,nu);
title('The Order Parameter in Conical Hole','fontsize',16);
xlabel('theta','fontsize',16); ylabel('\nu','fontsize',16);
```

% The M matrix loads up the first term, then add the second term.
% The boundary condition requires that at $\mu = 0$, the conducting wall is fixed at a constant potential.
%This is achieved by setting M(1,1)=0, M(1,:)=M(:,1)=0;
% The eigenvalue equation is solved by the solver for the sparse matrix eigs.
% The problem is solved for the conical angle opening ranged from 30° to 90° at 5° interval.
% The plot can be compared with Fig 3.6 on P107 in Jackson's book.

Homework 2.13:   Laplace Transform — a damped oscillator:
Solve in MATLAB the following equation $d^2x/dt^2 + \lambda dx/dt + x = 0$ given the initial values: $x(t = 0) = 1$ and $x'(t = 0) = 0$, making use of the Laplace transform.

```
function LaplaceSolver
syms t x s f F v0 x0 xp0;
L=0.1; x0=1; xp0=0;
X=((s-
L)*x0+xp0)/(s^2+L*s+1);
x=ilaplace(X,s,t)
t=0:0.1:100;
y=eval(x);
plot(t,y);
title('Laplace Transform – a
damped
oscillator','fontsize',16);
xlabel('t','fontsize',16);
ylabel('y','fontsize',16);
```

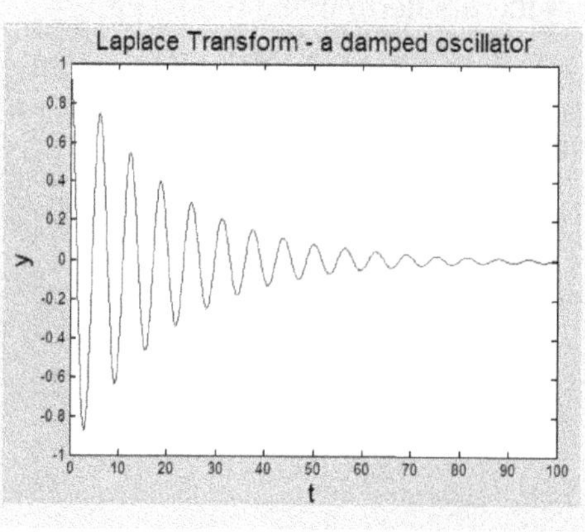

Homework 2.15: With the use of MATLAB limit function to find the following integral:

$$\int_1^x \left(1+\frac{1}{t}\right)^t dt \quad \textbf{for } x \gg 1$$

to leading order.

```
function h2p15
syms x t;
f=(1+1/t)^t;
I=int(f,t,1,x),
A=limit(f,t,inf)
I=int(A,t,1,x),
```

I=e*(x-1)

Homework 2.17: Find the roots of $\cos x = x \sin x$ to leading order.

$$|x| \gg 1\text{: } x \approx (-1)^n n\pi, n \neq 0$$
$$x \sim O(1)\text{: } x \approx \pm\sqrt{2/3}$$
$$|x| \ll 1\text{: no solution}$$

Homework 2.19: Compare the sizes of these terms:

$$N^N, \quad N!, \quad N^C, \quad C^N, CN \quad \text{for } N \gg C \gg 1.$$

$N \gg C \gg 1.$
$N^N \gg N! \gg C^N \gg N^C \gg CN.$

Take logarithmic operation on all terms and compare.

Homework 2.21: Find the asymptotic leading order term to Eq. (2.27) for $\mu \gg 1$.

$$A(\mu) \equiv \frac{\mu}{K_2(\mu)} \int_1^\infty \frac{(\gamma^2 - 1)^{3/2}}{\gamma} e^{-\mu\gamma}\, d\gamma\,.$$

For $\mu \gg 1$, take $\xi = x + 1$, and $x \ll 1$ in Taylor expansion,

$$K_2(\mu) = \frac{\sqrt{\pi}\,\mu^2}{4\Gamma\left(\frac{5}{2}\right)} \int_1^\infty e^{-\mu\xi}(\xi^2 - 1)^{\frac{3}{2}} d\xi \xrightarrow{\mu \gg 1} \sqrt{\frac{\pi}{2\mu}} e^{-\mu}.$$

$$A(\mu) \approx \mu^{3/2}\sqrt{\frac{2}{\pi}} e^{\mu} \int_1^\infty \frac{(\gamma^2 - 1)^{3/2}}{\gamma} e^{-\mu\gamma}\, d\gamma$$

$$\approx \frac{4\mu^{3/2}}{\sqrt{\pi}} \int_0^\infty x^{3/2} e^{-\mu x} dx = \frac{4}{\sqrt{\pi}\,\mu} \int_0^\infty z^{3/2} e^{-z} dz = \frac{3}{\mu},$$

where $x \equiv \gamma - 1$, and $z \equiv \mu x$.

Chapter Three

Single Particle

"We can learn from anyone; and we learn from everyone."

— David A. Paterson, New York Governor

A good starting point to explore a complex plasma phenomenon is often the single particle behavior, which can be the significant first step to gain insight into the relevant plasma physics. While the collective plasma phenomena and the nonlinear many-body physics can deviate from the overly simplified single particle picture, and wave, fluid, or kinetic and statistical descriptions are indispensable, physical effects, as a rule, have to originate from the single particle motions in a consistent way. Moreover, treating all the particles self-consistently, as, in particle simulation, unmistakably reveals the underlying physics. This Chapter will introduce the single particle physics and the related numerical algorithms.

3.1 Drifts

A particle under a strong uniform magnetic field, is characterized by the following equation of motion,

$$m\frac{d\vec{v}}{dt} = e\frac{\vec{v}}{c} \times \vec{B}. \tag{3.1}$$

The trajectory in the plane perpendicular to the B field is circular motion, while the parallel motion is free streaming,

$$\vec{v} = v_{0||}\widehat{b} + \vec{v}_0 \cos\Omega t + \vec{v}_0 \times \widehat{b} \sin\Omega t, \tag{3.2}$$

where $\Omega \equiv eB/mc$ is the gyrofrequency, and $\widehat{b} \equiv \vec{B}/B$ is the unit vector pointing along the magnetic field, and $\vec{v}_0$ is the initial velocity. We may adopt a straightforward finite difference integration, accurate only to the first order of time step size, $O(dt)$, as in the case of gyroorbit.m of Chap. 2 to obtain a reasonable solution of gyromotion for the particle as shown below. Here, we are concerned with the particle motion perpendicular to the field. Without elaborate effort on the numerical algorithm, it is clear that the error can be significant as indicated from the thickness of trajectory line.

```
function gyration0                              % first order finite difference
dt=0.001; OMEGA=1;                              scheme
Nt=50000;
vx=(1:Nt)*0; x=vx; y=x; vy=x;
vx(1)=-1; x(1)=0; y(1)=-1;                      % The algorithm follows the leap
x(1)=x(1)+vx(1)*dt/2;                           frog scheme and takes the leading
y(1)=y(1)+vy(1)*dt/2;                           order in the expansion, known as the
E0=(vx(1)^2+vy(1)^2)/2;                         Euler's finite difference method.
for it=2:Nt
    vx(it)=vx(it-1)+vy(it-1)*dt*OMEGA;
    vy(it)=vy(it-1)-vx(it-1)*dt*OMEGA;
    x(it)=x(it-1)+vx(it-1)*dt;
    y(it)=y(it-1)+vy(it-1)*dt;
end;
figure, plot(x,y); axis equal;
xlabel('x'); ylabel('y'); axis([-1 1 -1 1]);
dE=abs(sum(vx.^2+vy.^2)/2/Nt-E0)/E0,            % The error is dE=0.0254
```

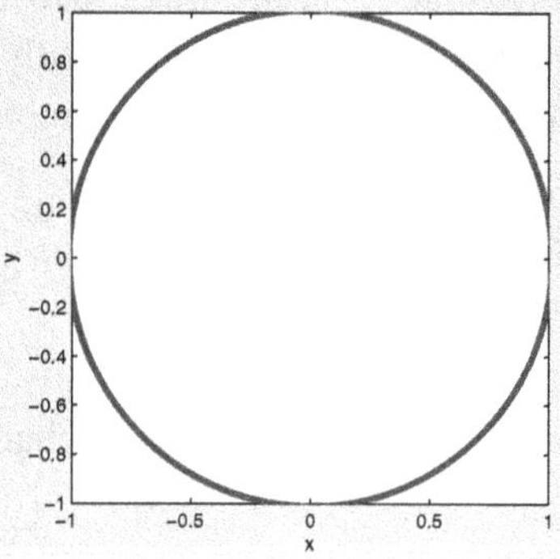

To improve the numerical accuracy, we may adopt the Taylor expansion to reduce the error to $O(dt^4)$, the fourth order in time step size. This can speed up the computation for the same elapsed time of particle motion, since a much larger time step is now workable to achieve similar or better accuracy. The energy, however, is still slightly non-conserved.

```
function Taylor4             Vx
syms dt vx vy v;             =v*cos(theta0)-v*sin(theta0)*dt-1/2*v*cos(theta0)*dt^2+
syms theta theta0;           1/6*v*sin(theta0)*dt^3
theta=theta0+dt;             Vy
vx=v*cos(theta);             =v*sin(theta0)+v*cos(theta0)*dt-1/2*v*sin(theta0)*dt^2-
vy=v*sin(theta);             1/6*v*cos(theta0)*dt^3
Vx=taylor(vx,dt,4)           V2 =1/36*v^2*(-3*dt^4+dt^6+36)
Vy=taylor(vy,dt,4)
V2=Vx^2+Vy^2;
V2=simplify(V2)
```

```
function gyration1
% Taylor expansion to the order of dt^4
dt=0.001; OMEGA=1; dT=dt*OMEGA;
Nt=50000;
vx=(1:Nt)*0; x=vx;y=x; vy=x;
vx(1)=-1; x(1)=0; y(1)=-1;
x(1)=x(1)+vx(1)*dt/2;
y(1)=y(1)+vy(1)*dt/2;
E0=(vx(1)^2+vy(1)^2)/2;
for it=2:Nt

vx(it)=vx(it-1)*(1-dT^2/2)+vy(it-1)*dT*(1-dT^2/6);

vy(it)=vy(it-1)*(1-dT^2/2)-vx(it-1)*dT*(1-dT^2/6);
    x(it)=x(it-1)+vx(it-1)*dT;
    y(it)=y(it-1)+vy(it-1)*dT;
end;
plot(x,y); axis equal;
axis([-1 1 -1 1]);
xlabel('x'); ylabel('y');
dE=abs(sum(vx.^2+vy.^2)/2/Nt-E0)/E0,
```

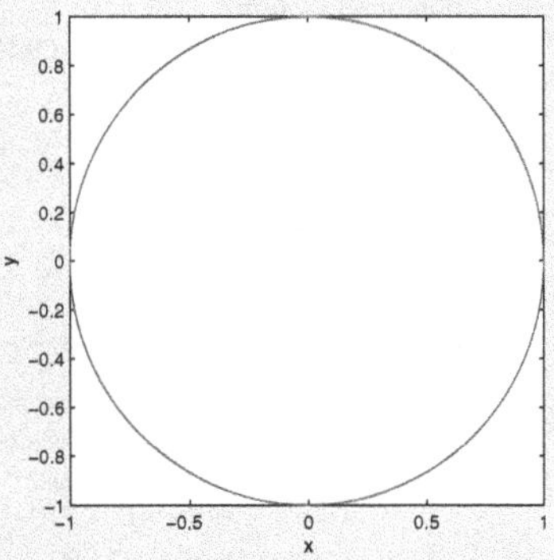


```
% The error is given by
  dE =2.0812e-009
```

There is an easy way to have energy conservation to all orders by following *the physical characteristics*. We may rotate the particle trajectory, thus obtained at its gyrofrequency to trace the exact trajectory. This is possible due to the conservation of particle kinetic energy, since the force is always perpendicular to the velocity. It can be shown by dotting Eq. (3.1) with $\vec{v}$ to obtain $m\vec{v} \cdot (d\vec{v}/dt) = d(\frac{1}{2}m\vec{v}^2)/dt = 0$ so that v is a constant of motion. By writing the solution as

$$\vec{v} = v_0 \cos\theta \hat{e}_x + v_0 \sin\theta \hat{e}_y, \tag{3.3a}$$

it is straightforward to find that

$$\frac{d\theta}{dt} = \Omega. \tag{3.3b}$$

The following program utilizes Eqs. (3.3) to advance particle trajectory. It is exact and incurs no error in the energy value. This same algorithm was conveniently used to find the test particle diffusion in Chap. 1 by adding the pitch angle scattering in the particle trajectory. It confirms the classical diffusion law of particle across the magnetic field to scale as $1/B^2$.

```
function gyration2
dt=0.025; OMEGA=1; dT=dt*OMEGA;
Nt=500;
vx=(1:Nt)*0; x=vx;y=x; vy=x;
vx(1)=-1; theta=-90*pi/180;
x(1)=cos(theta)+vx(1)*dT/2;
y(1)=sin(theta)+vy(1)*dT/2;
E0=(vx(1)^2+vy(1)^2)/2;
v0=sqrt(2*E0);
for it=2:Nt
    theta=theta+dT;
    vx(it)=v0*sin(theta);
    vy(it)=v0*cos(theta);
    x(it)=x(it-1)+vx(it-1)*dT;
    y(it)=y(it-1)+vy(it-1)*dT;
end;
figure
plot(x,y); axis equal;
axis([-1 1 -1 1]);
dE=abs(sum(vx.^2+vy.^2)/2/Nt-E0)/E0,
T=(0:Nt-1)*dT;
VX=sin(T-90*pi/180);
VY=cos(T-90*pi/180);
plot(x,y); axis equal;
figure;
plot(T,vx,'*g',T,vy,'xk',T,VX,' -r',T,VY,' -r');
title('Vx and Vy');
xlabel('time'); ylabel('v');
```

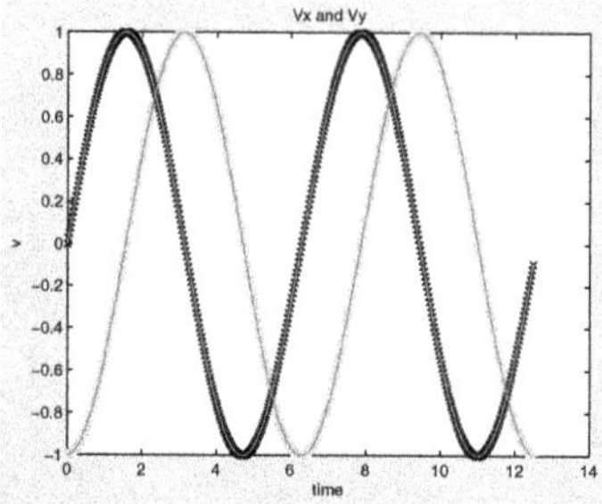

% The thin red lines are the analytical

shows Vx and black Vy.

dE =

0

3.1.1 *ExB Drift*

Consider a particle in a strong magnetic field and a dc electric field,

$$m\frac{d\vec{v}}{dt} = e\frac{\vec{v}}{c} \times \vec{B} + e\vec{E}. \tag{3.4}$$

This equation can be solved, in addition to the homogeneous solution of Eq. (3.2), the particular solution which is found by balancing the terms on the right hand side (RHS) to provide

$$\vec{v}_E = c\frac{\vec{E} \times \vec{B}}{B^2}. \tag{3.5}$$

Since the electric field is time independent, $d\vec{v}_E/dt = 0$. Equation (3.5) characterizes the ExB drift. It is an important effect in plasma physics that may be generalized for arbitrary force,

$$m\frac{d\vec{v}}{dt} = e\frac{\vec{v}}{c} \times \vec{B} + \vec{F}, \tag{3.6}$$

to obtain

$$\vec{v} = v_{\|}\hat{b} + \vec{v}_0 \cos\Omega t + \vec{v}_0 \times \hat{b} \sin\Omega t + c\frac{\vec{F} \times \vec{B}}{eB^2}. \tag{3.7}$$

For a charge independent force, such as the gravitation $\vec{F} = m\vec{g}$, the drift velocity $cm\ \vec{g} \times \vec{B}/eB^2$ is charge dependent. It can cause charge separation and consequently the **ambipolar potential**. The ambipolar potential can pull back the fast moving species by the slow moving counterparts. On the other hand, the ExB drift is charge independent. Therefore, electrons and ions could move out of the magnetic confinement together. This can be a cause of concern for any low frequency electric field fluctuations, thermal or otherwise, to result in significant plasma transport. Since it is not due to the classical collision, it is a kind of **anomalous transport**. By anomalous transport, we refer to transport that is not due to the classical collisions of particles. A case in point for the latter was demonstrated in the program TPD.m. The ExB drifts have been shown to yield the Bohm-like diffusion of $1/B$ scaling due to the long coherence time of electric field that allows virtually free streaming of particles across the field lines.

Homework 3.1: Write the MATLAB code to calculate the single particle trajectory in a strong magnetic field and a dc electric field.

Homework 3.2: Write the MATLAB code to calculate the single particle in the magnetic field of the form, $\vec{B} = B_0\vec{e}_z(1 + y/L_B)$.

3.1.2 *Curvature and Gradient-B Drifts*

The general description for a single particle trajectory under the electromagnetic force can be derived from the Hamiltonian formulation given by

$$H = \frac{1}{2m}\left(\vec{p} - \frac{e}{c}\vec{A}\right)^2 + e\phi. \tag{3.8}$$

Here the magnetic field $\vec{B}$ is defined from the vector potential $\vec{A}$ through $\nabla \times \vec{A} = \vec{B}$ and the electric field is $\vec{E} = -(1/c)\partial\vec{A}/\partial t - \nabla\phi$. The expression $\vec{P} \equiv \vec{p} - e\vec{A}/c$ is often termed as the **canonical momentum**. It is the generalized momentum for a particle in the presence of magnetic field, because the equations of motion for a single particle are given by

$$\dot{\vec{q}} = \frac{\partial H}{\partial \vec{p}} = \frac{1}{m}\left(\vec{p} - \frac{e\vec{A}}{c}\right) = \frac{\vec{P}}{m}, \tag{3.9a}$$

and

$$\dot{\vec{p}} = -\nabla H = \dot{\vec{q}} \times \frac{e}{c}\vec{B} + \dot{\vec{q}} \cdot \nabla \frac{e}{c}\vec{A} - e\nabla\phi. \tag{3.9b}$$

To arrive at Eq. (3.9b), we have utilized Eq. (3.9a) and the vector identity:

$$\nabla(\vec{A}\cdot\vec{B}) = \vec{A}\times\nabla\times\vec{B} + \vec{B}\times\nabla\times\vec{A} + \vec{A}\cdot\nabla\vec{B} + \vec{B}\cdot\nabla\vec{A}.$$

The Hamiltonian description is easier to generalize to the quantum and relativistic regimes. The classical single particle motion in the presence of electric and magnetic fields can therefore be obtained by taking the time derivative of Eq. (3.9a) to give

$$\ddot{\vec{q}} = \frac{\dot{\vec{p}}}{m} - \frac{e\dot{\vec{A}}}{mc} = \frac{e\vec{E}}{m} + \dot{\vec{q}} \times \frac{e}{mc}\vec{B}, \tag{3.10}$$

where Eq. (3.9b) and $\dot{\vec{A}} = d\vec{A}/dt = \partial\vec{A}/\partial t + \dot{\vec{q}}\cdot\nabla\vec{A}$ have been applied to arrive at the right hand expression. By defining $\vec{v} \equiv \dot{\vec{q}} = \vec{P}/m$, Eq. (3.10) can be expressed as

$$m\frac{d\vec{v}}{dt} = e\left(\vec{E} + \frac{\vec{v}}{c}\times\vec{B}\right). \tag{3.11}$$

The equation of motion along the magnetic field can be obtained by multiplying Eq. (3.11) with $\widehat{e}_{||} \equiv \vec{B}/B$ to give

$$\frac{d\vec{v}}{dt}\cdot\widehat{e}_{||} = \frac{e}{m}E_{||} = \frac{dv_{||}}{dt} - \frac{d\widehat{e}_{||}}{dt}\cdot\vec{v}.$$

Here, $E_{||} = \vec{E}\cdot\vec{B}/B$is the parallel electric field along the magnetic field. The perpendicular motion is given by

$$\begin{aligned}\frac{d\vec{v}}{dt} - \frac{d\vec{v}}{dt}\cdot\widehat{e}_{||}\widehat{e}_{||} &= \frac{d(\vec{v}_\perp + \vec{v}\cdot\widehat{e}_{||}\widehat{e}_{||})}{dt} - \frac{d\vec{v}}{dt}\cdot\widehat{e}_{||}\widehat{e}_{||}\\ &= \frac{e}{m}\left(\vec{E}_\perp + \frac{\vec{v}_\perp}{c}\times\vec{B}\right) + \vec{v}\cdot\frac{d\widehat{e}_{||}}{dt}\widehat{e}_{||} + v_{||}\frac{d\widehat{e}_{||}}{dt}.\end{aligned} \tag{3.12}$$

Since $d\widehat{e}_{||}/dt = \partial\widehat{e}_{||}/\partial t + \vec{v}\cdot\nabla\widehat{e}_{||}$, an important term from $v_{||}d\widehat{e}_{||}/dt$ surviving the cyclotron average is $v_{||}^2\partial\widehat{e}_{||}/\partial s$. Assuming that both $\vec{E}_\perp$ and $\vec{B}$ have weak temporal variation, the terms on the RHS of Eq. (3.12) get balanced to obtain the *ExB* drift and the **curvature drift**:

$$\vec{V}_c = \frac{v_{||}^2}{\Omega}\widehat{e}_{||}\times\frac{\partial\widehat{e}_{||}}{\partial s}. \tag{3.13}$$

The curvature drift is charge dependent and should as a result cause charge separation.

Another important drift in the inhomogeneous magnetic field is the **grad-B drift**. By expressing the particle position in terms of the guiding center coordinate $\vec{R}_{gc}$ and the gyrating orbital $\vec{\rho}$, so that $\vec{r} \equiv \vec{R}_{gc} + \vec{\rho}$, the time derivative of $\vec{r}$ gives $\dot{\vec{r}} = \vec{v} \equiv \vec{V}_{gc} + \vec{u} = \vec{V}_{gc} + \vec{\rho} \times \vec{\Omega}$, where we have defined the time derivative of $\vec{\rho}$ to be $\dot{\vec{\rho}} \equiv \vec{u} \equiv \vec{\rho} \times \vec{\Omega}$ and $\vec{\rho} \cdot \vec{\Omega} = 0$. Therefore, $\ddot{\vec{r}} = \dot{\vec{V}}_{gc} + \dot{\vec{u}} = \dot{\vec{V}}_{gc} + \dot{\vec{\rho}} \times \vec{\Omega} + \vec{\rho} \times \dot{\vec{\Omega}}$ $= \vec{v} \times \vec{\Omega} = (\vec{V}_{gc} + \vec{u}) \times \vec{\Omega}$. Thus,

$$\vec{V}_{gc} \times \vec{\Omega} = \dot{\vec{V}}_{gc} + \vec{\rho} \times \dot{\vec{\Omega}} = \dot{\vec{V}}_{gc} + \vec{\rho} \times (\vec{V}_{gc} + \vec{u}) \cdot \nabla\vec{\Omega}. \tag{3.14}$$

The gyrating $\vec{\rho}$would vanish after average over the cyclotron period. If we consider the low frequency phenomena so that $\dot{\vec{V}}_{gc} \approx 0$, the surviving terms in Eq. (3.14) after the gyro-average denoted by the bracket $\langle\,\rangle$ are: $\vec{V}_{gc} \times \vec{\Omega} \approx \langle \vec{\rho} \times \vec{u} \cdot \nabla\vec{\Omega} \rangle$. We end with $\vec{V}_{gc} = \langle \vec{\Omega} \times (\vec{\rho} \times \vec{u} \cdot \nabla\vec{\Omega})/\Omega^2 \rangle = \langle \vec{\rho}\vec{u} \cdot \nabla\Omega^2/2\Omega^2 \rangle = \langle (\vec{u} \times \vec{\Omega})(\vec{u} \cdot \nabla\Omega)/\Omega^3 \rangle$. Thus, a net drift due to the magnetic field gradient results,

$$\vec{V}_D = \frac{1}{2}\rho^2 \widehat{e}_{\parallel} \times \nabla\Omega, \tag{3.15}$$

where $\rho \equiv u_0/\Omega$ is the gyroradius.

There is also the polarization drift that results from the time variation of the electric field. From the equation of motion, Eq. (3.11), to lowest order, we have $\vec{v} = c\vec{E} \times \vec{B}/B^2$. To the next order, $c\dot{\vec{E}} \times \vec{B}/B^2 = e\vec{V}_p \times \vec{B}/c$. Therefore,

$$\vec{V}_p = \frac{c\dot{\vec{E}}}{B\Omega}. \tag{3.16}$$

3.1.3 *Runge Kutta Algorithm*

A robust algorithm is desirable that can solve similar problems in general so a new algorithm need not be developed every time. To calculate the particle trajectory, a popular algorithm is the Runge Kutta method. The second order Runge Kutta method has the following recipe: (1) Find the displacement of the final position with the use of the slope at the initial position. (2) Take the middle point at half of the displacement. (3) Find the slope at the middle point. (4) Advance from the initial position to the final position with the use of the slope at the middle point. In mathematical terms, given $y = y_0$ at the initial time $t = t_0$ and the governing equation $y' = f(t, y)$, the initial displacement is given by (1)$d_0 = f(t_0, y_0)\Delta t$. Thus, the middle point is given by (2)$y_{\frac{1}{2}} \approx y_0 + \frac{1}{2}d_0$. The slope at this middle point will be (3)$k_{\frac{1}{2}} = f(t_0 + \frac{1}{2}\Delta t, y_{\frac{1}{2}})$. The next point is then found to be (4)$y_1 = y_0 + k_{\frac{1}{2}}\Delta t$.

While advantageous for its speed, the second order Runge Kutta does not always have the accuracy for long time evolution. In case accuracy rules, the fourth order Runge Kutta (RK4) is the choice that runs somewhat slower, but compensated by a larger workable step size. The RK4 method is characterized by the following equations:

$$y_1 = y_0 + \frac{1}{6}\Delta t(k_0 + 2k_{\frac{1}{2}} + 2\kappa_{\frac{1}{2}} + k_1), \tag{3.17}$$

to advance from the time t_0 to $t_1 = t_0 + \Delta t$. Here, $k_0 = f(t_0, y_0)$ is the slope at the initial position, $k_{\frac{1}{2}} = f(t_0 + \frac{1}{2}\Delta t, y_0 + \frac{1}{2}k_0\Delta t)$ is the slope at the middle point determined from the initial slope, $\kappa_{\frac{1}{2}} = f(t_0 + \frac{1}{2}\Delta t, y_0 + \frac{1}{2}k_{\frac{1}{2}}\Delta t)$ is the slope at the new middle point determined from the middle slope, and $k_1 = f(t_0+\Delta t, y_0+\kappa_{\frac{1}{2}}\Delta t)$ is the slope at the final point determined by the new middle slope. The final position is determined by the displacement that gives greater weight to the middle points. The RK4 method is a fourth-order method, meaning that the error per step is on the order of $(\Delta t)^5$, while the total accumulated error is of order $(\Delta t)^4$.

```
function  yout=rk4(y,t,dt)                % y can be a vector of variables
k0=dt*force(t,y);
k=dt*force(t+dt/2,y+k0/2);
kappa=dt*force(t+dt/2,y+k/2);
k1=dt*force(t+dt,y+kappa);
yout=y+k0/6+k/3+kappa/3+k1/6;
```

Homework 3.3: Apply the Runge Kutta fourth order algorithm to the single particle trajectory in a uniform magnetic field. Take dt to be 0.1, 0.01, and 0.001 and check how good their energies are conserved.

3.2 The Adiabatic Invariants

The mechanical law for a dynamical system is to follow the path of the least action $A = \oint \vec{p} \cdot d\vec{q}$. Action represents the cumulative energy through time. The dynamical trajectory may go through higher energy states some of the time, but in doing so, the dynamical system maintains the lowest cumulative energy in the periodic process and admits an adiabatic invariant of least action. This leads to Sommerfeld's quantization rule in the old quantum mechanics. In applying to plasma physics, this is particularly helpful for tracking particles in a slowly varying magnetic field for example.

3.2.1 *The First Adiabatic Invariant*

The periodic circulation perpendicular to the magnetic field gives a current loop that has its **magnetic moment** as **the first adiabatic invariant**:

$$\mu = \frac{1}{4\pi}\oint \vec{v}_\perp \cdot d\vec{r} = \frac{1}{4\pi}\oint \vec{v}_\perp^{\,2} dt \equiv \frac{1}{2}\frac{\langle v_\perp^2 \rangle}{\Omega}. \tag{3.18}$$

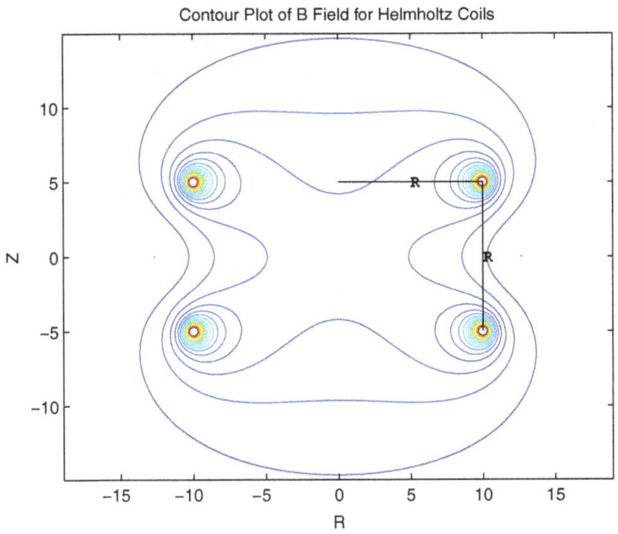

Fig. 3.1. Mod B surfaces of Helmholtz coil. The central region has high uniformity in magnetic field.

In a magnetic field slowly varying in space, the particle would gradually tilt its velocity with respect to the magnetic field and maintains the magnetic moment as an invariant. Into the stronger magnetic field region, the perpendicular particle energy will increase while the parallel velocity decreases. This was the underlying principle for the mirror machine with stronger magnetic fields at the ends to confine plasmas. However, particles in some part of velocity space having enough parallel velocity will escape. It will also cause the **loss cone instability** that, together with others, such as the interchange instability, made the mirror machine unfit to be a confinement device.

As an example we consider particle motion in a **Helmholtz coil** to demonstrate the first adiabatic invariant. A Helmholtz pair consists of two identical circular current coils, say, of radius R, placed symmetrically along z axis, and separated by a distance equal to the radius of the coils as shown in Fig. 3.1. Each coil carries an equal electrical current flowing in the same direction. The Helmholtz coil produces a nearly uniform magnetic field in the central region.

Homework 3.4: The Helmholtz Coil: Two identical circular current coils of radius R, placed symmetrically along z axis, and separated by a distance equal to the radius of the coils as shown in Fig. 3.1. Each coil carries an equal electrical current flowing in the same direction. Find the maximum field error near the center within the spherical volume of radius $R/4$. Also explain how this uniformity in magnetic field occurs.

Homework 3.5: The Helmholtz Coil: Write a MATLAB code to calculate the magnetic field of the Helmholtz coils. Also make the contour plot of the magnetic field.

We place a test particle near the center of a Helmholtz coil to examine the time evolution of its magnetic moment μ. As shown in the program DriftsHelmholtz.m

```
function getMU(mu,Nt,dt)
%get the mean and variance of the magnetic moment
MU=mean(mu),
muvar=var(mu/MU)*100,
time=(1:Nt)*dt;
[AX,H1,H2]=plotyy(time,(mu-mu(1))/mu(1)*100,time,cumsum(mu)./(1:Nt));
set(get(AX(1),'Ylabel'),'String','d\mu(%)','fontsize',10,'fontname','verdana','fontweight','bold')
set(get(AX(2),'Ylabel'),'String','<\mu>','fontsize',10,'fontname','verdana','fontweight','bold')
title('The First Adiabatic Invairant'); xlabel('time');

function b=getB(R)
% find the b-field at particle location from the Helmholtz routine
global a h;
x=R(1); y=R(2); z=R(3);
r=sqrt(x^2+y^2); theta=atan2(y,x);
[Br,Bz]=Helmholtz(a, h, r, z);
Bx=Br*sin(theta); By=Br*cos(theta);
b=[Bx By Bz];

function dy=force(t,R)
% get force term from vxB
global bx by bz N MU;
b=getB(R);
B=norm(b);
V=[R(4);R(5);R(6)];
dy=cross(V,b); dy=[V' dy];
Vp=dot(V,b)/B; % parallel velocity
MU=(dot(V,V)-Vp^2)/B; % mu
```

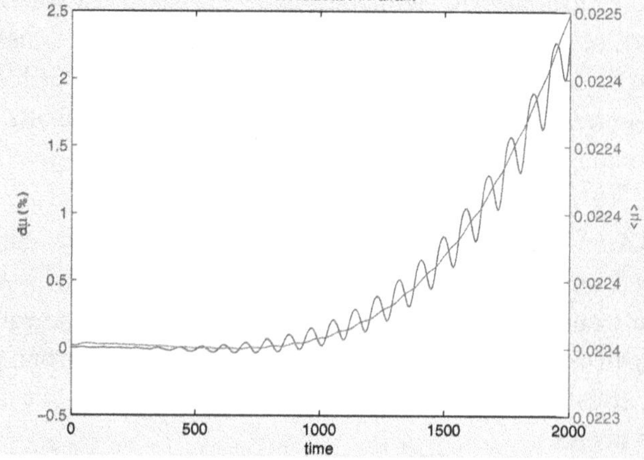

listed herein, μ oscillates around a mean value during the cyclotron period, and its mean value holds nearly constant until the particle drifts to the outer region of stronger field gradient where the adiabatic invariant fails to hold it. The green line shows the cumulative averaged μ.

```
function DriftsHelmholtz                      % main program for particle drifts in
close all; clc; clear all;                    Helmholtz coils
global bx by bz N MU;
global a h;                                   % a is the coil radius, h the distance
a=10; h=a/2; dt=0.05;                         between the two coils.
y0=[0.04 0 0 0.04 0.0 -0.002];
Nt=40000;                                     % y0 initial position near the
Xp=(1:Nt)*0;  Yp=Xp;  Zp=Xp;  mu=Xp;          geometry center.
PHI=mu;
y=y0;
for i=1:Nt;
    t=dt*i;
    y=rk4(y,t,dt);                            % Trajectory is advanced by the
    Xp(i)=y(1); Yp(i)=y(2); Zp(i)=y(3);       Runge-Kutte 4th order.
    mu(i)=MU;
    B=getB(y);
    r=[y(1);y(2);y(3)];
    v=[y(4);y(5);y(6)];
    PHI(i)=dot(B',cross(r,v));
end;
plot3(Xp,Yp, Zp);
hold on;
plot3(0.05,0,0,'.g','MarkerSize',36);
xlabel('X'); ylabel('Y'); zlabel('Z'); grid on;
title('Drifts in Helmholtz Coil');
getMU(mu,Nt,dt);                              % print out the conservation of
delE=norm([y(4),y(5),y(6)])^2/norm([y0(4),    magnetic moment and energy
y0(5),y0(6)])^2-1,
```

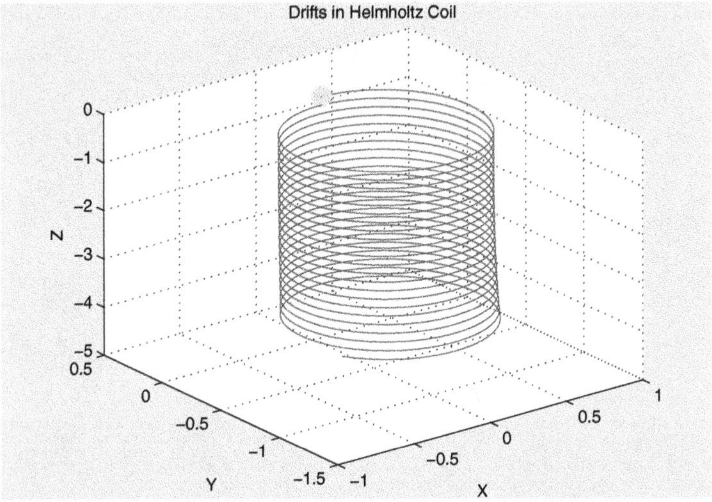

Homework 3.6: Time varying magnetic field: Write a MATLAB code to investigate the time evolution of the energy and the magnetic moment of a test particle in the magnetic field that increases at a rate $\gamma \equiv \partial B/\partial t(eB^2/mc) \gg 1$.

3.2.2 *The Second Adiabatic Invariant*

The second adiabatic invariant is the **J-invariant**:

$$J = \oint v_{||} ds \equiv \langle v_{||}^2 \rangle \tau_b, \tag{3.19}$$

which arises from the oscillation along the field lines in the mirror like magnetic field. The bounce period is denoted as τ_b. The magnetosphere has plasma particles bouncing between the north and the south poles. Due to the J-invariant, the particles maintain their attitude and *drift across the longitude eastwardly.* This can be visualized in the movie clip when running the program of DriftsInMagnetosphere.m. The magnetic field is approximated by a dipole field. The particle, as demonstrated, holds the J-invariant very well as shown in the dJ plot where the area underneath each periodic structure represents the J value.

Homework 3.7: Write the MATLAB code to calculate the earth magnetic field by treating the earth as a magnetic dipole.

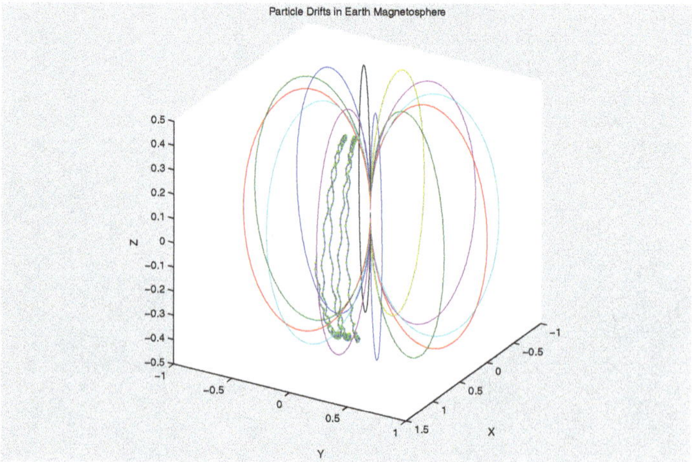

```
function DriftsInMagnetosphere
close all; clear all; clc;
global vpara;
%fig=figure('position', [50 50 650 600]);
%xlabel('X'); ylabel('Y'); zlabel('Z');
EarthMagneticField; hold on;
dt=0.01;
%y0=[1 0 0 0.025 0.0 -0.05];
y0=[1 0 0 -0.01 0.0 -0.025];
Nt=30000;
X=(1:Nt)*0; Y=X; Z=X; dJ=X;
y=y0; PHI=X;
for i=1:Nt;
   t=dt*i;
   y=rk4(y,t,dt);
   X(i)=y(1); Y(i)=y(2); Z(i)=y(3);
   dJ(i)=vpara^2*dt;
   B=getB(y);
   PHI(i)=dot(B',cross([y(1);y(2);y(3)],[y(4);y(5);y(6)]));
end;
delE=(y(4)^2+y(5)^2+y(6)^2-y0(4)^2-y0(5)^2-...
y0(6)^2)/(y0(4)^2+y0(5)^2+y0(6)^2),
plot3(X,Y,Z); hold on;
j=0;
for i=1:Nt/100
   plot3(X(i*100),Y(i*100),Z(i*100),'.g','MarkerSize',6);
   view(-2*i,25);
    title('Particle Drifts in Earth Magnetosphere');
   j=j+1,
   M=getframe;
end;
r=sqrt(X.^2+Y.^2+Z.^2);
rmean=mean(r),
rvar=var(r),
figure;
plot(1:Nt,dJ);
title('J Invariance - dJ as function of time');
plotPHI(PHI,Nt);       % plot the magnetic flux

function  yout=rk4(y,t,dt)
k1=dt*force(t,y);
k2=dt*force(t,y+k1/2);
k3=dt*force(t+dt/2,y+k2/2);
k4=dt*force(t+dt,y+k3);
yout=y+k1/6+k2/3+k3/3+k4/6;

function dy=force(t,R)    % get the force term
global vpara;
V=[R(4);R(5);R(6)];
B=getB(R);
dy=cross(V,B);
dy=[V' dy];
vpara=dot(V,B)/norm(B);
```

% test particle drifts in the earth magnetic field
% rmean =0.8671 rvar =0.0097
% around 1% variance in the radial distance

% the earth magnetic field is modeled by dipole field

% Rugge-Kutta 4th order scheme to advance the particle

% differential J value

% record down the magnetic flux

% the variance in the radial distance

```
function B=getB(R)                          % get B from the earth
omega0=1;                                   magnetic field
x=R(1); y=R(2); z=R(3);
s=[R(1);R(2);R(3)];
V=[R(4);R(5);R(6)];
r=sqrt(sum(s.^2,1));
ro=sqrt(x^2+y^2);
omega=omega0/r^3;
Bz=(3*z^2/r^2-1)*omega;
Bx=omega0*3*z/r*sqrt(1-z^2)/r*x/ro;
By=omega0*3*z/r*sqrt(1-z^2)/r*y/ro;
B=[Bx By Bz];
```

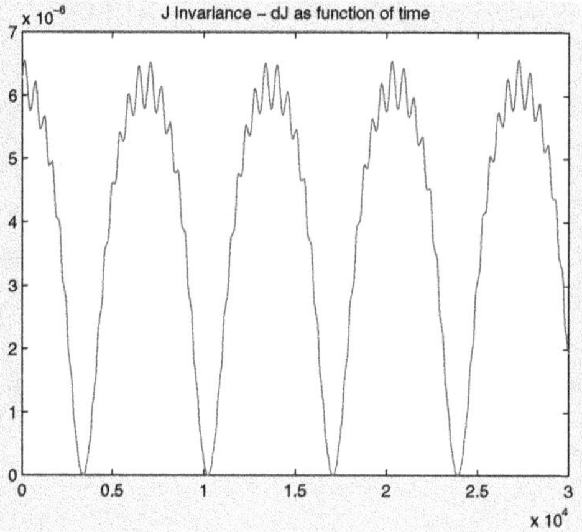

3.2.3 *The Third Adiabatic Invariant*

The third adiabatic invariant is the magnetic flux. Over the area as enclosed by the trajectory projected onto the plane perpendicular to the magnetic field, the magnetic flux,

$$\Phi = \int \vec{B} \cdot d\vec{A}, \tag{3.20}$$

is an adiabatic invariant. For a single particle motion, this may be approximated as

$$\Phi = \oint \vec{B} \cdot \vec{r} \times \vec{v} dt \approx \oint \vec{B} \cdot (\vec{e}_{\|} \times \vec{v}/\Omega) \times \vec{v} dt = -\mu(mc/e),$$

where $\vec{r}(t)$ is the particle position, $\vec{v}(t)$ is the particle velocity, and $\vec{B}$ is evaluated at the $\vec{r}$ location. Adding the line of coding plotPHI(PHI,Nt) in DriftsInMagnetosphere.m to plot the magnetic flux with the following function, we have the magnetic flux evaluated as an accumulated average, which appears to settle into a fairly constant value.

```
function plotPHI(PHI,Nt)                                    %phimean = -0.0021
figure(3); grid on;                                         %phivar = 2.1687e -005
title('Magnetic Flux');
phimean=mean(PHI),
phivar=var(PHI),
phi=cumsum(PHI);
[AX,H1,H2]=plotyy(1:Nt,PHI,1:Nt,phi./(1:Nt));
set(get(AX(1),'Ylabel'),'String','d\phi','fontsize',10,...
'fontname','verdana','fontweight','bold') ;
set(get(AX(2),'Ylabel'),'String','<\phi>','fontsize',10,...
'fontname','verdana','fontweight','bold');
```

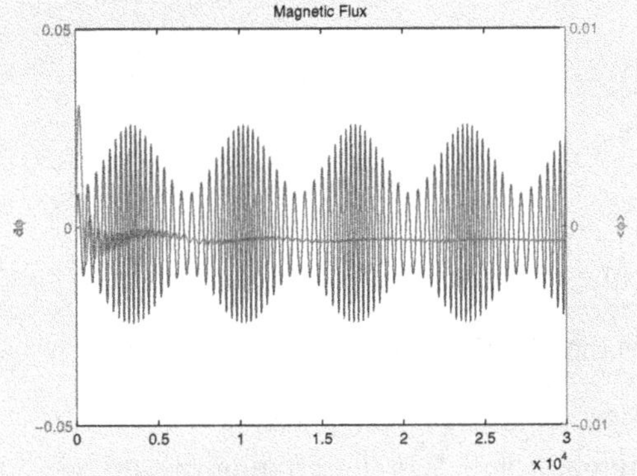

3.2.4 *Guiding Center Particle Trajectory in a Strong Magnetic Field*

Without loss of generality, we may divide the particle motion in a strong magnetic field into two parts, the **guiding center** and the gyration namely, $\vec{r} \equiv \vec{R}_{gc} + \vec{\rho}$ and $\vec{v} \equiv \vec{V}_{gc} + \vec{u}$. These parameters are defined by

$$d\vec{R}_{gc}/dt \equiv \vec{V}_{gc}, \tag{3.21}$$

$$d\vec{\rho}/dt \equiv \vec{u}, \tag{3.22}$$

$$\frac{d\vec{u}}{dt} \equiv \vec{u} \times \vec{\omega}. \tag{3.23}$$

Therefore

$$\frac{d\vec{V}_{gc}}{dt} = \vec{V}_{gc} \times (\vec{\omega} + \delta\vec{\Omega}) + \vec{u} \times \delta\vec{\Omega}. \tag{3.24}$$

Here $\vec{\omega} \equiv \vec{\Omega}(\vec{R}_{gc}) \equiv \vec{\Omega}(\vec{r}) - \delta\vec{\Omega}$ is the magnetic field evaluated at the guiding center instead at the physical position since the difference $\delta\vec{\Omega}$ is of higher order, and moreover, it keeps the particle gyration frequency to only vary on a slow time scale. The gyrating velocity is given by

$$\vec{u} = \vec{v}_0 \cos\omega t + \vec{v}_0 \times \hat{e}_{\|} \sin\omega t, \tag{3.25}$$

where $\hat{e}_{||} \equiv \vec{\omega}/\omega$. The gyroradius $\vec{\rho} = \hat{e}_{||} \times \vec{u}/\omega$ is much smaller than the guiding center displacement R_{gc} and the magnetic field scale length $L_m \sim R_{gc} \gg \rho$ so that $\delta\Omega/\Omega \sim \rho/R_c \sim \rho/L_m \sim O(\delta) \ll 1$.

It is clear that the guiding center velocity to lowest order is along the field line so that $\vec{V}^{(0)}_{gc} \times \vec{\omega} = 0$ in Eq. (3.24) . Proceed further by taking the inner product of Eq. (3.24) with $\vec{V}^{(0)}_{gc}$ to give

$$\vec{V}^{(0)}_{gc} \cdot d\vec{V}^{(0)}_{gc}/dt = \vec{V}^{(0)}_{gc} \cdot (\vec{u} \times \delta\vec{\Omega}) = \delta\vec{\Omega} \cdot (\vec{V}^{(0)}_{gc} \times \vec{u}).$$

Recognizing that $\delta\vec{\Omega} = \vec{\rho} \cdot \nabla\vec{\omega}$ to lowest order, we obtain

$$\frac{dV^{(0)}_{gc}}{dt} = \delta\vec{\Omega} \cdot \vec{\rho}\omega = (\vec{\rho} \cdot \nabla\vec{\omega}) \cdot \vec{\rho}\omega. \tag{3.26}$$

Since $\vec{\rho}$ is perpendicular to the magnetic field, we have $\oint d(\omega t/2\pi)\vec{\rho}\vec{\rho} = \frac{1}{2}(v_0^2/\omega^2)$ $(\vec{\vec{I}} - \hat{e}_{||}\hat{e}_{||})$. Here $\vec{\vec{I}}$ is the identity matrix. After taking an average over the cyclotron period, Eq. (3.26) gives $dV^{(0)}_{gc}/dt = \mu(\vec{\vec{I}} - \hat{e}_{||}\hat{e}_{||}) \cdot \nabla\vec{B} = \mu(\nabla \cdot \vec{B} - \partial B/\partial s)$. Here μ is the magnetic moment and S is the guiding center coordinate along the field line. Therefore, to the lowest order

$$\frac{dV^{(0)}_{gc}}{dt} = -\mu\frac{\partial B}{\partial s}. \tag{3.27}$$

This could have also been shown from the conservation of the energy $d\varepsilon/dt = d(\frac{1}{2}v_{||}^2 + \mu B)/dt = v_{||}dv_{||}/dt + \mu dB/dt = v_{||}dv_{||}/dt + \mu v_{||}\partial B/\partial s = 0$, with μ held as an adiabatic invariant. On the other hand, Eq. (3.27) and the energy conservation would prove that the magnetic moment is an adiabatic invariant. In the perpendicular direction, there are the grad-B and the curvature drifts as derived in Section 3.1:

$$\vec{V}^{(1)}_{gc} = \frac{\mu}{\Omega}\hat{e}_{||} \times \nabla\Omega + \frac{v_{||}^2}{\Omega}\hat{e}_{||} \times \frac{\partial\hat{e}_{||}}{\partial s} = \frac{1}{\Omega}\left(\mu + \frac{v_{||}^2}{\Omega}\right)\hat{e}_{||} \times \nabla\Omega. \tag{3.28}$$

Homework 3.8: Tokamak Physics: In the tokamak geometry (donut shape) with the magnetic field given by the toroidal magnetic field $\vec{B}_T = \hat{e}_\Phi B_0 R_0/R$ and the poloidal magnetic field $\vec{B}_p = \hat{e}_\theta b_0 r/a$ for $r < a$ and $\vec{B}_p = \hat{e}_\theta b_0 a/r$ for $r > a$ due to a uniform toroidal current taken in the cylindrical approximation. (R, Φ, Z) are the cylindrical coordinates with $\hat{e}_Z$ pointing normal to the donut plane, and $\hat{e}_\Phi$ pointing along the major circle of the donut. The (r, θ, z) are the cylindrical coordinates in the minor cross section with $\widehat{e}_z$ coinciding with $\widehat{e}_\Phi$ locally and $\widehat{e}_r$ pointing radially outward. Write a MATLAB computer code to investigate the following:

1. Make a contour plot for magnetic surfaces.
2. Trace the magnetic field lines starting from $R = R_0 + a$, $R_0 + a/2$, $R_0 + a/4$ and $\Phi = Z = 0$.
3. Find a few different particle trajectories in this magnetic field configuration. Indentify the passing, banana and potato orbitals.
4. Add a small radial electric field and find the modifications to 3.

Homework 3.9: Electron trajectory circulating two fixed ions: Write a MATLAB code to investigate the time evolution of an electron which is circulating around two fixed ions on the plane. This is in complete analogy to the *planet trajectory circulating the binary sun.*

Further Reading

Review the principle of the least action that leads to the equation of motion in the Lagrangian and Hamiltonian formulations of classical mechanics. See for example, Landau and Lifshitz, Mechanics (1969), or Thornton and Marion, Classical Dynamics of Particles and Systems (2008),

If you are a novice in the numerical algorithm, the book on Numerical Recipes — the Art of Scientific Computing by Press, Teukolsky, Vetterling, and Flnnery, can be a great source of self-learning.

Applications of single particle physics made great contributions to technology development. A few examples are listed below:

Free Electron Laser was first conceived by John Madey (1971), and reported in *J. Appl. Phys.* titled "Stimulated Emission of Bremsstrahlung in a Periodic Magnetic Field".

Gyrotron was first conceived by A. Kupiszezwki (1979) as "A High Frequency Microwave Amplifier". Its backward wave amplification was discovered from both particle simulation and experimental confirmation by Chu, Chen, Hung, Chang, Barnett, Chen, and Yang (1998). A good review on "The Electron Cyclotron Maser" is available in Rev. Modern Phys. by K. R. Chu (2004).

Plasma Accelerator was conceived by Tajima and Dawson, in Laser Electron Accelerator (1979). It was observed that 5 MeV electrons were accelerated to 55 MeV in 20 μm (Hsieh *et al.*, 2006).

Nonneutral plasma confinement and its applications can be seen in Physics Today article, "Trapped Plasmas with a Single Sign of Charge", by O'Neil (1999).

Homework Hints

Homework 3.1: Write the MATLAB code to calculate the single particle trajectory in a strong magnetic field and a dc electric field.

```
function ExB                                      %the energy difference  dE=
v0=1; x0=0; y0=1; p=0.05;                         1.1669e-005
dt=0.001; OMEGA=1; dT=dt*OMEGA;
Nt=100000;
vx=(1:Nt)*0; x=vx;y=x; vy=x;
vx(1)=v0; theta=0;
x(1)=vx(1)*dt/2;
y(1)=vy(1)*dt/2;
vt=OMEGA; r=sqrt(x(1)^2+y(1)^2);
v=sqrt(vx(1)^2+vy(1)^2);
for it=2:Nt
    theta=theta-dT;
    vx(it)=v*cos(theta);
    vy(it)=-v*sin(theta)+p*dT;
    x(it)=x(it-1)+vx(it)*dT;
    y(it)=y(it-1)+vy(it)*dT;
    v=sqrt(vx(it)^2+vy(it)^2);
end;
fplot(x,y);
title('EXB drift','fontsize',16);
xlabel('X','fontsize',16); ylabel('Y','fontsize',16);
axis equal;
E0=sum(vx(1:50000).^2+vy(1:50000).^2)/50000;
E=sum(vx.^2+vy.^2)/Nt;
dE=abs(E-E0)/E0,
```

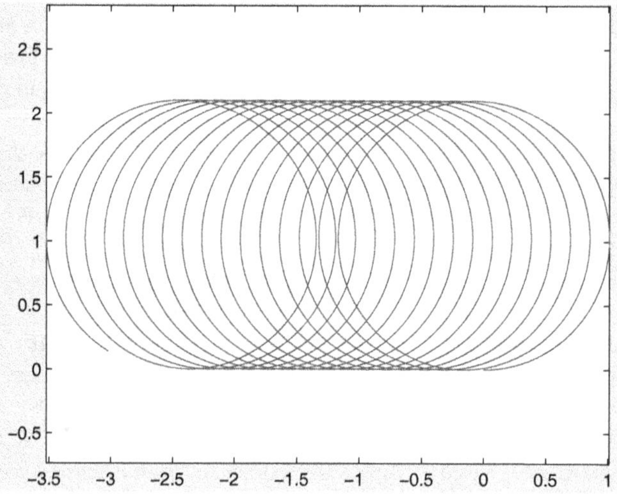

Homework 3.3: Apply the Runge Kutta fourth order algorithm to the single particle trajectory in a uniform magnetic field. Take dt to be 0.1, 0.01, and 0.001 and check how good their energies are conserved.

```
function gyration3(dt)                         % Runge Kutta fourth order
OMEGA=1; dT=dt*OMEGA;
Nt=50000;
vx=(1:Nt)*0; x=vx; y=x; vy=x;
vx(1)=-1; x(1)=0; y(1)=-1;
E0=(vx(1)^2+vy(1)^2)/2;
Z=[x(1) y(1) vx(1) vy(1)];
for i=1:Nt;
t=dt*i;
Z=rk4(Z,t,dt);
x(i)=Z(1); y(i)=Z(2); vx(i)=Z(3); vy(i)=Z(4);
end;
plot(x,y); axis equal;
axis([-1 1 -1 1]);
xlabel('x'); ylabel('y');                      dE =  0 for dt=0.001
dE=abs(sum(vx.^2+vy.^2)/2/Nt-E0)/E0,           dE=  3.4726e-010 for dt=0.01
%-----------------------------------------     dE=  3.4671e-004 for dt=0.1
function dy=force(t,R)
dy=[R(3), R(4), R(4), -R(3)];
```

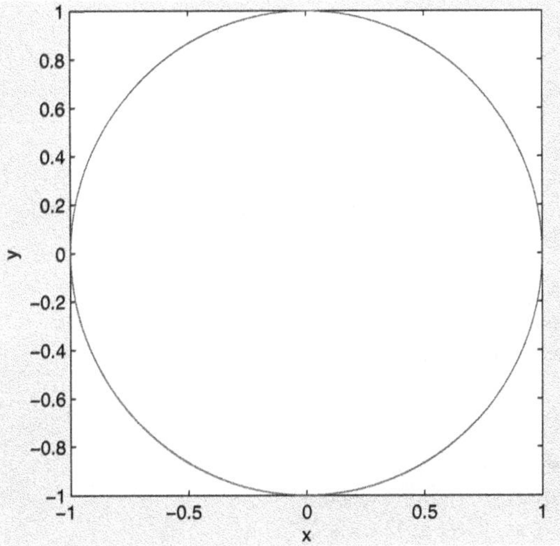

Homework 3.5: Helmholtz Coil: Write a MATLAB code to calculate the magnetic field of the Helmholtz coils. Also make the contour plot of the magnetic field.

```
function HelmholtzCoil
a=10;
h=a/2; dx=0.1; b=1.5*a;
N=fix(2*b/dx);
Br=zeros(N,N); Bz=Br;
x=-b+dx/2:dx:b-dx/2;
z=b-dx/2:-dx:-b+dx/2;
for i=1:length(x)
   X=x(i);
   for j=1:length(z)
      Z=z(j);
      [Br(j,i),Bz(j,i)]=Helmholtz(a,h,abs(X),Z);
   end;
end;
B=sqrt(Br.^2+Bz.^2);
Bmax=max(max(B,[],1),[],2); Bmin=min(min(B,[],1),[],2);
dB=(Bmax-Bmin)/100;
value=[Bmin:dB:Bmax/4];
figure;
R=repmat(x,N,1); Z=repmat(z',1,N);
contour(R,Z,B,value);
xlabel('R');ylabel('Z');
axis equal;
title('Contour Plot of B Field for Helmholtz Coils');
x=0:0.01:10; z=x*0+5;
line(x,z,'color','k');
z=-5:0.01:5; x=0*z+10;
line(x,z,'color','k')
text(5,5,'R','color','k','fontsize',10,'fontname','verdana','fontweight','bold');
text(10,0,'R','color','k','fontsize',10,'fontname','verdana','fontweight','bold');
save Helmholtz Br Bz R Z;

function [Br,Bz]=Helmholtz(a, h, r, z)
[br1,bz1]=Coil(a, h, r, z);
[br2,bz2]=Coil(a, -h, r, z);
Br=br1+br2;
Bz=bz1+bz2;

function [Br,Bz]=Coil(a, h, r, z)
k=sqrt(4*a*r./((r+a).^2+(z-h).^2+eps));
B=1/2/pi./sqrt((r+a).^2+(z-h).^2+eps);
[K,E]=ellipke(k^2);
Br=-(z-h).*(K-(1-k.^2/2)./(1-k.^2+eps).*E)./(r+eps);
Bz=K+(k.^2./2.*(1+a/(r+eps))-1).*E./(1-k.^2+eps);
Br=B.*Br;
Bz=B.*Bz;
if(r<0.1) Br=0; Bz=1/2/a./(1+((z-h)/a).^2)^(3/2); end;
```

```
function [Br,Bz]=Coil(a, h, r, z)
k=sqrt(4*a*r./((r+a).^2+(z-h).^2+eps));
B=1/2/pi./sqrt((r+a).^2+(z-h).^2+eps);
[K,E]=ellipke(k^2);
Br=-(z-h).*(K-(1-k.^2/2)./(1-k.^2+eps).*E)./(r+eps);
Bz=K+(k.^2./2.*(1+a/(r+eps))-1).*E./(1-k.^2+eps);
Br=B.*Br;
Bz=B.*Bz;
if(r<0.1) Br=0; Bz=1/2/a./(1+((z-h)/a).^2)^(3/2); end;

function [E,MU,PHI]=getEMUPHI(y,b)
r=[y(1) y(2) y(3)];
v=[y(4) y(5) y(6)];
B=norm(b);
Vp=dot(v,b)/B;
MU=(dot(v,v)-Vp^2)/B/2;
E=dot(v,v)/2;
PHI=dot(b',cross(r,v));
```

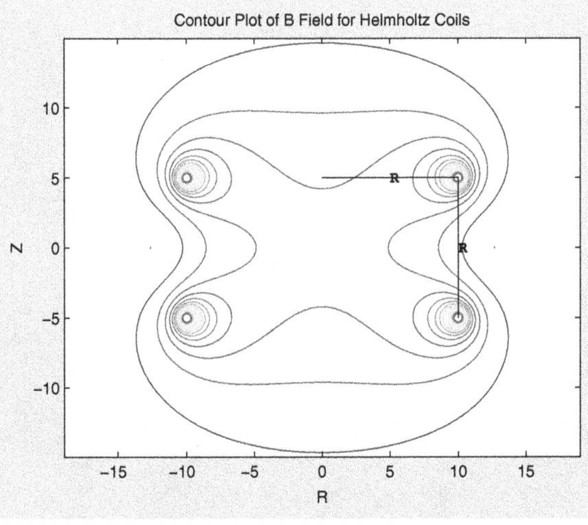

Homework 3.7: Write the MATLAB code to calculate the earth magnetic field.

```
function [Br,Bt]=EarthMagneticField
Nphi=12;
dphi=2*pi/Nphi;
phi=dphi*(1:Nphi);
N=300;
index=1:N;
dt=pi/N;
theta=dt*index;
r0=1;
r=r0*sin(theta);
z=r.*cos(theta);
Z=repmat(z',1,Nphi);
ro=r.*sin(theta);
X=ro'*cos(phi);
Y=ro'*sin(phi);
fig=figure('position', [50 50 600 600],'color','white');
plot3(X,Y,Z);
xlabel('X'); ylabel('Y'); zlabel('Z');
```

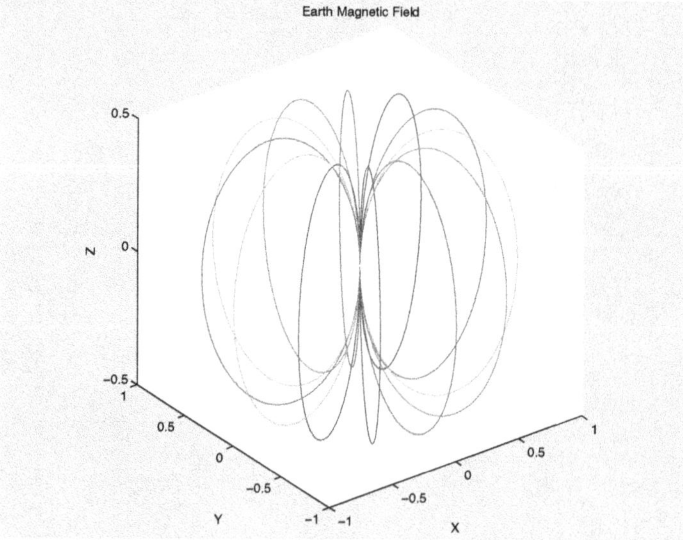

Homework 3.9: Electron trajectory circulating two fixed ions: Write a MATLAB code to investigate the time evolution of an electron which is circulating around two fixed ions on the plane. This is in complete analogy to the **planet trajectory circulating the binary sun.**

```
function BinaryStar                    % Potential energy of binary sun in the space
close all; clc; clear all;
syms X Y U D1 D2 ;
syms GD M1 M2 m R1 R2;

G=6.67e-11;
M1=2e30; M2=2e30; m=4e24; D=30e10;
X=repmat(linspace(-1.0e12, 1.0e12, 401), 401, 1);
Y=repmat(linspace(-1.0e12, 1.0e12, 401)', 1, 401);
D1=M2/(M1+M2)*D;   D2=M1/(M1+M2)*D;

U=-G*M1*m*((X-D1).^2+Y.^2+eps).^(-1/2)-G*M2*m*(((X+D2).^2+Y.^2+eps)).^(-1/2);
%contour(X, Y, U, [-1e35:2e33:0]);    % Motion of planet
syms u x y Sx Sy dt fx fy vx vy p;
u=-G*M1*m*((x-D1)^2+y^2+eps)^(-1/2)-G*M2*m*(((x+D2)^2+y^2+eps))^(-1/2);
fx=-diff(u,x); fy=-diff(u,y);
dt=684000; %dt=700000;
vx=20000; vy=-5000;
p=[-30e10 -30e10]; x=-3.0e11; y=-3.0e11;
Sx=[]; Sy=[];
for i=1:10000
    x=x+vx*dt;
    y=y+vy*dt;
    vx=vx+eval(fx)/m/2*dt;
    vy=vy+eval(fy)/m/2*dt;
    if(mod(i,10)==0)
        Sx=[Sx,x];
        Sy=[Sy,y];
    end;
end;

RANGE=-2e35./(0.01:10:500);
figure(1)
plot(Sx,Sy,'g-');
axis equal;
hold on;
contour(X, Y, U, RANGE);
hold off;
```

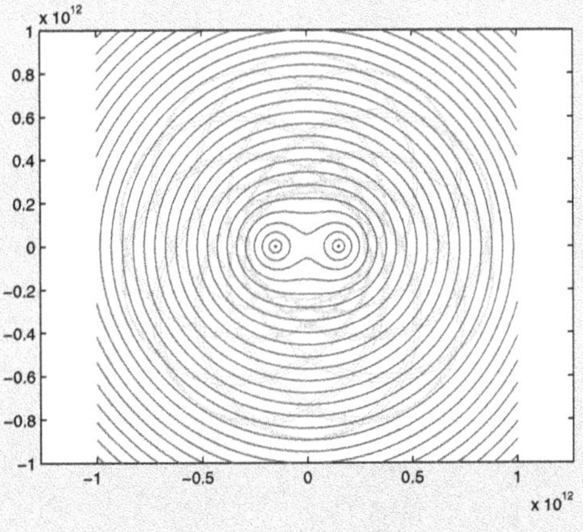

Chapter Four

Single Fluid

"Make everything as simple as possible, but not simpler."

Albert Einstein (1879–1955)

The single particle picture, while providing the first glance at the truth, is insufficient to describe the collective phenomena of many-body plasma physics. Generalizing it to keep track of all particles would be highly desirable. However, since plasma physics often involves a great number of particles, to follow their dynamics at least $6N$ variables for N particles in phase space are needed, not to mention the electric field, the magnetic field, and many other quantities in the configuration space. Thus, it is often too complicated to follow the kinematics of all particles analytically or numerically. On the other hand, the smoothed and averaged properties of wave, transport, and fluid, may be revealed by a few local variables in the configuration space such as density, momentum, energy, and current, to understand the governing principles by simplification with clarification. Many effects from the particle discreteness such as correlation, velocity anisotropy, runaway, kinetic effects, among others, may be neglected as a start. Therefore, resorting to a fluid model is naturally the next step in our exploration of plasma physics.

4.1 From Particle to Fluid

If we take a single particle motion as described by the Newton's second law, and generalize it to particles of species σ at the same velocity, we have the momentum equation

$$n_\sigma m_\sigma \frac{d\vec{v}_\sigma}{dt} = n_\sigma q_\sigma \left(\vec{E} + \frac{\vec{v}_\sigma}{c} \times \vec{B} \right).$$

Here the time derivative following the particles, often termed as the material derivative, is the sum of local and convective derivatives, namely $d/dt = \partial/\partial t + \vec{v}_\sigma \cdot \nabla$. In the fixed space description or the lab frame of reference, we have

$$n_\sigma m_\sigma \left(\frac{\partial \vec{v}_\sigma}{\partial t} + \vec{v}_\sigma \cdot \nabla \vec{v}_\sigma \right) = n_\sigma q_\sigma \left(\vec{E} + \frac{\vec{v}_\sigma}{c} \times \vec{B} \right). \tag{4.1}$$

Similarly, assuming these particles move in the positive direction without loss of generality, the particles that enter the local spot from the left would increase the density, while the particles that leave the local cell would deprive the density. We therefore have

$$\frac{dn_\sigma}{dt} = -\frac{n_\sigma v_{\sigma+} - n_\sigma v_{\sigma-}}{\Delta} = -n_\sigma \frac{v_{\sigma+} - v_{\sigma-}}{\Delta} = -n_\sigma \frac{\partial v_\sigma}{\partial x}.$$

Back to the laboratory frame, we have the continuity equation,

$$\frac{\partial n_\sigma}{\partial t} + \vec{v}_\sigma \cdot \nabla n_\sigma = -n_\sigma \nabla \cdot \vec{v}_\sigma. \tag{4.2}$$

If $\nabla \cdot \vec{v}_\sigma = 0$, the flow is said to be incompressible since the fluid density is unchanged, or frozen, with the flow, $dn_\sigma/dt = \partial n_\sigma/\partial t + \vec{v}_\sigma \cdot \nabla n_\sigma = 0$. In the steady state, Eq. (4.2) gives $\nabla \cdot (n_\sigma \vec{v}_\sigma) = 0$, which means that the particle flux enters a region must equal to the particle flux leaves the region.

When thermal motions are taken into account, we may write $\vec{v}_\sigma = \vec{V}_\sigma + \tilde{v}_\sigma$. Here $\tilde{v}_\sigma$ is the random fluctuations that is ensemble averaged to vanish, namely, $\langle \tilde{v}_\sigma \rangle = 0$, and the ensemble averaged velocity gives the flow velocity $\vec{V}_\sigma \equiv \langle \vec{v}_\sigma \rangle$. The pressure tensor is defined as $P_{\sigma_{ij}} \equiv n_\sigma m_\sigma \langle \tilde{v}_{\sigma_i} \tilde{v}_{\sigma_j} \rangle$, where i *and* j are the indices of the components. Therefore,

$$\left\langle n_\sigma m_\sigma \left(\frac{\partial \vec{v}_\sigma}{\partial t} + \vec{v}_\sigma \cdot \nabla \vec{v}_\sigma \right) \right\rangle = n_\sigma m_\sigma \left(\frac{\partial \vec{V}_\sigma}{\partial t} + \vec{V}_\sigma \cdot \nabla \vec{V}_\sigma + \left\langle \sum_i \tilde{v}_{\sigma i} \cdot \nabla \tilde{v}_{\sigma_j} \right\rangle \right)$$

$$= n_\sigma q_\sigma \left(\vec{E} + \frac{\vec{v}_\sigma}{c} \times \vec{B} \right).$$

We end with

$$n_\sigma m_\sigma \left(\frac{\partial \vec{V}_\sigma}{\partial t} + \vec{V}_\sigma \cdot \nabla \vec{V}_\sigma \right) = -\nabla \cdot \overline{\vec{P}}_\sigma + n_\sigma q_\sigma \left(\vec{E} + \frac{\vec{V}_\sigma}{c} \times \vec{B} \right). \tag{4.3}$$

This gives us the fluid description for the single species. The hydrodynamics deals with neutral fluid, that reduces Eq. (4.3) to

$$\frac{\partial \vec{V}}{\partial t} + \vec{V} \cdot \nabla \vec{V} = -\frac{1}{\rho} \nabla \cdot \overline{\vec{P}}, \tag{4.4}$$

where $\rho = nm$ is the mass density, and the subscript σ is omitted. Similarly, Eq. (4.2) can be cast into the continuity equation,

$$\frac{\partial \rho}{\partial t} + \nabla \cdot \rho \vec{V} = 0. \tag{4.5}$$

It is noteworthy that the nonlinear term in the momentum equation of Eq. (4.4),

$$\vec{V} \cdot \nabla \vec{V} = \frac{1}{2} \nabla \vec{V}^2 + (\nabla \times \vec{V}) \times \vec{V}, \tag{4.6}$$

is physically most complex. It advects the fluid, channels the vortex, steepens the mass, and makes object buoyant. Fluid dynamics is very much responsible for tornado, hurricane, typhoon, tsunami, microburst, etc. The **Computational Fluid Dynamics (CFD)** is a subject being actively investigated with ever-improved heavy computing.

An incompressible flow has the density kept intact along the fluid motion. It gives $dn/dt = \partial n/\partial t + \vec{V} \cdot \nabla n = 0$ that implies $\nabla \cdot \vec{V} = 0$, no density accumulated or dispersed in the volume following the flow. Thus, in the 2d flow, the velocity may be expressed as $\vec{V} = \nabla \times (\psi \hat{e}_z) = \hat{e}_z \times \nabla \psi$. Here, ψ is called the stream function. Any variable which only varies with the stream function will have no spatial gradient along the flow velocity since $\vec{V} \cdot \nabla \psi = 0$. Consequently, in steady state, density will be function of ψ only since $\vec{V} \cdot \nabla n = 0$. Dotting Eq. (4.4) with $\vec{V}$ eliminates the vortex term in Eq. (4.6) and leads to the steady state solution of $\vec{V} \cdot \nabla(p/\rho + V^2/2) = 0$ for the isotropic fluid, $\overline{\overline{P}} = p\overline{\overline{I}}$. Therefore, we may write $p/\rho + V^2/2 = \Phi(\psi)$, that is also a function of ψ.

4.1.1 *Potential Flow*

When the flow is **irrotational**, i.e.,$\nabla \times \vec{V} = 0$, then $\vec{V} = \nabla \phi$, free of **vortex flow**. Together with the **incompressibility** where $\nabla \cdot \vec{V} = 0$, it leads to $\nabla^2 \phi = 0$. This is the Laplace equation governing the potential flow. The **potential flow** described in Chapter Two is re-examined herein. Assume the fluid moves across an infinitely long cylinder of radius a. We specify the boundary conditions within a box of width $L \gg a$ and height $h \gg a$ by $\phi|_{z=\pm L} = \pm v_0 L$, $\vec{V}|_{z=\pm L} = \vec{V}|_{x=\pm h} = v_0 \hat{e}_z$, $\phi|_{\rho=a} = 2v_0 \cos\theta$. It has the following solution:

$$\phi = v_0 \left(\rho + \frac{a^2}{\rho}\right) \cos\theta. \tag{4.7}$$

The velocities are given by $\vec{V} = \nabla \phi = \hat{e}_\rho v_0 (1 - a^2/\rho^2) \cos\theta - \hat{e}_\theta v_0 (1 + a^2/\rho^2) \sin\theta$. The fluid circulates around the cylinder with the velocity $\vec{V}|_{\rho=a} = -2\hat{e}_\theta v_0 \sin\theta$. The velocity $\vec{V}$ vanishes at the stagnation points, $\theta = 0$ and $\theta = \pi$ on the surface of the cylinder.

4.1.2 *Laminar Flow and Turbulent Flow*

The potential flow is a good example of the **laminar flow**. The blood flow in the human body and the normal traffic flow are laminar flows. Laminar flows, however, can be disrupted and become turbulent. When this occurs, blood does not flow smoothly in adjacent layers, and can cause great congestion. A supersonic flow with **Mach number** exceeds unity, meaning the flow velocity exceeds the sound speed,

```
function PotentialFlow
a=1; L=10; h=5; v0=1;
dx=0.025; z=-L:dx:L; x=-h:dx:h;
m=length(x); n=length(z);
Z=repmat(z,m,1);
X=repmat(x',1,n);
r=Z.^2+X.^2;
PHI=Z.*v0.*(1+a^2./(r+eps));
[vz,vx]=gradient(PHI);
vx=vx.*(r>1); vz=vz.*(r>1);
PHI=PHI.*(r>1);
d=10; N=1:10;
V=-2^5./2.^N;
v=[V,0,-V(10:-1:1)];
contourf(Z,X,PHI,v);
axis equal; view(0,90); shading flat;
hold on;
quiver(Z(1:d:m,1:d:n),X(1:d:m,1:d:n),...
vz(1:d:m,1:d:n),vx(1:d:m,1:d:n));
title('Potential Flow','fontsize',16);
xlabel('X','fontsize',16); ylabel('Y','fontsize',16);
hold off;
```

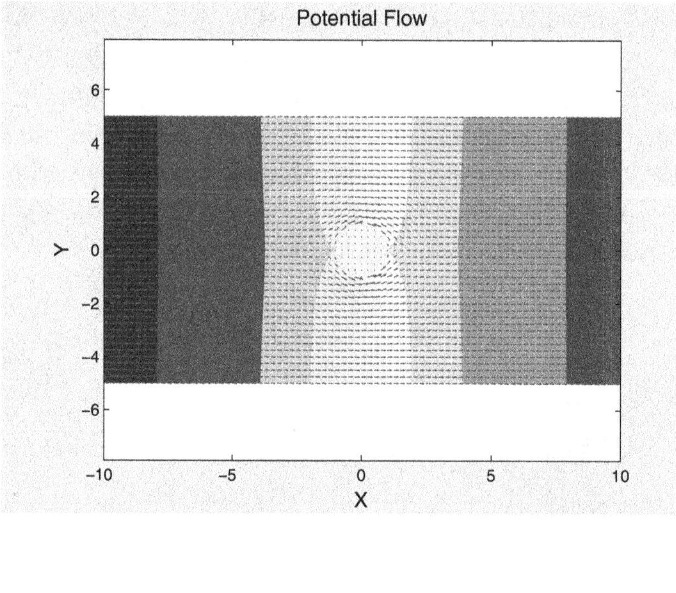

could pump energy into the sound wave. It will cause greater energy consumption by the enhanced viscosity on the supersonic jet, and can result in the acoustic boom as well as the **turbulent flow** behind the jet. Turbulent flow also occurs in large arteries at branch points, in diseased and narrowed arteries, and across stenotic heart valves. Turbulence increases the energy required to drive blood flow as it loses energy due to friction.

4.2 Ideal Fluid Equations

We develop the formal theory for the fluid equations from the first principles in this section. It is important to keep track of what has been neglected or assumed to arrive at the fluid description since at times the fluid model may be inadequate to describe the correct physics, and the equations can be ill posed in that regard.

4.2.1 *Local Variables of Velocity Moments in the Configuration Space*

The local variables of the velocity moments can be determined by summing up the contributions from the individual particles. Consider only one single species to begin with. The zeroth moment is the particle population,

$$n(\vec{x}, t) \equiv \sum_i \delta(\vec{x} - \vec{x}_i(t)), \tag{4.8}$$

the first moment is the **particle flux**,

$$\vec{\Gamma}(\vec{x}, t) \equiv n(\vec{x}, t)\vec{v}(\vec{x}, t) \equiv \sum_i \vec{w}_i(t)\delta(\vec{x} - \vec{x}_i(t)), \tag{4.9}$$

and the second moment is the **momentum flux dyadic**,

$$\overset{\rightharpoonup}{\vec{K}}(\vec{x}, t) \equiv \sum_i m\vec{w}_i(t)\vec{w}_i(t)\delta(\vec{x} - \vec{x}_i(t)), \tag{4.10}$$

where $\vec{w}_i$ is the velocity of ith particle, and the summations are taken over all the particles at position $\vec{x}$. We may further define the pressure tensor which is the momentum flux in the moving frame of the fluid,

$$\overset{\rightharpoonup}{\vec{P}}(\vec{x}, t) = \sum_i m(\vec{w}_i(t) - \vec{v}(\vec{x}, t))(\vec{w}_i(t) - \vec{v}(\vec{x}, t))\delta(\vec{x} - \vec{x}_i(t)). \tag{4.11}$$

Therefore,

$$\overset{\rightharpoonup}{\vec{K}} = \overset{\rightharpoonup}{\vec{P}} + nm\vec{v}\vec{v}. \tag{4.12}$$

Note that multiplying the trace of the momentum flux by $\frac{1}{2}$ gives the total random and ordered energies $\in (\vec{x}, t) \equiv \frac{1}{2}\sum_i m\vec{w}_i^2\delta(\vec{x} - \vec{x}_i) = \frac{1}{2}nm\vec{v}^2 + \frac{1}{2}trace(\overset{\rightharpoonup}{\vec{P}})$. Here $\vec{v}(\vec{x}, t)$ is the flow velocity at the local position $\vec{x}$ at time t, and $\frac{1}{2}trace(\overset{\rightharpoonup}{\vec{P}})$ is the pressure energy. We further define the **energy flux**

$$\vec{H}(\vec{x}) \equiv \frac{1}{2}\sum_i m\vec{w}_i w_i^2 \delta(\vec{x} - \vec{x}_i), \tag{4.13}$$

and the **heat flux**, which is the energy flux in the flow frame,

$$\vec{Q}(\vec{x}) \equiv \frac{1}{2}\sum_i m(\vec{w}_i - \vec{v})|\vec{w}_i - \vec{v}|^2 \delta(\vec{x} - \vec{x}_i). \tag{4.14}$$

Therefore, we have the following relation between the heat flux and the energy flux,

$$\vec{H} = \vec{Q} + \frac{1}{2}nm\vec{v}v^2 + \overset{\Rightarrow}{P}\cdot\vec{v} + \frac{1}{2}\vec{v}trace(\overset{\Rightarrow}{P}) = \vec{Q} + \vec{v} \in + \overset{\Rightarrow}{P}\cdot\vec{v}. \tag{4.15}$$

These somewhat involved formulae translate the single particle properties in the configuration space to local variables of physical significance. The approach here abstains from the phase space treatment as routinely done otherwise, and retains the **particle and fluid duality** in a convenient formulation that may have advantages in both analytical treatment and computational simulation.

4.2.2 *Conservation Laws of Particle, Momentum and Energy*

The fluid equations, as a manifestation of the conservation laws of particle, momentum and energy, can be found by taking the time derivative of corresponding variables. In this section, we will treat only a single species.

The time change of the density is given by

$$\frac{\partial}{\partial t}n(\vec{x},t) = \sum_i \frac{\partial}{\partial t}\delta(\vec{x} - \vec{x}_i(t)) = \sum_i \frac{\partial}{\partial \vec{x}_i}\delta(\vec{x} - \vec{x}_i(t)) \cdot \frac{\partial \vec{x}_i(t)}{\partial t}$$
$$= -\nabla\cdot\sum_i \vec{w}_i\delta(\vec{x} - \vec{x}_i) = -\nabla\cdot(n\vec{v}).$$

This results in the continuity equation

$$\frac{\partial}{\partial t}n + \nabla\cdot(n\vec{v}) = 0, \tag{4.16}$$

provided that there is no particle source or sink. Equation (4.16) represents **the conservation law of particle density**. Similarly, taking the time derivative on the particle flux gives

$$\frac{\partial n\vec{v}}{\partial t} = \sum_i \frac{\partial}{\partial t}[\vec{w}_i(t)\delta(\vec{x} - \vec{x}_i(t))] = \sum_i \vec{a}_i\delta(\vec{x} - \vec{x}_i) - \nabla\cdot\sum_i \vec{w}_i\vec{w}_i\delta(\vec{x} - \vec{x}_i).$$

Here, $\vec{a}_i \equiv d\vec{w}_i(t)/dt$ is the acceleration on the ith particle. The time derivative gives rise to a higher moment, the momentum flux dyadic $\overset{\Rightarrow}{K}$ as defined in Eq. (4.12).

Therefore,

$$\frac{\partial nm\vec{v}}{\partial t} + \nabla \cdot \vec{\vec{K}} = \vec{F} \equiv \sum_i m\vec{a}_i \delta(\vec{x} - \vec{x}_i). \tag{4.17}$$

The force $\vec{F} = \vec{F}^{ext} + \vec{F}^{int}$ is due to both the external fields, and internal interactions that may result in redistribution of the flow momentum. The left hand side of Eq. (4.17) when integrated over the entire volume represents **the conservation law of momentum** in the absence of external force. With use of the vector identity, $\nabla \cdot (\vec{A}\vec{B}) = (\nabla \cdot \vec{A})\vec{B} + \vec{A} \cdot \nabla \vec{B}$, the momentum flux dyadic can be split into the following

$$\nabla \cdot \vec{\vec{K}} = \nabla \cdot (n\vec{v})m\vec{v} + nm\vec{v} \cdot \nabla \vec{v} + \nabla \cdot \vec{\vec{P}}. \tag{4.18}$$

Thus, utilizing the continuity equation to eliminate $\partial n/\partial t$, we have the equation of motion,

$$nm\frac{\partial \vec{v}}{\partial t} + nm\vec{v} \cdot \nabla \vec{v} = -\nabla \cdot \vec{\vec{P}} + \vec{F}. \tag{4.19}$$

This can be cast into the Navier-Stokes equation by separating the pressure tensor into $\vec{\vec{P}} = p\vec{\vec{I}} + \vec{\vec{\sigma}}$, where $\vec{\vec{\sigma}}$ may be interpreted as the stress tensor to give

$$\rho\left(\frac{\partial \vec{v}}{\partial t} + \vec{v} \cdot \nabla \vec{v}\right) = -\nabla p + \nabla \cdot \vec{\vec{\sigma}} + \vec{F}. \tag{4.20}$$

Here, ρ is the mass density. The stress tensor resulting from transport effects will be discussed in Chap. 8.

Taking the time derivative on the total energy $\in (\vec{x}, t) = \frac{1}{2}\sum_i m\vec{w}_i^2 \delta(\vec{x} - \vec{x}_i)$ gives

$$\frac{\partial \in (\vec{x}, t)}{\partial t} = \sum_i m\vec{w}_i \cdot \vec{a}_i \delta(\vec{x} - \vec{x}_i) - \nabla \cdot \left(\frac{1}{2}\sum_i m\vec{w}_i^2 \vec{w}_i \delta(\vec{x} - \vec{x}_i)\right)$$

and results in,

$$\frac{\partial}{\partial t} \in (\vec{x}, t) = \vec{F}^{ext} \cdot \vec{v} + \varsigma^{int} - \nabla \cdot \vec{H}. \tag{4.21}$$

The first term on the RHS is the power density caused by the external force $\vec{F}^{ext}$ acting on the flow motion. A good example is the ohmic heating $\vec{J} \cdot \vec{E}$. The internal power consumption $\zeta^{int} \equiv \sum_i m\vec{a}_i^{int} \cdot \vec{w}_i \delta(\vec{x} - \vec{x}_i)$ may, for example, redistribute the energy and relax the fluid toward the maximum entropy state. Equation (4.21) represents **the conservation law of energy.**

Dotting Eq. (4.18) with $nm\vec{v}$, dotting Eq. (4.17) with $\vec{v}$, and adding them together, we arrive at the evolution equation for the flow energy,

$$\frac{\partial \in_f}{\partial t} = \frac{\partial \frac{1}{2} nm\vec{v}^2}{\partial t} = -\vec{v} \cdot \nabla \cdot \vec{\vec{K}} + \vec{v} \cdot \vec{F} + \frac{1}{2} m\vec{v}^2 \nabla \cdot (n\vec{v}), \tag{4.22}$$

that may be simplified to give

$$\frac{\partial \frac{1}{2} nm\vec{v}^2}{\partial t} + \nabla \cdot \left(\frac{1}{2} nm\vec{v}^2 \vec{v} \right) = -\vec{v} \cdot \nabla \cdot \vec{\vec{P}} + \vec{v} \cdot \vec{F}. \tag{4.22a}$$

Somewhere along this line of attack, we need a closure on the formulation so that we have a complete description of all variables. Therefore, a truncation at certain transport effect is necessary. We may for example, neglect the heat flux in Eq. (4.21), or describe the pressure by the adiabatic response $pn^{-\gamma} = \text{const}$ to bypass Eq. (4.21) in its entirety, or by the isothermal condition $p = nk_BT$ with a prescription of the temperature.

Homework 4.1: The particles are given small perturbations so that $n(\vec{r}, t) = \sum_i \delta(\vec{r} - \vec{x}_i(t) - \varepsilon \vec{\chi}_i(t))$, where $\varepsilon \ll 1$ and all the other quantities are considered order unity. Make a series expansion of $n(\vec{r}, t)$ to the second order in ε.

Homework 4.2: Find the transport equations for the flow energy $\frac{1}{2} nmv^2$ and the pressure energy $\frac{3}{2} p = \frac{1}{2} \sum_i m(\vec{w}_i(t) - \vec{v}(\vec{x}, t))^2 \delta(\vec{x} - \vec{x}_i(t))$, respectively. Add them together to compare with Eq. (4.21).

4.2.3 *The Mixed Fluid*

The formulation of ideal fluid equations will now be generalized to describe the mixed fluid. Applying the single-fluid model to multiple species loses track of substantial information. We will therefore restrict ourselves to electron and single ion species with reference to the hydrogen plasma for simplicity.

The conservation of particles in Eq. (4.16) implies the mass conservation for each species,

$$\frac{\partial}{\partial t} \rho_\sigma + \nabla \cdot (\rho_\sigma \vec{v}_\sigma) = 0, \tag{4.16a}$$

where $\rho_\sigma \equiv n_\sigma m_\sigma$. Defining the mass density for the multi-species plasma as,

$$\rho(\vec{x}, t) \equiv \sum_\sigma \rho_\sigma(\vec{x}, t), \tag{4.23}$$

and the flow momentum as,

$$\rho(\vec{x},t)\vec{U}(\vec{x},t) \equiv \sum_{\sigma} \rho_\sigma(\vec{x},t)\vec{v}_\sigma(\vec{x},t), \tag{4.24}$$

and summing up all the species from Eq. (4.16a), we find the continuity equation,

$$\frac{\partial}{\partial t}\rho + \nabla \cdot (\rho \vec{U}) = 0, \tag{4.25}$$

that represents the **conservation law of mass density**. Since the individual species satisfies its own continuity equation, it is not surprising that their sum would. The other continuity equation of the two-species plasma can be recovered by the equation for the charge density as defined by $\rho_q(\vec{x},t) \equiv \sum_{i,\sigma} q_\sigma \delta(\vec{x} - \vec{x}_{i\sigma}(t)) = \sum_\sigma q_\sigma n_\sigma(\vec{x},t)$. Again, the conservation of particles in Eq. (4.16) implies the charge conservation for each species,

$$\frac{\partial}{\partial t} n_\sigma q_\sigma + \nabla \cdot (n_\sigma q_\sigma \vec{v}_\sigma) = 0. \tag{4.16b}$$

Summing up all the species from Eq. (4.16b) yields the charge-conserving continuity equation,

$$\frac{\partial}{\partial t}\rho_q(\vec{x},t) = -\nabla \cdot \vec{J}, \tag{4.26}$$

where $\vec{J}(\vec{x},t) \equiv \sum_\sigma q_\sigma n_\sigma(\vec{x},t)\vec{v}_\sigma(\vec{x},t)$ is the current density.

Taking the time derivative on the flow momentum of Eq. (4.24) and applying Eq. (4.17) gives the following:

$$\frac{\partial \rho \vec{U}}{\partial t} = \vec{F} - \nabla \cdot \sum_{\sigma} \overleftrightarrow{K}_\sigma, \tag{4.27}$$

where the first term on the RHS is the force term given by $\vec{F} \equiv \sum_\sigma \rho_\sigma \vec{a}_\sigma$ that has both the external and internal forces. In the presence of electromagnetic fields, the external force is given by $\vec{F}^{ext} = \rho_q \vec{E} + \vec{J} \times \vec{B}/c$. The internal force $\vec{F}^{int}$ due to its long range nature of plasma interaction may cause damping and equilibration across the space. The $\sum_\sigma \overleftrightarrow{K}_\sigma$ term is the momentum flux dyadic given by

$$\overleftrightarrow{K}(\vec{x},t) \equiv \sum_{i,\sigma} m\vec{w}_{i\sigma}(t)\vec{w}_{i\sigma}(t)\delta(\vec{x} - \vec{x}_{i\sigma}(t)) = \sum_{\sigma} \overleftrightarrow{p}_\sigma(\vec{x},t) + \sum_{\sigma} \rho_\sigma \vec{v}_\sigma \vec{v}_\sigma. \tag{4.28}$$

To quantify the momentum flux dyadic, we need to define the total pressure tensor. The pressure energy is referred to as the random energy of the particles. As such, *pressure tensor of individual species is more appropriate*, and the total pressure of mixed fluid is simply the sum of the pressures of individual species, $\overleftrightarrow{P} \equiv \sum_\sigma \overleftrightarrow{p}_\sigma$. Thus,

$$\frac{\partial \rho \vec{U}}{\partial t} + \nabla \cdot \sum_\sigma \rho_\sigma \vec{v}_\sigma \vec{v}_\sigma = -\nabla \cdot \overleftrightarrow{P} + \rho_q \vec{E} + \frac{1}{c}\vec{J} \times \vec{B} + \vec{F}^{int}. \tag{4.27a}$$

To combine into a single-fluid description, we need to express the drift velocities in terms of the single-fluid flow $\vec{U} = (n_p M_p \vec{v}_p + n_e m_e \vec{v}_e)/(n_p M_p + n_e m_e)$ and the current $\vec{J} = q(n_p \vec{v}_p - n_e \vec{v}_e)$. We will take advantage of the light electron mass with respect to the ion mass and make an expansion accordingly. We find,

$$\begin{aligned}\vec{v}_e &= \left(1 + \frac{n_p - n_e}{n_e}\frac{M_p}{M_p + m_e}\right)\vec{U} - \frac{\vec{J}}{n_e q}\frac{M_p}{M_p + m_e} \\ &\approx \vec{U} - \frac{\vec{J}}{nq}\left(1 - \frac{m_e}{M_p}\right) + O(m_e^2),\end{aligned} \tag{4.29}$$

$$\begin{aligned}\vec{v}_p &= \left(1 + \frac{n_e - n_p}{n_p}\frac{m_e}{M_p + m_e}\right)\vec{U} + \frac{m_e}{M_p + m_e}\frac{\vec{J}}{n_p q} \\ &\approx \vec{U} + \frac{m_e}{M_p}\frac{\vec{J}}{nq} + O(m_e^2).\end{aligned} \tag{4.29a}$$

We will impose the charge neutrality condition $n_p \approx n_e = n$ and retain accuracy in terms of the mass ratio expansion in the subsequent treatment. We find,

$$\sum_\sigma \rho_\sigma \vec{v}_\sigma \vec{v}_\sigma \approx \rho \vec{U}\vec{U} + \frac{m_e}{nq^2}\vec{J}\vec{J} + O(m_e^2).$$

Therefore,

$$\rho \frac{\partial \vec{U}}{\partial t} + \rho \vec{U} \cdot \nabla \vec{U} = -\nabla \cdot \overleftrightarrow{P} + \frac{1}{c}\vec{J} \times \vec{B} + \vec{F}^{int} + O(m_e). \tag{4.30}$$

Another momentum equation to uncover is the time evolution of the electrical current. Utilizing Eq. (4.17), we find

$$\frac{\partial \vec{J}}{\partial t} = -\sum_\sigma \nabla \cdot (\overleftrightarrow{P}_\sigma + \rho_\sigma \vec{v}_\sigma \vec{v}_\sigma)\frac{q_\sigma}{m_\sigma} + \sum_\sigma \vec{F}_\sigma \frac{q_\sigma}{m_\sigma}. \tag{4.31}$$

Homework 4.3: Apply Eq. (4.31) to electron-proton plasma with use of the light electron mass expansion to find the leading order terms.

We arrive at the equation of current transport by keeping only the lowest order terms (cf. Homework 4.3),

$$\frac{\partial \vec{J}}{\partial t} \approx \frac{\omega_{pe}^2}{4\pi}\left(\vec{E}+\frac{\vec{U}}{c}\times\vec{B}\right)-\frac{q\vec{J}}{m_e c}\times\vec{B}+\nabla\cdot\left(\frac{q\overline{\overline{P}}_e}{m_e}-\vec{U}\vec{J}-\vec{J}\vec{U}+\frac{\vec{J}\vec{J}}{nq}\right)-\vec{F}_e^{int}\frac{q}{m_e}+O(m_e). \tag{4.31a}$$

We may assume the low frequency activities relative to both the electron and ion cyclotron frequencies so that the fluid motion will maintain $\vec{E}+\vec{U}\times\vec{B}/c\approx 0$. The first term on the RHS of Eq. (4.31a) vanishes. Assuming equipartition in the pressure energy with $\frac{1}{2}k_BT$ in each degree of freedom, we expect $\overline{\overline{P}}_e\sim\overline{\overline{P}}_p\sim\frac{1}{2}\overline{\overline{P}}$. The $\vec{J}\times\vec{B}$ force causes the current channel to self pinch. The mechanism is of great interest in the plasma confinement study of reversed field pinch, θ-pinch, z-pinch, and the like. The internal force term has an obvious component from the electrical resistance $-\vec{J}/\sigma$ with σ the electrical conductance.

> **Homework 4.4:** Derive the energy equation by summing up the species with use of Eq. (4.21). Apply it to electron-proton plasma with use of the light electron mass expansion.

Summing up different species in Eq. (4.21) and expanding in the mass ratio, we find

$$\frac{\partial}{\partial t}\in(\vec{x},t)+\nabla\cdot(\vec{U}\in)=\vec{J}\cdot\vec{E}-\nabla\cdot(\vec{Q}+\overline{\overline{P}}\cdot\vec{U})+\nabla\cdot\left((\overline{\overline{P}}_e+\in_e\overline{\overline{I}})\cdot\frac{\vec{J}}{nq}\right)+\varsigma^{int}+O(m_e), \tag{4.32}$$

where $\varsigma^{int}\equiv\varsigma_e^{int}+\varsigma_p^{int}$ is the total internal energy consumption, and $\overline{\overline{I}}$ is the identity matrix. Equation (4.32) represents the conservation law of energy.

From Eqs. (4.21) and (4.22), the total pressure energy is given by,

$$\frac{\partial\frac{3}{2}p}{\partial t}+\nabla\cdot\left(\frac{3}{2}\vec{U}p\right)=-\nabla\cdot\vec{Q}-\overline{\overline{P}}:\nabla\vec{U}+\overline{\overline{P}}_e:\nabla\frac{\vec{J}}{nq}+\nabla\cdot\left(\frac{3}{2}p_e\frac{\vec{J}}{nq}\right)-\vec{F}^{\mathrm{int}}\cdot\vec{U}+\varsigma^{\mathrm{int}}+\vec{F}_e^{\mathrm{int}}\cdot\frac{\vec{J}}{nq}+O(m_e). \tag{4.33}$$

The external force does not pump energy directly to the thermal energy, whereas the internal power consumption, $\sum_i m\vec{a}_i^{\mathrm{int}}\cdot(\vec{w}_i-\vec{v})\delta(\vec{x}-\vec{x}_i)=\varsigma_e^{int}+\varsigma_p^{int}-(\vec{F}_e^{int}+$

$\vec{F}_p^{int}) \cdot \vec{U} + \vec{F}_e^{int} \cdot \vec{J}/nq$, may contribute to equilibrate the thermal pressure. The pressure can alter the flow energy, and vice versa. The internal force can be long range and may not be nullified locally. The heat flux $\vec{Q}$, a higher velocity moment, is the sum of the individual heat flux, and will require an additional equation to describe. If we take the pressure tensor to be isotropic and equipartition, $\vec{\vec{P}} = p\vec{\vec{I}}$ and $p_e = p_p = \frac{1}{2}p$, Eq. (4.33) becomes

$$\frac{\partial p}{\partial t} + \left(\vec{U} - \frac{1}{2}\frac{\vec{J}}{nq}\right) \cdot \nabla p = -\frac{2}{3}\nabla \cdot \vec{Q} - \frac{5}{3}p\nabla \cdot \left(\vec{U} - \frac{1}{2}\frac{\vec{J}}{nq}\right)$$
$$- \frac{2}{3}\vec{F}^{int} \cdot \vec{U} + \frac{2}{3}\varsigma^{int} + \frac{2}{3}\vec{F}_e^{int} \cdot \frac{\vec{J}}{nq} + O(m_e). \tag{4.33a}$$

Since typically, $v_e \gg v_p$ that implies $J/nq \gg U$, the current can transport the pressure more quickly than the flow velocity.

4.3 The Magnetohydrodynamics

The **magnetodydrodynamics** is a one-fluid description that assumes the inseparable motion of electrons and ions at the $\vec{E} \times \vec{B}$ drifts across the field lines, and maintains the charge neutrality since the fast electron motion along the field lines may ensure its validity. Therefore, MHD theory is limited to activities of lower frequency than the ion cyclotron frequency and the electron transit time, and longer wavelength than the ion gyroradius and the Debye length. The collisional effect is considered to be weak and negligible. We thus have the ideal MHD model.

4.3.1 *The Ideal MHD Equations*

The MHD description serves many useful purposes such as: establish global understanding of (fusion) plasma, make zeroth order quantitative description, allow intuitive account of the experimental observables, describe large scale instabilities which severely limit the achievable plasma parameters, and enable new design concept. The governing equations are:
the continuity equation:

$$\frac{\partial}{\partial t}\rho + \nabla \cdot (\rho\vec{U}) = 0, \tag{4.25}$$

the momentum equation:

$$\rho\frac{\partial \vec{U}}{\partial t} + \rho\vec{U} \cdot \nabla\vec{U} = -\nabla p + \frac{1}{c}\vec{J} \times \vec{B}, \tag{4.30a}$$

the Ampere's law:

$$\nabla \times \vec{B} = \frac{4\pi}{c}\vec{J}, \tag{4.34}$$

the Faraday's law:

$$\nabla \times \vec{E} = -\frac{1}{c}\frac{\partial}{\partial t}\vec{B}, \tag{4.35}$$

the ExB drifts:

$$\vec{E} + \frac{1}{c}\vec{U} \times \vec{B} = 0, \tag{4.36}$$

the adiabatic law:

$$p\rho^{-\gamma} = \Theta = const., \tag{4.37}$$

where $\gamma = 5/3$. The last equation also leads to $dp/dt = \Theta\gamma\rho^{\gamma-1}d\rho/dt = -\Theta\gamma\rho^{\gamma}\nabla \cdot \vec{U}$. Thus, an alternative equation can be cast as the pressure equation:

$$\frac{\partial p}{\partial t} + \vec{U} \cdot \nabla p = -\frac{5}{3}p\nabla \cdot \vec{U}. \tag{4.38}$$

On the other hand, Eq. (4.33a) would modify Eq. (3.28) by replacing $\vec{U}$ with $\vec{U} - \frac{1}{2}\vec{J}/nq$. Equations (4.35) and (4.36) may be combined to give

$$\frac{\partial}{\partial t}\vec{B} = \nabla \times (\vec{U} \times \vec{B}). \tag{4.39}$$

The magnetic field also satisfies $\nabla \cdot \vec{B} = 0$ so that $\vec{B} = \nabla \times \vec{A}$, consistent with the absence of magnetic monopole.

4.3.2 *The Static MHD Equations*

The MHD equilibrium is a plasma state of great interest especially for fusion plasma confinement, and is governed by the time independent equations of the above ideal MHD equations. Thus, we have

$$\nabla \cdot (\rho\vec{U}) = 0, \tag{4.25a}$$

$$\rho\vec{U} \cdot \nabla\vec{U} = -\nabla p + \frac{1}{c}\vec{J} \times \vec{B}, \tag{4.30b}$$

$$\nabla \times \vec{B} = \frac{4\pi}{c}\vec{J}, \tag{4.34}$$

$$\nabla \times (\vec{U} \times \vec{B}) = 0, \tag{4.39a}$$

$$\vec{U} \cdot \nabla p = -\gamma p\nabla \cdot \vec{U}. \tag{4.38a}$$

Equation (4.34) implies $\nabla \cdot \vec{J} = 0$. In the case of vortex flow only, $\vec{U} = \nabla \times \vec{\Omega}$, so that the flow is incompressible with $\nabla \cdot \vec{U} = 0$, we have $\vec{U} \cdot \nabla\rho = \vec{U} \cdot \nabla p = 0$

that implies both density and pressure are constant along the flow. Since electron motions are on the faster time scale, it is conceivable that $\vec{J} \cdot \nabla \rho = \vec{J} \cdot \nabla p = 0$. When the flow is absent, the static equations are then reduced to

$$\nabla p = \frac{1}{c} \vec{J} \times \vec{B}, \tag{4.40}$$

$$\nabla \times \vec{B} = \frac{4\pi}{c} \vec{J}. \tag{4.41}$$

They lead to $\vec{B} \cdot \nabla p = \vec{J} \cdot \nabla p = 0$ so that the pressure is constant along both the magnetic field and current directions. Replacing $\vec{J}$ in Eq. (4.40) by Eq. (4.41), we have

$$\nabla \left(p + \frac{1}{8\pi} B^2 \right) - \frac{1}{4\pi} \vec{B} \cdot \nabla \vec{B} = 0. \tag{4.42}$$

The concept of the magnetic bottle can then be understood from the first term by setting $\vec{B} \cdot \nabla \vec{B} = 0$ to find $p + B^2/8\pi = const$. For the outer strong magnetic field it would be able to contain hot plasma pressure within, although in reality the magnetic configuration will be very complex to require the helical winding or the averaged magnetic well.

Homework 4.5: A force free MHD equilibrium is a plasmas with $p = 0$ so that $\vec{J} \times \vec{B} = 0$. Take the simplest case such that $\vec{J} = \frac{c}{4\pi} \lambda \vec{B}$ where λ is a constant. Find the equilibrium solution in the cylindrical geometry.

4.4 Fluid Dynamics

Fluid dynamics is an important discipline that is closely related to hydrodynamics and aerodynamics. It is applicable to weather prediction, aviation design, and natural resources management such as river, dam, estuary, seabed, ocean erosion, and biomedical implications. Its fundamental principles have been applied to traffic control, blood circulatory system, star evolution, to name a few. As magneto-hydrodynamics is, in principle, a superset of hydrodynamics, it is easy to recognize that the basic physics of fluid dynamics must be the integral part of the plasma physics discipline.

4.4.1 *Raleigh Taylor Instability*

In environmental hydrodynamics the gravitational force helps sediment transport in forming delta, wet land, channel diversion, etc. The gravitational acceleration $\vec{g}$

can be added to the momentum equation of Eq. (4.4) to study these phenomena. When a heavy fluid sits on top of the light fluid, the gravitational force will pull down the heavy fluid that is the so called **Raleigh Taylor instability** (**RTI**). The nature of the phenomena is that due to the higher potential energy of the heavy fluid, it is to exchange the position with the light fluid so to finally reach its minimum energy state. Thus, RTI is often referred to as the **exchange instability** in plasma physics. To account for the mechanism, a linear stability analysis is often performed which is, however, insufficient to do justice to the underlying physics as the sinking velocity of the heavy fluid is minimal at the beginning, but picking up the speed after acceleration with its terminal velocity determined by the fluid viscosity. Like dropping a ball to the ground, the potential energy will convert to kinetic energy. If the friction is not too strong the ball will bounce a few times before comes to rest. The RTI essentially behaves the same, but with more twists and turns, as fluid is complicated by its soft matter nature, that is capable to advect ($\vec{V} \cdot \nabla\rho, \vec{V} \cdot \nabla\vec{V}$), compress ($(d\rho/dt)/\rho = -\nabla \cdot \vec{V} \neq 0$), conduit ($\nabla \times \vec{V}) \times \vec{V}$, diffuse ($D\nabla^2\rho$), levitate ($-\nabla p/\rho - \nabla V^2/2$), and poise ($\mu\nabla^2\vec{V}$), contrary to a rigid body.

To examine the RTI, the equilibrium is first established. By placing the heavier fluid upon the light fluid with an interface layer that has the smooth transition of the density gradient, the pressure is balanced with the gravitational force, $\nabla p_0 = -\rho_0 g\hat{e}_y$. Nullified at the top of the container where $y = h$, it is found through the formulae,

$$p(\vec{r}, t) = \int_y^h dy' \rho(x, y', t) g. \tag{4.43}$$

We will assume that the density follows an incompressible flow, i.e., $\nabla \cdot \vec{V} = 0$, so that

$$\frac{\partial\rho}{\partial t} + \vec{V} \cdot \nabla\rho = 0. \tag{4.44}$$

The advection of density is solved by taking the density at the prior position before it arrives at the grid point,

$$\rho(\vec{X}, t) = \rho\left(\vec{X} - \int_0^t \vec{V}(X(\tau), \tau) d\tau\right). \tag{4.45}$$

Similarly, the advection of the velocity through $\partial\vec{V}/\partial t + \vec{V} \cdot \nabla\vec{V}$ is solved by the solution,

$$\vec{V}(\vec{X}, t) = \vec{V}\left(\vec{X} - \int_0^t \vec{V}(X(\tau), \tau) d\tau\right). \tag{4.46}$$

In the listed program Advection.m, the boundary conditions are chosen to be periodic in the X direction. At the top and bottom of the computational domain along the Y direction, the outward flow is prohibited as the fluid is confined in a container.

In the numerical scheme, false boundary is applied to ensure that the velocity in the Y direction will be turned back. This is done by setting the velocities at the false boundary to have the opposite velocity to its immediate neighbor on the physical boundary, while the density at the false boundary is kept the same so to conserve the density.

Without prescription on the compressibility, the numerical scheme can go awry easily. An important theorem is introduced here, which states that any vector field can be decomposed into the longitudinal and the transverse components, that is,

$$\vec{V} = \nabla\Phi + \nabla \times \vec{A}, \tag{4.47}$$

known as the **Helmholtz-Hodge Decomposition**. We thus eliminate the longitudinal component of the velocity by subtracting away $\nabla\Phi$, where Φ is solved from $\nabla^2\Phi = \nabla \cdot \vec{V}$ during the time evolution.

Homework 4.6: Prove the theorem of Helmholtz-Hodge Decomposition, namely, any vector can be expressed as $\vec{V} = \nabla\Phi + \nabla \times \vec{A}$.

The Navier Stokes equation

$$\frac{\partial \vec{V}}{\partial t} + \vec{V} \cdot \nabla \vec{V} = -\frac{1}{\rho}\nabla p - g\hat{e}_y + \mu \nabla^2 \vec{V}. \tag{4.48}$$

has a few components that have to be treated differently. The diffusion/viscous damping process is treated by the implicit scheme. The advection of flow velocity in the left hand side of Eq. (4.48) is treated the same way as Eq. (4.44) prescribes, while the acceleration by the pressure and the gravitation are advanced in time by finite difference scheme.

The movie clip of density contour from the numerical run shows that as the RTI develops, the fluid begins to bubble down and up. The fluid conduits itself in space by the forces of gravitation and velocity shear, and the density conservation is reasonable. The counter-streaming flows are not strong enough to trigger the vortex street formation.

```
function T=Advection(type)          %Periodic BC in X and fixed BC in Y
global X Y RO Vx Vy dx dt N MU;     %type refers to 1. Density, 2. Vx, 3. Vy.
isign=1;
if(type==1) T=RO; elseif(type==2) T=Vx; else T=Vy; isign=-1; end;
T0=T;
xmin=X(1,1);              xmax=X(N,N);
ymin=Y(1,1);              ymax=Y(N,N);
dX=Vx*dt;                 dY=Vy*dt;
Xold=X-xmin-dX;           Yold=Y -ymin-dY;
Iold=floor(Xold/dx)+1;    Jold=floor(Yold/dx)+1;
fx=Xold/dx-Iold+1;        fy=Yold/dx-Jold+1;
F1=(1-fx).*(1-fy);        F2=fx.*(1-fy);
F3=fx.*fy;                F4=(1-fx).*fy;
Iold=Iold.*(Iold<=N)+N.*(Iold>N); Jold=Jold.*(Jold<=N)+N.*(Jold>N);
for ix=1:N
```

```
for iy=1:N
    ix1=Iold(iy,ix); iy1=Jold(iy,ix);
    if(ix1>0&&ix1<N&&iy1>0&&iy1<N)
        ro1=T0(iy1,ix1); ro2=T0(iy1,ix1+1);
        ro3=T0(iy1+1,ix1+1); ro4=T0(iy1+1,ix1);
        end;
    if(ix1<=0)
        if(iy1<N&&iy1>0)
            ro1=T0(iy1,N); ro2=T0(iy1,1);
            ro3=T0(iy1+1,1); ro4=T0(iy1+1,N);
            end;
        if(iy1<=0)
            ro3=T0(1,1); ro4=T0(1,N);
            ro1=isign*ro4; ro2=isign*ro3;
            end;
        if(iy1>=N)
          ro1=T0(N,N); ro2=T0(N,1);
          ro3=isign*ro2; ro4=isign*ro1;
          end;
        end;
    if(ix1>=N)
        if(iy1<N&&iy1>0)
            ro1=T0(iy1,N); ro2=T0(iy1,1);
            ro3=T0(iy1+1,1); ro4=T0(iy1+1,N);
            end;
        if(iy1<=0)
            ro3=T0(1,1); ro4=T0(1,N);
            ro1=isign*ro4; ro2=isign*ro3;
            end;
        if(iy1>=N)
            ro1=T0(N,N); ro2=T0(N,1);
            ro3=isign*ro2; ro4=isign*ro1;
            end;
        end;
    if(iy1<=0&&ix1>0&&ix1<N)
        ro3=T0(1,ix1); ro4=T0(1,ix1+1);
        ro1=isign*ro4; ro2=isign*ro3;
        end;
    if(iy1>=N&&ix1>0&&ix1<N)
        ro1=T0(N,ix1); ro2=T0(N,ix1+1);
        ro4=isign*ro1; ro3=isign*ro2;
        end;
    T(iy,ix)=ro1*F1(iy,ix)+ro2*F2(iy,ix)+ro3*F3(iy,ix)+ro4*F4(iy,ix);
    end;
end;
```

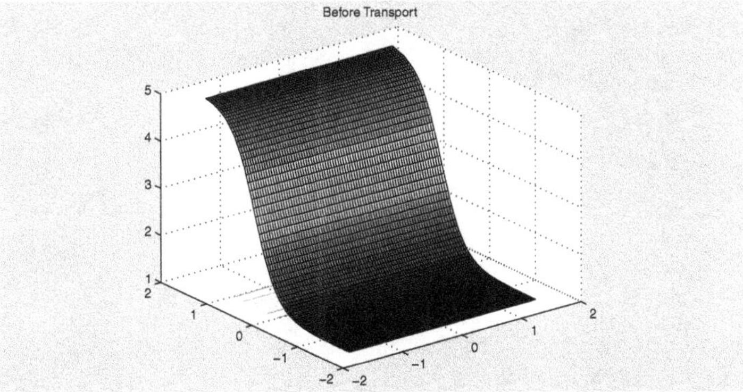

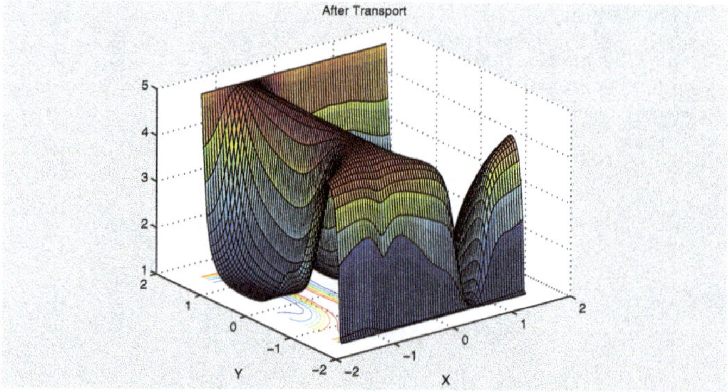


```
function [Ux,Uy]=HodgeDecomposition(X,Y,Vx,Vy,N,dx)
  %Helmholtz Decomposition
dxVx=([Vx(:,2:N),Vx(:,1)]-[Vx(:,N),Vx(:,1:N-1)])/2/dx;
dyVy=([Vy(2:N,:);Vy(1,:)]-[Vy(N,:);Vy(1:N-1,:)])/2/dx;
divV=dxVx+dyVy;
%get divV
Ux=0*Vx; Uy=0*Vy;
u=1/dx/dx;
M=SetM(N,u);
divV=reshape(divV,N^2,1);
PHI=M\divV;
%solve for PHI
PHI=reshape(PHI,N,N);
Ux=([PHI(:,2:N),PHI(:,1)]-[PHI(:,N),PHI(:,1:N-1)])/2/dx;
%get U=-grad PHI
Ux=Vx-Ux;
% eliminate the compressible flow
Uy=([PHI(2:N,:);PHI(1,:)]-[PHI(N,:);PHI(1:N-1,:)])/2/dx;
Uy(1,:)=0; Uy(N,:)=0;
Uy=Vy-Uy;
  % eliminate the compressible flow

function A=setA(N,u)
A=sparse(N);
A=A*0+diag(-4*u+0*(1:N),0);
A=A+diag(u+0*(1:N-1),1)+diag(u+0*(1:N-1),-1);

function B=setB(N,u)
B=sparse(N);
B=B*0+diag(u+0*(1:N));

function M=SetM(N,u)
M=sparse(N*N);
A=setA(N,u);
B=setB(N,u);
for i=1:N
    M(1+(i-1)*N:i*N,1+(i-1)*N:i*N)=A;
    if(i~=N) M(1+(i-1)*N:i*N,1+i*N:(i+1)*N)=B; end;
    if(i~=1) M(1+(i-1)*N:i*N,1+(i-2)*N:(i-1)*N)=B; end;
end;
M(1+(N-1)*N:N^2,1:N)=B;
M(1:N,1+(N-1)*N:N^2)=B;
```

```
function RTI(RERUN)
close all; clc
global X Y RO Vx Vy dx dy dt N MU;
MU=1.0e-3;  gamma=5/3; g=1.0e-3;
N=75; L=pi/2;
dx=2*L/N; N=N+1; dy=dx;
x=-L:dx:L; y=x;
X=repmat(x,N,1); Y=repmat(y',1,N);
ro1=5; ro2=1; cm=0.5;
RO=(ro1+ro2)/2+(ro1-ro2)/2*tanh(2*(Y-cm));
P=getP(RO,N,g,dx);
dt=0.02; TIME=0; iframe=0; Tc=50;
PSI=1.0e-2*cos(X*2).*exp(-4*abs(Y-cm));
Vx=getDy(PSI,N,dx); Vy=-getDx(PSI,N,dx);
V2=(Vx.^2+Vy.^2)/2;
if(RERUN) load RTI RO P Vx Vy X Y L dx dy dt N TIME iframe; end;
RO0=RO;
figure(1); surfc(X,Y,RO); title('Before Transport');
dt=0.01; RO=Advection(1); dt=2*dt;
tic, datestr(now),
for it=0:5000
   if(mod(it,Tc)==0) it,
      iframe=iframe+1; TIME=TIME+dt*Tc*(it>0);
      figure(1); contourf(X,Y,RO,0.0:0.25:5);
      axis([-L L -L L]);
str=sprintf('Rayleigh Taylor Instability @ t=%5.1f',TIME);
      title(str); xlabel('X'); ylabel('Y');
      M(iframe)=getframe;
      TotalDensity=sum(sum(RO,1),2)/N/N,
   end;
   Vy=Vy-g*dt-getDy(P,N,dx)./RO*dt;
   Vx=Vx-getDx(P,N,dx)./RO*dt;
   [Vx,Vy]=HodgeDecomposition(X,Y,Vx,Vy,N,dx,dy);
   Vx=Advection(2); Vy=Advection(3);
   Vy(N,:)=0; Vy(1,:)=0;
   RO0=RO; P0=P;
   RO=Advection(1);
   P=getP(RO,N,g,dx);
end;
toc, datestr(now),
figure(2);
RO=reshape(RO,N,N);
surfc(X,Y,RO);
title('After Diffusion'); xlabel('X'); ylabel('Y');
save RTI RO P Vx Vy X Y L dx dy dt N TIME iframe;
d=10;
figure(3);
quiver(Vx(1:d:N,1:d:N),Vy(1:d:N,1:d:N));
```

```
function P=getP(RO,N,g,dx)
for i=1:N
   P(:,i)=-Simpson(RO(:,i),N,dx)*g;
   P(:,i)=P(:,i)-P(N,i);
end;

function divP=getDIV(Px,Py,N,dx)
divP=getDx(Px,N,dx)+getDy(Py,N,dx);

function dxP=getDx(P,N,dx)
dxP=([P(:,2:N),P(:,1)]-[P(:,N),P(:,1:N-1)])/2/dx;

function dyP=getDy(P,N,dx)
dyP=([P(2:N,:);0*(1:N)]-P)/dx;

function d2P=getDEL(P,N,dx)
d2P=-4*P+[P(:,2:N),P(:,1)]+[P(:,N),P(:,1:N-1)];
d2P=d2P+[P(2:N,:);0*(1:N)']+[P(1,:);P(1:N-1,:)];
d2P=d2P/dx/dx;
```

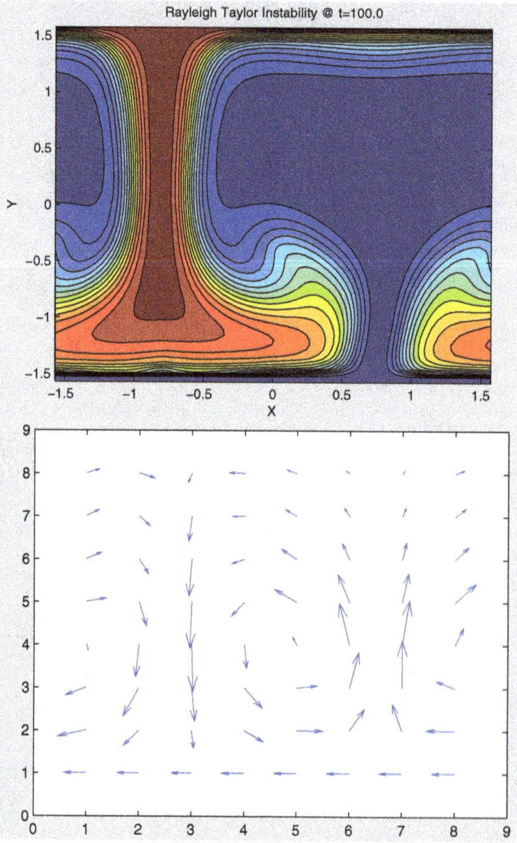

4.4.2 *Kelvin Helmholtz Instability*

Flow velocity can be a serious free energy to excite instabilities. The garden hose instability is a good example that flow within a not so rigid boundary can cause kinky twists. But the uniform flow can be stable in the infinite domain, since by sitting in the center of mass frame the free energy no longer exists. This however cannot be properly arranged for two-stream instability where the free energy cannot be transformed away. The two stream instability can be a kinetic instability in the sense that in the configuration space there is no net flow, and yet the instability is explosive to result in the relaxation from the narrowly peaked velocity distribution function. The **Kelvin Helmholtz instability (KHI)** can occur when two streams of flows are physically apart but with enough flow velocity gradient across the interface, or when sufficient velocity shear is present within a continuous fluid. Vortex structures generated in river water and by air flow, have also been observed in space and laboratory plasmas. Right after the Japan Sandai earthquake of year 2011, gigantic vortexes in the ocean were noted.

The MATLAB program KHIfft.m, as listed below, solves the Navier Stokes equation for the two diemensional flow,

$$\frac{\partial \vec{v}}{\partial t} + \vec{v} \cdot \frac{\partial \vec{v}}{\partial \vec{x}} = -\frac{1}{\rho}\nabla p + \mu \nabla^2 \vec{v}.$$

A vortex street developed in the contact layer between the two flow streams. The code adopts the Fast Fourier Transform to ensure the accuracies in the spatial derivatives that appear to ease the numerical instability.

```
function [dNx,dNy]=getGrad(n)
N=length(n);
Fn=fft2(n);
m=0:N-1;
M=repmat(m,N,1);
dNx=ifft2(-j*Fn.*M)/N;
dNy=ifft2(-j*Fn.*M')/N;
dNx=dNx+conj(dNx);
dNy=dNy+conj(dNy);
function divV=getDiv(U,V)
N=length(U);
fU=fft2(U);
m=0:N-1;
M=repmat(m,N,1);
dUx=ifft2(-j*fU.*M)/N;
fU=fft2(V);
dVy=ifft2(-j*fU.*M')/N;
divV=dUx+conj(dUx)+dVy+conj(dVy);
```

```
function OMEGA=getVortex(U,V)
N=length(U);
fU=fft2(V);
m=0:N-1;
M=repmat(m,N,1);
dVx=ifft2(-j*fU.*M)/N;
fU=fft2(U);
dUy=ifft2(-j*fU.*M')/N;
OMEGA=dVx+conj(dVx)-dUy-conj(dUy);
```


```
function OMEGA=getVortex(U,V)
N=length(U);
fU=fft2(V);
m=0:N-1;
M=repmat(m,N,1);
dVx=ifft2(-j*fU.*M)/N;
fU=fft2(U);
dUy=ifft2(-j*fU.*M')/N;
OMEGA=dVx+conj(dVx)-dUy-conj(dUy);
```


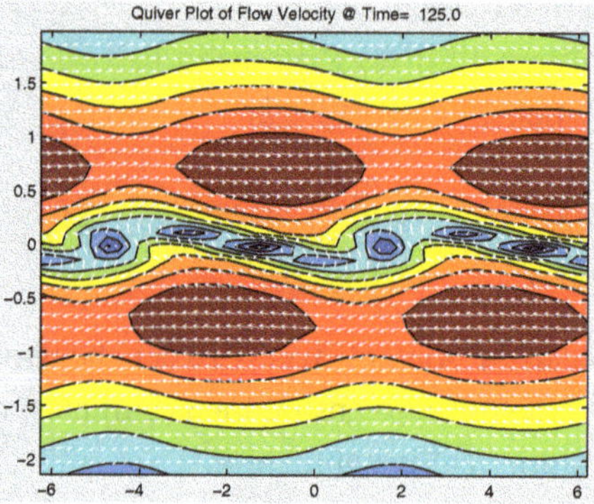


```
function init
global A B n N dx dy dt nt U V X Y ;
global K T NU NUn Omega Density Theta;
u0=0.5;     l=1;     nt=4000;
n=150;      N=2*n+2;
dy=4*pi/N; dt=0.05; dx=dy;
y=(-n-1/2:n+1/2)*dy; L=max(y),    x=y;
X=repmat(x,N,1); Y=repmat(y',1,N);
R=sqrt(X.^2+Y.^2);
u=u0*tanh(y'/l).*exp(-abs(y'/l));
U=repmat(u,1,N);
v=cos(x)*u0*0.025;
V=repmat(v,N,1);
```

```
B=fft2(V);
Omega=curl(X,Y,U,V);
Theta=0.*(Y>0)+pi.*(Y<0);
K=U.^2+V.^2;
Density=U*0+0.5;
T=1.0;     NUn=0.005;     NU=0.005;
Omega=getVortex(U,V);
[U,V]=AdvanceV(U,V,Density,T,NU,dt/2);

function Compressibility=getComp(U,V)
N=length(U);
divV=getDiv(U,V);
Compressibility=sum(sum(divV,1),2)/N^2;
```

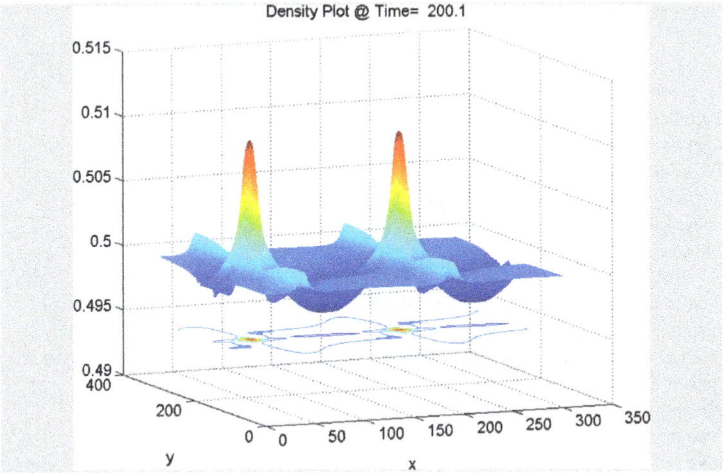


```
function KHIfft(RERUN)   % 2nd order explicit scheme FFT gradient
global A B n N dx dy dt nt U V X Y ;
global K T NU NUn Omega Density Theta;
clc; close all;
datestr(now),
init; it=0;
if(RERUN) load KHI U V Density it; end;
fig=figure('position', [50 50 950 650],'color','white');
tic; j=0;
while(it<=nt)
    if(mod(it,10)==0)
        j=j+1;
        plotQuiver(it);
        M(j)=getframe(gcf);
```

```
        end;
        Density=AdvanceN(U,V,Density,T,NUn,NU,dt);
        [U,V]=AdvanceV(U,V,Density,T,NU,dt);
        it=it+1;
    end;
    save KHI U V Density it;
    figure(2);
    surfc(Density);
    str=sprintf('Density Plot @ Time=%7.1f',it*dt);
    title(str);
    xlabel('x'); ylabel('y');
    toc;
    datestr(now),
    clear all;
```

```
function plotQuiver(it)
global A B n N dx dy dt nt U V X Y ;
global K T NU NUn Omega Density Theta;
it, n1=fix(N/3); n2=fix(N*2/3); m=3;
ParticleDensity=sum(sum(Density,1),2)/N^2.;
Compressibility=getComp(U,V),
FlowEnergy=sum(sum(K,1),2)/N^2/2,
Vorticity=sum(sum(Omega,1),2)/dx/N^2/2/dx,
value=[0:0.01:0.15];
FlowV=sqrt(U.^2+V.^2);
x= X(n1:m:n2,1:2*m:N); y= Y(n1:m:n2,1:2*m:N);
contourf(x,y,FlowV(n1:m:n2,1:2*m:N),value);
hold on;
u=U(n1:m:n2,1:2*m:N); v= V(n1:m:n2,1:2*m:N);
quiver('x,y,u,v,' color','w');
str=sprintf('Quiver Plot of Flow Velocity @ Time=%7.1f',it*dt);
title(str);
hold off;
getframe;
clear K;
```

```
function Density=AdvanceN(U,V,Density,T,NUn,NU,deltaT)
[dNx,dNy]=getGrad(Density);
d2N=getDiv(dNx,dNy);
divV=getDiv(U,V);
NN=-deltaT*(U.*dNx+V.*dNy+divV -NUn.*d2N);
[dUx,dUy]=getGrad(U);
[dVx,dVy]=getGrad(V);
d2U=getDiv(dUx,dUy);
d2V=getDiv(dVx,dVy);
UU=-deltaT*(U.*dUx+V.*dUy+T*dNx -NU*d2U);
VV=-deltaT*(U.*dVx+V.*dVy+T*dNy -NU*d2V);
[dNNx,dNNy]=getGrad(NN);
d2NN=getDiv(dNNx,dNNy);
NNN=-deltaT/2*(UU.*dNx+VV.*dNy+U.*dNNx+V.*dNNy+getDiv(UU,VV) -NUn.*d2NN);
Density=Density+NN+NNN;
```

```
function [U,V]=AdvanceV(U,V,Density,T,NU,deltaT)
[dNx,dNy]=getGrad(Density);
divV=getDiv(U,V);
NN=-deltaT*(U.*dNx+V.*dNy+divV);
[dUx,dUy]=getGrad(U);
[dVx,dVy]=getGrad(V);
d2U=getDiv(dUx,dUy);
d2V=getDiv(dVx,dVy);
UU=-deltaT*(U.*dUx+V.*dUy+T*dNx -NU*d2U);
VV=-deltaT*(U.*dVx+V.*dVy+T*dNy-NU*d2V);
[dUUx,dUUy]=getGrad(UU);
[dVVx,dVVy]=getGrad(VV);
UUU=-deltaT/2*(U.*dUUx+V.*dUUy+UU.*dUx+VV.*dUy+T*dNx);
VVV=-deltaT/2*(U.*dVVx+V.*dVVy+UU.*dVx+VV.*dVy+T*dNy);
U=U+UU+UUU;
V=V+VV+VVV;
```

```
function [dNx,dNy]=getGrad(n)
N=length(n);
Fn=fft2(n);
m=0:N-1;
M=repmat(m,N,1);
dNx=ifft2(-j*Fn.*M)/N;
dNy=ifft2(-j*Fn.*M')/N;
dNx=dNx+conj(dNx);
dNy=dNy+conj(dNy);
```

4.4.3 *Vortex Dynamics*

Recall that the Taylor-Green vortex (Sec. 1.1.4) is governed by the pressure energy $p = \frac{1}{4}\rho(\cos 2mx + \cos 2my)F^2(t)$, and the vortex $\vec{\Omega} = \nabla \times \vec{v} = 2m \sin mx \sin my F(t)\hat{e}_z$. The pressure energies are averaged in space to zero, and the flow energies are $\frac{1}{2}\langle \rho v^2 \rangle = \frac{1}{4}\rho F^2(t)$. Thus, these vortex states are degenerate as their energy contents are identical. Therefore, while the higher m modes of vortices may decay faster, there is no lower energy state for the Taylor-Green vortex to relax to.

If the vortex flow is the major component, $\vec{v} = v_\theta \hat{e}_\theta + v_r \hat{e}_r$, that implies $v_\theta \gg v_r$, we have,

$$\frac{\partial v_\theta}{\partial t} + v_\theta \frac{1}{r}\frac{\partial v_\theta}{\partial \theta} + v_r \frac{\partial v_\theta}{\partial r} \approx \mu \nabla^2 v_\theta. \tag{4.48}$$

The steady state to lowest order can be achieved when $v_\theta = v(r)$ only. This has the implication that vortices of the same polarity in close proximity would tend to merge into one, as aided by the expansive force term $\vec{\Omega} \times \vec{v}$.

```
function Vortex(RERUN)
close all; clc
global X Y P RHO Vx Vy dx dt N MU;
MU=0.0e-6; N=100; L=2; d=5;
dx=2*L/N; N=N+1;
x=-L:dx:L; y=x;
X=repmat(x,N,1); Y=repmat(y',1,N);
%4 0.3PSI 0.6 cm N=150 %2 0.2PSI 0.3 cm N=150
RHO=1.0+0*Y;
dt=0.025; cm=0.6;
PSI= 0.1*(exp(-4*sqrt((Y-cm).^2+X.^2))+exp(-4*sqrt((Y+cm).^2+X.^2)));
PSI=PSI+0.1*(exp(-4*sqrt((X-cm).^2+Y.^2))+exp( -4*sqrt((X+cm).^2+Y.^2)));
Vx=getDy(PSI,N,dx); Vy=-getDx(PSI,N,dx);
V2=Vx.^2+Vy.^2;
Temperature=1.0,
P=RHO.*Temperature;
TIME=0; iframe=0; Tc=20;
if(RERUN) load VORTEX RHO P Vx Vy X Y L dx dt  N TIME iframe; end;
RHO0=RHO;
TotalDensity=sum(sum(RHO,1),2)/N/N,
dt=dt/2; RHO=AdvectionFB(1); dt=dt*2;
tic, datestr(now),
values=0.0:0.025:0.35;
figure(3);
quiver(X(1:d:N,1:d:N),Y(1:d:N,1:d:N),Vx(1:d:N,1:d:N),Vy(1:d:N,1:d:N));
for it=0:4000
      P=RHO*Temperature;
  if(mod(it,Tc)==0) it,
    iframe=iframe+1; TIME=TIME+dt*Tc*(it>0);
    V=sqrt(Vx.^2+Vy.^2);
    figure(1);
    contourf(X,Y,V);
    axis([-L L -L L]); axis equal;
    str=sprintf('Vortex Coalease @ t=%5.1f',TIME);
    title(str); xlabel('X'); ylabel('Y');
    M(iframe)=getframe;
    TotalDensity=sum(sum(RHO,1),2)/N/N,
    MaxV=max(max(V,[],2),[],1),
  end;
  Vy=Vy-getDy(P,N,dx)./RHO*dt;
  Vx=Vx-getDx(P,N,dx)./RHO*dt;
  Vx1=AdvectionFB(2);
  Vy=AdvectionFB(3);
  Vx=Vx1;
  [Vx,Vy]=HodgeDecomposition(X,Y,Vx,Vy,N,dx);
  RHO=AdvectionFB(1);
end;
toc, datestr(now),
save VORTEX RHO P Vx Vy X Y L dx dt N TIME iframe;
toc, datestr(now),
figure(3);
quiver(X(1:d:N,1:d:N),Y(1:d:N,1:d:N),Vx(1:d:N,1:d:N),Vy(1:d:N,1:d:N));
```

```
function M=SetM(N,u)
M=sparse(N*N);
A=setA(N,u);
B=setB(N,u);
for i=1:N
   M(1+(i-1)*N:i*N,1+(i-1)*N:i*N)=A;
   if(i~=N) M(1+(i-1)*N:i*N,1+i*N:(i+1)*N)=B; end;
   if(i~=1) M(1+(i-1)*N:i*N,1+(i-2)*N:(i-1)*N)=B; end;
end;
if(i==1||i==N)
   M(1+(i-1)*N:i*N,1+(i-1)*N:i*N)=M(1+(i-1)*N:i*N,1+(i-1)*N:i*N)-
diag(u+0*(1:N));
end;
M(1+(N-1)*N:N^2,1+(N-1)*N:N^2)=diag(1+0*(1:N),0);
M(1:N,1:N)=diag(1+0*(1:N),0);
```

```
function P=getP(RHO,N,g,dx)
for i=1:N
   P(:,i)=-Simpson(RHO(:,i),N,dx)*g;
   P(:,i)=P(:,i)-P(N,i);
end;

function divP=getDIV(Px,Py,N,dx)
divP=getDx(Px,N,dx)+getDy(Py,N,dx);

function dxP=getDx(P,N,dx)
dxP=([P(:,2:N),P(:,N-1)]-[P(:,2),P(:,1:N-1)])/dx/2;

function dyP=getDy(P,N,dx)
dyP=([P(2:N,:);P(N-1,:)]-[P(2,:);P(1:N-1,:)])/dx/2;

function d2P=getDEL(P,N,dx)
d2P=-4*P+[P(:,2:N),0*(1:N)']+[0*(1:N)',P(:,1:N-1)];
d2P=d2P+[P(2:N,:);0*(1:N)']+[0*(1:N)';P(1:N-1,:)];
d2P=d2P/dx/dx;

function RHO=Diffusion(RHO,D,N)
RHO=reshape(RHO,N*N,1);
RHO=D*RHO;
RHO=reshape(RHO,N,N);

function D=getD
global X Y P RHO Vx Vy dx dt N MU;
u=MU*dt/dx/dx/2;
M=SetM(N,u);
W=SetM(N,-u);
D=M\W;

function A=setA(N,u)
A=sparse(N);
A=A*0+diag(1+4*u+0*(1:N),0);
A=A+diag(-u+0*(1:N-1),1)+diag(-u+0*(1:N-1),-1);
A(1,1)=A(1,1)-u; A(N,N)=A(N,N)-u;
function B=setB(N,u)
B=sparse(N);
B=B*0+diag(-u+0*(1:N));
```

The capability of **vortex coalesce** is demonstrated in the following numerical code. The flow is confined within the fixed boundary domain. Notice that ∇v^2 points outward at the geometrical center, the $\vec{v}_{drift}$ is rotating counterclockwise. The pressure is also imposed without affecting the center of mass motion or distorting the structures of the vortices. It does not alter the physical picture of the vortex coalesce.

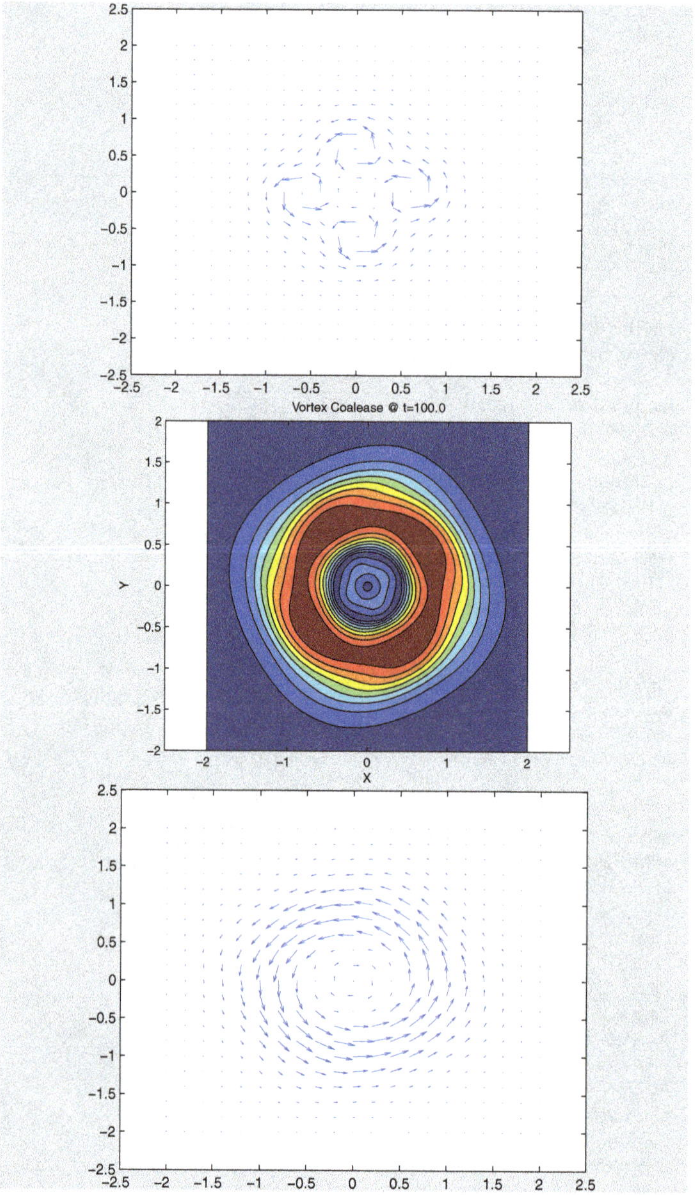

```
function AdvectionFB(type) %Both X and Y are Fixed Boundary condition
global X Y RO Vx Vy dx dt N MU;
isign=-1;
if(type==1) T=RO; isign=1; elseif(type==2) T=Vx; else T=Vy; end;T0=T;
xmin=X(1,1);        xmax=X(N,N);
ymin=Y(1,1);        ymax=Y(N,N);
dX=Vx*dt;        dY=Vy*dt;
Xold=X-xmin-dX;     Yold=Y -ymin-dY;
Iold=floor(Xold/dx)+1;  Jold=floor(Yold/dx)+1;
fx=Xold/dx-Iold+1;     fy=Yold/dx -Jold+1;
F1=(1-fx).*(1-fy);     F2=fx.*(1-fy);
F3=fx.*fy;           F4=(1-fx).*fy;
Iold=Iold.*(Iold<=N)+N.*(Iold>N);
Jold=Jold.*(Jold<=N)+N.*(Jold>N);
for ix=1:N
  for iy=1:N
    ix1=Iold(iy,ix); iy1=Jold(iy,ix);
    if(ix1>0&&ix1<N&&iy1>0&&iy1<N)
      ro1=T0(iy1,ix1); ro2=T0(iy1,ix1+1);
      ro3=T0(iy1+1,ix1+1); ro4=T0(iy1+1,ix1);
      end;
    if(ix1<=0)
      if(iy1<N&&iy1>0)
        ro2=T0(iy1,1); ro3=T0(iy1+1,1);
        ro1=isign*ro2; ro4=isign*ro3;
        end;
      if(iy1<=0)
        ro3=T0(1,1); ro4=isign*ro3;
        ro1=isign*ro3; ro2=isign*ro3;
        end;
      if(iy1>=N)
        ro2=T0(N,N); ro2=isign*ro2;
        ro3=isign*ro2; ro4=isign*ro2;
        end;
      end;
    if(ix1>=N)
      if(iy1<N&&iy1>0)
        ro1=T0(iy1,N); ro4=T0(iy1+1,N);
        ro3=isign*ro4; ro2=isign*ro1;
        end;
      if(iy1<=0)
        ro4=T0(1,N); ro3=isign*ro4;
        ro1=isign*ro4; ro2=isign*ro4;
        end;
      if(iy1>=N)
        ro1=T0(N,N); ro2=isign*ro1;
        ro3=isign*ro1; ro4=isign*ro1;
        end;
      end;
    if(iy1<=0&&ix1>0&&ix1<N)
      ro4=T0(1,ix1); ro3=T0(1,ix1+1);
      ro1=isign*ro4; ro2=isign*ro3;
      end;
    if(iy1>=N&&ix1>0&&ix1<N)
      ro1=T0(N,ix1); ro2=T0(N,ix1+1);
      ro4=isign*ro1; ro3=isign*ro2;
      end;
    T(iy,ix)=ro1*F1(iy,ix)+ro2*F2(iy,ix)+ro3*F3(iy,ix)+ro4*F4(iy,ix);
    end;
  end;
```

We may treat the center of the vortex like the center of mass that is subject to the pressure gradient of a longer length scale. The vortex would move as a whole by the pressure gradient. The vortex flow, like the spin of the rigid body, can be studied in terms of its 'principal axes' on the body frame. By assuming $\vec{v} = \vec{V}_{cm} + \vec{u}$, and separating the momentum equation of motion into the slow time and long length scales,

$$\frac{\partial \vec{V}_{cm}}{\partial t} \approx -\frac{1}{\rho}\nabla p + \mu \nabla^2 \vec{V}_{cm}, \tag{4.48}$$

and the fast time, short length scales,

$$\frac{\partial \vec{u}}{\partial t} + \vec{u} \cdot \frac{\partial \vec{V}_{cm}}{\partial \vec{x}} + \vec{V}_{cm} \cdot \frac{\partial \vec{u}}{\partial \vec{x}} + \vec{u} \cdot \frac{\partial \vec{u}}{\partial \vec{x}} = \mu \nabla^2 \vec{u}. \tag{4.49}$$

Taking an average over the volume of the entire vortex flow, then $\langle \vec{u} \cdot \partial \vec{V}_{cm}/\partial \vec{x} \rangle = \langle \vec{V}_{cm} \cdot \partial \vec{u}/\partial \vec{x} \rangle \approx 0$, we end with

$$\frac{\partial \vec{u}}{\partial t} + \vec{u} \cdot \frac{\partial \vec{u}}{\partial \vec{x}} = \mu \nabla^2 \vec{u}. \tag{4.50}$$

The accelerating force due to $\vec{u} \cdot \nabla \vec{u} = \frac{1}{2}\nabla u^2 + (\nabla \times \vec{u}) \times \vec{u}$ is short ranged and effective only when flow velocities are significant. When two vortices are close enough, by taking $\vec{u} \cdot \nabla \vec{u} \approx 0$, the gradient of flow energy can cause a drift velocity $\vec{v}_d = (\vec{\Omega} \times \frac{1}{2}\nabla u^2)/\Omega^2$, for strong vorticity $\vec{\Omega} \equiv \nabla \times \vec{u}$. This is similar to the ExB drift for the charged particles in EM fields. To the next order, there is an analogous polarization drift, $\vec{v}_p = (\dot{\vec{\Omega}} \times \vec{v}_d)/\Omega^2$, that tends to smear out the gradient of flow energy.

Homework 4.7: Modify the program vortex.m to initialize two vortices of opposite polarities and observe the subsequent evolution. Report the differences of its physical effects in comparison with that of the same polarities.

Homework 4.8: Initiate a large vortex in the center of the computational domain, and add a few smaller vortexes by modifying the program vortex.m. Report the evolution.

Further Reading

To obtain the fluid equations, the derivation presented in this chapter by defining the fluid properties in the configuration space is much simpler than that of Klimotovich (1982), who sums up particles in the phase space. The complete set of transport coefficients in the fluid equations will further be derived in Chap. 9.

The fluid equations were routinely obtained by taking velocity moments of the Boltzmann equation or the Vlasov equation. See for example, the textbooks by Chen, Boyd and Sanderson, or Krall and Triverlpiece.

For a comprehensive review on fluid mechanics, the book by Landau and Lifshitz is recommended.

Small vortexes of the same polarity would coalesce into a larger vortex that was demonstrated in the work by Montgomery *et al.* (1992).

Homework Hints

Homework 4.1: The particles are given small perturbations so that $n(\vec{r}, t) = \sum \delta(\vec{r} - \vec{x}_i(t) - \varepsilon\vec{\chi}_i(t))$, where $\varepsilon \ll 1$ and all the other quantities are considered order unity. Make a series expansion of $n(\vec{r}, t)$ to the second order in ε.

$$\begin{aligned}
n(\vec{r}, t) &= \sum_i \delta(\vec{r} - \vec{x}_i(t) - \varepsilon\vec{\chi}_i(t)) \\
&= \sum_i \delta(\vec{r} - \vec{x}_i(t)) - \sum_i \varepsilon\vec{\chi}_i(t) \cdot \nabla\delta(\vec{r} - \vec{x}_i(t)) \\
&\quad + \frac{1}{2}\sum_i \varepsilon^2\vec{\chi}_i(t)\vec{\chi}_i(t) : \nabla\nabla\delta(\vec{r} - \vec{x}_i(t)) + \cdots \\
&= \sum_i \delta(\vec{r} - \vec{x}_i(t)) - \varepsilon\nabla \cdot \sum_i \vec{\chi}_i(t)\delta(\vec{r} - \vec{x}_i(t)) \\
&\quad + \frac{1}{2}\varepsilon^2\nabla\nabla : \sum_i \vec{\chi}_i(t)\vec{\chi}_i(t)\delta(\vec{r} - \vec{x}_i(t)) + \cdots
\end{aligned}$$

Homework 4.3: Apply Eq. (4.31) to electron-proton plasma with use of the light electron mass expansion to find the leading order terms.

Evaluating the individual terms, we have

$$\begin{aligned}
\sum_\sigma \rho_\sigma\vec{v}_\sigma\vec{v}_\sigma\frac{q_\sigma}{m_\sigma} &\approx -nq\left[\vec{U} - \frac{\vec{J}}{nq}\left(1 - \frac{m_e}{M_p}\right)\right]\left[\vec{U} - \frac{\vec{J}}{nq}\left(1 - \frac{m_e}{M_p}\right)\right] \\
&\quad + nq\left[\vec{U} + \frac{m_e}{M_p}\frac{\vec{J}}{nq}\right]\left[\vec{U} + \frac{m_e}{M_p}\frac{\vec{J}}{nq}\right] \\
&\approx \left[\vec{U}\vec{J} + \vec{J}\vec{U} - \frac{\vec{J}\vec{J}}{nq}\left(1 - \frac{2m_e}{M_p}\right)\right] + O(m_e^2),
\end{aligned}$$

and

$$\sum_\sigma \vec{F}_\sigma \frac{q_\sigma}{m_\sigma} = \sum_\sigma n_\sigma \frac{q_\sigma^2}{m_\sigma}\left(\vec{E} + \vec{v}_\sigma \times \frac{\vec{B}}{c}\right) - \vec{F}_e^{int}\frac{q}{m_e} + \vec{F}_p^{int}\frac{q}{M_p}$$

$$\approx \frac{\omega_{pe}^2 + \omega_{pi}^2}{4\pi}\left(\vec{E} + \frac{\vec{U}}{c} \times \vec{B}\right) - \frac{q\vec{J}}{m_e c} \times \vec{B}\left(1 - \frac{m_e}{M_p}\right)$$

$$- \vec{F}_e^{int}\frac{q}{m_e} + \vec{F}_p^{int}\frac{q}{M_p} + O(m_e^2)$$

We arrive at the equation of current transport by keeping only the lowest order terms,

$$\frac{\partial \vec{J}}{\partial t} \approx \frac{\omega_{pe}^2}{4\pi}\left(\vec{E} + \frac{\vec{U}}{c} \times \vec{B}\right) - \frac{q\vec{J}}{m_e c} \times \vec{B}$$

$$+ \nabla \cdot \left(\frac{q\vec{\vec{P}}_e}{m_e} - \vec{U}\vec{J} - \vec{J}\vec{U} + \frac{\vec{J}\vec{J}}{nq}\right) - \vec{F}_e^{int}\frac{q}{m_e} + O(m_e) \qquad (4.31a)$$

Homework 4.5: A force free MHD equilibrium is a plasmas with $p = 0$ so that $\vec{J} \times \vec{B} = 0$. Take the simplest case such that $\vec{J} = \frac{c}{4\pi}\lambda\vec{B}$ where λ is a constant. Find the equilibrium solution in the cylindrical geometry.

$$\nabla \times \vec{B} = \frac{4\pi}{c}\vec{J} = \lambda\vec{B} \qquad B \sim e^{i(m\phi - kz)},\, k \sim \frac{n}{L}$$

$$\frac{1}{r}\frac{\partial B_z}{\partial \phi} - \frac{\partial B_\phi}{\partial z} = \lambda B_r \qquad \frac{im}{r}B_z + ikB_\phi = \lambda B_r$$

$$\frac{\partial B_r}{\partial z} - \frac{\partial B_z}{\partial r} = \lambda B_\phi \quad \Rightarrow \quad -ikB_r - \frac{\partial B_z}{\partial r} = \lambda B_\phi \quad \Rightarrow$$

$$\frac{1}{r}\frac{\partial r B_\phi}{\partial r} - \frac{\partial B_r}{r\partial \phi} = \lambda B_z \qquad \frac{1}{r}\frac{\partial r B_\phi}{\partial r} - \frac{im}{r}B_r = \lambda B_z$$

$$(k^2 - \lambda^2)B_\phi = \frac{-mk}{r}B_z + \lambda\frac{\partial B_z}{\partial r}$$

$$\lambda\frac{1}{r}\frac{\partial r B_\phi}{\partial r} + \frac{mk}{r}B_\phi = \left(\lambda^2 - \frac{m^2}{r^2}\right)B_z \quad \Rightarrow$$

$$\frac{1}{r}\frac{\partial}{\partial r}\left(r\frac{\partial B_z}{\partial r}\right)+\left(\lambda^2-k^2-\frac{m^2}{r^2}\right)B_z=0$$

$$\therefore\quad \frac{\partial^2 B_z}{\partial\rho^2}+\frac{1}{\rho}\frac{\partial B_z}{\partial\rho}+\left(1-\frac{m^2}{\rho^2}\right)B_z=0,\quad \rho\equiv\sqrt{k^2-\lambda^2}r$$

$$B_z=b_{mn}J_m(\sqrt{\lambda^2-k^2}r)e^{im\phi-ikz}$$

Homework 4.7: Modify the program vortex.m to initialize two vortices of opposite polarities and observe the subsequent evolution. Report the differences of its physical effects in comparison with that of the same polarities.

Here is the coding for the two vortices of opposite polarities:

```
cm2=0.9; PSI=0.1*(exp(-4*sqrt((Y-cm2).^2+X.^2))-
exp(-4*sqrt((Y+cm2).^2+X.^2)));
```

The twin vortexes of opposite polarity do not merge but retain their separate entity.

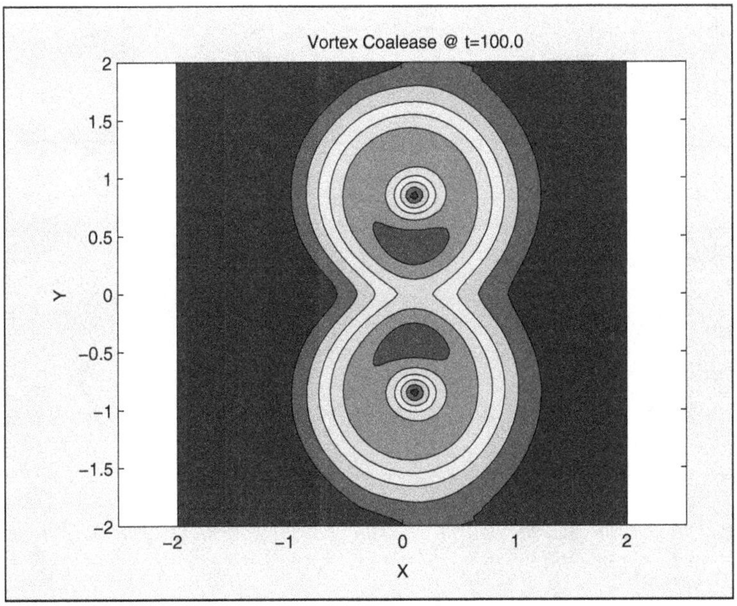

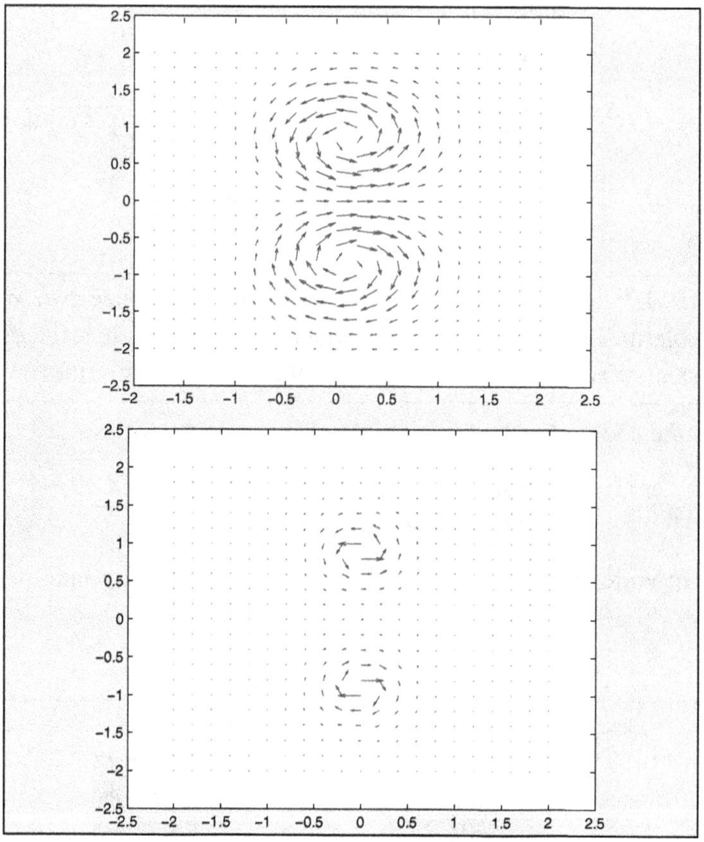
2.5
2
1.5
1
0.5
0
−0.5
−1
−1.5
−2
−2.5
−2 −1.5 −1 −0.5 0 0.5 1 1.5 2 2.5
2.5
2
1.5
1
0.5
0
−0.5
−1
−1.5
−2
−2.5
−2.5 −2 −1.5 −1 −0.5 0 0.5 1 1.5 2 2.5

Chapter Five

Single Wave

"Why is it that you physicists always require so much expensive equipment? Now the Department of Mathematics requires nothing but money for paper, pencils, and erasers... and the Department of Philosophy is better still. It doesn't even ask for erasers."

University President

Confined plasma, as a rule, has plenty of free energy due to its spatial inhomogeneity, external heating, or fusion reactions. This makes plasma rich in wave phenomena whether in the controlled fusion device or otherwise such as in the **solar flare**. The long-range nonlinear many-body interactions allow waves of various wavelengths and frequencies to occur. Moreover, the plasma fluid model of MHD is a superset of hydrodynamics. As such, it exhibits wave physics of hydrodynamics and beyond. Therefore, plasma has many kinds of waves and instabilities. Knowing wave physics is an indispensable discipline for plasma physicists.

The wave physics is one of the tractable many body descriptions and a comprehensible analysis of collective plasma phenomena. It is also an appropriately chosen view, more convenient and closer to the truth. The distinctions between the particle and wave pictures lie in the fact that the emphasis of the former tends to be on localized variables in phase space or configuration space alike, while the wave model is nonlocal and long range through the collective response. It goes beyond the local variables such as the density, temperature, or current, and is suitable to describe large scale dynamics and transport. A **sawtooth oscillation** due to the internal kink instability, for example, would turn the inner region of a tokamak inside out, and a **whistler wave** could carry the lightening disturbance to the end of the ionosphere.

In a single wave analysis, we often investigate a natural mode in a system under consideration, and may follow its nonlinear evolution to fully explore its long time behavior. Beyond the single wave picture, **wave-particle** and **wave-wave interactions** are the next level of complications. The mechanism of wave-particle interaction is the underlying principle to make gyrotron, free electron laser (FEL) or **particle accelerator**. It may also lead to **strange attractor**, chaotic behavior, and **stochastic heating**. Wave-wave interaction may give rise to mode conversion (two wave coupling), parametric decay/instability (three wave coupling), **energy cascading**, **self organization**, **dynamo effect** (many wave coupling), turbulence spectrum (beyond wave recognition), and much more. They are important as to impact many applications and to explain many natural phenomena.

Before we start examining the individual natural mode in the plasma, note that linear waves may share components by superposition. Analyzing linear wave theory for instability or otherwise without proper orthogonal wave functions is redundant at best and misleading for new physics at worst. The very basic components of a wave should be examined accordingly, and not to be confused with the 'borrowed' properties. It makes no sense to superimpose linear waves to lead to a "scientific discovery". To recognize the nature of a wave, it is important to examine the least effects that support such a wave. And to fully understand the physics, the nonlinear or statistical analysis, in general, has to follow.

5.1 Longitudinal Waves

Since any vector can be expressed in terms of the longitudinal and transverse components by the Helmholtz-Hodge decomposition, the wave electric field can be classified into the longitudinal and the transverse waves. The wave electric field may be written as $\vec{E} = -\nabla\Phi - \frac{1}{c}\partial\vec{A}/\partial t$. A **longitudinal wave** in plasma is an electrostatic wave, with the vector potential $\vec{A} = 0$, while a **transverse wave** has $\Phi = 0$, characterized with the magnetic field component. A longitudinal wave would have the density perturbation causing compressible flow, and the wave vector and the wave electric field aligned in the same direction. The transverse waves are the orthogonal wave basis functions to the longitudinal waves, and their electric fields are perpendicular to the wave vector. We will start with the electrostatic waves.

5.1.1 *Langmuir Plasma Wave*

The **Langmuir wave** is a density wave associated with the electron oscillation in the plasma. It readily exists in the absence of an external magnetic field. To describe a density wave, we may start with the electron continuity equation given by Eq. (4.16) $\partial n/\partial t + \nabla \cdot n\vec{v} = 0$. By assuming a uniform background density n_0, the linearized equation is,

$$\frac{\partial \delta n}{\partial t} + n_0 \nabla \cdot \delta\vec{v} = 0. \tag{5.1}$$

Either the electron motion generates the density response or the density perturbation induces the particle motion are the two sides of the same coin. The electron momentum equation given by Eq. (4.19) without the thermal effect, is needed to elucidate the electron motion. Its linearized version is

$$\frac{\partial \delta\vec{v}}{\partial t} = \frac{e\delta\vec{E}}{m_e}. \tag{5.2}$$

A description of the electric field is required to complete the story. At this juncture, we do not choose the wave to be electromagnetic governed by the Faraday's law,

$\nabla \times \vec{E} = -(1/c)\partial\vec{B}/\partial t$, for it would need another equation, namely, the Ampere's law to describe $\vec{B}$. Moreover, it would, in fact, describe the plasma density modified electromagnetic wave branch, instead. We choose the Poisson's equation to examine the electrostatic branch. The electromagnetic branch tends to be more energetic since there is the tendency of equipartition in the electric and the magnetic field energies. Therefore, the electrostatic branch would more easily be excited.

We further make the assumption of immobile uniform ion background since we are looking at the physics on the electron time scale. It results in the following

$$\nabla \cdot \delta\vec{E} = -\nabla^2 \delta\Phi = 4\pi e \delta n. \tag{5.3}$$

Taking time differentiation on Eq. (5.1), and eliminating $\delta\vec{v}$ by utilizing Eq. (5.2), then substituting the electric field by Eq. (5.3), we end with

$$\frac{\partial^2 \delta n}{\partial t^2} + \omega_{pe}^2 \delta n = 0. \tag{5.4}$$

We have derived the electron plasma frequency $\omega_{pe} = \sqrt{4\pi n_0 e^2/m_e}$, a very fundamental property of plasma physics. Equation (5.4) gives a pure plasma oscillation.

By retaining the thermal pressure in the momentum equation of (5.2), we have

$$\frac{\partial \delta\vec{v}}{\partial t} = \frac{e\delta\vec{E}}{m_e} - \frac{\nabla \delta n k_B T_e}{n_0 m_e}, \tag{5.2a}$$

Equation (5.4a) is modified to give the following wave equation,

$$\frac{\partial^2 \delta n}{\partial t^2} + \nabla \cdot n_0 \frac{e\delta\vec{E}}{m_e} = \frac{\partial^2 \delta n}{\partial t^2} + \omega_{pe}^2 \delta n = V_e^2 \nabla^2 \delta n. \tag{5.4a}$$

That has the dispersion relation after **Fourier transform,**

$$\omega^2 = \omega_{pe}^2 + k^2 V_e^2. \tag{5.4b}$$

The Fourier transform is defined in this book as follows:

$$f(x) = \int_{-\infty}^{\infty} dk\, f_k e^{ikx}, \quad f_k = \frac{1}{2\pi} \lim_{L\to\infty} \int_{-L/2}^{L/2} dx\, f(x) e^{-ikx}. \tag{5.5}$$

The electron thermal velocity $V_e = \sqrt{\frac{3}{2} k_B T_e / m_e}$ enables the wave to propagate and disperse accordingly.

The counterpart of this electrostatic plasma wave is the plasma density modified electromagnetic wave having the dispersion relation $\omega^2 = \omega_{pe}^2 + k^2 c^2$. Both have the cutoff at the plasma density where $\omega = \omega_{pe}$ and the wave number goes to zero. Beyond the cutoff point, the wave number becomes imaginary, the wave amplitude is exponentially decreased, and the wave is termed to be **evanescent**. It would be reflected as demonstrated in the program EMPW in Chap. 1.

The following program shows a monochromatic electron plasma wave propagating in the uniform plasma. The wave frequency is slightly higher than the plasma frequency. The differentiation is implemented with the two-point central difference scheme. The time advancement is carried out by the leapfrog method.

```
function [n,v,e]=init
global N dz dt L z;
global w k;
w=1.1; k=sqrt(w^2-1);
L=4*pi/k; N=3000;
dz=L/(N+1); dt=dz/1.5;
z=0.5*dz:dz:(N)*dz;
v=0*z; e=v;
iterations=250;
for i=1:iterations
   amplitude=2.15e-3*i/iterations;
   n=1+amplitude*sin(k*z-w*i*dt);
   e=amplitude/k*cos(k*z-w*i*dt);
   v=w/k*amplitude*sin(k*z-w*i*dt);
   dn=D(n);
end;
```

% initialize the program
% The system is periodic of length L.
% N is the number of grids, dz and dt are the space and time steps, z is the space grid coordinates, w is the frequency normalized to unity, k is the wave vector
% v is the wave phse velocity, e is the electric field amplitude,

% Set up the wave in the first 250 time steps.
% initialize the amplitude
% n is the desist
% e is the electric field
% v is the phase velocity
% dn is the density gradient

```
function PW
clear all; close all; clc;
global N dz dt L z;
global w k;
[n,v,e]=init;
dt=dt/20;
iterations=40000;
[v,dn]=getv(n,v,e);
dt=dt*2;
figure(1);
for i=1:iterations
   n=getn(n,v);
   e=gete(n);
   [v,dn]=getv(n,v,e);
   if(mod(i,100)==0)
      plot(z,dn,'cd',z,v,'r-',z,n-1,'g*',z,e,'b-');
      axis([0 L -1.0e-2 1.0e-2]);
      s=sprintf('Propagating Plasma Wave @time=%4.1f',dt*i);
      title(s,'fontsize',14);
      xlabel('Z','fontsize',14); ylabel('amplitude','fontsize',14);
      getframe;
   end;
end;
text(4,0.008,'---v','color','r','fontsize',14);
text(9,0.008,'---E','color','b','fontsize',14);
text(14,0.008,'---n-1','color','g','fontsize',14);
text(19,0.008,'---dn/dx','color','c','fontsize',14);
```

% main program for the Plasma Wave

% leap frog, take half time step to advance v and dn

% get n and e at integer time steps

% get v and dn at half integer time steps.

% every 100 steps plot the figure

Plotted are the perturbed velocity, the electric field, the perturbed density, and the density gradient.

```
function dn=D(n)       % take first
derivative
global N dz dt L z;
np=[n(2:N),n(1)];
nn=[n(N),n(1:N-1)];
dn=(np-nn)/2/dz; %two point
central difference

function n=getn(n,v)
global N dz dt L z;
dv=D(v);
n=n-dv*dt;                %Eq(5.1)

function [v,dn]=getv(n,v,e)
global N dz dt L z;
dn=D(n);
v=v+dt*(-e-dn);           %Eq(5.2a)
                                        %the ion density profile treated as uniform.
function e=gete(n)
global w k;
ep=[e(2:N),e(1)];
en=[e(N),e(1:N-1)];
phi=(1-n)/k^2;            %Eq(5.3)
e=-D(phi);
```

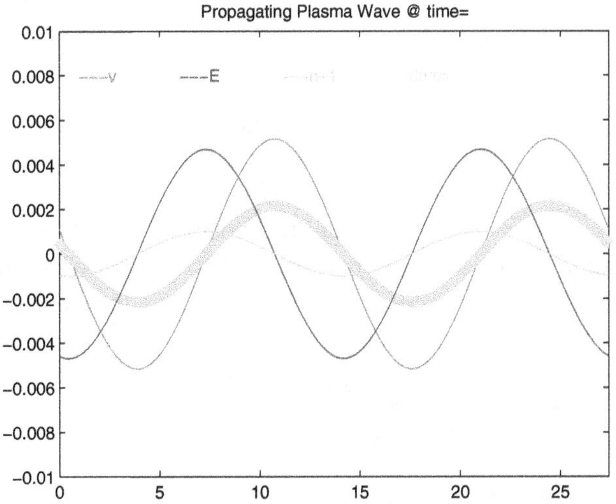

Homework 5.1: Carry out the description of the electromagnetic wave that is modified by the plasma.

5.1.2 *Ion Sound Wave*

The **ion sound wave** is a low frequency longitudinal wave associated with the ion density disturbance. The wave perturbs the ion density governed by the continuity equation,

$$\frac{\partial}{\partial t}\delta n_i + \nabla \cdot (n_{0i}\delta \vec{v}_i) = 0, \tag{5.6}$$

and the ion flow is governed by the momentum equation,

$$n_{0i}m_i\frac{\partial \delta \vec{v}_i}{\partial t} = -\nabla \delta p_i + n_{0i}q_i\delta \vec{E}. \tag{5.7}$$

Assuming that the flow is mainly electrified and not pressured so that δp_i is negligible, which is valid for a relatively cold ions compared to the electrons, namely the electron temperature is much greater than the ion temperature, $T_e \gg T_i$, we then have

$$\frac{\partial^2}{\partial t^2}\delta n_i + \frac{n_{0i}q_i}{m_i}\nabla \cdot \delta \vec{E} = 0 = \frac{\partial^2}{\partial t^2}\delta n_i + \frac{4\pi n_{0i}q_i}{m_i}(\delta n_i q_i + \delta n_e e), \tag{5.6a}$$

where the electrostatic field has been replaced with use of the Poisson's equation:

$$\nabla \cdot \delta \vec{E} = -\nabla^2 \delta\phi = 4\pi(\delta n_i q_i + \delta n_e e). \tag{5.8}$$

The electrons serves as the dynamical background and achieves the energy equilibration on the electron time scale by balancing the terms on the RHS of Eq. (5.2a),

$$\delta n_e = -n_{0e}\frac{e\delta\phi}{k_B T_e}, \tag{5.9}$$

which simply represents a balance of the electron thermal pressure with the electrostatic potential. We will assume, for simplicity, the hydrogen plasma, $q_e = e = -q_i < 0$, with the equilibrium density in charge neutrality, $n_{0e} = n_{0i} = n_0$. Combining Eqs. (5.8) and (5.9) gives

$$\nabla^2\delta\phi - \frac{4\pi n_{0e}e^2}{k_B T_e}\delta\phi = -4\pi q_i \delta n_i. \tag{5.10}$$

As the ions move along, the background electrons are light, mobile, responsive, and forming a shielding cloud. This contrasts with the electron plasma wave where the ions are heavy, immobile, irresponsive and serving as the uniform background. It is clear that the electric potential of a bare ion in Eq. (5.10) is limited to a Debye sphere at radius of Debye length $\lambda_D = \sqrt{k_B T_e/4\pi n_{0e}e^2}$by the electron screening, and can be rewritten as,

$$\lambda_D^2 \nabla^2 \Phi - \Phi = -\xi, \tag{5.10a}$$

where we have defined $\Phi \equiv q_i \delta\phi / T_e$ and $\xi \equiv \delta n_i / n_{0i}$. Similarly, rewriting Eq. (5.6a) gives

$$\frac{\partial^2}{\partial t^2}\xi + \omega_{pi}^2 (\xi - \Phi) = 0. \tag{5.6b}$$

Here, $\omega_{pi} \equiv \sqrt{4\pi n_{0i} q_i^2 / m_i}$ is the ion plasma frequency. The dispersion relation is obtained by the Fourier transform of Eqs. (5.6b) and (5.10a) to give

$$\omega^2 = \omega_{pi}^2 \frac{k^2 \lambda_D^2}{1 + k^2 \lambda_D^2} = \frac{T_e}{m_i} \frac{n_{0i} q_i^2}{n_{0e} e^2} \frac{k^2}{1 + k^2 \lambda_D^2} = C_s^2 \frac{q_i}{|e|} \frac{k^2}{1 + k^2 \lambda_D^2}, \tag{5.11}$$

where the **ion sound speed** is defined as $C_s \equiv \sqrt{T_e / m_i}$, and the charge neutrality condition has been imposed. The long wavelength spectrum ($k\lambda_D \ll 1$) of the wave is nondispersive, while the short wavelength spectrum ($k\lambda_D \gg 1$) remains as simple oscillation at the ion plasma frequency.

The Debye screening can be understood by posting the **Green's function** solution to Eq. (5.10a) cast in the following form,

$$\nabla^2 G(\vec{r}) - G(\vec{r}) = -4\pi \delta(\vec{r}). \tag{5.12}$$

This equation can be solved by letting $G(\vec{r}) = f(\vec{r})/r$ with the condition that $f(\vec{r}) = 1$ at $r = 0$. Applying the prior knowledge that $\nabla^2(1/r) = -4\pi\delta(\vec{r})$, we end with $d^2 f/dr^2 - f = 0$. Therefore, $G(\vec{r}) = \exp(-r)/r$ is the solution to Eq. (5.12). Equation (5.10a) can then be solved in terms of the Green's function by taking $r \to |\vec{r} - \vec{r}'|/\lambda_D$,

$$\Phi(\vec{r}) = -\int \frac{d\tau'}{4\pi \lambda_D^2} \xi(\vec{r}') \frac{e^{-|\vec{r} - \vec{r}'|/\lambda_D}}{|\vec{r} - \vec{r}'|}.$$

Homework 5.2: Assume that an ion sound wave is excited as a Gaussian packet, $\delta n_k = \delta n \lambda_D^3 \pi^{-3/2} e^{-k^2 \lambda_D^2}$. Its amplitude δn is 1% of the plasma density n_0. Calculate its energy flux for a hydrogen plasma of density $n_0 \sim 10^{13}$/cc, and temperature $T \sim 1$ Kev.

5.1.3 *Ion Cyclotron Wave*

The above derivation of the ion sound wave may be generalized to describe the **ion cyclotron wave** when an external magnetic field is imposed. The wave has the frequency near the ion cyclotron frequency $\Omega_{ci} = q_i B_0/m_i c$. As such, it is a powerful mechanism to transfer the wave energy to heat the ions by the ion cyclotron resonance, and vice versa, energetic fusion plasma may excite ion cyclotron waves through the same. We limit ourselves to the electrostatic branch, and the continuity equation of Eq. (5.6) is valid to describe the ion density perturbations. The equation of ion motion is given by

$$\frac{\partial \delta \vec{v}_i}{\partial t} = -\frac{q_i}{m_i}\nabla \delta\phi + \frac{q_i}{m_i c}\delta\vec{v}_i \times \vec{B}_0, \tag{5.13}$$

where the external magnetic field is included and the less significant thermal pressure is excluded. Depending on how the ion cyclotron period is compared with the transient time of longitudinal electron motion, we may assume that electrons equilibrate with the electrostatic potential governed by Eq. (5.9), provided that $\omega \sim \Omega_{ci} \ll k_{\|} v_{Te}$. Taking the time derivative on Eq. (5.10) gives

$$\nabla^2\frac{\partial \delta\phi}{\partial t} - \frac{4\pi n_0 e^2}{T_e}\frac{\partial \delta\phi}{\partial t} = -4\pi q_i \frac{\partial \delta n_i}{\partial t} = 4\pi n_0 q_i \nabla\cdot\delta\vec{v}_i. \tag{5.14}$$

These partial differential equations can be conveniently solved by the Fourier transform so that

$$-i\omega\delta\vec{v}_i = -\frac{q_i}{m_i} i\vec{k}\delta\phi + \delta\vec{v}_i \times \vec{\Omega}_{ci}, \tag{5.13a}$$

and

$$\left(k^2 + \frac{1}{\lambda_D^2}\right)\omega\delta\phi = 4\pi n_0 q_i \vec{k}\cdot\delta\vec{v}_i. \tag{5.14a}$$

Once we eliminate $\delta\phi$ from Eq. (5.13a) with use of Eq. (5.14a), we end with only the variable $\delta\vec{v}_i$,

$$\omega\delta\vec{v}_i = \frac{T_e}{m_i}\frac{\vec{k}}{\omega}\frac{\vec{k}\cdot\delta\vec{v}_i}{1+k^2\lambda_D^2} + i\delta\vec{v}_i \times \vec{\Omega}_{ci}. \tag{5.15}$$

Dotting Eq. (5.15) with $\vec{k}$ gives

$$\vec{k}\cdot\delta\vec{v}_i = \frac{T_e}{m_i}\frac{k^2}{\omega^2}\frac{\vec{k}\cdot\delta\vec{v}_i}{1+k^2\lambda_D^2} + i\frac{\vec{k}}{\omega}\cdot\delta\vec{v}_i \times \vec{\Omega}_{ci}. \tag{5.16}$$

Dotting Eq. (5.15) with $\vec{\Omega}_{ci} \times \vec{k}$ eliminates the first term on the RHS, and leads to

$$\begin{aligned}\vec{\Omega}_{ci} \times \vec{k}\cdot\delta\vec{v}_i &= \frac{i}{\omega}(\vec{\Omega}_{ci}\times\vec{k})\cdot(\delta\vec{v}_i \times \vec{\Omega}_{ci}) \\ &= \frac{i}{\omega}\vec{\Omega}_{ci}\cdot\delta\vec{v}_i\vec{k}\cdot\vec{\Omega}_{ci} - \frac{i}{\omega}\Omega_{ci}^2\vec{k}\cdot\delta\vec{v}_i.\end{aligned} \tag{5.17}$$

Combining these two equations by imposing the two-dimensional wave propagation on the plane perpendicular to the magnetic field so that the first term on the RHS of

Eq. (5.17) drops out, we find

$$\vec{\Omega}_{ci} \times \vec{k} \cdot \delta\vec{v}_i = -\frac{i}{\omega}\Omega_{ci}^2 \vec{k} \cdot \delta\vec{v}_i. \tag{5.17a}$$

Replacing the last term on the RHS of Eq. (5.16) with use of Eq. (5.17a), we then have the dispersion for the electrostatic ion cyclotron wave,

$$\omega^2 = \frac{k^2 C_s^2}{1 + k^2\lambda_D^2} + \Omega_{ci}^2. \tag{5.18}$$

The propagation along the field lines will pick up the ion sound wave as shown in Eq. (5.16) by taking $\vec{k} = k_{\parallel}\hat{e}_{\parallel}$ to yield $\omega^2 = k^2 C_s^2/(1 + k^2\lambda_D^2)$. When the wave propagates across the field lines, the ion cyclotron motion will modify the wave characteristics to result in the ion cyclotron wave.

Homework 5.3: Show that the ion cyclotron wave governed by Eq. (5.18) has a vanishing group velocity as the wave approaches the ion cyclotron resonance layer, where $\omega = \Omega_{ci}$.

5.1.4 *Electron Cyclotron Wave*

Analogous to the ion cyclotron wave, the **electron cyclotron wave** is characterized by the electron cyclotron frequency $\Omega_{ce} = eB_0/m_e c$ with $e < 0$, and therefore may tap the electron energy or accelerate electrons through the electron cyclotron resonance. We will here examine for simplicity the electrostatic electron cyclotron wave. In the presence of the wave electric field, the equation of motion for electrons is given by

$$\frac{\partial\delta\vec{v}}{\partial t} = \frac{e\delta\vec{E}}{m_e} + \frac{e\delta\vec{v} \times \vec{B}_0}{m_e c}. \tag{5.19}$$

Taking the Fourier transform in time gives $(-i\omega + \vec{\Omega}_{ce}\times)\delta\vec{v} = e\delta\vec{E}/m_e$. Therefore,

$$\delta\vec{v}_{\perp} = \frac{e}{m_e}\frac{i\omega\delta\vec{E}_{\perp} + \vec{\Omega}_{ce} \times \delta\vec{E}_{\perp}}{\omega^2 - \Omega_{ce}^2}, \tag{5.20}$$

and

$$\delta\vec{v}_{\parallel} = \frac{e}{m_e}\frac{i\delta\vec{E}_{\parallel}}{\omega}. \tag{5.21}$$

Homework 5.4: Carry out the algebra to derive Eq. (5.20) for the general response of electron to an electric field perturbation under the uniform external magnetic field.

In the low frequency limit, $\omega \ll \Omega_{ce}$, $\delta\vec{v}_\perp \approx c\delta\vec{E}_\perp \times \hat{e}_z/B_0 + c\delta\dot{\vec{E}}_\perp/B_0\Omega_{ce}$. They are simply the $E \times B$ drift and the polarization drift, respectively, thus giving us an independent verification of the formula. This general solution to Eq. (5.19) does reveal the electron cyclotron resonance in the denominator of Eq. (5.20). The continuity equation gives, $\omega\delta n = n_0\vec{k} \cdot \delta\vec{v}$. Eliminating $\delta\vec{v}$ with use of Eq. (5.20) and replacing the electric field by the Gauss's law

$$\nabla \cdot \delta\vec{E} = 4\pi e\delta n = i\vec{k} \cdot \delta\vec{E} = 4\pi n_0 e\vec{k} \cdot \delta\vec{v}/\omega$$

gives

$$i\vec{k} \cdot \delta\vec{E} = \frac{\omega_{pe}^2}{\omega^2}\left(ik_{||}\delta\vec{E}_{||} + \omega\vec{k} \cdot \frac{i\omega\delta\vec{E}_\perp + \vec{\Omega}_{ce} \times \delta\vec{E}_\perp}{\omega^2 - \Omega_{ce}^2}\right).$$

Since the wave is longitudinal and $\delta\vec{E} = -\nabla\delta\Phi$, we find

$$k^2 = \frac{\omega_{pe}^2}{\omega^2}\left(k_{||}^2 + \frac{\omega^2 k_\perp^2}{\omega^2 - \Omega_{ce}^2}\right). \tag{5.22}$$

Equation (5.22) gives a simple plasma oscillation when $k_\perp = 0$, and an oscillation at **upper hybrid frequency**, $\omega = \sqrt{\Omega_{ce}^2 + \omega_{pe}^2}$, when $k_{||} = 0$. It demonstrates that when the electron plasma wave propagates across the magnetic field lines, the electron cyclotron motion does modify the plasma wave to become the upper hybrid wave.

Homework 5.5: Derive the electromagnetic branch of the electron cyclotron waves by considering the current perturbations instead of the density perturbations.

5.1.5 *Lower Hybrid Wave*

The **lower hybrid wave** has its frequency lying between the electron cyclotron frequency and the ion cyclotron frequency. We will start with the flow velocities of the electrons and ions, which has the same general form as given in Eqs. (5.20) and (5.21),

$$\delta\vec{v}_{\perp\sigma} = \frac{q_\sigma}{m_\sigma}\frac{i\omega\delta\vec{E}_\perp + \vec{\Omega}_\sigma \times \delta\vec{E}_\perp}{\omega^2 - \Omega_\sigma^2}, \tag{5.20a}$$

$$-i\omega\delta v_z = \frac{q_\sigma}{m_\sigma}\delta E_z. \tag{5.21a}$$

The density perturbations are given by

$$\omega\delta n_\sigma = n_0\vec{k}\cdot\delta\vec{v}_\sigma = \frac{n_0 q_\sigma i}{m_\sigma}\left(\frac{k_{||}\delta E_{||}}{\omega} + \frac{\omega\vec{k}_\perp\cdot\delta\vec{E}_\perp}{\omega^2-\Omega_\sigma^2}\right). \tag{5.23}$$

Thus, we have the electron density perturbation

$$\delta n_e = \frac{n_0 e i}{m_e}\left(\frac{\vec{k}_\perp\cdot\delta\vec{E}_\perp}{\omega^2-\Omega_{ce}^2} + \frac{k_{||}\delta E_{||}}{\omega^2}\right), \tag{5.24}$$

and the ion density perturbation,

$$\delta n_i = \frac{n_0 q i}{m_i}\left(\frac{\vec{k}\cdot\delta\vec{E}}{\omega^2-\Omega_{ci}^2} + \frac{k_{||}\delta E_{||}}{\omega^2}\right), \tag{5.25}$$

Due to the relatively low frequency nature of the wave, the electrons are mobile enough to maintain the charge neutrality, $\delta n_e = \delta n_i$. Therefore,

$$\left(\frac{\vec{k}_\perp\cdot\delta\vec{E}_\perp}{\omega^2-\Omega_{ce}^2} + \frac{k_{||}\delta E_{||}}{\omega^2}\right)\frac{m_i}{m_e} = \left(\frac{\vec{k}_\perp\cdot\delta\vec{E}_\perp}{\omega^2-\Omega_{ci}^2} + \frac{k_{||}\delta E_{||}}{\omega^2}\right),$$

that leads to

$$\frac{k_\perp^2}{\omega^2-\Omega_{ce}^2} + \frac{k_{||}^2}{\omega^2} \approx -\frac{m_e}{m_i}\frac{k_\perp^2}{\omega^2-\Omega_{ci}^2}, \tag{5.26}$$

where the ion motion along the field line has been neglected due to the mass ratio. When $k_{||} = 0$, we have the dispersion relation, $\omega^2 - \Omega_{ci}^2 = -m_e(\omega^2-\Omega_{ce}^2)/m_i$. Taking the limit, $\Omega_{ce} \gg \omega \gg \Omega_{ci}$, we end with

$$\omega^2 = \frac{\Omega_{ci}^2 + \Omega_{ci}\Omega_{ce}}{1+(m_e/m_i)} \approx \Omega_{ci}\Omega_{ce}. \tag{5.27}$$

The wave frequency happens to be the geometric mean of the two cyclotron frequencies.

> **Homework 5.6:** Show that the particle trajectories in the lower hybrid waves of Eq. (5.27) are elliptical and the major radius of the electron orbital equals the minor radius of the ion orbital. Take $k_{||} = 0$ for simplicity.

Suppose the charge neutrality is violated when the plasma density is relatively low, or when the electrons could not move along the field lines fast enough, we may explore the full Poisson's equation but keep $k_{||} = 0$ for simplicity,

$$i\vec{k} \cdot \delta\vec{E} = 4\pi e(\delta n_i - \delta n_e) = \omega_{pe}^2 \frac{i\vec{k} \cdot \delta\vec{E}}{\omega^2 - \Omega_{ce}^2} + \omega_{pi}^2 \frac{i\vec{k} \cdot \delta\vec{E}}{\omega^2 - \Omega_{ci}^2}.$$

The dispersion relation is then given by

$$1 = \frac{\omega_{pe}^2}{\omega^2 - \Omega_{ce}^2} + \frac{\omega_{pi}^2}{\omega^2 - \Omega_{ci}^2} \approx \omega_{pi}^2 \left[\frac{-1}{\Omega_{ci}\Omega_{ce}} + \frac{1}{\omega^2 - \Omega_{ci}^2} \right], \tag{5.28}$$

where ω^2 is neglected in comparison with Ω_{ce}^2. Therefore,

$$\omega^2 = \Omega_{ci}^2 + \frac{\omega_{pi}^2 \Omega_{ci}\Omega_{ce}}{\omega_{pi}^2 + \Omega_{ci}\Omega_{ce}} \approx \begin{cases} \Omega_{ci}\Omega_{ce} & \text{for } \omega_{pi}^2 \gg \Omega_{ci}\Omega_{ce}. \\ \Omega_{ci}^2 + \omega_{pi}^2 & \text{for } \omega_{pi}^2 \ll \Omega_{ci}\Omega_{ce}. \end{cases} \tag{5.29}$$

At high density, the lower hybrid wave goes to the geometrical mean of the electron and ion cyclotron frequencies. When the plasma density is low, $\omega_{pi}^2 \ll \Omega_{ci}\Omega_{ce}$, it goes to $\sqrt{\Omega_{ci}^2 + \omega_{pi}^2}$. If $\Omega_{ce} \gg \omega_{pi} \gg \Omega_{ci}$, the wave is a simple oscillation at the *ion plasma frequency*.

5.2 Transverse Waves

The electrostatic modes discussed in the last section are longitudinal waves. Their transverse counterparts are often the electromagnetic waves in the vacuum but modified by the plasma dielectric response. The exceptions are the **Alfvén wave** and **the Whistler mode**. They are transverse waves in their own right but exist only inside the plasma medium.

5.2.1 *Alfvén Wave*

The Alfvén wave has a frequency lower than the ion cyclotron frequency. It causes the magnetic oscillation that could infringe the magnetic confinement, but also has the potential of magnetic pumping. This wave is analogous to the spring oscillator with the ion mass as the inertia and the magnetic field the restoring tensile force. If the wave propagates along the field line with the magnetic perturbations perpendicular to the equilibrium magnetic field, it is a **shear Alfvén wave**. Otherwise, it is a **compressional Alfvén wave**. Given the same magnitude in the magnetic perturbations, the shear Alfvén wave has less energy content than the compressional Alfvén wave (cf. Homework 1.1). Therefore, it tends to be excited first. Unless the plasma is highly energetic, the compressional Alfvén wave will be less significant.

It is more convenient to resort to the MHD equations to describe this wave, as historically so discovered by Alfvén, thus often referred to as the **magnetosonic waves** in analogy to the ion sound wave. For simplification, we restrict ourselves to the uniform external magnetic field. Taking from Eq. (4.30) without the pressure effect or internal friction, we have

$$\rho\left(\frac{\partial \vec{U}}{\partial t}+\vec{U}\cdot\nabla\vec{U}\right)=\frac{1}{c}\vec{j}\times\vec{B}. \tag{5.30}$$

The wave frequency is much lower than both the electron and the ion cyclotron frequencies so that both the electrons and ions are moving at the $\vec{E}\times\vec{B}$ drifts and the Faraday's law can be written as,

$$\frac{1}{c}\frac{\partial \vec{B}}{\partial t}=-\nabla\times\vec{E}=\nabla\times\left(\frac{\vec{U}}{c}\times\vec{B}\right). \tag{5.31}$$

Linearizing Eqs. (5.30) and (5.31) results in

$$\rho_0\frac{\partial\delta\vec{u}}{\partial t}=\frac{1}{4\pi}(\nabla\times\delta\vec{b})\times\vec{B}_0, \tag{5.30a}$$

$$\frac{\partial\delta\vec{b}}{\partial t}=\nabla\times(\delta\vec{u}\times\vec{B}_0). \tag{5.31a}$$

Taking the Fourier transform of Eqs. (5.30) and (5.31) gives

$$-\rho_0\omega\delta\vec{u}=\frac{1}{4\pi}(\vec{k}\times\delta\vec{b})\times\vec{B}_0, \tag{5.30b}$$

$$-\omega\delta\vec{b}=\vec{k}\times(\delta\vec{u}\times\vec{B}_0). \tag{5.31b}$$

Eliminating $\delta\vec{u}$ from Eq. (5.31b) with use of Eq. (5.30b), we arrive at

$$\begin{aligned}\rho_0\omega^2\delta\vec{b}&=\frac{1}{4\pi}\vec{k}\times[((\vec{k}\times\delta\vec{b})\times\vec{B}_0)\times\vec{B}_0]\\&=\frac{B_0^2}{4\pi}((\delta\vec{b}k_z-\vec{k}\delta b_z)k_z-\widehat{e}_z(\vec{k}\cdot\delta\vec{b}k_z-k^2\delta b_z)).\end{aligned} \tag{5.32}$$

The wave with $\vec{k}=k_z\widehat{e}_z$ and $\vec{k}\cdot\delta\vec{b}/k=\delta b_z=0$ is governed by

$$\rho_0\omega^2\delta\vec{b}_\perp=\frac{B_0^2}{4\pi}\delta\vec{b}_\perp k_z^2. \tag{5.33}$$

This is the shear Alfvén wave with the dispersion relation $\omega=k_zV_A$, where $V_A=B_0/\sqrt{4\pi\rho_0}$ is the Alfvén velocity. It has the wave magnetic component perpendicular to the external magnetic field, propagates along the field line, and commands the incompressible flow since the flow is along $\delta\vec{b}$ as given in Eq. (5.30a), and $\vec{k}\cdot\delta\vec{u}=0$.

The other wave with $\vec{k} \cdot \widehat{e}_z = 0$ and $\delta\vec{b} = \delta b_z \widehat{e}_z$ is governed by

$$\rho_0 \omega^2 \delta b_z \widehat{e}_z = \frac{B_0^2}{4\pi} \widehat{e}_z k_\perp^2 \delta b_z. \tag{5.33a}$$

This is the compressional Alfvén wave with the dispersion relation $\omega = k_\perp V_A$ that has the magnetic component parallel and the wave vector perpendicular to the external magnetic field. Its fluid element has a velocity $\delta\vec{u} = \vec{k}\delta\vec{b} \cdot \vec{B}_0/(4\pi\rho_0\omega)$ that causes compressible flow, and is thus modified by the ion sound wave when the pressure is included.

> **Homework 5.7:** Show that there is an equipartition between the fluid kinetic energy and the magnetic field energy in the shear Alfvén wave.

> **Homework 5.8:** Derive the dispersion relation for the Alfvén wave by including the pressure term $p = \rho C_s^2$ and assuming the density compressibility.

5.2.2 *Whistler Wave*

The Alfvén wave at higher frequency regime turns into a **whistler wave**. This wave, as it travels along the field lines, has the unique feature of making the whistle-like acoustic sound. It is often triggered by the lightening, and propagates mainly along the geomagnetic field lines. It coincides with the same frequency regime as the lower hybrid wave, namely, $\Omega_{ce} \gg \omega \gg \Omega_{ci}$, and may be considered as its electromagnetic counterpart. The theory can be described as in the following. The perpendicular velocity responses of electrons and ions are given by Eq. (5.20). The wave is electromagnetic in nature, therefore governed by the Faraday's law to give $\vec{k} \times \delta\vec{E} = \omega\delta\vec{B}/c$, and the Ampere's law gives $\vec{k} \times \delta\vec{B} = 4\pi\delta\vec{J}/ic - \omega\delta\vec{E}/c$. Combining them together to eliminate the magnetic perturbation $\delta\vec{B}$ gives

$$\vec{k} \times (\vec{k} \times \delta\vec{E}) = \frac{\omega}{c}\left(\frac{4\pi}{ic}\delta\vec{J} - \frac{\omega}{c}\delta\vec{E}\right). \tag{5.34}$$

The current perpendicular to the external magnetic field, $\delta\vec{J}_\perp = n_0 e \delta\vec{v}_\perp$can be obtained by substituting the perpendicular velocities from electrons and ions

$$\delta\vec{J}_\perp = \frac{n_0 e^2}{m_i}\left(\frac{i\omega\delta\vec{E}_\perp + \vec{\Omega}_{ci} \times \delta\vec{E}_\perp}{\omega^2 - \Omega_{ci}^2}\right) + \frac{n_0 e^2}{m_e}\left(\frac{i\omega\delta\vec{E}_\perp + \vec{\Omega}_{ce} \times \delta\vec{E}_\perp}{\omega^2 - \Omega_{ce}^2}\right). \tag{5.35}$$

The current parallel to the external magnetic field, $\delta\vec{J}_{\|} = n_0 q \delta\vec{v}_{\|}$ can be obtained from Eq. (5.2)

$$\delta\vec{v}_{\|} = ie\delta\vec{E}/m/\omega,$$

$$\delta\vec{J}_{\|} = \left(\frac{n_0 q_i^2}{m_i} + \frac{n_0 e^2}{m_e}\right)\frac{i\delta\vec{E}_{\|}}{\omega}. \tag{5.36}$$

By replacing the current components from the above two equations, Eq. (5.34) yields the wave equation,

$$\begin{aligned}
&\vec{n} \times (\vec{n} \times \delta\vec{E}) + \delta\vec{E} \\
&= -\frac{\omega_{pi}^2}{\omega}\left[\left(\frac{\omega\delta\vec{E}_\perp - i\vec{\Omega}_{ci} \times \delta\vec{E}_\perp}{\omega^2 - \Omega_{ci}^2}\right) + \frac{m_i}{m_e}\left(\frac{\omega\delta\vec{E}_\perp - i\vec{\Omega}_{ce} \times \delta\vec{E}_\perp}{\omega^2 - \Omega_{ce}^2}\right)\right] \\
&\quad + \frac{\omega_{pi}^2}{\omega^2}\delta\vec{E}_{\|}\left(1 + \frac{m_i}{m_e}\right) \\
&\approx -\frac{\omega_{pi}^2}{\omega}\left[\left(\frac{\omega\delta\vec{E}_\perp - i\vec{\Omega}_{ci} \times \delta\vec{E}_\perp}{\omega^2}\right) + \frac{m_i}{m_e}\left(\frac{\omega\delta\vec{E}_\perp - i\vec{\Omega}_{ce} \times \delta\vec{E}_\perp}{-\Omega_{ce}^2}\right)\right] \\
&\quad + \frac{\omega_{pe}^2}{\omega^2}\delta\vec{E}_{\|}.
\end{aligned}$$

Here the **index of refraction** $\vec{n} = c\vec{k}/\omega$ is used. Rewrite it in the matrix form, we have

$$\begin{pmatrix} S - n_z^2 & iD & n_x n_z \\ -iD & S - n^2 & 0 \\ n_x n_z & 0 & P - n_x^2 \end{pmatrix}\begin{pmatrix} \delta E_x \\ \delta E_y \\ \delta E_z \end{pmatrix} = 0, \tag{5.37}$$

where we have defined $S \equiv 1 - \omega_{pi}^2/\omega^2 + \omega_{pi}^2/\Omega_{ce}\Omega_{ci}$, $D \equiv (\omega_{pi}^2/\omega^2)(\Omega_{ci}/\omega) + (\omega_{pi}^2/\omega\Omega_{ci})$, and $P \equiv 1 - \omega_{pe}^2/\omega^2$. The wave of interest has the characteristics in that $n_{\|} \gg n_\perp \sim O(1)$. As a consequence, its $\delta E_z \to 0$ due to the transverse nature of the wave. The wave dispersion relation is the governed by

$$(S - n_{\|}^2)(S - n^2) \approx D^2, \tag{5.38}$$

and $D \approx \omega_{pi}^2/\omega\Omega_{ci}$. For $\omega^2 \sim \Omega_{ce}\Omega_{ci}$, we have $S \sim O(1)$, and in the limit, $\omega_{pi}^2 \gg \omega\Omega_{ci}$, $D \gg 1$. Therefore, the wave dispersion function becomes

$$n^2 \approx \left|\frac{\omega_{pi}^2}{\omega\Omega_{ci}}\frac{1}{\cos\theta}\right| \gg 1, \tag{5.39}$$

where $\theta \to 0$.

Homework 5.9: Assume the earth magnetic field is 0.3 Gauss, and nitrogen is the major ion species in the air. Show that the whistler wave has a frequency in the KHz range.

5.3 Instabilities

We may identify the existence of an instability by examining the virtual displacement of relevant parameter to see whether the system would then lower its energy as a result. This is often referred to as the energy principle. Then the corresponding perturbation will cause the system to become unstable. Another method is to identify the mode frequency to see whether it has the imaginary part. This is often referred to as the normal mode analysis. Since the system variables are in general real, the solution of the frequency analysis would yield complex conjugates if an imaginary solution in the frequency is found. One of the modes then grows with time. Typically, we make discrete or continuous Fourier transforms of the system variables. Take the following representation as the example,

$$F(x,t) = \sum_{k,\omega} f_{k\omega} e^{i(kx-\omega t)}. \tag{5.40}$$

It is clear that if $\omega = \omega_r + i\gamma$, with $\gamma > 0$, we would have $F(x,t) \sim f_{k\omega_r} e^{i(kx-\omega_r t)+\gamma t}$. The wave grows exponentially. Note that the linear theory easily breaks down even before the instability amplitude becomes comparable with the equilibrium quantities, since as a rule, the linear theory assumes the instability as a small perturbation to begin with. Extrapolate the linear instability theory to go beyond a few e-folding time of the growth rate can be misleading and erroneous.

5.3.1 *Drift Wave*

The simplest form of a **drift wave** is an electrostatic, low frequency wave, driven by the free energy associated with the density gradient. In general, any free energy, be it with the spatial gradient of temperature, current, or trapped particle population, for examples, has the potential of exciting some kind of drift waves, that may lead to weak turbulence. When there are stabilizing or restoring mechanisms to balance out such as the pinch effect or the averaged magnetic well, it would enhance transport and result in confinement degradation. In the extreme case, the argument by quasilinear theory would require the gradient to be evened out. That could mean a more severe confinement loss.

Consider the drift waves driven by the density-gradient. The equilibrium is established from balancing the magnetic confinement with the electron pressure gradient: $ne\vec{v} \times \vec{B}/c = \nabla p$. Therefore, a zeroth order diamagnetic drift arises,

$$\vec{V}_D = \frac{c}{ne}\frac{\vec{B} \times \nabla p}{B^2}. \tag{5.41}$$

We may assume a constant $\vec{V}_D$ in an isothermal plasma $p = nT_e$ for simplicity. We also define the drift frequency by $\omega* = \vec{k} \cdot \vec{V}_D$. We will neglect the electron inertia but impose a phenomenological damping term. Since the electron/ion cyclotron frequency is much greater than the collisional frequency ν_c, we will neglect the collisional effect in the perpendicular velocities. The equation of motion for electrons is then modeled as follows,

$$e\delta\vec{E} + \frac{e\delta\vec{v} \times \vec{B}}{c} + \frac{\delta ne\vec{V}_D \times \vec{B}}{nc} - \frac{T_e \nabla \delta n}{n} - \nu_c m_e \delta\vec{v}_{||} = 0, \tag{5.42}$$

which is solved to give,

$$\delta\vec{v}_\perp = \frac{c\delta\vec{E} \times \vec{B}}{B^2} + \frac{cT_e}{ne}\frac{\vec{B} \times \nabla\delta n}{B^2} - \vec{V}_D\frac{\delta n}{n} - \frac{ci\vec{k} \times \vec{B}T_e}{eB^2}(\tilde{\phi} + \tilde{n}) - \vec{V}_D\tilde{n}, \tag{5.43}$$

$$\delta v_{||} = -\frac{T_e}{m_e}\frac{\partial_{||}\delta n}{n\nu_c} + \frac{e\delta E_{||}}{\nu_c m_e} = -\frac{i}{\mu}\frac{\omega}{k_{||}}(\tilde{n} + \tilde{\phi}), \tag{5.44}$$

where we have defined the dimensionless variables of potential $\tilde{\phi} \equiv e\phi/T_e$, density $\tilde{n} \equiv \delta n/n$, and damping factor $\mu \equiv \omega\nu_c/k_{||}^2 V_T^2$, with $V_T \equiv \sqrt{T_e/m_e}$ the electron thermal velocity. Substituting the velocities into the linearized electron continuity equation,

$$\partial\delta n/\partial t + \nabla \cdot (n\delta\vec{v} + \vec{V}_D\delta n) = 0 = -i\omega\delta n + ik_{||}n\delta v_{||} + (\delta\vec{v}_\perp + \vec{V}_D\tilde{n}) \cdot \nabla n,$$

gives

$$\tilde{n} + \frac{i}{\mu}(\tilde{n} + \tilde{\phi}) + \frac{\omega*}{\omega}(\tilde{n} + \tilde{\phi}) = 0. \tag{5.46}$$

Solving for $\tilde{n}$ from Eq. (5.46) gives

$$\tilde{n} = -\frac{1 + i/\mu}{1 + \omega * /\omega + i/\mu}\tilde{\phi}, \tag{5.46a}$$

The ions are treated as cold by taking $T_i = 0$. They execute the ExB and the polarization drifts perpendicular to the magnetic field , $\delta\vec{v}_{i\perp} = c\delta\vec{E}_\perp \times \vec{B}/B^2 +$

$c\delta \dot{\vec{E}}_{\perp}/B\Omega_i$, and the simple oscillation along the field lines: $\delta\vec{v}_{i||} = iq_i\delta\vec{E}_{||}/m_i\omega$. The density perturbation from the continuity equation gives

$$i\delta n_i\omega = \nabla\cdot(n\delta\vec{v}_i) = \delta\vec{v}_{i\perp}\cdot\nabla n + n\nabla\cdot\delta\vec{v}_{i\perp} + inq_i\nabla_{||}\cdot\delta\vec{E}_{||}/m_i\omega. \quad (5.47)$$

Thus, the ion density perturbation is related to the electrostatic potential by

$$\tilde{n}_i = \frac{\omega*}{\omega}\tilde{\phi} - k_\perp^2\rho_s^2\tilde{\phi} + \frac{k_{||}^2C_s^2}{\omega^2}\tilde{\phi}, \quad (5.48)$$

where the ion gyroradius is defined at the ion sound speed as $\rho_s \equiv C_s/\Omega_{ci}$. Imposing the charge neutrality condition, we arrive at the drift wave dispersion relation,

$$\frac{1-i\mu}{1-i\mu(1+\omega*/\omega)} - \frac{\omega*}{\omega} + k_\perp^2\rho_s^2 - \frac{k_{||}^2C_s^2}{\omega^2} = 0. \quad (5.49)$$

A characteristic frequency can readily be identified as $\omega \approx \omega*$ by nullifying the damping effect, viz., setting $\mu = 0$ and dropping the third term, which is the finite Larmor radius correction, and the last term, which is the coupling to the ion sound wave. Note that we have defined $\omega* = \vec{k}\cdot\vec{V}_D = (k_y cT_e/neB^2)(dn/dx)$ with the electron charge $e < 0$, by taking the external magnetic field $\vec{B} = B\hat{e}_z$, and the equilibrium density $n = n(x)$. We are interested in the drift wave and not the ion sound wave modified by the drift wave characteristics. So the mode of interest is $\omega = \omega_0 + i\gamma \approx \omega*(1-k_\perp^2\rho_s^2) + k_{||}^2C_s^2/\omega* + i\mu\omega*$. Therefore,

$$\omega_0 \approx \omega*(1-k_\perp^2\rho_s^2) + k_{||}^2C_s^2/\omega*,$$
$$\gamma \approx \mu\omega*. \quad (5.50)$$

The drift wave becomes unstable with the growth rate at $\mu\omega*$ when there is a dissipation mechanism to tap the free energy.

Homework 5.10: Show that the drift wave is a backward wave in that its group velocity and the phase velocity are travelling opposite to each other in the direction perpendicular to the magnetic field.

Homework 5.11: Derive the dispersion relation for the Alfvén wave as modified by the drift wave characteristics.

5.3.2 *Two-Stream Instability*

The **two-stream instability** has its hydrodynamic analogy known as the Kelvin Helmholtz instability (KHI) that may lead to severe weather condition including tornado, and hurricane. Its free energy is due to the flow velocity that cannot be transformed away by Galilean transformation. The two stream instability is a kinetic instability in the sense that in the configuration space there can be no net flow, and yet the instability is explosive to result in the relaxation from the narrowly peaked velocity distribution function. By contrast, the KHI arises from the free energy of velocity shear in the configuration space with the two streams of flows adjacent but apart, and dominates at the largest velocity gradient.

Consider the electrons with the flow velocities $\pm V_0$. Taking the density and the velocity perturbations in the continuity equation,

$$\frac{\partial \delta n}{\partial t} + \nabla \cdot (n_0 \delta \vec{v}) + \nabla \cdot (\delta n \vec{V}_0) = 0, \tag{5.51}$$

and in the momentum equation,

$$\frac{\partial \delta \vec{v}}{\partial t} + \vec{V}_0 \cdot \nabla \delta \vec{v} = \frac{e \delta \vec{E}}{m}, \tag{5.52}$$

after the Fourier transform in space and time, we find

$$\delta n = n_0 \frac{\vec{k} \cdot \delta \vec{v}}{\omega - \vec{k} \cdot \vec{V}_0}, \tag{5.53}$$

and

$$\delta \vec{v} = \frac{e}{m} \frac{i \delta \vec{E}}{\omega - \vec{k} \cdot \vec{V}_0}. \tag{5.54}$$

Combining the last two equations together, we have

$$\delta n = \frac{n_0 e}{m} \frac{i \vec{k} \cdot \delta \vec{E}}{(\omega - \vec{k} \cdot \vec{V}_0)^2}. \tag{5.55}$$

The Poisson's equation $\nabla \cdot \delta \vec{E} = 4\pi e(\delta n_+ + \delta n_-)$ gives

$$1 = \frac{1}{2} \omega_{pe}^2 \left[\frac{1}{(\omega - \vec{k} \cdot \vec{V}_0)^2} + \frac{1}{(\omega + \vec{k} \cdot \vec{V}_0)^2} \right], \tag{5.56}$$

where the density perturbations due to the $\pm V_0$ have been summed up. Eq. (5.56) can be rewritten as

$$1 = \frac{1}{(x - \Omega)^2} + \frac{1}{(x + \Omega)^2}, \tag{5.56a}$$

where $x \equiv \sqrt{2}\omega/\omega_{pe}$, $\Omega \equiv \sqrt{2}\vec{k} \cdot \vec{V}_0/\omega_{pe}$. Equation (5.56a) has four roots given by

$$x = \pm\sqrt{\Omega^2 + 1 \pm \sqrt{4\Omega^2 + 1}} \rightarrow \begin{cases} \pm\sqrt{2}\left(1 + \frac{3}{4}\Omega^2\right) + \cdots \\ \pm i\Omega + \cdots \end{cases}, \tag{5.57}$$

There are two plasma oscillations, one purely growing, and one purely damping. The explosive instability has the growth rate $\gamma = \vec{k} \cdot \vec{V}_0$. The following MATLAB program shows that the growth rate goes linearly with Ω at small value, consistent with the serial expansion, and peaks at around $\sqrt{2}\vec{k} \cdot \vec{V}_0 \sim \omega_{pe}$. Beyond $\vec{k} \cdot \vec{V}_0 > \omega_{pe}$, the mode is stable. This indicates that the two stream instability will not excite waves with wavelength shorter than V_0/f_{pe}.

```
function TwoStream
syms x Omega;
f=1-1/(x+Omega)^2-1/(x-Omega)^2;
A=solve(f,x),
KV=[]; gamma=[];
for Omega=0:0.01:2
   KV=[KV,Omega];
   a=eval(A(3));
   gamma=[gamma,a];
end;
figure(1);
plot(KV,real(gamma),'b -',KV,imag(gamma),'r*-');
```

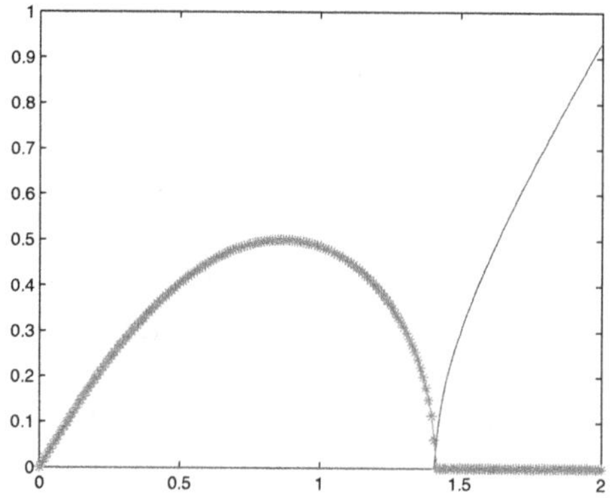

There are a few instabilities in close analogy to the two-stream instabilities. The trapped particle instability, for example, has the counter streaming particles of banana orbits. That free energy can drive the trapped particle instability. The particles in the oppositely propagating travelling waves can execute chaotic motions due to the stochastic instability.

Homework 5.12: Instead of cold beams of counter streaming particles, assume the particles have a thermal spread in the beam distribution. Derive the growth rate and find when the explosive instability ceases to be unstable.

Homework 5.13: The Rayleigh-Taylor instability has a growth rate given by $\gamma = \sqrt{gLk_z^2/(k_x^2L^2 + k_y^2L^2 + \frac{1}{4})}$, where L is the density scale length, g is the gravitational acceleration along the z direction, and k_x and k_y are the wave numbers in the plane perpendicular to the gravitational force. Suppose an additional effect is found to give $\gamma_A = \sqrt{gLk_z^2/(k_x^2L^2 + k_y^2L^2 + \frac{1}{4}) - k_x^2L^2A}$. Does this imply that the RTI can be stabilized by A?

5.4 Wave Characteristics

Wave categorization helps the understanding of wave characteristics, and provides the insight into a wave before great effort for detailed studies.

5.4.1 *Electromagnetic and Electrostatic*

From the foregoing examples, we recognize that the electrostatic wave is longitudinal, $\vec{k} \cdot \delta\vec{E} \neq 0$ and $\vec{k} \times \delta\vec{E} = 0$, while the electromagnetic wave is transverse with $\vec{k} \cdot \delta\vec{E} = 0$ and $\vec{k} \times \delta\vec{E} \neq 0$. There are always these two branches of waves in the same frequency range. The electromagnetic wave tends to be of longer wavelength than its electrostatic counterpart since the index of refraction $\vec{n} \approx \upsilon\vec{k}/\omega$, as a rule, is associated with the speed of light, $\upsilon \sim O(c)$, while the electron branch of the electrostatic wave is associated with the electron thermal speed, $\upsilon \sim O(\upsilon_e)$. Naturally, the wavelength becomes successively shorter, as $\lambda \sim 2\pi\, \upsilon/\omega$ at the same frequency. The wavelength mismatch does make the mode conversion from the electromagnetic branch to the electrostatic branch not so easily occur.

5.4.2 *Instability and Wave Damping*

A wave can become unstable and grow into a serious instability at the expense of confined plasma energy. Long range and large scale transport can occur provided that there is the free energy available and a mechanism to tap that free energy. On the other hand, if the particles are taking energy away from the wave, the wave will be damped while the particles are heated up and driven along. Heating and current drive by waves are important topics for fusion study.

5.4.3 *Cutoff and Resonance*

A wave could be a local oscillation that goes nowhere, or propagates in a region, but becomes evanescent beyond the cutoff layer. When a wave approaches a **cutoff** layer, its wave number along the propagation direction is reducing to zero, its wave speed is slowing down, subsequently the wave reflects back. When a wave approaches a **resonance** layer, its wave number increases, and the wavelength becomes very short, it can effectively transfers its energy to the particles through wave-particle resonance, or to the short wavelength ES modes through the mode conversion. When the cutoff and resonance occurs in close proximity, energy can tunnel through the cutoff layer to the resonance and effectively damp the wave.

5.4.4 *Ordinary and Extraordinary*

A wave is called **ordinary** if its dispersion relation does not depend on the magnetic field. This is the case when the wave electric field is along the magnetic field lines and particle response does not involve the magnetic field. The **extraordinary** wave would be the opposite to have the dispersion relation depend on the magnetic field, and in general its wave electric field comprises a component perpendicular to the magnetic field. Quite often we refer specifically to waves near the electron cyclotron frequency (cf. Homework 5.5), where the ordinary (O) wave has the dispersion $n_\perp^2 = P \approx 1 - \omega_{pe}^2/\omega^2$, and the extraordinary (X) wave has $n_\perp^2 = (S^2 - D^2)/S$, (setting $n_z = 0$ in Eq. (5.38)).

5.4.5 *Polarization*

A transverse wave is **linearly polarized** if the electric field does not rotate as the wave travels along. It is **circularly** polarized if the electric field rotates but does not change its strength. When the orthogonal components of the electric field rotate at different phase angular speeds, the tip of the electric field vector can form an ellipse, the wave is **elliptically polarized**. Right handed rotation of the electric field vector as the wave propagates is termed the right **polarization**, and left handed rotation the **left polarization**. Usually, a wave associated with electron characteristics will have the right polarization, a consequence of the right circulation of electrons in the magnetic field, and ion characteristics the left polarization.

5.4.6 *Phase Velocity and Group Velocity*

As demonstrated in the drift wave case (cf. Homework 5.10), the **phase velocity** and the **group velocity** can differ, and may even be moving in the opposite directions in the case of a backward wave. The phase velocity is defined as $V_{ph} \equiv \omega/k$ for the particular wave number k given the wave frequency ω. A pure k mode has to extend to infinite domain and does not exist in reality, the very reason for the uncertainty principle. Wave will travel as a wave packet with many k modes packed together. Its energy propagates at the group velocity $V_g \equiv \partial\omega/\partial k$.

5.4.7 *Fast Wave and Slow Wave*

The **fast and slow waves** are often referred to the magnetosonic waves. The shear Alfvén wave travels slower than the compressional Alfvén wave, is thus termed the slow wave. The compressional Alfvén wave with more energy content than the shear Alfvén wave (cf. Homework 1.1) due to its magnetic component along the field lines is referred to as the fast wave. But the terms may also be applicable to the waves at the same frequency, where the fast wave tends to be the transverse wave of electromagnetic structure, and slow wave the electrostatic.

5.4.8 *MHD and Kinetic Modes*

Beyond what has been discussed so far, there are also the **kinetic modes** such as the Bernstein wave, the **MHD modes** such as the kink instability and the interchange instability, and their more elaborate siblings: resistive tearing mode and the ballooning mode, and many more. The categorization is somewhat historical, but indicates that the irreducible properties of the waves identified from the named plasma model are well pronounced thereof.

5.4.9 *Linear Wave versus Nonlinear Wave*

When the wave amplitude becomes large, it may induce many nonlinear effects, and the physics can be very complex, especially when a serious instability is at work. Not only the particles will have to be examined for the nonlinear wave-particle interactions, the waves may also induce nonlinear mode-mode coupling. It is important to recognize that the linear theory provides very helpful first insight, but its conclusions cannot be all trustworthy until the long term, fully nonlinear evolution is examined, or statistical significance is assessed. One celebrated strength and unassuming weakness in the classical mechanics is its solubility to the initial and boundary value problem leading to a deterministic conclusion. Yet, given the example of the identical twins to arrive in this world only a minute apart, and to grow up in the identical environment, their life expectations can be drastically different. A specific instance that may be totally correct in its own right, cannot necessarily predict the same for another instance, and fails more often than not to represent the majority and may thus lose its physical meaning in its entirety. This draws attention to even the minute aberration, the **butterfly effect**, that could result in quite different long term behavior in the nonlinear physics. Statistical model could often come closer to the truth than a deterministic linear theory or its nonlinear extrapolation, of an initial value problem would.

5.4.10 *Wave Normal Surface*

While waves are often examined for parallel or perpendicular propagations to the externally imposed uniform magnetic field, the oblique propagation in general can be visualized by the **wave normal surface**. By designating the external magnetic field along the z-axis, the wave normal surface depicts the phase velocity of the wave as the radial distance in the polar plane panned out by the θ angle between $\widehat{e}_z$ and the radial $\widehat{e}_\rho$. The wave normal surface of whistler wave is plotted in the following program. Since its dispersion relation is $n^2 \approx |D/\cos\theta|$, and the phase velocity is $\vec{V}_p = \omega\vec{k}/k^2 = c\vec{n}/n^2$, we have $V_{p,x} = cn\sin\theta/n^2$, and $V_{p,z} = cn\cos\theta/n^2$. We set $c = 1$, $D = 1$ as the numerical units. The plot imitates a wave propagating outward, and presents its wave normal surface per fixed time interval.

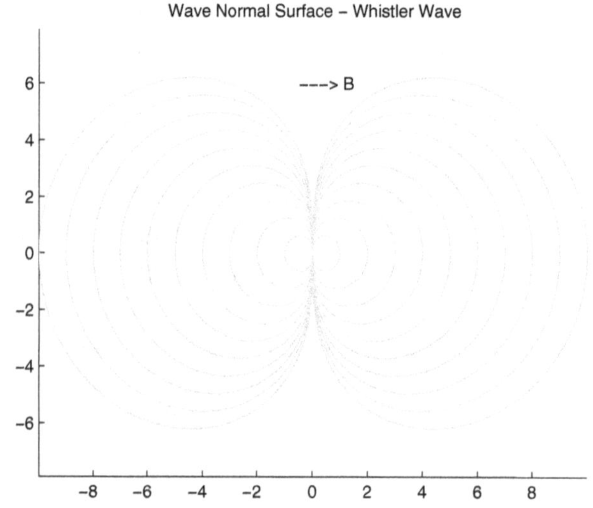

```
function WNSwhistler
close all; clear all; clc;
D=1;
theta=0:0.01:2*pi;
N=length(theta);
cost=cos(theta);
sint=sin(theta);
n2=abs(D./(cost+eps));
n=sqrt(n2);
Vz=n.*cost./n2;
Vx=n.*sint./n2;
x=0*theta; z=0*theta;
dt=0.1;
figure(1);
axis equal;
hold on;
for t=1:100
   x=x+Vx*dt;
   z=z+Vz*dt;
   if(mod(t,10)==0)
      plot(z,x,'g-');
   end;
end;
title('Wave Normal Surface - Whistler Wave');
str=sprintf('---> B');
text(-0.5,6,str); axis equal;
hold off;
```

Homework 5.14: Modify WNWwhistler.m to plot the wave normal surface for the compressional Alfvén wave. Its dispersion relation is given by $\omega^2 = C_s^2k_x^2 + V_A^2k^2 + C_s^4k_x^2k_z^2/(k^2V_A^2 + C_s^2k_x^2 - C_s^2k_z^2)$, where V_A is the Alfvén speed and C_s the ion sound speed.

5.5 The Cold Plasma Dielectric Response

To complete the linear theory of the full wave spectrum in plasmas, we need the general dielectric response. It can be derived from the Ampere's law that relates the displacement current to the free current,

$$\nabla \times \vec{B} = \frac{4\pi}{c}\vec{j} + \frac{1}{c}\frac{\partial \vec{E}}{\partial t} = \frac{1}{c}\frac{\partial \vec{D}}{\partial t}. \tag{5.58}$$

In the Fourier components, the electric displacement $\vec{D}$ and the electric field $\vec{E}$ are related by the dielectric tensor $\vec{\vec{\varepsilon}}$,

$$\vec{D}(\omega, \vec{k}) = \vec{\vec{\varepsilon}}(\omega, \vec{k}) \cdot \vec{E}(\omega, \vec{k}) = \vec{E}(\omega, \vec{k}) + \frac{4\pi i}{\omega}\vec{j}(\omega, \vec{k}) = \frac{c\vec{k}}{\omega} \times \vec{B}(\omega, \vec{k}).$$

Defining the conductivity tensor $\overleftrightarrow{\sigma}$, we have $\overleftrightarrow{\varepsilon}(\omega,\vec{k}) = \overleftrightarrow{I} + \sum_s 4\pi i\overleftrightarrow{\sigma}_s(\omega,\vec{k})/\omega$. By summing up the currents from the individual species $\vec{j}_s = n_s q_s \vec{v}_s = \overleftrightarrow{\sigma}_s(\omega,\vec{k})\cdot\vec{E}(\omega,\vec{k})$, the total current is given by $\vec{j} = \sum_s \vec{j}_s$. Here, $\overleftrightarrow{\sigma}_s$ is the conductivity tensor of the individual species. The Faraday's law gives us $\vec{k}\times\vec{E}(\omega,\vec{k}) = \omega\vec{B}(\omega,\vec{k})/c$. Thus, the wave equation can be cast into

$$\vec{n}\times(\vec{n}\times\vec{E}) + \overleftrightarrow{\varepsilon}\cdot\vec{E} = 0, \tag{5.59}$$

where $\vec{n} \equiv c\vec{k}/\omega$ is the refractive index. The remaining task is to evaluate $\overleftrightarrow{\sigma}_s$.

5.5.1 *Linear Plasma Response*

Assuming the magnetic field is uniform and along the $\widehat{e}_z$direction, $\vec{B}_0 = B_0\widehat{e}_z$, we will treat the electrons and ions in the cold plasma model by neglecting the pressure term in Eq. (4.3) or Eq. (5.19). Therefore, the velocity components are given by Eqs. (5.20) and (5.21)

$$\delta v_{\sigma,x} = \frac{q_\sigma}{m_\sigma}(\omega^2 - \Omega_{c\sigma}^2)^{-1}(i\omega\delta E_x - \Omega_{c\sigma}\delta E_y), \tag{5.60}$$

$$\delta v_{\sigma,y} = \frac{q_\sigma}{m_\sigma}(\omega^2 - \Omega_{c\sigma}^2)^{-1}(i\omega\delta E_y + \delta E_x\Omega_{c\sigma}), \tag{5.60a}$$

$$\delta v_{\sigma,z} = \frac{iq_\sigma}{m_\sigma\omega}\delta E_z. \tag{5.60b}$$

The velocity of the $\widehat{e}_z$ component oscillates in the electric field, while the perpendicular velocities give rise to the E × B and polarization drifts in the low frequency limit, but retains the cyclotron resonance in its general form. The conductivity tensor can then be expressed as follows:

$$\delta\vec{J} = \sum_\sigma n_\sigma q_\sigma \delta\vec{v}_\sigma = \overleftrightarrow{\sigma}\cdot\delta\vec{E} = \sum_\sigma \frac{i\omega_{p\sigma}^2}{4\pi}\begin{pmatrix} \dfrac{\omega}{\omega^2-\Omega_{c\sigma}^2} & \dfrac{i\Omega_{c\sigma}}{\omega^2-\Omega_{c\sigma}^2} & 0 \\ \dfrac{-i\Omega_{c\sigma}}{\omega^2-\Omega_{c\sigma}^2} & \dfrac{\omega}{\omega^2-\Omega_{c\sigma}^2} & 0 \\ 0 & 0 & \dfrac{1}{\omega} \end{pmatrix}\begin{pmatrix}\delta E_x \\ \delta E_y \\ \delta E_z\end{pmatrix}. \tag{5.61}$$

Without loss of generality, we will assume $\vec{k} = k_x\widehat{e}_x + k_z\widehat{e}_z$. Equation (5.59) can then be cast into

$$\overleftrightarrow{M}\cdot\vec{E} = \begin{pmatrix} S-n_z^2 & -iD & n_xn_z \\ iD & S-n^2 & 0 \\ n_xn_z & 0 & P-n_x^2 \end{pmatrix}\cdot\begin{pmatrix}E_x \\ E_y \\ E_z\end{pmatrix} = 0, \tag{5.62}$$

where

$$P \equiv 1 - \sum_\sigma \frac{\omega_{p\sigma}^2}{\omega^2}, \tag{5.63}$$

$$R \equiv 1 - \sum_\sigma \frac{\omega_{p\sigma}^2}{\omega(\omega + \Omega_{c\sigma})}, \tag{5.63a}$$

$$L \equiv 1 - \sum_\sigma \frac{\omega_{p\sigma}^2}{\omega(\omega - \Omega_{c\sigma})}, \tag{5.63b}$$

$$S \equiv \frac{1}{2}(R + L) = 1 - \sum_\sigma \frac{\omega_{p\sigma}^2}{\omega^2 - \Omega_{c\sigma}^2}, \tag{5.63c}$$

$$D \equiv \frac{1}{2}(R - L) = \sum_\sigma \frac{\omega_{p\sigma}^2 \Omega_{c\sigma}}{(\omega^2 - \Omega_{c\sigma}^2)\omega}. \tag{5.63d}$$

If $M \cdot E = 0$, for a nontrivial E, we need $M = 0$. If M is a matrix, for nontrivial solutions, the determinant of $\overleftrightarrow{M}$ has to vanish so that the equations are linearly dependent and any value of $\vec{E}$can be supported. Therefore, the dispersion relation is given by det $\overleftrightarrow{M} = 0$.

5.5.2 *Some Properties of the Dispersion Matrix*

[Propagation along the field lines] Consider waves propagating along the field lines. The dispersion relation is much simpler, given by det $\overleftrightarrow{M} = P((S - n_z^2)^2 - D^2) = 0$ from Eq. (5.62). The three roots are: the plasma oscillation $P = 0$, the right-polarized $n_z^2 = R$, and the left-polarized $n_z^2 = L$. The following program makes the contour plots of the index of refraction n_z on the plane of ω_{pe} and ω_{ce} normalized to the wave frequency.

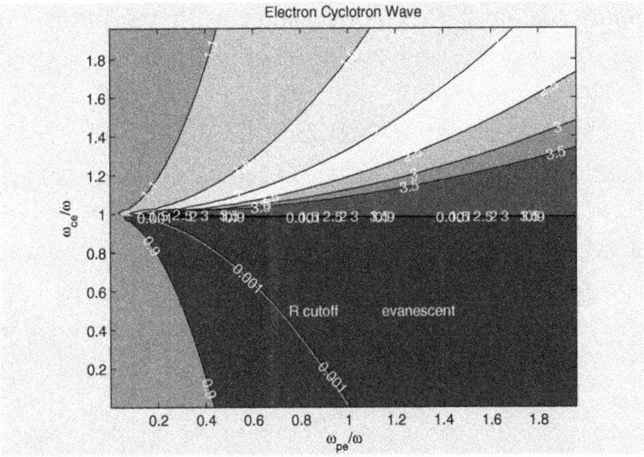

```
function KzElectron
c=1;omega=1;Mi=1837;
N=500; Ndata=1:N; dw=2e-5;
wpe=dw*1.02325.^(Ndata);
wce=dw*1.02325.^(Ndata);
Wpe=repmat(wpe,N,1);
Wpi=Wpe/sqrt(Mi);
Wce=repmat(wce',1,N);
Wci=Wce/Mi;
R=1-Wpe.^2./(1-Wce)-Wpi.^2./(1+Wci);
K2=R;
value=[0.001,0.9,1.1,1.5,2:0.5:3.5];
[c,h]=contourf(Wpe,Wce,sqrt(K2),value);
clabel(c,h,'color','w');
str=sprintf('R cutoff        evanescent');
text(0.75,0.5,str,'color','w');
hold on;
y=1-wpe.^2;
plot(wpe,y,'y-');
xlabel('\omega_p_e/\omega');
ylabel('\omega_c_e/\omega');
title('Electron Cyclotron Wave');
```

% The n_z values are marked on the contour plot.
Near the electron cyclotron resonance n_z varies drastically. The evanescent region is shown in the low field high density side marked by the yellow line.

The electron branch of the right-polarized wave, $n_z^2 \approx 1-\omega_{pe}^2/\omega(\omega-|\Omega_{ce}|)$, has the electron cyclotron resonance at $\omega \sim |\Omega_{ce}|$. It is evanescent in the high density, low field region due to the $R = 0$ cutoff, as traced by $y = 1 - x^2$ in yellow line in the program where $x \equiv \omega_{pe}/\omega$ and $y \equiv \Omega_{ce}/\omega$. In the low frequency range where $\Omega_{ce} > \omega$, the right polarized wave does not resonate with the ions, and also has no density cutoff. It is in fact the whistler wave. By taking $\omega \ll |\Omega_{ce}|$ and consider $n_z \gg 1$ in the high density regime so that $n_z^2 \approx \omega_{pe}^2/\omega|\Omega_{ce}|$ that is Eq. (5.39) at $\theta = 0$.

The left polarized wave does not resonate with the plasma in the electron frequency range, but has the ion cyclotron resonance at $\omega \sim \Omega_{ci}$ in the ion frequency range. It has the $L \approx 1 - \omega_{pe}^2/\omega|\Omega_{ce}| - \omega_{pi}^2/\omega(\omega - \Omega_{ci}) = 0$ cutoff, and thus difficult to access the ion cyclotron resonance layer from the low field side. By defining $x \equiv \omega_{pi}/\omega$ and $y \equiv \Omega_{ci}/\omega$, the cutoff frequency occurs at $x^2 + (\frac{1}{2} - y)^2 = \frac{1}{4}$ which is the semi-circle around the origin as plotted in the KzIon program. At the low frequency range, viz.,$\omega < \Omega_{ci}$, the wave is propagative without a density cutoff. This is in fact the shear Alfvén wave. By taking $\omega \ll \Omega_{ci}$in $n_z^2 = L \approx 1-\omega_{pi}^2/\Omega_{ci}(\omega-\Omega_{ci}) \approx 1+\omega_{pi}^2/\Omega_{ci}^2 \approx c^2/V_A^2$, it recovers the dispersion relation $\omega = k_z V_A$ for the shear Alfvén wave.

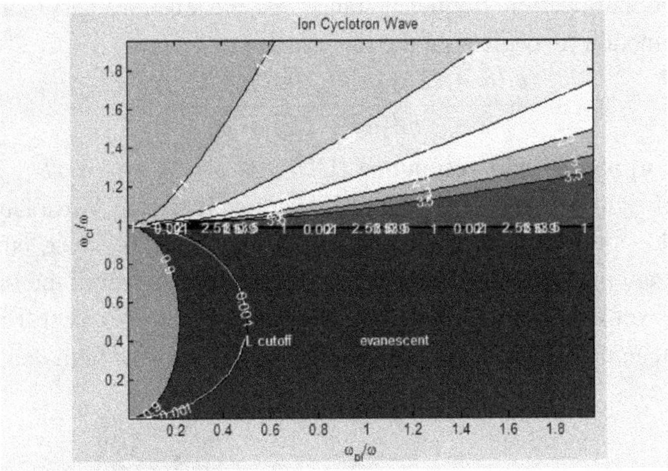


```
function KzIon
c=1;omega=1;Mi=1837;
format short e;
N=250; Ndata=1:N; dw=2e-5;
wpi=dw*1.02325.^(Ndata),
wci=dw*1.02325.^(Ndata);
Wpi=repmat(wpi,N,1);
Wpe=Wpi*sqrt(Mi);
Wci=repmat(wci',1,N);
Wce=Wci*Mi;
L=1-Wpe.^2./(1+Wce)-Wpi.^2./(1-Wci);
K2=L;
value=[0.001,0.9,1,1.1,1.5,2:0.5:3.5];
[c,h]=contourf(Wpi,Wci,sqrt(K2),value);
clabel(c,h,'color','w');
str=sprintf('L cutoff
evanescent');
text(2,0.5,str,'color','w');
hold on;
[x,index]=max(wpi.*(wpi<=1/2));
x=wpi(1:index); y=sqrt(1/4-x.^2);
z=1/2-y; y=1/2+y; plot(x,y,'y-',x,z,'y-');
xlabel('\omega_p_i/\omega');
ylabel('\omega_c_i/\omega');
title('Ion Cyclotron Wave');
hold off;
```

% The n_z values are marked on the contour plot. Near the ion cyclotron resonance, n_z varies drastically. The evanescent region is shown in the low field and high density side, which has a propagative region in the shape of a semi-circle, as indicated by the yellow line.

[Propagation perpendicular to the field lines] At the perpendicular propagation when $n_z = 0$, the cold plasma dispersion function has two roots: the ordinary mode $n_x^2 = P$, and the extraordinary mode $n_x^2 = RL/S$. The ordinary wave has the plasma cutoff at $P = 0$.

Consider the electron frequency range for the extraordinary wave with the dispersion function by neglecting the ion effect,

$$n_\perp^2 \approx \frac{(\omega(\omega+\Omega_{ce})-\omega_{pe}^2)(\omega(\omega-\Omega_{ce})-\omega_{pe}^2)}{\omega^2(\omega^2-\Omega_{ce}^2-\omega_{pe}^2)}, \tag{5.64}$$

that has the **upper hybrid resonance** (UHR) at $\omega^2 \approx \omega_{pe}^2+\Omega_{ce}^2$, but also the $R=0$ cutoff right before the resonance in the low field side. Wave launched from the high field side will have good accessibility to the UHR layer, but due to the narrow separation of cutoff-resonance pair, wave tunneling from the low field side to the UHR layer where mode conversion to electrostatic wave can still be a working scenario. The upper hybrid wave also has the $L=0$ cutoff at high density.

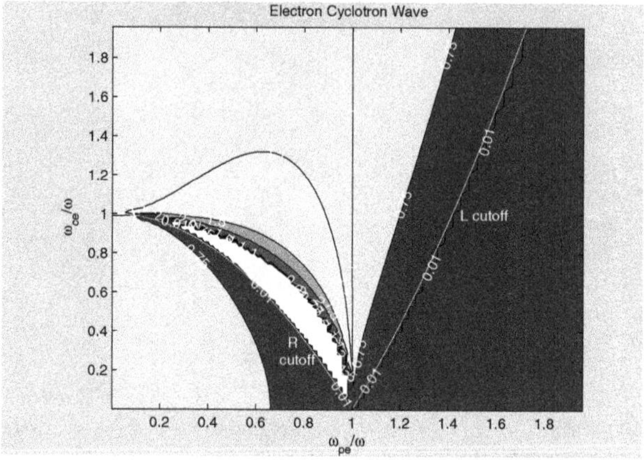

Homework 5.15: Make the contour plot for the extraordinary waves in the electron frequency range by modifying KzElectron.m.

The ion branch of the extraordinary wave has the dispersion function approximated by

$$n_\perp^2 \approx \frac{((\omega+\Omega_{ci})\Omega_{ci}+\omega_{pi}^2)((\omega-\Omega_{ci})\Omega_{ci}-\omega_{pi}^2)}{\left((\omega^2-\Omega_{ci}^2)\Omega_{ci}^2+\omega_{pi}^2\left[\frac{m_e}{m_i}(\omega^2-\Omega_{ci}^2)-\Omega_{ci}^2\right]\right)}. \tag{5.65}$$

The denominator has the lower hybrid resonance at $\omega=\sqrt{\Omega_{ci}^2+\omega_{pi}^2}$ in the low density regime. It well connects to $\omega \sim \sqrt{\Omega_{ci}\Omega_{ce}}$ in the high density (cf. Eq. (5.27)) where $\omega_{pi} \gg \Omega_{ci}$ by balancing the two terms $(m_e/m_i)\omega^2 \approx \Omega_{ci}^2$ in the denominator of Eq. (5.65). The numerical solution of $n_x^2 = RL/S$ is plotted in the following program, KxIon, in the ion frequency range. The lower hybrid resonance (LHR) is marked by the yellow line that traces from $\omega=\sqrt{\Omega_{ci}^2+\omega_{pi}^2}$ to $\omega \sim \sqrt{\Omega_{ci}\Omega_{ce}}$. There is a large evanescent region at the low field side. Thus, to access the lower hybrid resonance layer, it is favorable to launch the wave from the high field side.

```
function KxIon
c=1;omega=1;M=1837;
N=500; Ndata=1:N; dw=2e-5;
wpi=dw*1.02325.^(Ndata);
wci=dw*1.02325.^(Ndata);
Wpi=repmat(wpi,N,1);
Wci=repmat(wci',1,N);
Wpe=Wpi*sqrt(M);
Wce=Wci*M;
L=1-Wpe.^2./(1+Wce)-Wpi.^2./(1-Wci);
R=1-Wpe.^2./(1-Wce)-Wpi.^2./(1+Wci);
S=(R+L)/2;
K2=R.*L./S;
value=[0.1,0.9,0.99,1.01,1.1,1.25,1.5,2:5];
[c,h]=contourf(Wpi,Wci,sqrt(K2),value);
hold on;
x=wpi;
y=(2^(1/2)*((M^4*x.^4 - 2*M^4*x.^2 + M^4
+ 2*M^3*x.^4 + 2*M^3*x.^2 + M^2*x.^4 +
2*M^2*x.^2 - 2*M^2 - 2*M*x.^2 + 1).^(1/2)
- M*x.^2 + M^2 - M^2*x.^2 +
1).^(1/2))/(2*M)
plot(wpi,y,'y-');
clabel(c,h,'color','w');
str=sprintf('L\ncutoff \n\nevanescent');
text(0.5,0.4,str,'color','w');
str=sprintf('LHR');
text(1.3,0.05,str,'color','y');
xlabel('\omega_p_i/\omega');
ylabel('\omega_c_i/\omega');
title('Ion Branch - Extraordinary Mode');
hold off;
```

```
M=1837; syms x y;
%x=wpi/w; y=wci/w;
A=solve('x^2*M/(1-
y^2*M^2)+x^2/(1-y^2)=1','y');

A(1) =(2^(1/2)*((M^4*x^4 -
2*M^4*x^2 + M^4 +
2*M^3*x^4 + 2*M^3*x^2 +
M^2*x^4 + 2*M^2*x^2 -
2*M^2 - 2*M*x^2 + 1)^(1/2) -
M*x^2 + M^2 - M^2*x^2 +
1)^(1/2))/(2*M)
```

% The n_z values are marked on the contour plot. An evanescent region is in the low field and high density side, next to a propagative region that is enclosed by a semi-circle. As traced by the yellow line, the lower hybrid resonance (LHR) is smoothly connected to the ion cyclotron resonance.

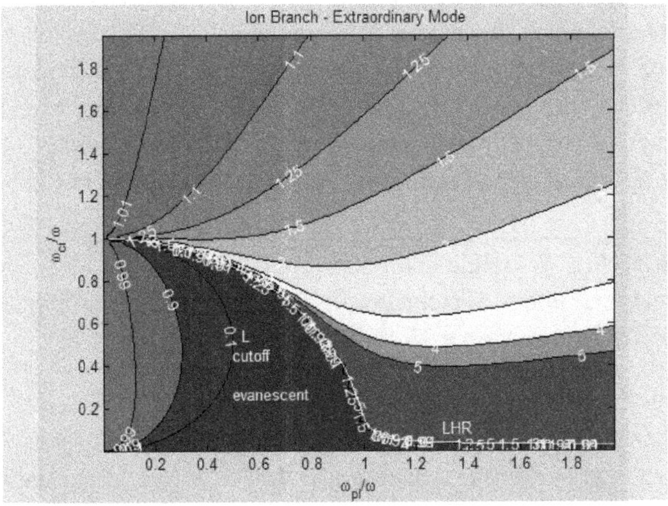

At the low frequency and high density regime, the wave is propagative without a density cutoff. By taking $\omega \ll \Omega_{ci}$ in Eq. (5.65), it gives $n_\perp^2 = RL/S \approx 1 + \omega_{pi}^2/\Omega_{ci}^2 \approx c^2/V_A^2$, and recovers the dispersion relation $\omega = k_\perp V_A$ as a compressional Alfvén wave.

[Oblique Propagation] The dispersion function from Eq. (5.62) for wave propagation in arbitrary directions can be cast into the following:

$$An^4 - Bn^2 + C = 0, \tag{5.66}$$

where $A = S\sin^2\theta + P\cos^2\theta$, $B = RL\sin^2\theta + PS(1+\cos^2\theta)$, and $C = PRL$. The two solutions are $n_\pm^2 = \frac{1}{2}(-B \pm \sqrt{B^2 - 4AC})/A$, referred to as $n_\pm$modes in the following discussions. The cutoffs occur when $C = 0$ at $P = 0$, $R = 0$, or $L = 0$, and the resonance at $A = S\sin^2\theta + P\cos^2\theta = 0$ that has $S = 0$ at $\theta = 90°$. The $P = 0$ at $\theta = 0°$ is not a resonance since it is factored out in Eq. (5.66) to give only the plasma oscillation.

It is instructive to examine waves propagating at $\theta = 45°$, as the two limits $\theta = 0°$ and $\theta = 90°$ have been discussed. In the electron frequency range, the n_+mode, with the $P = 0$ cutoff, is more of an ordinary mode but gains the $R = 0$ cutoff at the high density high field regime. The n_-mode is more of an extraordinary mode with the $R = 0$ cutoff before the upper hybrid resonance (UHR), retains the $L = 0$ cutoff in the high density side, but picks up the $P = 0$ cutoff in the high magnetic field region. The resonances of $S = 0$ at $\theta = 90°$ and $P = 0$ at $\theta = 0°$ are now replaced by $A = \frac{1}{2}S + \frac{1}{2}P = 0$ resonance that has two segments: $x > \sqrt{2}$ and $x < 1$, where $x \equiv \omega_{pe}/\omega$ and $y \equiv \Omega_{ce}/\omega$. The $x > \sqrt{2}$ branch, to be distinguished from the $x < 1$ branch of the UHR, will be denoted as the SP resonance. Both are plotted in yellow line as shown in the figure. It is interesting to note that at $\theta = 45°$, $A = 0$ occurs at $\omega = \Omega_{ce}$ for $\omega_{pe}/\omega \to 0$, but at $\omega = \sqrt{2}\Omega_{ce}$ for $\omega_{pe}/\omega \gg 1$. The SP resonance moves away to the very low frequency regime (upper right corner in the plot) as θ is increased. It moves to the electron cyclotron resonance and the Pcutoff when θ is reduced, and coincides with the $P = 0$ and $\Omega_{ce}/\omega \sim 1$ when $\theta \to 0$. This SP resonance is accessible only from the high field side.

> **Homework 5.16:** Write the computer program to plot the electron branch of the n_-mode. Locate the $A = 0$ contour for arbitrary θ value, and show that $P = 0$ is a cutoff-resonance pair when $\theta \to 0$.

In the ion frequency range, the n_+mode retains the $L = 0$ cutoff, and the ion cyclotron resonance appears to be pushed to the very high density side. Then_-mode, retains both the ion cyclotron resonance and the lower hybrid resonance, but has a small evanescent region at the plasma edge.

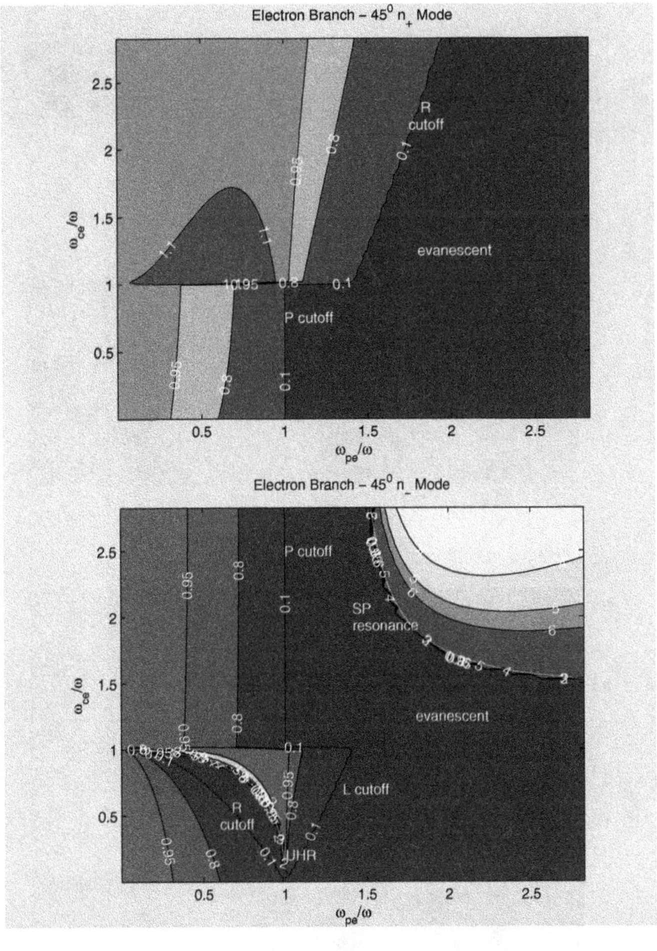

5.6 The Hot Plasma Dielectric Response

The cold plasma model only describes the oscillation frequencies of electrostatic waves. Since $\vec{n} \times \vec{E} = 0$ for electrostatic waves, it implies that $\vec{n} \times (\vec{n} \times \vec{E}) = \overleftrightarrow{\varepsilon} \cdot \vec{E} = 0$ from Eq. (5.59). As a consequence, no propagating electrostatic waves can be recovered, and the coupling of the electrostatic wave to its electromagnetic counterpart is absent. We now retain the electron temperature T_e through the pressure term, but keep the ions cold. We limit ourselves to waves with wavelength longer than the gyroradius, namely, $k\rho_{e,s} < 1$, since the short wavelength modes require kinetic approach to retain the **finite Larmor radius** (**FLR**) effect in terms of Bessel functions. Nonetheless, the current approach will reveal the prominent hot plasma

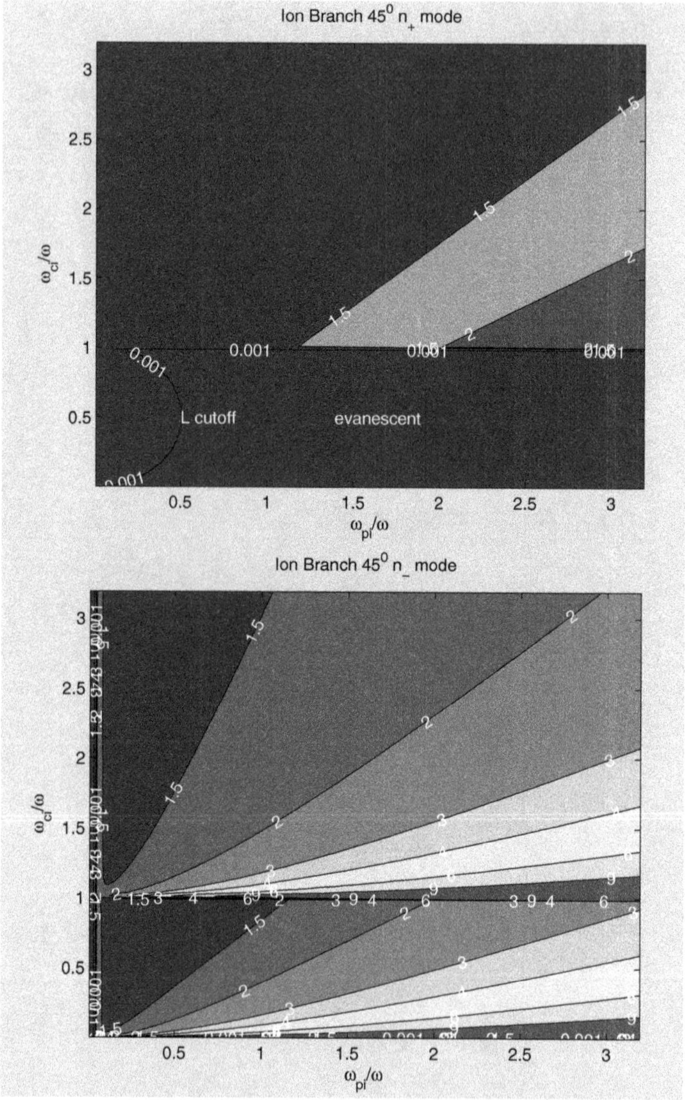
Ion Branch 45⁰ n+ mode
ωci/ω
ωpi/ω
L cutoff
evanescent
0.001
1.5
2
Ion Branch 45⁰ n− mode
ωci/ω
ωpi/ω

effect. Thus,

$$m\frac{\partial\delta\vec{v}_e}{\partial t} = -T_e\nabla\frac{\delta n_e}{n_0} + e\delta\vec{E} + e\frac{\delta\vec{v}_e \times \vec{B}_0}{c}, \tag{5.67}$$

which needs the continuity equation to account for the electron density perturbation,

$$\frac{\partial\delta n_e}{\partial t} + n_0\nabla\cdot\delta\vec{v}_e = 0. \tag{5.68}$$

Both the equilibrium density n_0 and external magnetic field $\vec{B}_0 = B_0\hat{e}_z$ are assumed uniform. Without loss of generality, we take $\vec{k} = k_x\hat{e}_x + k_z\hat{e}_z$. Equation (5.12a) becomes after Fourier transform,

$$-i\omega m\delta\vec{v}_e = -i\vec{k}T_e\frac{\vec{k}\cdot\delta\vec{v}_e}{\omega} + e\delta\vec{E} + e\frac{\delta\vec{v}_e \times \vec{B}_0}{c}. \tag{5.67}$$

Equation (5.12c) yields the following three electron velocities:

$$\delta v_z = \frac{k_z V_T^2}{\omega^2}\vec{k}\cdot\delta\vec{v}_e + i\frac{e\delta E_z}{m\omega} = (1 - n_z^2\beta)^{-1}(n_x n_z\beta\delta v_x + i\varepsilon_z), \tag{5.67a}$$

$$\delta v_y = i\varepsilon_y - i\Omega\delta v_x, \tag{5.67b}$$

$$\begin{aligned}\delta v_x &= i\varepsilon_x + n_x\beta\vec{n}\cdot\delta\vec{v}_e + i\Omega\delta v_y\\ &= (1 - \Omega^2 - n_x^2\beta)^{-1} \times (n_x n_z\beta\delta v_z - \Omega\varepsilon_y + i\varepsilon_x),\end{aligned} \tag{5.67c}$$

where we have defined $\beta \equiv V_T^2/c^2 \equiv T_e/m_e c^2$, $\vec{\varepsilon} \equiv e\delta\vec{E}/m\omega$, and $\Omega \equiv \Omega_{ce}/\omega$. After some more algebra, we find the three velocity components are given by,

$$\delta v_x = \frac{(i\varepsilon_x - \Omega\varepsilon_y)(1 - n_z^2\beta) + in_x n_z\beta\varepsilon_z}{1 - \Omega^2 - n^2\beta + n_z^2\beta\Omega^2}, \tag{5.68}$$

$$\delta v_y = \frac{i\varepsilon_y(1 - n^2\beta) + \Omega\varepsilon_x(1 - n_z^2\beta) + \Omega n_x n_z\beta\varepsilon_z}{1 - \Omega^2 - n^2\beta + n_z^2\beta\Omega^2}, \tag{5.68a}$$

$$\delta v_z = \frac{i\varepsilon_z(1 - \Omega^2 - n_x^2\beta) + n_z n_x\beta(i\varepsilon_x - \Omega\varepsilon_y)}{1 - \Omega^2 - n^2\beta + n_z^2\beta\Omega^2}. \tag{5.68b}$$

Homework 5.17: Prove that Eqs. (5.68) are the solutions to Eq. (5.67).

Following the same derivations as Eqs. (5.61) and (5.62), the current density $\delta\vec{J}$, and the conductivity tensor $\overset{=}{\sigma}$ can be found. The matrix $\overset{=}{M}$ can be cast into

$$\overleftrightarrow{M} = \begin{pmatrix} 1 - \frac{\omega_{pi}^2}{\omega^2 - \Omega_{ci}^2} - \frac{\omega_{pe}^2}{\omega^2} \frac{1 - n_z^2 \beta}{1 - \Omega^2 - n^2 \beta + \Omega^2 n_z^2 \beta} - n_z^2 & -i \frac{\omega_{pi}^2}{\omega^2 - \Omega_{ci}^2} \frac{\Omega_{ci}}{\omega} - i \frac{\omega_{pe}^2}{\omega^2} \frac{\Omega (1 - n_z^2 \beta)}{1 - \Omega^2 - n^2 \beta + \Omega^2 n_z^2 \beta} & n_z n_x - \frac{\omega_{pe}^2}{\omega^2} \frac{n_z n_x \beta}{1 - \Omega^2 - n^2 \beta + \Omega^2 n_z^2 \beta} \\ i \frac{\omega_{pi}^2}{\omega^2 - \Omega_{ci}^2} \frac{\Omega_{ci}}{\omega} + i \frac{\omega_{pe}^2}{\omega^2} \frac{\Omega (1 - n_z^2 \beta)}{1 - \Omega^2 - n^2 \beta + \Omega^2 n_z^2 \beta} & 1 - \frac{\omega_{pi}^2}{\omega^2 - \Omega_{ci}^2} - \frac{\omega_{pe}^2}{\omega^2} \frac{1 - n^2 \beta}{1 - \Omega^2 - n^2 \beta + \Omega^2 n_z^2 \beta} - n^2 & i \frac{\omega_{pe}^2}{\omega^2} \frac{n_z n_x \beta \Omega}{1 - \Omega^2 - n^2 \beta + \Omega^2 n_z^2 \beta} \\ n_z n_x - \frac{\omega_{pe}^2}{\omega^2} \frac{n_z n_x \beta}{1 - \Omega^2 - n^2 \beta + \Omega^2 n_z^2 \beta} & -i \frac{\omega_{pe}^2}{\omega^2} \frac{n_z n_x \beta \Omega}{1 - \Omega^2 - n^2 \beta + \Omega^2 n_z^2 \beta} & 1 - \frac{\omega_{pi}^2}{\omega^2} - \frac{\omega_{pe}^2}{\omega^2} \frac{(1 - \Omega^2 - n_x^2 \beta)}{1 - \Omega^2 - n^2 \beta + \Omega^2 n_z^2 \beta} - n_x^2 \end{pmatrix} \tag{5.69}$$

5.6.1 *The Hot Plasma Electromagnetic Waves*

[Propagation along the field lines] For the ES wave propagates along the field lines with $n_x = 0$, the $P = 0$ plasma oscillation is modified into a propagating wave but has a cutoff at the plasma frequency:

$$1 - \frac{\omega_{pi}^2}{\omega^2} - \frac{\omega_{pe}^2}{\omega^2}\frac{1}{1 - n_z^2\beta} = 0. \tag{5.70}$$

The EM waves of both the right polarized $n_z^2 = R$ and the left polarized waves $n_z^2 = L$ are not modified by the finite electron temperature, since the β factor is completely eliminated.

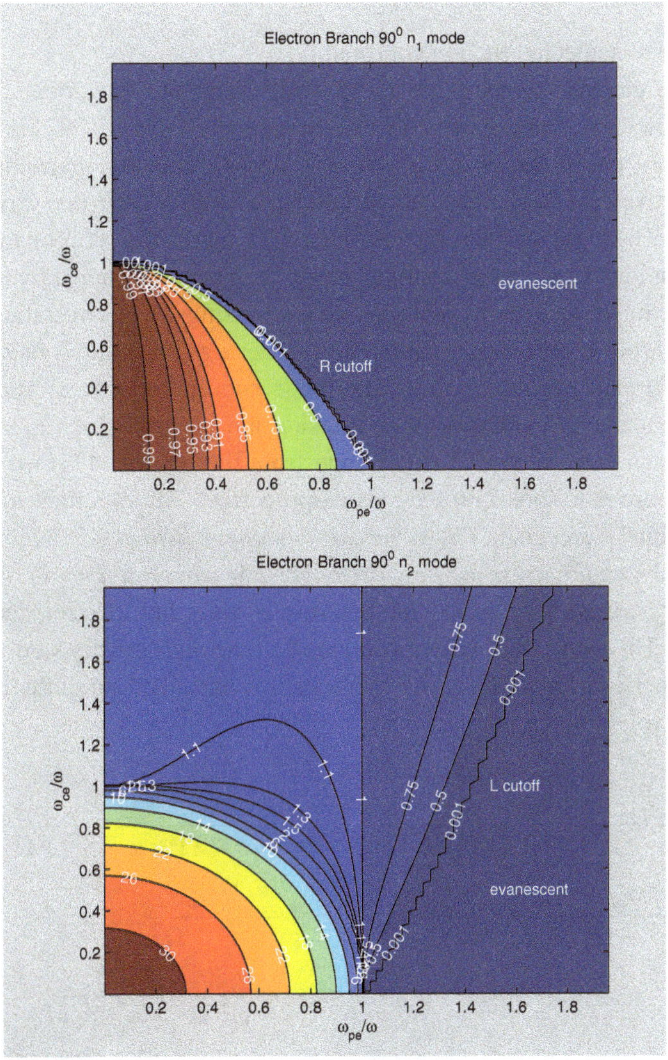

[Propagation perpendicular to the field lines] When the EM wave propagates in the perpendicular direction to the magnetic fields with $n_z = 0$, the ordinary mode is not modified by the hot plasma effect since the β factor is completely eliminated in the ε_{zz} term. The electron branch of the extraordinary mode has the following dispersion relation,

$$\det \begin{pmatrix} 1 - \dfrac{\omega_{pe}^2}{\omega^2} \dfrac{1}{1 - \Omega^2 - n_x^2\beta} & -i\dfrac{\omega_{pe}^2}{\omega^2} \dfrac{\Omega}{1 - \Omega^2 - n_x^2\beta} \\ i\dfrac{\omega_{pe}^2}{\omega^2} \dfrac{\Omega}{1 - \Omega^2 - n_x^2\beta} & 1 - \dfrac{\omega_{pe}^2}{\omega^2} \dfrac{1 - n_x^2\beta}{1 - \Omega^2 - n_x^2\beta} - n_x^2 \end{pmatrix} = 0, \qquad (5.71)$$

where we have neglected the ion contribution,

We take electron temperature to be 5 KeV, and the numerical result shows there are now two branches of the electromagnetic waves. The left plot in the accompanying figure shows that in the low density and low magnetic field side, the propagative EM wave has the R cutoff, beyond which the wave is totally evanescent. The right plot shows the other EM wave, that is propagative in the high field but low density side, continues to travel beyond the upper hydrid resoance (UHR) with much shorter wavelength, characterizing an electrostatic wave. This shows *the EM wave arriving at the UHR from the high field side is mode converted to the electrostatic cyclotron wave.* The mode conversion process is reversible in the sense that the ES cyclotron waves inside the plasma when travelling to the UHR layer can convert to EM and escape from the plasma. This may serve as a noninvasive probe to look into the plasma properties. An important implication is for the earthquake precursor. *Given the earth magnetic field at 0.25~0.65 Gauss, the tectonic plates under stress may emit stronger EM waves at 4.4~11.4 MHz range of electron cyclotron frequency.* Note that this assumes low electron density so that $\Omega_{ce} \gg \omega_{pe}$. This wave does have the L cutoff at the high density side,

The dispersion function for the ion branch is expanded up to the first order in m_e/m_i to give,

$$\begin{pmatrix} 1 - \dfrac{\omega_{pi}^2}{\omega^2 - \Omega_{ci}^2} + \dfrac{\omega_{pi}^2}{\omega^2} \dfrac{1}{\Omega_{ci}^2 M + k_x^2 C_s^2} & -i\dfrac{\omega_{pi}^2}{\omega^2 - \Omega_{ci}^2} \dfrac{\Omega_{ci}}{\omega} - i\dfrac{\omega_{pi}^2}{\omega^2 \Omega_{ci}} \\ i\dfrac{\omega_{pi}^2}{\omega^2 - \Omega_{ci}^2} \dfrac{\Omega_{ci}}{\omega} + i\dfrac{\omega_{pi}^2}{\omega^2 \Omega_{ci}} & \begin{array}{l} 1 - \dfrac{\omega_{pi}^2}{\omega^2 - \Omega_{ci}^2} \\ -\dfrac{\omega_{pi}^2}{\omega^2} \dfrac{k_x^2 C_s^2}{\Omega^2 + k_x^2 C_s^2 / M} - n_x^2 \end{array} \end{pmatrix} = 0. \qquad (5.72)$$

The following plot at the top is the cold plasma solution, and at the bottom the hot plasma. The only modification comes from the finite Larmor radius correction due to the thermal effect. It makes little difference to the index of refraction, and there is no additional wave branch in the perpendicular propagation for the ion waves.

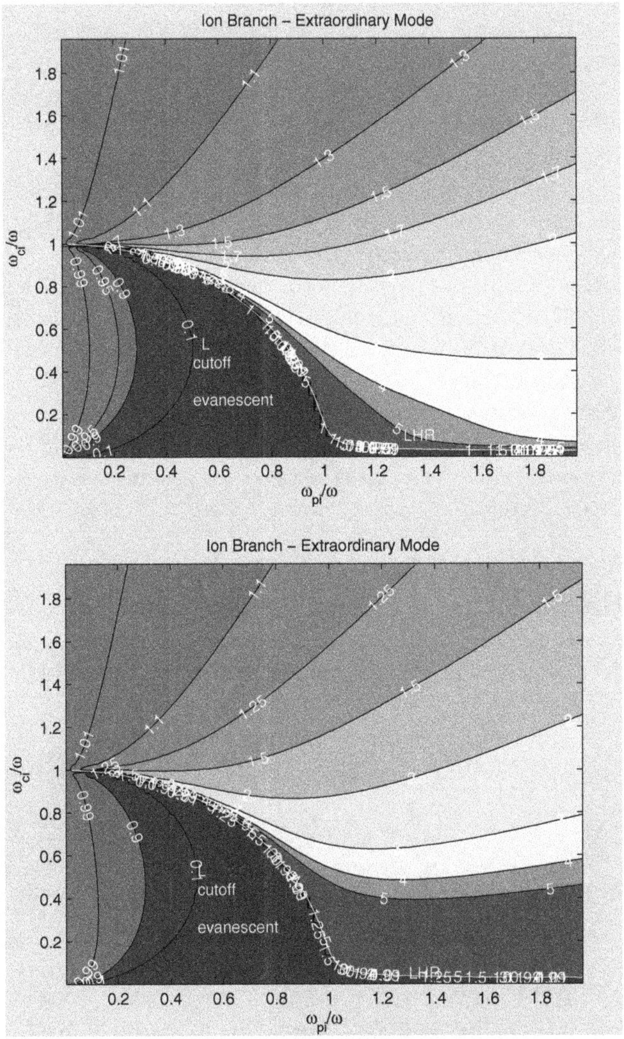

Further Reading

A few good textbooks on plasma waves are readily available. To gain more physical understanding, F. Chen's book is a good start. For more elaborate formulation, consult books by Stix or Swanson.

The original paper by H. Alfven on the "Existence of Electromagnetic-hydrodynamic Waves" appears in Nature (1942). Mode conversion was first recognized in the paper by Stix (1965).

Relativistic formulation of electron cyclotron wave was developed by Matsuda *et al.* (1990). For literatures in the **electron cyclotron resonance heating (ECRH)**, check into the review article on Electron cyclotron resonance heating and current drive in toroidal fusion plasmas by Erckmann and Gasparino (1994). A summary report on ECRH experiments can be found in Llyod (1998). Yoshimura *et al.* (2013) reported the ECRH via electron Bernstein wave in **Large Helical Device (LHD)**.

For the **ion cyclotron resonance heating (ICRH)**, a good review on "Heating tokamaks via the ion-cyclotron and ion-ion hybrid resonances" by Perkins (1977). Mode conversion to ion Bernstein wave in the mixed species plasma was reported by Lee *et al.* (1983).

Momentum and energy flux transported by waves was included in the fluid equations and applied to the solar atmosphere and solar wind (Jacques 1977). While wave heating is to ignite plasma, it may also modify the transport property (Hsu *et al.*, 1984).

Large-scale resonant modification of the polar ionosphere by electromagnetic waves was reported (Wong *et al.*, 1989). It is interesting to note that ionospheric precursors before strong earthquakes were noticed (Pulinets 2003). Liu *et al.* (2000) reported the seismo-ionospheric signatures prior to M > 6 Taiwan earthquakes.

For the wave and beam interaction, Liu and Tripathi, (1994) had a summary review. Experimentalists tend to believe that magnetic or velocity shear is a stabilizing effect on instabilities and turbulences (Burrell 1997). Waves given whatever configurations, however, may degenerate into more complex forms.

Homework Hints

Homework 5.1: Carry out the description of the electromagnetic wave that is modified by the plasma.

To describe the electromagnetic wave, we start from the Maxwell equations,

$$\nabla \times \vec{E} = -\frac{1}{c}\frac{\partial \vec{B}}{\partial t}, \quad \text{and} \quad \nabla \times \vec{B} = \frac{4\pi}{c}\vec{j} + \frac{1}{c}\frac{\partial \vec{E}}{\partial t}. \tag{5.1.1}$$

Eliminating the magnetic field from the two equations, we end with

$$\frac{\partial^2 \vec{E}}{\partial t^2} + 4\pi ne\frac{\partial \vec{v}}{\partial t} + c^2\nabla \times (\nabla \times \vec{E}) = 0, \tag{5.1.2}$$

where the current is given by $\vec{j} = ne\vec{v}$. Replacing the velocity by $\partial\vec{v}/\partial t = e\vec{E}/m$, we find

$$\frac{\partial^2 \vec{E}}{\partial t^2} + \omega_{pe}^2\vec{E} + c^2\nabla \times (\nabla \times \vec{E}) = 0. \tag{5.1.3}$$

Taking the Fourier transform gives

$$\vec{n} \times (\vec{n} \times \vec{E}) + (1 - \frac{\omega_{pe}^2}{\omega^2})\vec{E} = 0, \tag{5.1.4}$$

where $\vec{n} = c\vec{k}/\omega$ is the index of refraction. Note that Eq. (5.1.4) gives $\vec{n} \cdot \vec{E} = 0$. Therefore,

$$-n^2\vec{E} + (1 - \frac{\omega_{pe}^2}{\omega^2})\vec{E} = 0, \tag{5.1.5}$$

which has the dispersion relation, $\omega^2 = \omega_{pe}^2 + k^2c^2$.

Homework 5.3: Show that the ion cyclotron wave governed by Eq. (5.18) has a vanishing group velocity as the wave approaches the ion cyclotron resonance layer, where $\omega = \Omega_{ci}$.

Take the partial derivative (5.18) with respect to k gives the group velocity

$$v_g = \frac{\partial\omega}{\partial k} = \frac{kC_s^2}{\omega(1 + k^2\lambda_D^2)^2}. \tag{5.3.1}$$

When $\omega \to \Omega_{ci}$, the wave vector vanishes, $k \to 0$. Therefore, $v_g \to 0$.

Homework 5.5: Derive the electromagnetic branch of the electron cyclotron waves by considering the current perturbations instead of the density perturbations.

Starting from Eq. (5.1.1), we find

$$\vec{k} \times (\vec{k} \times \delta\vec{E}) = \frac{\omega}{c}\left(\frac{4\pi}{ic}\delta\vec{J} - \frac{\omega}{c}\delta\vec{E}\right), \tag{5.5.1}$$

where the current perturbation is given by $\delta\vec{J} = ne\delta\vec{v}$. Adding Eq. (5.20) with the parallel motion we have

$$\delta\vec{v} = \frac{e}{m}\frac{i\omega\delta\vec{E}_\perp + \vec{\Omega}_{ce} \times \delta\vec{E}}{\omega^2 - \Omega_{ce}^2} + \frac{e}{m}\frac{i\delta\vec{E}_\parallel}{\omega}. \tag{5.5.2}$$

Eliminating $\delta\vec{J}$ in favor of $\delta\vec{E}$ gives

$$\vec{n} \times (\vec{n} \times \delta\vec{E}) + \delta\vec{E} = -\frac{\omega_{pe}^2}{\omega}\left(\frac{\omega\delta\vec{E}_\perp - i\vec{\Omega}_{ce} \times \delta\vec{E}_\perp}{\omega^2 - \Omega_{ce}^2}\right) + \frac{\omega_{pe}^2}{\omega^2}\delta\vec{E}_\parallel. \tag{5.5.3}$$

Writing in the matrix form, Eq. (5.4.3) gives

$$\begin{pmatrix} 1 - n_z^2 - \dfrac{\omega_{pe}^2}{\omega^2 - \Omega_{ce}^2} & i\dfrac{\omega_{pe}^2}{\omega^2 - \Omega_{ce}^2}\dfrac{\Omega_{ce}}{\omega} & n_x n_z \\ -i\dfrac{\omega_{pe}^2}{\omega^2 - \Omega_{ce}^2}\dfrac{\Omega_{ce}}{\omega} & 1 - n^2 - \dfrac{\omega_{pe}^2}{\omega^2 - \Omega_{ce}^2} & 0 \\ n_x n_z & 0 & 1 - n_x^2 - \dfrac{\omega_{pe}^2}{\omega^2} \end{pmatrix} \begin{pmatrix} \delta E_x \\ \delta E_y \\ \delta E_z \end{pmatrix} = 0 \tag{5.5.4}$$

Consider the waves with $\delta E_z = 0$ and $n_z = 0$. Both the wave vector and the electric field are perpendicular to the magnetic field. This is the **extraordinary mode** with the following dispersion function:

$$\left[1 - \frac{\omega_{pe}^2}{\omega^2 - \Omega_{ce}^2}\right]\left[1 - n^2 - \frac{\omega_{pe}^2}{\omega^2 - \Omega_{ce}^2}\right] - \left[\frac{\omega_{pe}^2}{\omega^2 - \Omega_{ce}^2}\frac{\Omega_{ce}}{\omega}\right]^2 = 0. \qquad (5.5.5)$$

One other mode with $n_z = 0$ and both $\delta E_x = \delta E_y = 0$, has the dispersion relation given by

$$1 - n_x^2 - \frac{\omega_{pe}^2}{\omega^2} = 0, \qquad (5.5.6)$$

which is independent of the magnetic field. It is often termed as the **ordinary mode** since its electric field is along the field lines, and will not cause the particles to execute any magnetic field dependent motions such as ExB drift or polarization drift. It is a density modified EM wave that happens to have the frequency near the electron cyclotron frequency. It will reach the resonance layer without encountering a cutoff layer, where $k \to 0$, unlike the extraoridinary wave. Moreover, it will still heat electrons at $\omega = \Omega_{ce}$ since the electric field would appear as a dc field to the particles, while the wavelength in the x-direction tends to be relatively long in comparison with the electron gyroradius, unless it reaches the plasma density cutoff, $\omega = \Omega_{pe}$.

Homework 5.7: Show that there is an equipartition between the fluid kinetic energy and the magnetic field energy in the shear Alfvén wave.

After Fourier transform, Eq. (5.21a) gives $\delta\vec{u} = -k_{||}B_0\delta\vec{b}/(4\pi\omega\rho_0)$, and Eq. (5.22a) gives $\delta\vec{b} = -k_{||}B_0\delta\vec{u}/\omega$. Multiplying these two equations together yields,

$$\frac{k_{||}}{\omega}B_0|\delta\vec{u}|^2 = \frac{1}{4\pi}\frac{k_{||}}{\omega}\frac{B_0}{\rho_0}|\delta\vec{b}|^2,$$

which is simplified to,

$$\frac{1}{2}\rho_0|\delta\vec{u}|^2 = \frac{1}{8\pi}|\delta\vec{b}|^2. \qquad (5.7.1)$$

Homework 5.9: Assume the earth magnetic field is 0.3 Gauss, and nitrogen is the major ion species in the air. Show that the whistler wave has a frequency in the KHz range.

The electron cyclotron frequency is given by $f_{ce} = 2.8e^6\,\mathrm{BHz} \approx 8.4e^5\,\mathrm{Hz}$, and the ion cyclotron frequency, $f_{ci} = 1.5e^3 B/(m_N/m_H)\,\mathrm{Hz} \approx 32\,\mathrm{Hz}$. Here m_N is the nitrogen atomic mass and $m_H$is the hydrogen atomic mass. B is the magnetic field in the unit of gauss. Therefore, the whistler frequency is given by

$$f_{whistler} = \sqrt{f_{ci} f_{ce}} \approx 5\,\mathrm{KHz}.$$

Homework 5.11: Derive the dispersion relation for the Alfvén wave as modified by the drift wave characteristics.

By treating the ions as cold fluid, the ions execute the ExB and polarization drifts perpendicular to the magnetic field, and simple oscillations along the field lines. Therefore, the ion density perturbation is given by Eq. (5.41). Any electric field along the field line will be balanced out by the redistribution of electron pressure. Thus, by imposing the charge neutrality condition, the electric potentials are given by

$$\delta\psi = \frac{\vec{k}\cdot\vec{V}_D}{\omega}\delta\phi - k_\perp^2\rho_s^2\delta\phi + \frac{k_{||}^2 C_s^2}{\omega^2}\delta\psi. \tag{5.42a}$$

Here, $\delta\vec{E}_\perp \equiv -\nabla_\perp\delta\phi$, and $\delta\vec{E}_{||} \equiv -\nabla_{||}\delta\psi$ to accommodate the Faraday's law where $\delta\vec{b}_\perp = c\vec{k}\times\vec{k}_{||}(\delta\psi - \delta\phi)/i\omega$. It is noteworthy that the vector potential is given by $\delta\vec{A} = c\vec{k}_{||}(\delta\phi - \delta\psi)/\omega$. We take the density gradient along the $\hat{e}_x$ direction, the equilibrium magnetic field along the $\hat{e}_z$ direction, and $\vec{k} = k\hat{e}_y$. The charge neutrality at this low frequency ensures $\nabla\cdot\delta\vec{E} \approx 0$. Therefore, $\nabla\cdot\delta\vec{j} = 0$ by taking divergence on the Ampere's law,

$$\nabla\cdot(\nabla\times\delta\vec{b}_\perp) = \frac{1}{c}\nabla\cdot\left(4\pi\delta\vec{j} + \frac{\partial\delta\vec{E}}{\partial t}\right) = \frac{4\pi}{c}\nabla\cdot\delta\vec{j} = 0.$$

The parallel current is given from the Ampere's law,

$$\begin{aligned}\delta\vec{j}_{||} &= \frac{c^2}{4\pi\omega}\hat{e}_{||}\hat{e}_{||}\cdot\vec{k}\times(\vec{k}\times\vec{k}_{||})(\delta\psi - \delta\phi) + \frac{\omega k_{||}}{4\pi}\delta\psi \\ &= -\frac{c^2}{4\pi\omega}\vec{k}_{||}k_\perp^2(\delta\psi - \delta\phi) + \frac{\omega k_{||}}{4\pi}\delta\psi.\end{aligned}$$

The perpendicular current can be found by noting that the electrons have the ExB and diagmagnetic drifts. The ExB drifts of electrons and ions do not produce net current. Therefore,

$$\delta\vec{j}_\perp = n_0 q_i c\delta\dot{\vec{E}}_\perp/B\Omega_i + n_0 ec(\delta\vec{b}_\perp\times\nabla n_0 T_e + \vec{B}\times\nabla\delta n T_e)/B^2.$$

Only the first term of ion polarization drift survives in contributing to $\nabla \cdot \delta \vec{j}_\perp$. Thus,

$$-\frac{c^2}{4\pi \omega} k_{||}^2 k_\perp^2 (\delta\psi - \delta\phi) + \frac{\omega k_{||}^2}{4\pi}\delta\psi - \frac{\omega k_\perp^2 n_0 q_i c}{B\Omega_i}\delta\phi = 0.$$

It can be simplified into

$$\left(1 - \frac{k_{||}^2 C_s^2}{\omega^2}\right)\delta\psi = \left(\frac{\omega *}{\omega} - k_\perp^2 \rho_s^2\right)\delta\phi.$$

Therefore, the dispersion relation is found,

$$\left(1 - \frac{\omega^2}{k_{||}^2 V_A^2}\right)\left(1 - \frac{k_{||}^2 C_s^2}{\omega^2}\right) = \left(\frac{\omega *}{\omega} - k_\perp^2 \rho_s^2\right)\left(1 - \frac{\omega^2}{k_\perp^2 c^2}\right).$$

There are four types of waves in this dispersion function: the shear Alfvén wave ($\omega \approx k_{||} V_A$), the ion sound wave ($\omega \approx k_{||} C_s$), the drift wave ($\omega \approx \omega *$), and the electromagnetic wave ($\omega \approx k_\perp c$). The drift Alfvén wave is given by

$$\omega^2 \approx k_{||}^2 V_A^2 \left[1 + \left(\frac{\omega *}{k_{||} V_A} - k_\perp^2 \rho_s^2\right)\left(1 - \frac{k_{||}^2 V_A^2}{k_\perp^2 c^2}\right)\left(1 - \frac{C_s^2}{V_A^2}\right)^{-1}\right]$$

$$\to k_{||}^2 V_A^2 + \omega * k_{||} V_A, \qquad (5.43)$$

where for the drift Alfvén wave, it requires to order the frequency of interest lower than EM wave, $k_\perp c \gg \omega \sim k_{||} V_A \sim \omega *$, and small FLR effect, $1 \gg k_\perp \rho_s$.

Homework 5.13: The Rayleigh-Taylor instability has a growth rate given by $\gamma = \sqrt{gLk_z^2/(k_x^2 L^2 + k_y^2 L^2 + \frac{1}{4})}$, where L is the density scale length, g is the gravitational acceleration along the z direction, and k_x and k_y are the wave numbers in the plane perpendicular to the gravitational force. Suppose an additional effect is found to give $\gamma_A = \sqrt{gLk_z^2/(k_x^2 L^2 + k_y^2 L^2 + \frac{1}{4}) - k_x^2 L^2 A}$. Does this imply that the RTI can be stabilized by A?

The answer is no since the growth rate remains the same for modes with $k_x = 0$. Defining $\gamma_0 \equiv \sqrt{g/L}$, $\bar{\gamma} \equiv \gamma/\gamma_0$, and $\bar{k}_s \equiv k_s L$ for $s = x, y, z$, we have $\bar{\gamma} = \sqrt{\bar{k}_z^2/(\bar{k}_\perp^2 + \frac{1}{4})}$, and $\bar{\gamma}_A = \sqrt{\bar{k}_z^2/(\bar{k}_\perp^2 + \frac{1}{4}) - \bar{k}_x^2 A}$, where $\bar{k}_\perp^2 \equiv \bar{k}_x^2 + \bar{k}_y^2$. For modes with the same $\bar{k}_z$ and $\bar{k}_\perp$, the growth rate remains the same. Thus, the A effect is immaterial so long as the modes rotate their wave vector to the y direction in the plane perpendicular to the gravitational force.

Homework 5.15: Make the contour plot for the extraordinary waves in the electron frequency range by modifying the codes of KzElectron.m.

```
function KxElectron
c=1;omega=1;M=1837;
N=500; Ndata=1:N; dw=2e-5;
wpe=dw*1.02325.^(Ndata);
wce=dw*1.02325.^(Ndata);
Wpe=repmat(wpe,N,1);
Wpi=Wpe/sqrt(M);
Wce=repmat(wce',1,N);
Wci=Wce/M;
R=1-Wpe.^2./(1-Wce)-Wpi.^2./(1+Wci);
L=1-Wpe.^2./(1+Wce)-Wpi.^2./(1-Wci);
S=(R+L)/2;
K2=R.*L./S;
value=[0.01,0.75,1,1.1,1.5,2];
[c,h]=contourf(Wpe,Wce,sqrt(K2),value);
clabel(c,h,'color','w');
str=sprintf('evanescent');
text(2,0.5,str,'color','w');
text(1.45,1,'L cutoff','color','w');
str=sprintf('  R\ncutoff');
text(0.7,0.3,str,'color','w');
xlabel('\omega_p_e/\omega');
ylabel('\omega_c_e/\omega');
title('Electron Cyclotron Wave');
hold on;
[x,index]=max(wpe.*(wpe<=1));
x=wpe(1:index);
y=(1-x.^2); plot(x,y,'y-');
x=wpe(index+1:N);
z=(x.^2-1); plot(x,z,'y-');
hold off;
```

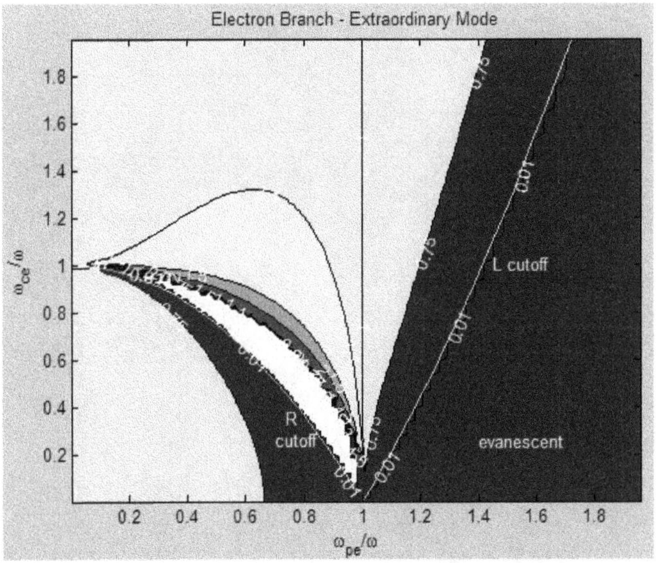

Homework 5.17: Prove that Eqs. (5.68) are the solutions to Eq. (5.67).

We take the solutions of Eqs. (5.68) and substitute into Eqs. (5.67) and prove that every equation is satisfied.

```
function WarmPlasmaVelocities
syms Ex Ey Ez nx nz Beta OMEGA;
n2=nx^2+nz^2;
denominator=1- OMEGA^2-n2*Beta+ OMEGA^2*nz^2*Beta;
Vx=(i*Ex-OMEGA*Ey)*(1-nz^2*Beta)+i*nx*nz*Beta*Ez;
Vy=i*Ey*(1-n2*Beta)+OMEGA*Ex*(1-
nz^2*Beta)+OMEGA*nx*nz*Beta*Ez;
Vz=i*Ez*(1-OMEGA^2-nx^2*Beta) +nx*nz*Beta*(i*Ex-
OMEGA*Ey);
Vx=Vx/denominator;
Vy=Vy/denominator;
Vz=Vz/denominator;
Eq5p67a=Vz-i*Ez-nz*Beta*(nx*Vx+nz*Vz);
Eq5p67b=Vy-i*Ey+i*OMEGA*Vx;
Eq5p67c=Vx-i*Ex-i*OMEGA*Vy-nx*Beta*(nx*Vx+nz*Vz);
Eqa=simplify(Eq5p67a),
Eqb=simplify(Eq5p67b),
Eqc=simplify(Eq5p67c),
```

```
Eqa =

0

Eqb =

0

Eqc =

0
```

Chapter Six

Analytical Chapter

"I must study politics and war that my sons may have liberty to study mathematics and philosophy."

— John Adams (1735–1826), Second President of the USA

Plasma parameters are very wide ranged. The electron density can be as low as 10^{-5}/cc in the interstellar, and as high as 10^{25}/cc in the center of the sun. The temperature could be as low as 10^{-2} eV, and as high as 10^{5} eV. Moreover, the mass disparity in the electron-to-ion ratio makes the characteristic frequencies and lengths vary by many orders of magnitude. Plasma physics thus offers great challenges in the uncompromising time and length scales for analysis or simulation.

Beyond the complexity of simultaneous and short range space-time interactions, many plasma phenomena are often of cumulative and long range effects. The lasting influence through space and time should be recognized as such and analyzed accordingly. Mixing long time scale with short time scale, or similarly, mixing the long range with boundary layer will only make the analysis difficult and the physics unclear. The proper representation would make the serial or asymptotic expansions rapidly convergent, and the incurred error vastly reduced.

6.1 Multiple Time Scales

The multiple time methodology tackling the temporal evolution with the mathematical rigorousness and the physical insightfulness is by itself an exercise for understanding a complex system. By making the single time coordinate τ into several orthogonal coordinates, $(\tau_1, \tau_2, \tau_3, \ldots)$, we would be able to describe events on different time scales that are separated through the orderings of variable sizes with respect to an expansion parameter $\varepsilon \ll 1$. Therefore, defining

$$\frac{d\tau_0}{d\tau} = 1, \quad \frac{d\tau_1}{d\tau} = \varepsilon, \quad \frac{d\tau_2}{d\tau} = \varepsilon^2, \ldots, \tag{6.1}$$

we treat the time variables $\tau_0, \tau_1, \tau_2, \ldots$ as independent. The time differentiation follows the partial derivative chain rule.

$$\frac{d}{d\tau} = \frac{\partial}{\partial\tau_0} + \varepsilon\frac{\partial}{\partial\tau_1} + \varepsilon^2\frac{\partial}{\partial\tau_2} + \cdots \tag{6.2}$$

And the second order time derivative is expressed as

$$\frac{d^2}{d\tau^2} = \frac{\partial^2}{\partial\tau_0^2} + 2\varepsilon\frac{\partial^2}{\partial\tau_0\partial\tau_1} + \varepsilon^2\left[\frac{\partial^2}{\partial\tau_1^2} + 2\frac{\partial^2}{\partial\tau_0\partial\tau_2}\right] + \cdots \tag{6.3}$$

We may formally expand the variable of interest into

$$x \approx x^{(0)}(\tau_0, \tau_1, \ldots) + \varepsilon x^{(1)}(\tau_0, \tau_1, \ldots) + \varepsilon^2 x^{(2)}(\tau_0, \tau_1, \ldots) + \cdots \tag{6.4}$$

6.1.1 *The Damped Oscillator*

```
function fcubic                    %Solve the equation
x=0; v=1;                          d2x/dt2+x+e*(dx/dt)^3=0
epsilon=0.025;                     %v=dx/dt; %dv/dt=-e *v^3-x
dt=0.0125;
N=2000000;
Idata=500;
X=(1:N/Idata)*0;
j=0; T=[];
for i=1:N
   t=i*dt;
   x=x+v*dt;
   v=v-x*dt-epsilon*v^3*dt;
   if(mod(i,Idata)==1) T=[T,t]; j=j+1; X(j)=x;
end;
end;
Y=sin(T)./sqrt(1+3/4*epsilon*T);
A=1./sqrt(1+3/4*epsilon*T);
plot(T,X,'g*',T,Y,'r -',T,A,'k-',T,-A,'b-')
xlabel('T');
title('Multiple Time Scale');
axis([0 dt*N -0.5 0.5]);
```

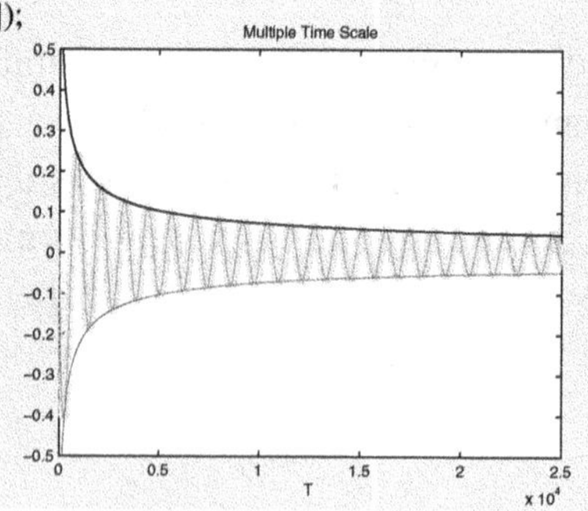

As a detailed example, we are to find the solution of the equation of a damped oscillator

$$\frac{d^2x}{d\tau^2} + x + \varepsilon\left(\frac{dx}{d\tau}\right)^3 = 0, \tag{6.5}$$

with use of multiple time scales, given the initial conditions: $x(0) = 0$, and $dx/d\tau|_0 = 1$, and $0 < \varepsilon \ll 1$. By inspecting the equation when setting $\varepsilon = 0$, it is clear that a simple harmonic oscillation is the primary phenomenon. The term $\varepsilon(dx/d\tau)^3$ provides the frictional damping. Since the damping effect is weak, it is natural to introduce the multiple time scales.

Accordingly, the equation can be expanded order by order and solved the same. To lowest order, O(1):

$$\frac{\partial^2 x^{(0)}(\tau_0, \tau_1, \ldots)}{\partial \tau_0^2} + x^{(0)}(\tau_0, \tau_1, \ldots) = 0 \tag{6.6}$$

The solution is given by

$$x^{(0)}(\tau_0, \tau_1, \ldots) = A(\tau_1, \ldots)\sin[\tau_0 + \phi(\tau_1, \ldots)]. \tag{6.7}$$

To satisfy the initial conditions, we will require $A(\tau_0 = 0) = 1$ and $\phi(\tau_0 = 0) = 0$. To next order, O(ε):

$$\frac{\partial^2 x^{(1)}(\tau_0, \tau_1, \ldots)}{\partial \tau_0^2} + x^{(1)}(\tau_0, \tau_1, \ldots)$$
$$= -2\frac{\partial^2 x^{(0)}(\tau_0, \tau_1, \ldots)}{\partial \tau_0 \partial \tau_1} - \left[\frac{\partial x^{(0)}(\tau_0, \tau_1, \ldots)}{\partial \tau_0}\right]^3. \tag{6.8}$$

It is now important to recognize that there could be unbounded solutions when solving the equations order by order. The physical solution, as we know it, has to be bounded, otherwise we would have an **ill-posed** equation, meaning that the equation does not correspondingly describe a physical system. This gives us an important tool to not only find the coefficients in the expansion scheme, but also retain the dominant effect, which tends to secularly behave in the longer time scale if not properly treated. The secularity can be observed by looking at the right hand side of the last equation, which has driving terms at the resonance frequency of the left hand equation, and would produce an unbounded solution. We have to make these secular terms vanish, By setting a time average over the resonance harmonics, it gives,

$$\oint d\tau_0 \begin{bmatrix} \cos \tau_0 \\ \sin \tau_0 \end{bmatrix} F(\tau_0, \tau_1, \ldots) = 0, \tag{6.9}$$

where F is the driving term on the right hand side of the said equation. We arrive at the following equations:

$$\frac{dA\cos\phi}{d\tau_1} = -\frac{3}{8}A^3\cos\phi, \quad \text{and} \quad \frac{dA\sin\phi}{d\tau_1} = -\frac{3}{8}A^3\sin\phi, \tag{6.10}$$

which readily infers that $d\phi/d\tau_1 = 0$ and $\phi = 0$. The other equation is given by

$$\frac{dA}{d\tau_1} = -\frac{3}{8}A^3. \tag{6.11}$$

We then arrive at the final solutions:

$$A^2 = 1/(1 + 3\tau_1/4), \quad x^{(0)} = \sin t\Big/\sqrt{1 + \frac{3}{4}\varepsilon t}. \tag{6.12}$$

The analytical result is compared below and shows consistency with the numerical solution.

Homework 6.1: Van der Pol equation: Solve the following equation with use of the multiple time scales,

$$\frac{d^2x}{d\tau^2} - \varepsilon(1 - \beta x^2)\frac{dx}{d\tau} + x = 0$$

where all quantities are dimensionless, $0 < \varepsilon \ll 1$, and $\beta > 0$ but $\beta \sim O(1)$.

6.1.2 *Frequency Mismatch and Nonlinear Resonance*

Resonance is an important topic in physics. The linear analysis of resonance does not necessarily do justice to its characteristics. **Frequency mismatch**, for example, may be compensated by the large amplitude of the pump wave to sustain the resonance (Hsu 1981). On the other hand, the energy pumping by resonance is by no means unlimited as the linear theory might suggest. In fact, saturation is readily reached unless, for example, the drainage of gained energy by an irreversible process is at work. Here the nonlinear theory at integer harmonics of the natural frequency of the system is presented with the following model equation,

$$\frac{d^2x}{d\tau^2} + x = \varepsilon a \sin(x - \nu\tau), \tag{6.13}$$

where the pump wave frequency normalized to the natural frequency is given by $\nu = n + \varepsilon\Delta$, with $\varepsilon\Delta$ the frequency mismatch from the nearest integer harmonic. To the lowest order in the multiple timescale, it is given by

$$\frac{\partial^2 x^{(0)}(\tau_0, \tau_1, \ldots)}{\partial \tau_0^2} + x^{(0)}(\tau_0, \tau_1, \ldots) = 0. \tag{6.14}$$

Its solution can be cast into

$$x^{(0)}(\tau_0, \tau_1, \ldots) = A(\tau_1, \ldots)\sin[\tau_0 + \Phi(\tau_1, \ldots)]. \tag{6.15}$$

To the next order,

$$\frac{\partial^2 x^{(1)}(\tau_0, \tau_1, \ldots)}{\partial \tau_0^2} + x^{(1)}(\tau_0, \tau_1, \ldots) + 2\frac{\partial^2 x^{(0)}(\tau_0, \tau_1, \ldots)}{\partial \tau_0 \partial \tau_1} = a \sin(x^{(0)} - \nu \tau_0). \tag{6.16}$$

By substituting Eq. (6.15) into the RHS of Eq. (6.16) and making use of the Bessel function expansion, viz., $\sin(\alpha \sin(x) + \varphi) = \sum_l J_l(\alpha) \sin(lx + \varphi)$, it can be expressed as

$$a \sin(A \sin(\tau_0 + \Phi) - \nu \tau_0) = a \sum_l J_l(A) \sin[(l - n)\tau_0 + l\Phi - \Delta \tau_1).$$

Thus, eliminating the secularity by taking out the resonance components

$$\oint d\tau_0 \begin{bmatrix} \cos \tau_0 \\ \sin \tau_0 \end{bmatrix} \left[2\frac{\partial}{\partial \tau_1} A \cos(\tau_0 + \Phi) \right.$$

$$\left. - a \sum_l J_l(A) \sin[(l - n)\tau_0 + l\Phi - \Delta \tau_1) \right] = 0,$$

we end with

$$\frac{dA}{d\tau_1} = \frac{na}{A} J_n(A) \sin(n\Phi - \Delta \tau_1), \tag{6.17a}$$

$$A\frac{d\Phi}{d\tau_1} = a J_n'(A) \cos(n\Phi - \Delta \tau_1), \tag{6.17b}$$

where these identities:

$$J_{n+1}(A) + J_{n-1}(A) = 2n J_n(A)/A$$

and

$$J_{n-1} - J_{n+1} = 2J_n'$$

have been applied. By defining the variable $\psi \equiv \Phi - \Delta \tau_1/n$, Eq. (6.17) can be solved to give this relationship between the amplitude and the phase angle, $J_n(A) \cos(n\Phi - \Delta \tau_1) - \Delta A^2/2na = C_0$, which is a constant of the motion if the frequency mismatch Δ is absent. We plot this constant of motion for the n = 1 and n = 2 harmonics in the following that shows distinct trajectory characteristics for the two harmonics. For very low velocity particles, they can be accelerated to a velocity toward the first zeros of the Bessel functions.

The following program solves the particle trajectory at given pump wave amplitude and frequency. The top left picture shows the **Poincare section plot** for $p = 0.25$ and $\nu = 1$, and to its right for $p = 0.25$ and $\nu = 2$. The bottom left for $p = 0.25$ and to its right for $p = 0.75$ both at $\nu = 1.5$. The frequency mismatch

```
function NonlinearResonanceCP
clear all; close all; clc;
dx=0.01;
N=2500;
x=-N/2*dx:dx:N/2*dx;
y=x;
N=N+1;
X=repmat(x,N,1);
Y=repmat(y',1,N);
R=sqrt(X.^2+Y.^2);
PHI=acos(X./R);
f=besselj(1,R).*cos(PHI);
V=-1:0.1:1;
figure(1);
contour(f,V)
axis equal;
f=besselj(2,R).*cos(2*PHI);
V=-1:0.1:1;
figure(2);
contour(f,V);
axis equal;
```

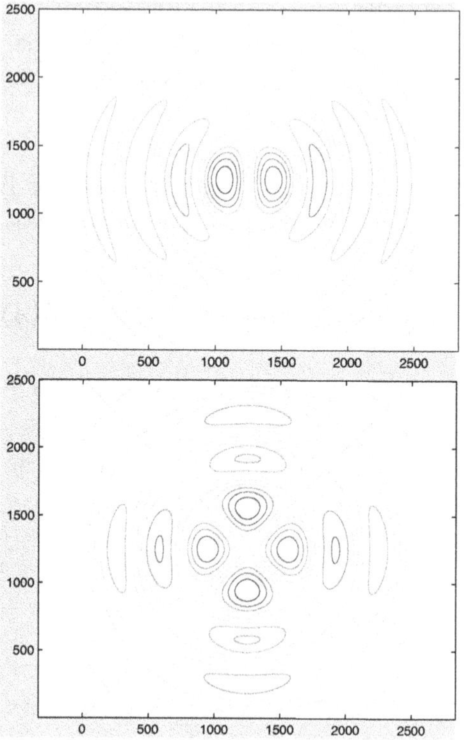

can be compensated by the large amplitude of the pump wave, and it leads to the stochastic heating when the stochastic threshold is exceeded.

Homework 6.2: Duffing oscillator: Given the following equation

$$\frac{d^2x}{d\tau^2} + x + \varepsilon x^3 = 0$$

with the initial conditions $x(0) = 1$ and $dx/d\tau|_0 = 0$ show that the nonlinear term causes a blue shift in the oscillation frequency. Here it is assumed that $0 < \varepsilon \ll 1$,

```
function NR(p,nu)
dt=2*pi/100; N=10^5;
v=0; x=rand(1);
X=[]; V=[];
J=0;
v=v+dt/2*p*sin(x)
for i=1:N
  x=x+v*dt;
  t=i*dt;
```

```
    v=v-x*dt+dt*p*sin(x-nu*t);
    if(mod(i,100)==0) J=J+1;  X=[X,x]; V=[V,v];  end;
end;
figure(1);
plot(X,V,'.g');
```

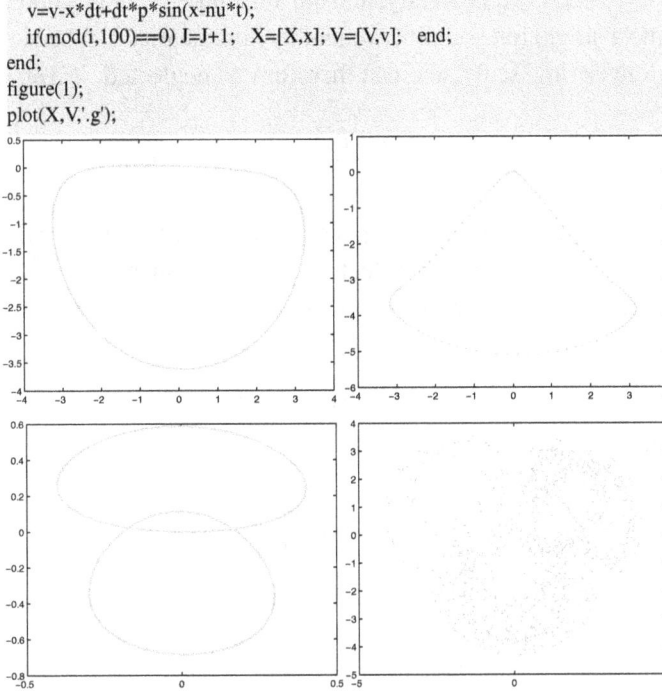

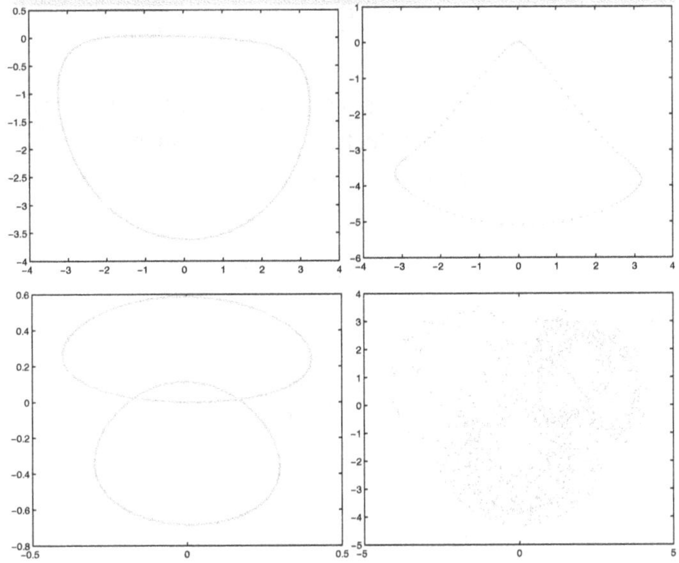

6.1.3 *Interaction of Two Plasma Waves with the Ion Sound Wave*

We now investigate the interaction of two plasma waves with the ion sound wave. To describe both the electron plasma wave and the ion sound wave, we resort to the fluid equations. The continuity equation is

$$\frac{\partial n_\sigma}{\partial t} + \nabla \cdot (n_\sigma \vec{v}_\sigma) = 0, \tag{6.18}$$

and the momentum equation,

$$\frac{\partial \vec{v}_\sigma}{\partial t} + \vec{v}_\sigma \cdot \nabla \vec{v}_\sigma = -\frac{1}{n_\sigma m_\sigma} \nabla p_\sigma + \frac{q_\sigma \vec{E}}{m_\sigma}. \tag{6.19}$$

Taking the time derivative on Eq. (6.18) and eliminating $\partial \vec{v}_\sigma / \partial t$ gives,

$$\frac{\partial^2 n_\sigma}{\partial t^2} = \nabla \cdot \left(n_\sigma \vec{v}_\sigma \cdot \nabla \vec{v}_\sigma + \frac{T_\sigma}{m_\sigma} \nabla n_\sigma - \frac{n_\sigma q_\sigma \vec{E}}{m_\sigma} + \vec{v}_\sigma \nabla \cdot (n_\sigma \vec{v}_\sigma) \right), \tag{6.20}$$

where we have assumed an isothermal plasma. The electric field is governed by the Poisson's equation that provides the closure to the system:

$$\nabla \cdot \vec{E} = 4\pi q (n_i - n_e). \tag{6.21}$$

The problem is treated in 1d geometry, and the ion temperature is ignored. Moreover, we will assume a single ion sound wave and its nonlinearity can be shown to occur on the much longer time scale, and can therefore be neglected. We arrive at

$$\frac{\partial^2 n_i}{\partial t^2} = -\nabla \cdot \frac{n_0 q \vec{E}}{m_i} = -\omega_{pi}^2 (n_i - n_e). \tag{6.22}$$

Defining $\varepsilon \equiv \sqrt{m_e/m_i} \ll 1$ as the expansion parameter, and normalize the space by the Debye length, λ_D, time by the electron plasma frequency, $1/\omega_{pe}$, and velocity by the electron thermal velocity, V_{th}, and applying the multiple time expansion,

$$\frac{d}{d\tau} = \frac{\partial}{\partial \tau_0} + \varepsilon \frac{\partial}{\partial \tau_1} + \varepsilon^2 \frac{\partial}{\partial \tau_2} + \cdots,$$

we may rewrite Eq. (6.22) as

$$\frac{\partial^2 n_i^{(1)}}{\partial \tau_1^2} + n_i^{(1)} - \oint \frac{d\tau_0}{T_0} n_e^{(1)} = 0, \tag{6.23}$$

where $\omega_{pi}\tau_1/\omega_{pe} \sim \tau_0$ the density has been normalized to the equilibrium density n_0. A good general solution to Eq. (6.23) is given by an ion sound wave $n_3 = A\sin(\omega_3\tau_1 - k_3\varsigma + \phi_3)$ plus the particular solution due to the electron response to the ion wave $n_p = \oint n_e^{(1)} d\tau_0/T_0$. Therefore,

$$n_i^{(1)} = n_3(\tau_1, \varsigma) + n_p(\tau_1, \varsigma). \tag{6.24}$$

To lowest order, the electron density from Eq. (6.20) is governed by

$$\frac{\partial^2 n_e^{(1)}}{\partial \tau_0^2} - \frac{\partial^2 n_e^{(1)}}{\partial \varsigma^2} + (n_e^{(1)} - n_i^{(1)}) = 0. \tag{6.25}$$

Two electron plasma waves are present and assumed to be given by $n_1 = a_1 \sin(\omega_1\tau_0 - k_1\varsigma + \varphi_1)$ and $n_2 = a_2 \sin(\omega_2\tau_0 - k_2\varsigma + \varphi_2)$, respectively. Thus,

$$n_e^{(1)} = a_1 \sin(\omega_1\tau_0 - k_1\varsigma + \varphi_1) + a_2 \sin(\omega_2\tau_0 - k_2\varsigma + \varphi_2) + n_p(\tau_1, \varsigma). \tag{6.26}$$

It is not surprising that the particular solutions for both ions and electrons are identical since it simply reflects the charge neutrality that any ion density perturbation is being neutralized by the fast motion of the electrons. The ion density perturbation does not vary on the fast time scale, and can be extracted through the particular solution of Eq. (6.25) to be $n_p = n_3/k_3^2$. Note that after the fast time integration $\oint d\tau_0/T_0$, both n_1 and n_2 vanish. It can be recognized that Eqs. (6.24) and (6.26) satisfy both Eqs. (6.23) and (6.25), and also give rise to the dispersion relation for the ion sound wave: $\omega_3^2 = k_3^2/(1 + k_3^2)$, and for the electron plasma waves: $\omega_1^2 = 1 + k_1^2$ and $\omega_2^2 = 1 + k_2^2$. We also have $n_i^{(1)} = n_3(1 + k_3^2)/k_3^2$.

The electric field to the first order can be found from $\partial E^{(1)}/\partial\varsigma = n_i^{(1)} - n_e^{(1)}$ to be

$$E^{(1)} = E_e^{(1)} + E_i^{(1)}, \tag{6.27a}$$

where

$$E_e^{(1)} = -\frac{a_1}{k_1}\cos(\omega_1\tau_0 - k_1\varsigma + \varphi_1) - \frac{a_2}{k_2}\cos(\omega_2\tau_0 - k_2\varsigma + \varphi_2), \tag{6.27b}$$

$$E_i^{(1)} = \frac{A}{k_3}\cos(\omega_3\tau_1 - k_3\varsigma + \varphi_3). \tag{6.27c}$$

The electron velocity to the first order governed by $\partial v_e^{(1)}/\partial\tau_0 = -\partial n_e^{(1)}/\partial\varsigma - E^{(1)}$ is given by

$$v_e^{(1)} = \frac{a_1}{k_1\omega_1}\sin(\omega_1\tau_0 - k_1\varsigma + \varphi_1) + \frac{a_2}{k_2\omega_2}\sin(\omega_2\tau_0 - k_2\varsigma + \varphi_2). \tag{6.28}$$

[The Resonance Condition] To the next order, Eq. (6.20) gives the electron response as in the following:

$$\frac{\partial^2 n_e^{(2)}}{\partial\tau_0^2} - \frac{\partial^2 n_e^{(2)}}{\partial\varsigma^2} + n_e^{(2)} - n_i^{(2)} + 2\frac{\partial^2 n_e^{(1)}}{\partial\tau_0\partial\tau_1} + n_e^{(1)}(n_e^{(1)} - n_i^{(1)})$$
$$= \frac{\partial}{\partial\varsigma}\left(2v_e^{(1)}\frac{\partial}{\partial\varsigma}v_e^{(1)}\right) - E^{(1)}\frac{\partial}{\partial\varsigma}n_e^{(1)}. \tag{6.29}$$

Only the beating of electron terms with the ion terms will drive the secularity, provided that the resonance conditions are met, namely,

$$\omega_1 - \omega_2 = \omega_3, \tag{6.30a}$$

and

$$k_1 - k_2 = k_3. \tag{6.30b}$$

Other conditions with sign change do not amount to new physics, so this particular set is chosen without loss of generality. Note that $n_e^{(1)}(n_e^{(1)} - n_i^{(1)}) = (n_1 + n_2 + n_p) \times (n_1 + n_2 - n_3)$, results in a secular term $(n_1 + n_2) \times (n_p - n_3)$. Ion flow velocity is immaterial to the first order. Thus, Eq. (6.29) retaining only resonance terms leads to

$$2\frac{\partial^2 n_e^{(1)}}{\partial\tau_0\partial\tau_1} + (n_1 + n_2)n_3\left(\frac{1}{k_3^2} - 1\right) = -E_e^{(1)}\frac{\partial}{\partial\varsigma}n_p - \vec{E}_i^{(1)}\frac{\partial}{\partial\varsigma}(n_1 + n_2). \tag{6.29a}$$

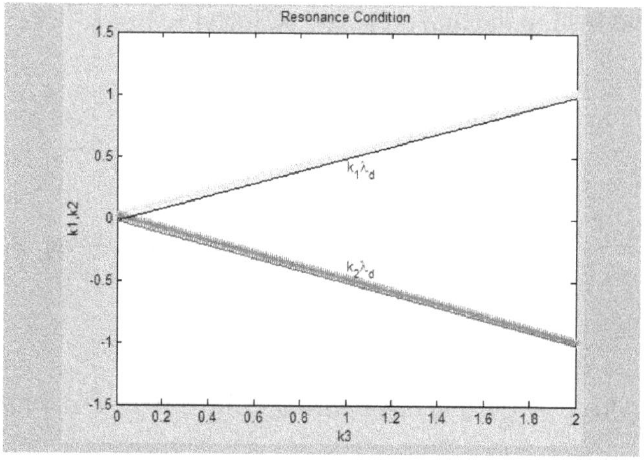

The resonance conditions of Eq. (6.30), translated to the dimensional variables, are given by $\omega_1 \to \omega_{pe}\sqrt{1+k_1^2\lambda_D^2}$, $\omega_2 \to \omega_{pe}\sqrt{1+k_2^2\lambda_D^2}$ and $\omega_3 \to k_3\lambda_D \times \omega_{pi}/\sqrt{1+k_3^2\lambda_D^2}$. By taking $\omega_{pi} = \omega_{pe}/\sqrt{1837}$for the hydrogen plasma, $k_1\lambda_D$ and $k_2\lambda_D$ can be solved in terms of $k_3\lambda_D$ and are plotted in the figure. There are two sets of solutions. Both electron plasma waves have almost twice the wavelength of the ion sound wave and are propagating in the opposite directions. Taking the case, $k_3\lambda_D = 1$, we find $k_1\lambda_D = 0.52$ and $k_2\lambda_D = -0.48$, and the frequencies are $1.126\omega_{pe}$ and $1.11\omega_{pe}$ for electron waves and $0.707\omega_{pi}$ for ion wave, respectively. The electron plasma waves are propagating in the opposite directions, and their wavelengths are about twice that of the ion sound wave.

[*The* ***Breather***] Eq. (6.29a), as extracted from putting together all the terms that drive the secularity, can be expanded to give the following:

$$\left(\frac{\partial a_1}{\partial \tau_1}\omega_1 \cos(\omega_1\tau_0 - k_1\varsigma + \varphi_1) - \frac{\partial \varphi_1}{\partial \tau_1} a_1\omega_1 \sin(\omega_1\tau_0 - k_1\varsigma + \varphi_1)\right) + (1 \to 2)$$

$$= \frac{Aa_1}{2k_3^2}(k_3^2 - 1)\sin(\omega_1\tau_0 - k_1\varsigma + \varphi_1)\sin(\omega_3\tau_1 - k_3\varsigma + \varphi_3)$$

$$+ \frac{Aa_1}{2k_3k_1}(k_1^2 - 1)\cos(\omega_1\tau_0 - k_1\varsigma + \varphi_1)\cos(\omega_3\tau_1 - k_3\varsigma + \varphi_3) + (1 \to 2).$$

Without loss of generality, we will set $\varphi_3 = 0$. Utilizing the resonance conditions, we arrive at

$$\frac{da_1}{d\tau_1} = \frac{Aa_2}{4k_3\omega_1}\frac{(k_2 - k_3)(k_2k_3 + 1)}{k_3k_2}\cos(\varphi_2 - \varphi_1) = \frac{Aa_2}{4k_3^2}\lambda_1 \cos\Phi, \qquad (6.30a)$$

$$\frac{d\varphi_1}{d\tau_1} = \frac{Aa_2}{4k_3a_1\omega_1}\frac{(k_2-k_3)(k_2k_3+1)}{k_3k_2}\sin(\varphi_2-\varphi_1) = -\frac{Aa_2}{4k_3^2a_1}\lambda_1\sin\Phi, \tag{6.30b}$$

$$\frac{da_2}{d\tau_1} = \frac{Aa_1}{4k_3\omega_2}\frac{(k_1+k_3)(k_1k_3-1)}{k_3k_1}\cos(\varphi_1-\varphi_2) = -\frac{Aa_1}{4k_3^2}\lambda_2\cos\Phi, \tag{6.30c}$$

$$\frac{d\varphi_2}{d\tau_1} = \frac{Aa_1}{4k_3a_2\omega_2}\frac{(k_1+k_3)(k_1k_3-1)}{k_3k_1}\sin(\varphi_1-\varphi_2) = -\frac{Aa_1}{4k_3^2a_2}\lambda_2\sin\Phi. \tag{6.30d}$$

where $\Phi \equiv \varphi_1 - \varphi_2$, $\lambda_1 \equiv (k_2-k_3)(k_2k_3+1)/\omega_1k_2$ and $\lambda_2 \equiv (k_1+k_3)(1-k_1k_3)/\omega_2k_1$. Subtracting (6.30d) from (6.30b), we have

$$\frac{d\Phi}{d\tau_1} = \frac{A}{4k_3^2}\left[\frac{a_1}{a_2}\lambda_2 - \frac{a_2}{a_1}\lambda_1\right]\sin\Phi. \tag{6.31}$$

Eliminate τ_1 in favor of Φ in Eqs. (6.30a) and (6.30c) with use of Eq. (6.31), we end with

$$\frac{da_1}{d\Phi} = \frac{a_1a_2^2\lambda_1}{a_1^2\lambda_2 - a_2^2\lambda_1}\cot\Phi, \tag{6.32a}$$

$$\frac{da_2}{d\Phi} = -\frac{a_2a_1^2\lambda_2}{a_1^2\lambda_2 - a_2^2\lambda_1}\cot\Phi. \tag{6.32b}$$

Further dividing these two equations to eliminate Φ, we end with

$$\frac{da_1^2}{da_2^2} = -\Lambda, \tag{6.33}$$

that implies $H \equiv a_1^2 + \Lambda a_2^2 = const.$ is a constant of the motion, where $\Lambda \equiv \lambda_1/\lambda_2$. Moreover, eliminating a_2^2 from Eq. (6.32a) with use of H, we have

$$\frac{da_1}{d\Phi} = \frac{a_1(H-a_1^2)}{2a_1^2 - H}\cot\Phi, \tag{6.34}$$

that can be integrated to give $W \equiv a_1^2(H-a_1^2)\sin^2\Phi = (H-\Lambda a_2^2)\Lambda a_2^2\sin^2\Phi = const.$, another constant of the motion.

We can now integrate the time evolution equation with use of $H$and W:

$$\frac{da_1}{d\tau_1} = \frac{Aa_2}{4k_3^2}\lambda_1\cos\Phi = \pm\alpha\sqrt{H - a_1^2 - \frac{W}{a_1^2}}, \tag{6.30a'}$$

which gives

$$a_1^2 = \frac{1}{2}H \pm \frac{1}{2}\sqrt{H^2-4W}\sin(\alpha\tau_1 \mp \theta_1), \tag{6.35}$$

where $\theta_1 = a\sin(H - 2a_1^2(0)/\sqrt{H^2 - 4W})$, and $\alpha \equiv A\sqrt{\lambda_1\lambda_2}/2k_3^2$. Similarly, we have

$$\frac{da_2}{d\tau_1} = -\frac{Aa_1\lambda_2}{4k_3^2}\cos\Phi = \mp\frac{\alpha}{\sqrt{\Lambda}}\sqrt{H - \Lambda a_2^2 - \frac{W}{\Lambda a_2^2}}, \tag{6.30c'}$$

which gives

$$\Lambda a_2^2 = \frac{1}{2}H \mp \frac{1}{2}\sqrt{H^2 - 4W}\sin(\alpha\tau_1 \pm \theta_2), \tag{6.35}$$

where $\theta_2 = a\sin(H - 2\Lambda a_2^2(0)/\sqrt{H^2 - 4W}) = -\theta_1$. We may choose one set of the solutions since the two solutions are complimentary to each other as the two electron waves are interchangeable. Thus, choosing the positive sign for a_1^2, we readily find from the definition of W,

$$\sin\Phi = \pm\frac{\sqrt{W}}{\sqrt{a_1^2(H - a_1^2)}} = \pm\frac{2\sqrt{W}}{\sqrt{H^2\cos^2(\alpha\tau_1 - \theta_1) + 4W\sin^2(\alpha\tau_1 - \theta_1)}}, \tag{6.36}$$

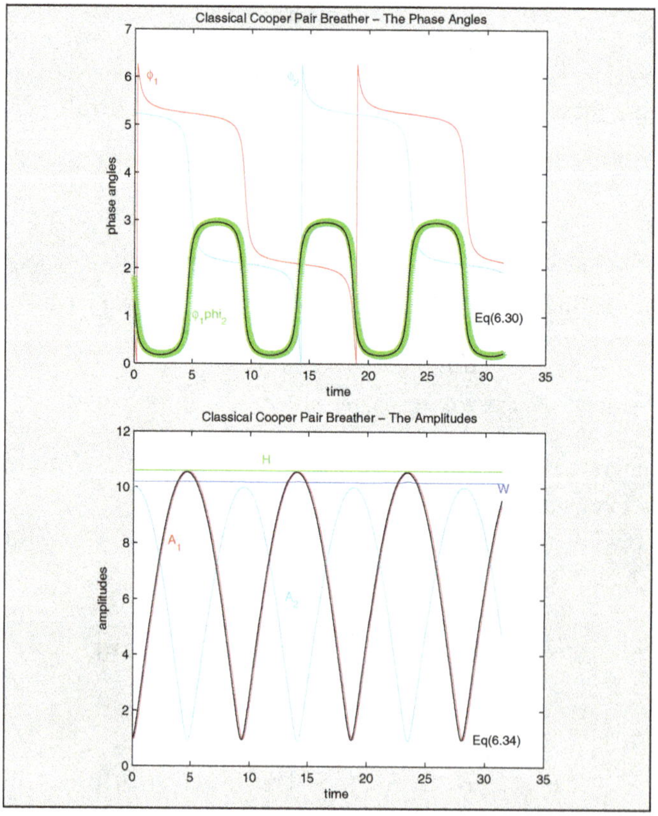

```
function breather
close all; clc; clear all;
A=1.0; a1=1.0; a2=10; ep=1/sqrt(1837);
k1=0.5184; k2=-0.4816; k3=1;
f1=1.1264; f2=1.1099; f3=0.0165;
phi1=pi/4; phi2=-pi/3;
A1=[]; A2=[]; PHI1=[]; PHI2=[];
N=200000; TIME=2*pi*5; dt=TIME/N;
for i=1:N
   F1=A*a2/4/k3/f1*(k2-k3)*(k2*k3+1)/k2/k3*cos(phi2-phi1);
   a1=a1+F1*dt;
   F2=A*a2/4/k3/f1/a1*(k2-k3)*(k2*k3+1)/k2/k3*sin(phi2-phi1);
   phi1=phi1+F2*dt;
   F3=A*a1/4/k3/f2*(k1+k3)*(k1*k3-1)/k1/k3*cos(phi1-phi2);
   a2=a2+F3*dt;
   F4=A*a1/4/k3/f2/a2*(k1+k3)*(k1*k3-1)/k1/k3*sin(phi1-phi2);
   phi2=phi2+F4*dt;
   if(mod(i,100)==0) A1=[A1,a1]; A2=[A2,a2]; PHI1=[PHI1,phi1];
                        PHI2=[PHI2,phi2]; end;
   end;
t=(1:N/100)*dt*100;
lambda1=(k2-k3)*(1+k2*k3)/f1/k2; lambda2=(k1+k3)*(1-k1*k3)/f2/k1;
LAMBDA=lambda1/lambda2,
a1=A1(1), a2=A2(1),
H=(a1^2+LAMBDA*a2^2),
B1=a1*sin(f1*t+phi1); B2=a2*sin(f2*t+phi2);
E=(max(B1)^2+max(B2)^2)/2,
energy=0.5*sum(A1.^2+A2.^2)/length(A1),
figure(1);  y=sqrt(H-A2.^2*LAMBDA);
plot(A2,A1,'*g',A2,y,'r-');
xlabel('A_2(t)'); ylabel('A_1(t)');
text(1,11,'A_1','color','g'); text(6,9,'sqrt(H-A_2^2\lambda)','color','r');
title('Amplitudes of Electron Waves');
figure(2);  W=A1.^2.*(H-A1.^2).*sin(PHI1-PHI2).^2;
alpha=A*sqrt(lambda1*lambda2)/2/k3^2;
s0=asin((H-2*a1.^2)./sqrt(H.^2-4*W));
A1Square=H/2+sqrt(H.^2/4-W).*sin(alpha*t-s0);
A1t=sqrt(A1Square);
plot(t,A1,'r-',t,A2,'c-',t,sqrt(H),'g-',t,(W).^(1/2),'b-',t,A1t,'k-');
text(3,8,'A_1','color','r'); text(13,6,'A_2','color','c'); text(11,11,'H','color','g');
text(31,10,'W','color','b');
text(29,1,'Eq(6.34)');
title('Classical Cooper Pair Breather - The Amplitudes');
xlabel('time'); ylabel('amplitudes');
PHIt=asin((sqrt(W(1))./sqrt(H-A1t.^2)./A1t).*(1-
2*([0,diff(A1t)]<0)))+pi*([0,diff(A1t)]<0);
figure(3); plot(t,mod(PHI1,2*pi),'r-',t,mod(PHI2,2*pi),'c-',t,PHI1-
PHI2,'g*',t,real(PHIt),'k-');
xlabel('time'); ylabel('phase angles');
```

```
title('Classical Cooper Pair Breather - The Phase Angles');
text(1,6,'\phi_1','color','r'); text(13,6,'\phi_2','color','c');
text(5,1,'\phi_1phi_2','color','g');text(29,1,'Eq(6.30)');
```

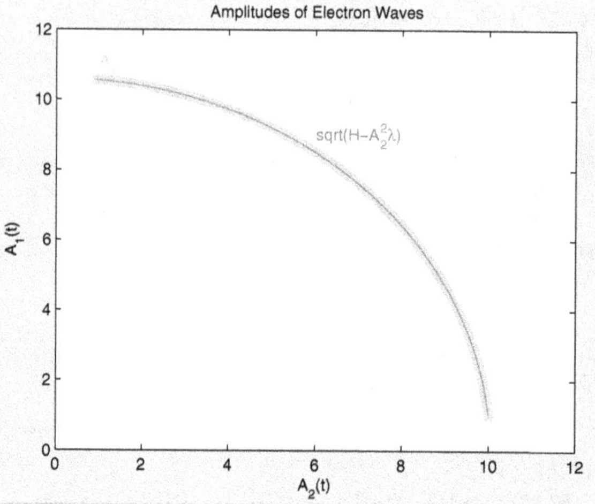

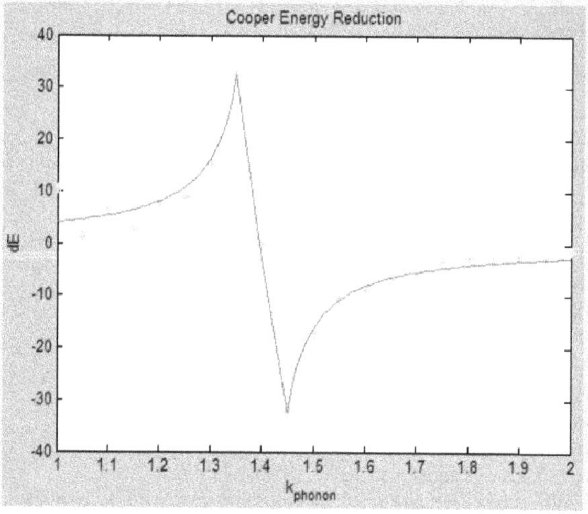

that gives the time evolution of Φ. The two electron plasma waves execute the breather oscillation. Their electric field energy without the mediation of the ion sound wave is $|E_e^{(1)}|^2 = \frac{1}{2}a_{10}^2/k_1^2 + \frac{1}{2}a_{20}^2/k_2^2$, while in the presence of the ion sound wave coupling, the energy is $E_{\text{int}} = \frac{1}{2}\langle a_1^2\rangle + \frac{1}{2}\langle a_2^2\rangle = \frac{1}{4}H(1 + 1/\Lambda) = \frac{1}{4}(a_{10}^2 \sin^2\varphi_1 + \Lambda a_{20}^2 \sin^2\varphi_2)(1 + 1/\Lambda)$. By assuming $a_{10} = a_{20} = 1$ to represent the equal amplitude of electron waves, the numerical solution shows that the energy transfer among the three waves goes both ways depending on the interaction conditions.

6.2 Multiple Space Scales

A boundary layer of small domain may alter the physics in its entirety in the major region. In fluid mechanics a turbulent layer often occurs at the channel wall where the viscosity is significant. While the major flow pattern could be laminar, the boundary layer can develop into the vortex street, which if allowed growing can cause turbulence to penetrate into the main flow channel, and disrupt the whole flow pattern. It has important implications in jet proposition, blood circulation, combustion engine efficiency, etc.

When a boundary layer is present, such as in the traffic merge, the blood fluid-vessel interface, the separatrix, the mode conversion, the tearing and reconnection, to name a few, the multiple space scales can be rather useful to further the analytical solution. We will demonstrate the mathematical apparatus, in particular, the **matching procedure** to this effect starting from the **singular perturbation** and the boundary layer.

6.2.1 *Singular Perturbation Problem and Matching Procedure*

A singular perturbation problem in general has a small higher order term that may be unimportant for some solutions but can be the source of an easily ignored but seriously principal solution.

Consider the following cubic equation, $\varepsilon x^3 + (x-2)^2 = 1$. The regular perturbation treatment would consider the first term as small, and the two roots are readily available. To lowest order, $x^{(0)} = 2 \pm 1$, and to next order, $x_1 \approx 2 + \sqrt{1-9\varepsilon}$ and $x_2 \approx 2 - \sqrt{1-\varepsilon}$. We missed one root since the problem is the so called singular perturbation problem in that the highest order term is the small term. To recover this root, we may order $x \gg 1$ and balance the two terms of the highest power in x to give $x_3 \approx -1/\varepsilon$.

Many problems in physics are singular perturbation problems, especially those with dissipation or viscosity. While these effects are weak, they could be crucial in the boundary layer and affect the solutions of boundary value problems to even yield bifurcated states, for example. The technique to treat a boundary layer problem is the matching procedure. Here is a good example: $\varepsilon y'' + y' = \alpha$ with the boundary conditions $y(0) = 0$ and $y(1) = 1$, where $\varepsilon \ll 1$. It has an exact solution given by $y = (1-\alpha)(1-e^{-x/\varepsilon})/(1-e^{-1/\varepsilon}) + \alpha x$. We may treat the space by two separate regions. The outer region has the solution $y_{out} = (1-\alpha) + \alpha x$. In the inner region near the origin, we define an inner variable z, $\varepsilon z \equiv x$ to have the following zeroth and first order equations:

$$\frac{d^2 y_{in}^{(0)}}{dz^2} + \frac{dy_{in}^{(0)}}{dz} = 0; \quad \frac{d^2 y_{in}^{(1)}}{dz^2} + \frac{dy_{in}^{(1)}}{dz} = \alpha. \tag{6.37}$$

Therefore, $y_{in} = A(e^{-z} - 1) + \alpha\varepsilon z = A(e^{-z} - 1) + \alpha\varepsilon z \xrightarrow{z\gg 1} -A + \alpha x$. While as $x \to 0$, the outer solution gives $y_{out} \to (1 - \alpha) + \alpha x$. It is clear that all coefficients of the two solutions must be consistent in the proper limits.

```
function SPP                                %Solve the singular perturbation
N=1000;                                     problem by the shooting method
dx=1/N;                                     and compare with the analytical
x=(0:N)*dx;                                 solution.
y=x*0;
z=y;
alpha=0.5;
epsilon=0.1;
z(1)=(1-alpha)/epsilon+0.5;
z(1),
for i=1:N
   y(i+1)=y(i)+z(i)*dx;
   z(i+1)=z(i)+(alpha-z(i))*dx/epsilon;
end;
y(N+1),
Y=(1-alpha)*(1-exp(-x/epsilon))+alpha*x;
plot(x,y,'g*',x,Y,'r-')
title('Shooting Method versus Multiple
Length Scale');
xlabel('x');
ylabel('y');
```

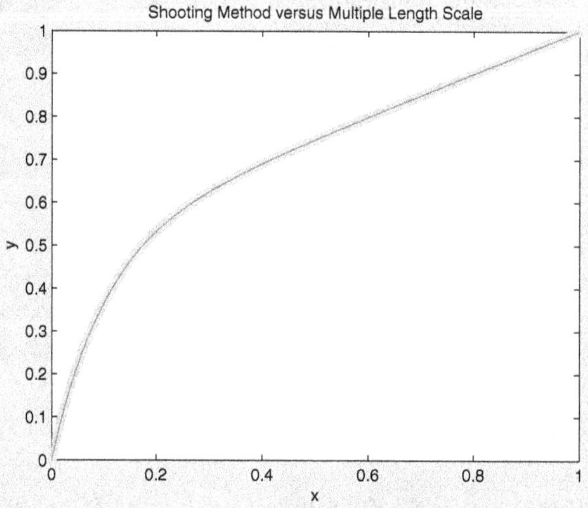

We have to assign $A = \alpha - 1$, which implies $y = (1 - \alpha)(1 - e^{-x/\varepsilon}) + \alpha x$. The difference between the exact solution and the matched solution is of order $O(e^{-1/\varepsilon})$.

Homework 6.3: Composite Solution: Determine a leading order composite solution for the boundary value problem

$$\varepsilon y'' + 2(2x - 1)(y' + 2y) = 0,$$

where $\varepsilon \ll 1$ and $y(0) = 1$, $y(1) = 2/e$. Construct a uniformly valid expansion for $y(x)$ and sketch the solution.

Homework 6.4: Multiple Length Scale: Determine the leading order terms of the outer and inner expansions for the problem

$$\varepsilon y'' - (2x + 1)y' + y = 0,$$

where $\varepsilon \ll 1$ and $y(0) = 1$, $y(1) = 0$. Construct a uniformly valid composite solution for $y(x)$ and sketch the result. Compare this solution with your numerical analysis.

6.2.2 *Separatrix Motion*

We are to solve the following equation,

$$\varepsilon^2 \frac{d^2 f}{dx^2} + f - f^3 = 0, \tag{6.38}$$

according to the boundary conditions: $f(0) = 0$ and $f(\infty) = 1$ with $\varepsilon \ll 1$. This nonlinear equation while describing a spatial function, is in analogy to the particle motion on the separatrix in the phase space if the x variable is interpreted as the time coordinate. It, however, can be tackled by the multiple space scales more easily and is a good example of demonstrating the technique to construct the composite solution, since the equation can be solved exactly.

In the outer region, we may take $f_{out} - f_{out}^3 = 0$. Thus, $f_{out} = 0$ or $f_{out} = 1$. Choose $f_{out} = 1$ as the solution that satisfies the boundary condition. It implies that $f_{out} \approx 1$ for $x \gg 1$. To find the higher order solution in the outer region we assume $f_{out} \approx 1 + w$ with $w \ll 1$, which is governed by $\varepsilon^2 d^2 w/dx^2 - 2w = 0$ and gives $w = A\exp(\sqrt{2}x/\varepsilon) + B\exp(-\sqrt{2}x/\varepsilon)$. It is clear that $A = 0$ so that $f_{out} \approx 1 + Be^{-\sqrt{2}x/\varepsilon}$.

The inner region solution is found by the series expansion so that the leading term is $f_{in}^{(0)} = \alpha s$, and to the next order,

$$\frac{d^2 f_{in}^{(1)}}{ds^2} + \alpha \varepsilon s = 0. \tag{6.39}$$

Thus, $f_{in}^{(1)} = -\frac{1}{3}\alpha\varepsilon s^3$ and

$$f_{in} \approx \alpha(s - \frac{1}{6}\varepsilon s^3) = \alpha\left(\frac{x}{\varepsilon} - \frac{x^3}{6\varepsilon^2}\right).$$

To match the inner solution to the outer solution $f_{out} \approx 1 + B - B\sqrt{2}x/\varepsilon$ for small x, it is clear that we will need $B = -1$, and $\alpha = -\sqrt{2}B = \sqrt{2}$. Therefore, $f_{out} \approx \sqrt{2}x/\varepsilon$, and $f_{in} \approx \sqrt{2}(x/\varepsilon - x^3/6\varepsilon^2)$. Thus, we may by observation conclude that a uniformly valid composite solution is given by

$$f \approx 1 - \exp(-\sqrt{2}x/\varepsilon). \tag{6.40}$$

To find the exact solution, we multiply the equation by df/dx

$$\varepsilon^2 \frac{d^2 f}{dx^2}\frac{df}{dx} + (f - f^3)\frac{df}{dx} = 0 = \frac{\varepsilon^2}{2}\frac{d}{dx}\left(\frac{df}{dx}\right)^2 + \frac{d}{dx}\left(\frac{f^2}{2} - \frac{f^4}{4}\right), \tag{6.41}$$

and obtain the "constant of motion",

$$\frac{\varepsilon^2}{2}\left(\frac{df}{dx}\right)^2 + \frac{f^2}{2} - \frac{f^4}{4} = E = \frac{1}{4}. \tag{6.42}$$

Therefore,

$$\varepsilon\left(\frac{df}{dx}\right) = \sqrt{\frac{1}{2} - f^2 + \frac{f^4}{2}} = \pm\sqrt{\frac{1}{2}}(1 - f^2). \tag{6.43}$$

Choosing the positive moving solution so that $\varepsilon df/dx = \sqrt{\frac{1}{2}}(1 - f^2)$, we find

$$\int_0^f \frac{df}{1 - f^2} = \int_0^x \sqrt{\frac{1}{2}}\frac{dx}{\varepsilon} = \sqrt{\frac{1}{2}}\frac{x}{\varepsilon} = \frac{1}{2}\ln\left|\frac{1+f}{1-f}\right|. \tag{6.44}$$

Therefore,

$$f = \frac{1 - e^{-\sqrt{2}x/\varepsilon}}{1 + e^{-\sqrt{2}x/\varepsilon}}. \tag{6.45}$$

The accompanying program compares the numerical solution, the exact solution and the composite solution. The error of the composite solution from the exact solution occurs in the transition region.

Homework 6.5: Composite Solution: A one dimensional heat transfer problem is governed by the dimensionless equations,

$$\varepsilon\frac{d^2T}{dx^2} + x\frac{dT}{dx} - xT = 0,$$

where $T(0) = T_0$ and $T(1) = T_1$ and T is the scaled temperature. Determine a uniformly valid leading order expansion of $T(x)$ for $\varepsilon \ll 1$ and sketch the results.

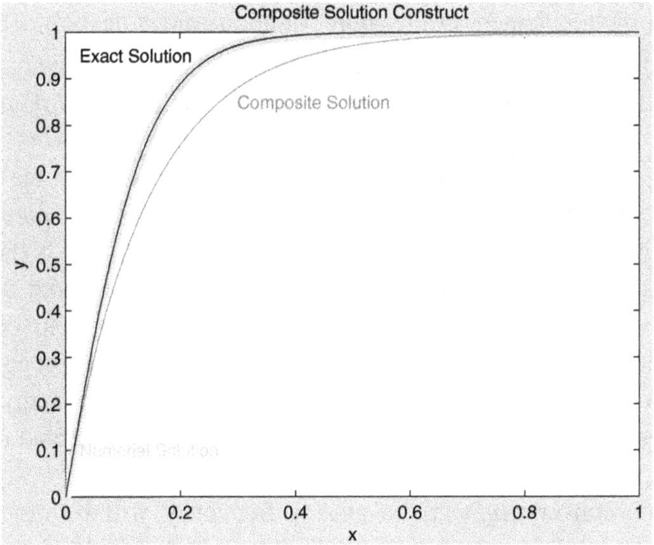


```
function Separatrix
dx=0.001;
N=1000; epsilon=0.1;
x=(0:N)*dx; y=x*0;
for i=1:N
   f=sqrt(1/2-y(i)^2+y(i)^4/2);
   y(i+1)=y(i)+f*dx/epsilon;
end;
y(N+1),
Y=1-exp(-x/sqrt(2)/epsilon);
Z=(1-exp(-sqrt(2)*x/epsilon))./(1+exp(-
sqrt(2)*x/epsilon));
plot(x,y,'g*',x,Y,'r- ',x,Z,'k-')
title('Composite Solution Construct');
xlabel('x'); ylabel('y');
text(0.3,0.85,'Composite Solution','color','r');
text(0.025,0.1,'Numerial Solution','color','g');
text(0.025,0.95,'Exact Solution');
```

6.2.3 *Mode Conversion*

Mode conversion occurs when two waves of different physical properties have the same frequency $\omega_1 = \omega_2$ and wave number $\vec{k}_1 = \vec{k}_2$ with the latter often compensated by the gradient of the equilibrium quantities, namely, $\vec{k}_1 = \vec{k}_2 + \vec{k}_L$, where $\vec{k}_L$ represents the effective wave number from the spatial gradient.

The mode conversion at the upper hybrid resonance was demonstrated in the previous Chapter, and the connection from the EM wave to the ES wave is imminent as the dispersion relation is a natural extension from the cold plasma model to the hot plasma model. Here, we will examine the cutoff-resonance pair. The wave penetrates through the evanescent layer to reach the resonance. The wave numbers are mismatched, but due to the cutoff to both the EM wave and the ES wave, they become easier to match with the help of the spatial gradient. Moreover, when the waves approach the cutoff, their propagation speed is reduced to stand still, that increases the interaction time for energy transfer. As the process is reversible, the mode conversion efficiency is expected to reach 50% at best.

The mode conversion near the plasma frequency will be examined in the following. A transverse wave that is modified by the plasma density matches with the longitudinal plasma wave at the mode conversion layer where $\omega = \omega_{pe}$. This phenomenon is important since it could be a signature from the earthquake, an absorption mechanism for photo cells, and an energy channel from thermal energy to blackbody radiation. In the nonlinear large amplitude regime, it can generate dc magnetic field and vortex flow, evidenced in the high power laser experiments and in the solar flare.

The occurrence of the mode conversion needs an electric field component along the density gradient since the EM wave has the electric field $\vec{E} \cdot \vec{k}_1 = 0$, while the electrostatic wave has $\vec{E} \times \vec{k}_2 = 0$. The electric field orientation matters for the conversion efficiency.

Combining the linearized continuity equation $\partial \delta n / \partial t + \nabla \cdot n_0 \delta \vec{v} = 0$ and the linearized momentum equation $n_0 m \partial \delta \vec{v} / \partial t = -\nabla \delta p + n_0 e \delta \vec{E}$ gives us the density equation

$$\frac{\partial^2 \delta n}{\partial t^2} - V_T^2 \nabla^2 \delta n + \omega_p^2 \delta n + \frac{e}{m} \delta \vec{E} \cdot \nabla n_0 = 0. \tag{6.46}$$

Similarly, combining the Ampere's law and the Faraday's law gives

$$\frac{\partial^2 \delta \vec{E}}{\partial t^2} + c^2 \nabla \times (\nabla \times \delta \vec{E}) = -4\pi \frac{\partial}{\partial t} \delta \vec{j} = 4\pi \frac{e}{m} (\nabla \delta p - n_0 e \delta \vec{E}), \tag{6.47}$$

where the current density $\delta \vec{j} = n_0 e \delta \vec{v}$ is applied. Without loss of generality, the EM wave vector is assumed to be $\vec{k} = k_x \hat{e}_x + k_z \hat{e}_z$ in the vacuum, where the $\hat{e}_z$

direction is along the density gradient. Since the electric field component that is perpendicular to both the density gradient and the wave vector does not contribute to the mode conversion process, we define two dimensionless variables:

$$\xi \equiv 4\pi\,\delta ne = (\partial\delta E_z/\partial z + ik_x\delta E_x), \tag{6.48}$$

that measures the electrostatic charge density, and

$$\varsigma \equiv \widehat{e}_y \cdot \nabla \times \delta\vec{E} = \partial\delta E_x/\partial z - ik_x\delta E_z, \tag{6.49}$$

that represents the electromagnetic wave. Eqs. (6.46) and (6.47) become,

$$\frac{\partial^2\xi}{\partial t^2} - V_T^2\nabla^2\xi + \omega_p^2\xi + \frac{\partial\omega_p^2}{\partial z}\delta E_z = 0, \tag{6.50}$$

$$\frac{\partial^2\varsigma}{\partial t^2} - c^2\nabla^2\varsigma + \omega_p^2\varsigma + \frac{\partial\omega_p^2}{\partial z}\delta E_x = 0. \tag{6.51}$$

The mode conversion arises from the coupling of the electric field with the density gradient in the last terms in both equations. It is noteworthy that the first three terms in Eqs. (6.50) and (6.51) give the dispersion relation of the electrostatic wave, $\omega^2 = \omega_p^2 + k^2V_T^2$, and the electromagnetic wave, $\omega^2 = \omega_p^2 + k^2c^2$, respectively. They would merge to one if V_T approaches the speed of light, and the conversion efficiency would continue to increase according to the nonrelativistic theory. The relativistic mass dependency, however, would eventually set in and make the conversion efficiency saturate. We will not treat the relativistic theory here by setting $\beta \equiv V_T^2/c^2 \ll 1$.

We assume the density profile to be given by $\omega_p^2 = \omega_{p0}^2(1 + \sigma\tanh(z/l))$, where $\omega^2 = \omega_{p0}^2$. The density reaches plateaus beyond the boundary layer of sharp density gradient around $-l < z < l$ with $(1-\sigma)\omega_{p0}^2$ to its left and $(1+\sigma)\omega_{p0}^2$ to the right. We will take $k_xl \ll 1$, $\xi = a(z)e^{ik_xx-i\omega t}$, $\varsigma = b(z)e^{ik_xx-i\omega t}$, and put the equations in the dimensionless form by the new variables $s \equiv z\omega/c$, $\varepsilon \equiv l\omega/c$, $\kappa \equiv k_xc/\omega$:

$$-\beta\frac{d^2a}{ds^2} + \beta\kappa^2a + a\sigma\tanh\frac{s}{\varepsilon} + \sigma\frac{E_z}{\varepsilon}\,\mathrm{sec}\,h^2\frac{s}{\varepsilon} = 0, \tag{6.50a}$$

$$-\frac{d^2b}{ds^2} + \kappa^2b + b\sigma\tanh\frac{s}{\varepsilon} + \sigma\frac{E_x}{\varepsilon}\,\mathrm{sec}\,h^2\frac{s}{\varepsilon} = 0. \tag{6.51a}$$

The electric fields are governed by

$$a = \frac{\partial E_z}{\partial s} + i\kappa E_x, \quad b = \frac{\partial E_x}{\partial s} - i\kappa E_z. \tag{6.52}$$

Taking the integration $\int_{0-}^{0+} ds$ of Eqs. (6.52) over the mode conversion layer shows the electric fields are continuous across the layer. Eqs. (6.52) can be cast into the following:

$$\frac{\partial^2 E_z}{\partial s^2} - \kappa^2 E_z = \frac{\partial a}{\partial s} - i\kappa b, \tag{6.52a}$$

$$\frac{\partial^2 E_x}{\partial s^2} - \kappa^2 E_x = \frac{\partial b}{\partial s} + i\kappa a. \tag{6.52b}$$

Thus, these Poisson's equations can be solved for the given a and b.

We first consider the simpler situation when $\varepsilon \to 0$. Utilizing $\int_{-\infty}^{\infty} \frac{1}{2\varepsilon} \sec h^2 (x/\varepsilon)dx = 1$,

$$\lim_{\varepsilon \to 0} \tanh(s/\varepsilon) = s/|s|,$$

and

$$\lim_{\varepsilon \to 0} \frac{1}{2\varepsilon} \sec h^2(x/\varepsilon) = \delta(x),$$

we will solve the problem for now with use of the jump conditions. The equations to be solved are,

$$-\beta\frac{d^2 a}{ds^2} + \beta\kappa^2 a + \frac{s}{|s|}\sigma a + 2\sigma E_z \delta(s) = 0, \tag{6.50b}$$

$$-\frac{d^2 b}{ds^2} + \kappa^2 b + \frac{s}{|s|}\sigma b + 2\sigma E_x \delta(s) = 0. \tag{6.51b}$$

Taking the integration $\int_{0-}^{0+} ds$ on both sides of Eqs. (6.50b) and (6.51b) gives the jump conditions:

$$\left[\frac{da}{ds}\right]_{s=0} = 2\frac{\sigma}{\beta} E_{z0}, \quad \left[\frac{db}{ds}\right]_{s=0} = 2\sigma E_{x0}. \tag{6.53}$$

Waves are evanescent in the $s > 0$ region so that

$$a_> = a_{0R} \exp(-s\sqrt{\sigma\beta^{-1} + \kappa^2}),$$
$$b_> = b_{0R} \exp(-s\sqrt{\sigma + \kappa^2}), \tag{6.54}$$

and propagative in the $s < 0$ region:

$$a_< = a_{0L} \exp(-is\sqrt{\sigma\beta^{-1} - \kappa^2}),$$
$$b_< = \exp(is\sqrt{\sigma - \kappa^2}) + b_{0L} \exp(-is\sqrt{\sigma - \kappa^2}). \tag{6.55}$$

Here we consider an incoming EM wave with unity amplitude and its reflected wave of amplitude b_{0L}, and the converted ES wave of amplitude a_{0L} that is outgoing from the resonance layer. We then readily find the electric fields due to both EM and ES waves from the particular solutions to Eq. (6.52):

$$E_{z>} = \frac{i}{\sigma}\left(i\beta\sqrt{\sigma\beta^{-1}+\kappa^2}a_{0R}e^{-s\sqrt{\sigma\beta^{-1}+\kappa^2}} - \kappa b_{0R}e^{-s\sqrt{\sigma+\kappa^2}}\right), \tag{6.56a}$$

$$E_{x>} = \frac{i}{\sigma}\left(\beta\kappa a_{0R}e^{-s\sqrt{\sigma\beta^{-1}+\kappa^2}} + i\sqrt{\sigma+\kappa^2}b_{0R}e^{-s\sqrt{\sigma+\kappa^2}}\right), \tag{6.56b}$$

$$E_{z<} = \frac{i}{\sigma}\left(a_{0L}\beta\sqrt{\sigma\beta^{-1}-\kappa^2}e^{-is\sqrt{\sigma\beta^{-1}-\kappa^2}} + (b_{0L}e^{-is\sqrt{\sigma-\kappa^2}} + e^{is\sqrt{\sigma-\kappa^2}})\kappa^2\right), \tag{6.56c}$$

$$E_{x<} = \frac{i}{\sigma}\left(-a_{0L}\beta\kappa e^{-is\sqrt{\sigma\beta^{-1}-\kappa^2}} + (b_{0L}e^{-is\sqrt{\sigma-\kappa^2}} - e^{is\sqrt{\sigma-\kappa^2}})\sqrt{\sigma-\kappa^2}\right). \tag{6.56d}$$

The continuity of electric fields across the boundary layer gives two equations for the unknown coefficients:

$$ia_{0R}\beta\sqrt{\sigma\beta^{-1}+\kappa^2} - b_{0R}\kappa = a_{0L}\beta\sqrt{\sigma\beta^{-1}-\kappa^2} + (b_{0L}+1)\kappa, \tag{6.57}$$

$$a_{0R}\beta\kappa + ib_{0R}\sqrt{\sigma+\kappa^2} = -a_{0L}\beta\kappa + (b_{0L}-1)\sqrt{\sigma-\kappa^2}. \tag{6.58}$$

The other two equations are obtained from the jump conditions of Eq. (6.53):

$$\left[\frac{da}{ds}\right]_{s=0} = 2\sigma E_{z0}\beta^{-1} = -a_{0R}\sqrt{\sigma\beta^{-1}+\kappa^2} + ia_{0L}\sqrt{\sigma\beta^{-1}-\kappa^2}, \tag{6.59}$$

$$\left[\frac{db}{ds}\right]_{s=0} = 2\sigma E_{x0} = -b_{0R}\sqrt{\sigma+\kappa^2} - i(1-b_{0L})\sqrt{\sigma-\kappa^2}. \tag{6.60}$$

Putting these four equations in the matrix form, we find

$$\begin{bmatrix} -i\beta\sqrt{\sigma\beta^{-1}+\kappa^2} & \kappa & \beta\sqrt{\sigma\beta^{-1}-\kappa^2} & \kappa \\ \beta\kappa & i\sqrt{\sigma+\kappa^2} & \beta\kappa & -\sqrt{\sigma-\kappa^2} \\ -\sqrt{\sigma\beta^{-1}+\kappa^2} & -2i\beta^{-1}\kappa & -i\sqrt{\sigma\beta^{-1}-\kappa^2} & 0 \\ 2i\beta\kappa & -\sqrt{\sigma+\kappa^2} & 0 & -i\sqrt{\sigma-\kappa^2} \end{bmatrix} \begin{bmatrix} a_{0R} \\ b_{0R} \\ a_{0L} \\ b_{0L} \end{bmatrix}$$

$$= \begin{bmatrix} -\kappa \\ -\sqrt{\sigma-\kappa^2} \\ 0 \\ -i\sqrt{\sigma-K_s^2} \end{bmatrix} \tag{6.61}$$

```
function ModeConversion(T)                          % T is in unit of keV.
close all; clc;
syms k S b;
m11=-i*b*sqrt(S/b+k^2); m12=k;
m13=b*sqrt(S/b-k^2);       m14=k;
m21=b*k;    m22=i*sqrt(S+k^2);
m23=b*k;    m24=-sqrt(S-k^2);
m31=-sqrt(S/b+k^2);  m32=-2*i*k/b;
m33=-i*sqrt(S/b-k^2);  m34=0;
m41=2*i*b*k;    m42=-sqrt(S+k^2);
m43=0;  m44=-i*sqrt(S-k^2);
M=[m11 m12 m13 m14;m21 m22 m23 m24;m31
m32 m33 m34;m41 m42 m43 m44];
Minv=inv(M)
V=[-m14; m24; 0; m44];
Coeff=Minv*V, n= -1; hold on;
b=T/500; % T the temperature in keV.
for S=0.25:0.25:0.75 % sigma value for density
Efficiency=[]; Eff=[]; n=n+1;
for angle=5:1:85,
  theta=pi*angle/180;
  k=sin(theta);
  Reflection=abs(eval(Coeff(4))).^2;
  ModeConversion=1-Reflection;
  Efficiency=[Efficiency,ModeConversion*100];
end;
Angle=5:1:85;
plot(Angle,Efficiency,'g*',Angle,Efficiency,'r -');
title('Mode Conversion at Plasma Frequency');
str=sprintf('temperature %d keV',T);
text(10,45,str); axis([0 90 0 50]);
xlabel('the incident angle'); ylabel('efficiency (%)');
str=sprintf('=%3.2f',S);
text(30+n*15,10,str); text(29+n*15,10,'\sigma');
getframe;
end;
hold off;
```

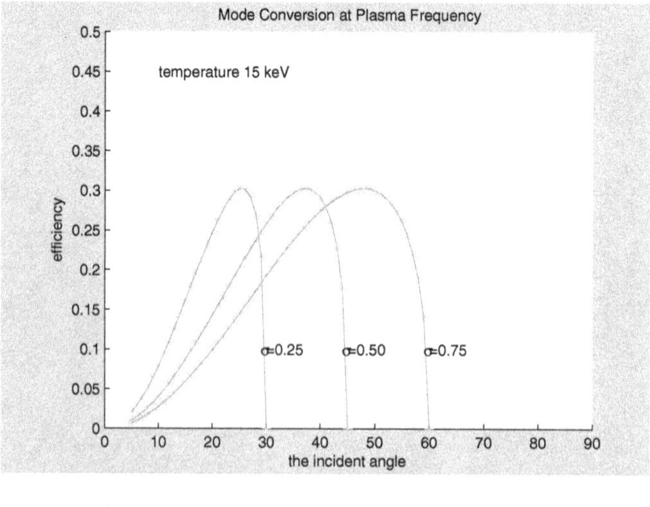

The matrix Eq. (6.61) is inverted symbolically on MATLAB , and the mode conversion efficiency given by $\eta \equiv 1 - |b_{0L}|^2$ is plotted in the program ModeConversion.m. The code allows the temperature and the density plateaus to be varied. The sharper gradient makes the efficiency weaker. Higher temperature tends to increase the mode conversion efficiency. The best efficiency is expected, however, to be around 50% since the EM and ES waves reach equilibration for this **reversible process** of mode conversion. As the problem is being solved by the Fourier transform, it finds only the steady state solution that would have allowed energy to transfer back to EM wave from the ES wave, contrary to the transient situation. The latter requires a different numerical scheme, such as the **finite difference time domain (FDTD)** approach. The mode conversion efficiency increases with the temperature. However the nonrelativistic formulation breaks down when the plasma thermal energy is a substantial fraction of the electron rest mass energy.

> **Homework 6.6: Jump Condition**: With use of the ModeConversion program, show that the values of ξ and ς are continuous, i.e., $a_{0R} = a_{0L}$, $b_{0R} = 1 + b_{0L}$.

6.2.4 *Mode Conversion by the Matching Procedure*

We want to solve Eqs. (6.50a) and (6.51a) by the multiple space scales. Outside the boundary layer, $s \sim O(1)$, to lowest order, we have

$$s > 0, \quad \frac{d^2 b_>}{ds^2} + \kappa^2 b + \sigma b_> = 0 = -\beta \frac{d^2 a_>}{ds^2} + \beta\kappa^2 a_> + \sigma a_>,$$

$$s < 0, \quad -\frac{d^2 b_<}{ds^2} + \kappa^2 b_< - \sigma b_< = 0 = -\beta \frac{d^2 a_<}{ds^2} + \beta\kappa^2 a_< - \sigma a_<.$$

The solutions are given by Eqs. (6.54) and (6.55). They have the following series expansions,

$$\begin{aligned} a_> &\xrightarrow{s\to 0} a_{0R}(1 - s\sqrt{\sigma\beta^{-1} + \kappa^2} + \cdots), \\ b_> &\xrightarrow{s\to 0} b_{0R}(1 - s\sqrt{\sigma + \kappa^2} + \cdots), \end{aligned} \tag{6.62a}$$

$$\begin{aligned} a_< &\xrightarrow{s\to 0} a_{0L}(1 - is\sqrt{\sigma\beta^{-1} - \kappa^2} + \cdots), \\ b_< &\xrightarrow{s\to 0} 1 + b_{0L} + is(1 - b_{0L})\sqrt{\sigma - \kappa^2} + \cdots \end{aligned} \tag{6.62b}$$

The electric fields are given by Eq. (6.56) and can be expanded as in the following:

$$E_{z>} \to \frac{i}{\sigma}\left(i\beta\sqrt{\sigma\beta^{-1} + \kappa^2}a_{0R}(1 - s\sqrt{\sigma\beta^{-1} + \kappa^2}) - \kappa b_{0R}(1 - s\sqrt{\sigma + \kappa^2})\right),$$

$$E_{x>} \to \frac{i}{\sigma}\left(\beta\kappa a_{0R}(1 - s\sqrt{\sigma\beta^{-1} + \kappa^2}) + i\sqrt{\sigma + \kappa^2} b_{0R}(1 - s\sqrt{\sigma + \kappa^2})\right),$$

$$E_{z<} \to \frac{i}{\sigma}\left(a_{0L}\beta\sqrt{\sigma\beta^{-1} - \kappa^2}(1 - is\sqrt{\sigma\beta^{-1} - \kappa^2})\right.$$
$$\left. + (1 + b_{0L})\kappa + (1 - b_{0L})is\sqrt{\sigma - \kappa^2}\kappa\right),$$

$$E_{x<} \to \frac{i}{\sigma}\left(-a_{0L}\beta\kappa(1 - is\sqrt{\sigma\beta^{-1} - \kappa^2})\right.$$
$$\left. + (b_{0L} - 1)\sqrt{\sigma - \kappa^2} - (b_{0L} + 1)is(\sigma - \kappa^2)\right). \quad (6.63)$$

In the boundary layer, where $s \sim O(\varepsilon)$ we may change the length variable to $s = \varepsilon\rho$, where $\rho \sim O(1)$. Therefore,

$$-\beta\frac{d^2a}{\varepsilon^2 d\rho^2} + \beta\kappa^2 a + \sigma a \tanh\rho + \sigma\frac{E_z}{\varepsilon}\,\mathrm{sec\,h}^2\rho = 0, \quad (6.64a)$$

$$-\frac{d^2b}{\varepsilon^2 d\rho^2} + \kappa^2 b + \sigma b \tanh\rho + \sigma\frac{E_x}{\varepsilon}\,\mathrm{sec\,h}^2\rho = 0. \quad (6.64b)$$

These two equations are to be solved to $O(\varepsilon)$. Before doing so, we need to have a handle on the electric fields which are governed by $a = \partial E_z/\partial s + iE_x$, $b = \partial E_x/\partial s - iE_z$, and can be obtained by the following series expansion:

$$E_z^{in} = E_{z0} - iE_{x0}s + a_0 s + O(\varepsilon^2), \quad (6.65a)$$

$$E_x^{in} = E_{x0} + iE_{z0}s + b_0 s + O(\varepsilon^2). \quad (6.65b)$$

Thus, we find,

$$a_{in} = a_0 + a_1\rho + \varepsilon\sigma\beta^{-1}E_{z0}\log(\cosh(\rho)) + O(\varepsilon^2), \quad (6.66)$$

$$b_{in} = b_0 + b_1\rho + \varepsilon\sigma\beta^{-1}E_{x0}\log(\cosh(\rho)) + O(\varepsilon^2). \quad (6.67)$$

Taking the limit of $|\rho| \gg 1$ so to match with the outer solutions, we find,

$$a_{in} \to \begin{cases} a_0 + a_1 s/\varepsilon + \sigma\beta^{-1}(s - \varepsilon\log 2)E_{z0} + \cdots, \\ a_0 + a_1 s/\varepsilon - \sigma\beta^{-1}(s + \varepsilon\log 2)E_{z0} + \cdots, \end{cases} \quad (6.68)$$

$$b_{in} \to \begin{cases} b_0 + b_1 s/\varepsilon + \sigma(s - \varepsilon\log 2)E_{x0} + \cdots, \\ b_0 + b_1 s/\varepsilon - \sigma(s + \varepsilon\log 2)E_{x0} + \cdots, \end{cases} \quad (6.69)$$

Matching a_{in} in Eq. (6.68) to $a_>$ and $a_<$ of Eq. (6.62) gives

$$a_{0R} = a_0 - \sigma\beta^{-1}\varepsilon\log 2E_{z0}, \quad -a_{0R}\sqrt{\sigma\beta^{-1} + \kappa^2} = \sigma\beta^{-1}E_{z0} + a_1/\varepsilon,$$

$$a_{0L} = a_0 - \sigma\beta^{-1}\varepsilon\log 2E_{z0}, \quad -ia_{0L}\sqrt{\sigma\beta^{-1} - \kappa^2} = -\sigma\beta^{-1}E_{z0} + a_1/\varepsilon.$$

Thus by eliminating both a_0 and a_1, we have

$$a_{0R} = a_{0L}, \tag{6.70}$$

and

$$2\sigma\beta^{-1}E_{z0} = -a_{0R}\sqrt{\sigma\beta^{-1}+\kappa^2} + ia_{0L}\sqrt{\sigma\beta^{-1}-\kappa^2}. \tag{6.71}$$

Equation (6.71) is identical to Eq. (6.50). Similarly by matching b_{in} in Eq. (6.69) to $b_>$ and $b_<$ of Eq. (6.62) gives

$$b_{0R} = b_0 - \sigma\varepsilon\log 2E_{x0} = 1 + b_{0L}$$

and

$$-b_{0R}\sqrt{\sigma+\kappa^2} = \sigma E_{x0} + b_1/\varepsilon, \quad i(1-b_{0L})\sqrt{\sigma-\kappa^2} = -\sigma E_{x0} + b_1/\varepsilon.$$

Thus, eliminating both b_0 and b_1 gives

$$b_{0R} = 1 + b_{0L}, \tag{6.72}$$

$$i(1-b_{0L})\sqrt{\sigma-\kappa^2} + b_{0R}\sqrt{\sigma+\kappa^2} = -2\sigma E_{x0}. \tag{6.73}$$

This is identical to Eq. (6.60).

Matching the electric fields in Eqs. (6.65) with Eqs. (6.63) gives from the constant terms,

$$\sigma E_{z0} = -\beta\sqrt{\sigma\beta^{-1}+\kappa^2}a_{0R} - i\kappa b_{0R} = ia_{0L}\beta\sqrt{\sigma\beta^{-1}-\kappa^2} + i(1+b_{0L})\kappa, \tag{6.74}$$

which is identical to (6.57), and

$$\sigma E_{x0} = i\left(\beta\kappa a_{0R} + i\sqrt{\sigma+\kappa^2}b_{0R}\right) = i\left(-a_{0L}\beta\kappa + (b_{0L}-1)\sqrt{\sigma-\kappa^2}\right), \tag{6.75}$$

that is identical to (6.58).

Eliminating the electric fields from Eqs. (6.71), (6.73)–(6.75), we arrive at

$$\begin{aligned}
&\frac{1}{2}a_{0L}(-\sqrt{\sigma\beta^{-1}+\kappa^2} + i\sqrt{\sigma\beta^{-1}-\kappa^2}) \\
&\quad = -\sqrt{\sigma\beta^{-1}+\kappa^2}a_{0L} - \beta^{-1}i\kappa(1+b_{0L}) \\
&\quad = ia_{0L}\sqrt{\sigma\beta^{-1}-\kappa^2} + i\beta^{-1}(1+b_{0L})\kappa \\
&\qquad -\frac{1}{2}\left(i(1-b_{0L})\sqrt{\sigma-\kappa^2} + (1+b_{0L})\sqrt{\sigma+\kappa^2}\right) \\
&\quad = i\left(\beta\kappa a_{0L} + i\sqrt{\sigma+\kappa^2}(1+b_{0L})\right) \\
&\quad = i(-a_{0L}\beta\kappa + (b_{0L}-1)\sqrt{\sigma-\kappa^2})
\end{aligned}$$

The terms on the LHS of the above two equations are simply the sum of the terms on the right but divided by 2, the redundancy unnecessary for finding a_{0L} and b_{0L}. The LHS terms can then be solved to find,

$$b_{0L} = \frac{i\left(\sqrt{\sigma - \kappa^2} + i\sqrt{\sigma + \kappa^2}\right)\left(\sqrt{\sigma\beta^{-1} + \kappa^2} + i\sqrt{\sigma\beta^{-1} - \kappa^2}\right) + 4\kappa^2}{i\left(\sqrt{\sigma - \kappa^2} - i\sqrt{\sigma + \kappa^2}\right)\left(\sqrt{\sigma\beta^{-1} + \kappa^2} + i\sqrt{\sigma\beta^{-1} - \kappa^2}\right) - 4\kappa^2}. \tag{6.76}$$

The solution to lowest order does not depend on the layer thickness $\varepsilon \equiv l\omega/c \ll 1$. Therefore, b_{0L} is identical to that obtained from Eq. (6.61) by the jump condition which can yield only to this order of accuracy, while the matching procedure, in principle, can achieve higher order accuracy if so desired.

Homework 6.7: Reflected Wave: With use of the ModeConversion program, verify that the values of b_{0L} as given by Eq. (6.76) is identical to that found from Eq. (6.61).

Homework 6.8: Budden Equation: Find the transmission and reflection coefficients with use of the multiple length scale for the Budden equation, $\frac{d^2\psi}{dx^2} + (1 + \frac{\delta}{x})\psi = 0$, where $0 < \delta \ll 1$.

Homework 6.9: Secularity and Singularity: Explain the meanings of secularity and singularity, give examples to elucidate their properties, and clarify their differences.

Further Reading

The multiple time scale is popularly adopted to solve plasma physics problems. A good reference book is R. C. Davidson (1972). Gyrokinetic theory applies the multiple time and multiple space scales to the low-frequency electromagnetic responses in general plasma equilibria (Frieman and Chen 1982). It was applied to the particle simulation model (Lee, 1987).

Tachikawa and Fujimoto (2007) found that diversification of the time scales emerges through the multi clustering process for the coupled oscillators. The model imitates the self-organizing phenomena in highly complex system such as living organisms. The stochasticity boundary indicates that heating can be effective at the three-halves harmonic, $\nu = 1.5$, with large amplitude pump wave (Hsu 1981).

The boundary layer theory was well documented by Schlichting (1979).

Homework Hints

Homework 6.1: Van der Pol equation: Solve the following equation with use of the multiple time scales,

$$\frac{d^2x}{d\tau^2} - \varepsilon(1 - \beta x^2)\frac{dx}{d\tau} + x = 0$$

where all quantities are dimensionless, $0 < \varepsilon \ll 1$, and $\beta > 0$ but $\beta \sim O(1)$.

```
function VanderPol
e=0.1;  beta=0.5;
dt=0.01; T=250; N=T/dt;
t=dt:dt:T;
v=t*0;
x=t*0;
x(1)=0.1; v(1)=0;
for i=2:N
v(i)=v(i-1)+dt*e*(1-beta*x(i-1)^2)*v(i-1)-x(i-1)*dt;
x(i)=x(i-1)+v(i)*dt;
end;
A=x(1)./sqrt(beta/4*x(1)^2*(1-exp(-e*t))+exp(-e*t));
plot(t,x,'b',t, A,'r-',t,-A,'r-');
title('Vander Pol Equation');
xlabel('time'); ylabel('Amplitude');
```

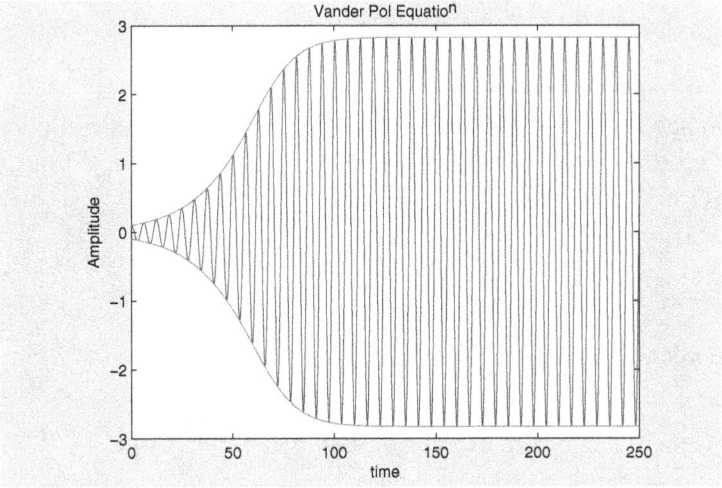

To lowest order, the solution is given by

$$x^{(0)}(\tau_0, \tau_1, \ldots) = A(\tau_1, \tau_2, \ldots)\cos[\tau_0 + \varphi(\tau_1, \tau_2, \ldots)]. \tag{6.1.1}$$

To next order,

$$\frac{\partial^2 x^{(1)}(\tau_0, \tau_1, \ldots)}{\partial \tau_0^2} + x^{(1)}(\tau_0, \tau_1, \ldots)$$
$$= -2\frac{\partial^2 x^{(0)}(\tau_0, \tau_1, \ldots)}{\partial \tau_0 \partial \tau_1} + [1 - \beta x^{(0)^2}(\tau_0, \tau_1, \ldots)]\frac{\partial x^{(0)}(\tau_0, \tau_1, \ldots)}{\partial \tau_0}. \tag{6.1.2}$$

Applying Eq. (6.9), we arrive at the following equations:

$$\frac{\partial \varphi_1}{\partial \tau_1} = 0, \quad \text{and} \quad \frac{\partial A}{\partial \tau_1} = \frac{A}{2}\left[1 - \frac{\beta A^2}{4}\right]. \tag{6.1.3}$$

The phase angle is constant in the longer time scale. By changing the variable to $z = A^2/2$, the governing equation is then given by $\partial z/\partial \tau_1 = z[1 - \beta z/2]$, which has the solution $\ln(z/z_0) - \ln[(\beta z/2 - 1)/(\beta z_0/2 - 1)] = \tau_1$, where z_0 is the initial value. Thus, $A = A_0[\frac{1}{4}\beta A_0^2(1 - e^{-\tau_1}) + e^{-\tau_1}]^{-1/2}$, $A = A_0$ at $\tau_1 = 0$, and the saturation amplitude is given by$A_\infty = 2/\sqrt{\beta}$ at $\tau_1 \to \infty$.

Homework 6.3: Composite Solution: Determine a leading order composite solution for the boundary value problem

$$\varepsilon y'' + 2(2x - 1)(y' + 2y) = 0,$$

where $\varepsilon \ll 1$ and $y(0) = 1$, $y(1) = 2/e$. Construct a uniformly valid expansion for $y(x)$ and sketch the solution.

It is important to observe that this problem is easier to solve by locating the boundary layer at $x = \frac{1}{2}$, where we define $z \equiv (x - \frac{1}{2})/\sqrt{\varepsilon}$, the governing equation is

$$\frac{d^2 y}{dz^2} + 4z\left(\frac{dy}{dz} + 2\sqrt{\varepsilon} y\right) = 0. \tag{6.3.1}$$

To lowest order, $O(1)$,

$$\frac{d^2 y_{in}^{(0)}}{dz^2} + 4z\frac{dy_{in}^{(0)}}{dz} = 0 = \frac{d}{dz}\left(e^{2z^2}\frac{dy_{in}^{(0)}}{dz}\right). \tag{6.3.2}$$

It has the solution $y_{in}^{(0)} = a + b\int_0^z \sqrt{8/\pi}\, dz' e^{-2z'2}$.

In the region to the right hand side of $x = \frac{1}{2}$, we have to lowest order, $dy_>^{(0)}/dz + 2y_>^{(0)} = 0$, that has the solution $y_>^{(0)} = 2e^{1-2x} \xrightarrow{x \to \frac{1}{2}} 2 + 2(1 - 2x) + \cdots = 2 - 4\varepsilon z$.

It satisfies the boundary condition $y(1) = 2/e$. To the left hand side of $x = \frac{1}{2}$,

$$y_<^{(0)} = e^{-2x} \xrightarrow{x \to \frac{1}{2}} e^{-1} - e^{-1}(2x-1) = e^{-1} - e^{-1}(2\varepsilon z). \tag{6.3.3}$$

By matching the asymptotic limits of the inner solution to the outer solutions, we may choose the composite solution,

$$y = a + b\int_0^z \sqrt{8/\pi}\, dz' e^{-2z'^2} - 2(1 - e^{1-2x}), \quad x > \frac{1}{2}, \tag{6.3.4a}$$

$$y = a + b\int_0^z \sqrt{8/\pi}\, dz' e^{-2z'^2} - \left(\frac{1}{e} - e^{-2x}\right), \quad x < \frac{1}{2}. \tag{6.3.4b}$$

Thus,

$$a + b\int_0^{\frac{1}{2\sqrt{\varepsilon}}} \sqrt{8/\pi}\, dz' e^{-2z'^2} - 2\left(1 - \frac{1}{e}\right) = \frac{2}{e},$$

and

$$a - b\int_0^{-\frac{1}{2\sqrt{\varepsilon}}} \sqrt{8/\pi}\, dz' e^{-2z'^2} - \left(\frac{1}{e} - 1\right) = 1.$$

Therefore, we have

$$a = 1 + \frac{1}{2e}, \quad b\int_0^{\frac{1}{2\sqrt{\varepsilon}}} \sqrt{8/\pi}\, dz' e^{-2z'^2} = 1 - \frac{1}{2e}. \tag{6.3.5}$$

The error is still substantial in the inner solution due to the fact that there is no boundary condition to be referenced. If the value at $x = 0.5$ is adjusted according to the exact solution, the consistency will be vastly improved.

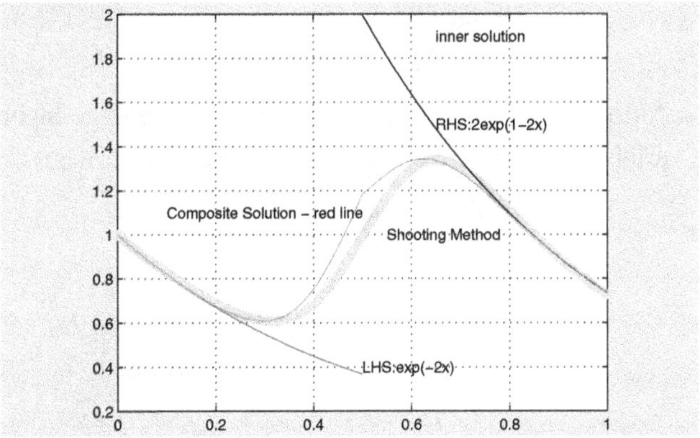

```
function HW6p3
clear all; close all; clc; % epsilon=0.1; z(1)= -1.804125;
N=1000; epsilon=0.05; rte=sqrt(epsilon); e=exp(1);
dx=(1-1.0e-4)/N;
x=1.0e-4:dx:1;
N=N+1; ih=(N+1)/2;
dz=dx/rte;
yp=2*exp(1-2*x).*(x>=0.5); yn=exp(-2*x).*(x<=0.5);
z=abs(x(ih:N)-0.5)/rte; X=2*z*rte;
W=sqrt(pi/8)*Simpson(exp(-z.^2*2),ih,dz);
a=1+1/2/e; b=(1-1/2/e)/W(ih);
Y=a+b*[-W(ih:-1:1),W(2:ih)];
y(1)=1; z(1)=-1.9043375;
Z=Y-(1/e-exp(-2*x)).*(x<=1/2)-2*(1-exp(1-2*x)).*(x>1/2);
for i=1:N-1
  y(i+1)=y(i)+z(i)*dx;
  z(i+1)=z(i)-2/epsilon*(2*x(i)-1).*(z(i)+2*y(i))*dx;
end;
y(N),
plot(x,y,'g*',x,Y,'c -',x(ih:N),yp(ih:N),'k-',x(1:ih-1),
yn(1:ih-1),'b-',x,Z,'r-');
grid on;
text(0.65,1.5,'RHS:2exp(1-2x)');
text(0.65,1.9,'inner solution');
text(0.5,0.4,'LHS:exp(-2x)');
text(0.55,1,'Shooting Method');
text(0.1,1.1,'Composite Solution - red line');
```

Homework 6.5: Composite Solution: A one dimensional heat transfer problem is governed by the dimensionless equations,

$$\varepsilon\frac{d^2T}{dx^2} + x\frac{dT}{dx} - xT = 0,$$

where $T(0) = T_0$ and $T(1) = T_1$ and T is the scaled temperature. Determine a uniformly valid leading order expansion of $T(x)$ for $\varepsilon \ll 1$ and sketch the results.

In the main region,

$$\frac{dT_{out}^{(0)}}{dx} - T_{out}^{(0)} = 0, \tag{6.5.1}$$

that gives the solution, $T_{out}^{(0)} = T_1\exp(x-1)$. In the boundary layer, by changing the coordinate variable to $z = x/\sqrt{\varepsilon}$, the governing equation is given by

$$\frac{d^2T}{dz^2} + z\frac{dT}{dz} - \sqrt{\varepsilon}zT = 0. \tag{6.5.2}$$

To lowest order, the solution is found from the first two terms, $T_{in}^{(0)} = T_0 + a\int_0^{x/\sqrt{\varepsilon}} dze^{-z^2/2}$.

$$T_{in}^{(0)} = T_0 + a\int_0^{x/\sqrt{\varepsilon}} dze^{-z^2/2} \to T_0 + a\sqrt{\frac{\pi}{2}}. \tag{6.5.3}$$

Therefore, $a = (T_1/e - T_0)\sqrt{2/\pi}$ so that the inner solution approaches that of the outer solution in the limit of $x \to 0$. The composite solution is given by

$$T = T_0 + \left(\frac{T_1}{e} - T_0\right)\sqrt{\frac{2}{\pi}}\int_0^{x/\sqrt{\varepsilon}} dz e^{-z^2/2} + \frac{T_1}{e}(e^x - 1). \tag{6.5.4}$$

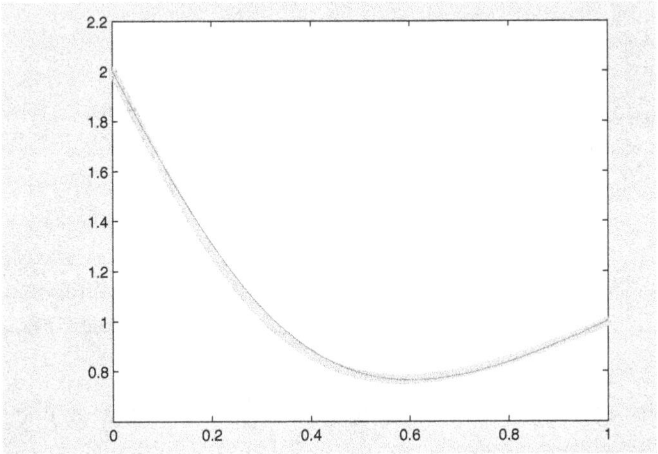


```
function HW6p5                                          %Solve the
N=1000; epsilon=0.1;                                    equation by the
dx=(1-1.0e-4)/N;                                        shooting method
x=1.0e-4:dx:1;                                          and compare with
N=N+1;                                                  the analytical
T0=2.0; T1=1.0;                                         solution.
Y=T0+T1*exp(-1)*(exp(x)-1);
Y=Y+(T1*exp(-1)-T0)*sqrt(2/pi)*Simpson(exp(-
x.^2/2/epsilon),N,dx)/sqrt(epsilon);
y(1)=T0; z(1)=(T1*exp(-1)-T0)*sqrt(pi/2)/sqrt(epsilon)+T1*exp(-1);
z(1)=z(1)+2.15;
for i=1:N-1
   y(i+1)=y(i)+z(i)*dx;
   z(i+1)=z(i)+x(i)*(y(i)-z(i))*dx/epsilon;
end;
y(N),
plot(x,y,'g*',x,Y,'r-');
```

Note that as $x \to 0$, $T \to T_0$, and as $x \to 1$,

$$T \to T_0 + \left(\frac{T_1}{e} - T_0\right) + \frac{T_1}{e}(e^1 - 1) \to T_1.$$

Homework 6.7: Reflected Wave: With use of the ModeConversion program, verify that the values of b_{0L} as given by Eq. (6.55) is identical to that found from Eq. (6.61).

```
%In the ModeConversion program after the
line:
Coeff=Minv*V, add these lines:
A=sqrt(S/b+k^2)+i*sqrt(S/b-k^2);
B=sqrt(S-k^2)+i*sqrt(S+k^2);
C=sqrt(S-k^2)-i*sqrt(S+k^2);
b0L=(i*A*B+4*k^2)/(i*A*C-4*k^2);
dB0=b0L-Coeff(4);
dB0=simplify(dB0),
```

```
%The program would give
dB0=0.
```

Homework 6.9: Secularity and Singularity: Explain the meanings of secularity and singularity, give examples to elucidate their properties, and clarify their differences.

Singularity refers to the pole of a term in a differential or integral equation that would give an unbounded value thereof. It has significant physical effect like damping as in Landau damping, resonance as in cyclotron resonance, or simply an unbounded potential origin like $1/r$ in Coulomb interaction. Singularity does not necessarily cause physical effect to be unbounded. Like delta function while the local value might be unbounded, the integrated total is not.

Secularity refers to the unbounded temporal evolution that increases with time. Removing the secularity in the multiple time analysis allows well behaved temporal solutions to be found in the long time scale since all the physical effects have to be finite and the corresponding equations have to be well posed. Secularity may come from the resonance effect that drives the amplitude to infinity if unchecked, for example. But it may thus bring in the other physical effects to quench the said secularity behavior, and bring out the correct physics by the renormalization process or otherwise.

Equations can have singularity but well posed. Secularity is itself unbounded and can't be left without proper treatment.

Chapter Seven

Numerical Algorithm

"If computers get too powerful, we can organize them into a committee — that will do them in."

Bradley's Bromide

Algorithm is the procedure for solving a mathematical problem in a finite number of steps. It is a deterministic automation, particularly suitable to program a computer to find the analytical or numerical solutions. Algorithm is central to all areas of scientific computing. It is the recipe that makes computer perform according to the knowledge, concept or expertise. It is capable of transferring the know-how to the novice for efficient and best-known procedures in problem solving. Algorithm is the spirit of computing. It can be said that where there is the algorithm, there is the solution. In truth, it is the solubility condition in computational science. If you love algorithm you may be amazed to find that algorithm does find love for some, by matching the personality traits and more.

Algorithm may come in all kinds of shapes, sizes and creativities. Students should avoid falling into the mind frame of the old saying, "If you are a hammer, everything else is a nail." The rightful problem solving mentality is to let the problem reveal where the solution might be, and study the clues to find the answers. A presumed methodology or a predefined tool, as a rule, is not always the best approach. Therefore, start from the asymptotic analysis once the problem is translated into a mathematical equation, and get all the hints possible before the inevitable numerical solution is explored.

7.1 Fundamentals in Problem Solving

A few fundamental problem solving techniques are in fact available, such as: **Divide-and-conquer**, **Elimination**, **Substitution**, and **Comparison (DESC)**. DESC can be effective when applied to debugging your computer program or devising the numerical algorithm. It may also help you to better understand the physical concept of a new subject, systematically resolve issues in a new research, and efficiently figure out flaws in a new approach.

7.1.1 *Divide-and-Conquer*

The concept of divide-and-conquer may be traced back to the ancient time, as documented in the **Art of War**, a military treatise of war strategies, written by **Sun Tzu** in the 6th century BC, some twenty six hundred years ago. The strategy of divide-and-conquer can be effective in devising a numerical solver. The algorithm of Fast Fourier transform by Danielson and Lanczos in 1942 is a good example. It further utilizes the recursive formulae, which not only reduces the coding lines but also is itself a powerful method of divide-and-conquer, and is commonly adopted to evaluate special functions.

Here we consider the **Tower of Hanoi** that involves 64 stone rings of varying sizes and three posts, designated as X, Y and Z. The task is to move the 64 stone rings of ascending sizes that are mounted on the post, say, X, to the post Y, with use of the post Z. The bigger stone ring is prohibited to place over the smaller one. Legend has it that the task was carried on by the monks, one move per day in an Indian temple, and when finished it will be the end of the world since it takes a total of $2^{64} - 1$ moves. The algorithm to solve this problem involves these steps:

Subroutine MOVE N from X to Y using Z
If $N = 1$ then output $X \rightarrow Y$.
Otherwise do the following:
Call MOVE $N-1$ from X to Z using Y;
Output $X \rightarrow Y$.
Call MOVE $N-1$ from Z to Y using X.

The idea is to place the largest stone on the final post Y, but the entire $N - 1$ stone rings needs be moved to the middle post first. Once this is accomplished, the problem is reduced from N to $N - 1$. Repeat the process while the source shifts between X and Z, and the intermediate post shifts between Y and X, so that one stone is conquered at a time.

7.1.2 *Elimination*

In the scientific pursue, it is always helpful to first get a simple but nontrivial solution. The spirit of "*make everything as simple as possible, but not simpler.*" implies eliminating as much complications as possible. We often balance only two leading terms to find the dominant solution in the asymptotic limit so to begin the understanding of a complex theory. Moreover, it is a general practice to start from something understandable. Take the picture on the next page as an example: If you could not see two ladies, try covering the eye of the lady you can recognize. This way you reduce the complication and you should be able to find what you have been missing: less is more!

The knowledge tree can be explosively divergent, and the alpha-beta pruning by reducing the number of nodes in the search tree is an effective algorithm commonly adopted in the field of **artificial intelligence (AI)** to find the optimized solution. The **genetic algorithm** discards the representative population of inferior traits, to improve the odds of creating children of better fitness.

7.1.3 *Substitution*

By placing an impurity of charge Z, the force balance requires $-\nabla\delta P - ne\delta\vec{E} = 0$, where the pressure perturbation is given by $\delta P = \delta n k_B T$. The Poisson's equation gives$\nabla \cdot \delta\vec{E} = 4\pi e[Z\delta(\vec{r}) - \delta n]$. Combining these equations, we end with $\nabla^2\delta n = -4\pi ne^2[Z\delta(\vec{r}) - \delta n]/k_B T$. Define $\lambda_D^2 \equiv k_B T/4\pi ne^2$, we have

$$\nabla^2\delta n - \frac{1}{\lambda_D^2}\delta n = -\frac{Z}{\lambda_D^2}\delta(\vec{r}). \tag{7.1}$$

The solution to the equation is (cf. Homework 8.8)

$$\delta n(\vec{r}) = \frac{Z}{4\pi r\lambda_D^2}e^{-r/\lambda_D}. \tag{7.2}$$

The electrostatic field of the impurity is screened to diminish at the e-folding length $\lambda_D = v_T/\omega_p$, the Debye length, and the electron density response to this electric field weakens accordingly. In the condensed matter, the electron velocity is determined by the Fermi energy, where the electrons are on the top of the energy band and are free electrons in a conductor. We may define the Fermi velocity $v_F = \sqrt{2\varepsilon_F/3m}$. Thus, the screening length is $\lambda_{TF} = v_F/\omega_p$, the Thomans-Fermi shielding length, instead of the Debye length. The Fermi energy in the condensed matter is typically much higher (7 eV in copper) than the thermal energy (25 meV at room temperature) so that the Debye length is negligible until the plasma temperature reaches tens of thousands of Kelvin. Thus, *upon a substitution*, in this case the effective velocity, *it allows us to understand a different topic in the light of a familiar one*.

In plasma physics, the magnetic curvature has the effect equivalent to the gravitational force. It may cause the charge separation in the 'bad' curvature region, and develop the interchange instability, commonly recognized in hydrodynamics as the Rayleigh Taylor instability. The heavy fluid cannot stay on top of the light fluid and an interchange of the fluids is predictable. This also causes sediment in river, lake, or dam, a very common experience to us. Thus substituting with the familiar or definitive effect, the unfamiliar factor may lead to the expected and understandable results. We often replace a component to diagnose a problem in networking, computer system, or an application program. Substitution allows us to quickly identify the source of the problem.

7.1.4 *Comparison*

Making observations by comparing the features of seemingly entirely different entities may lead to the global understanding than otherwise. Darwin and many biologists apply comparison to reach definitive answers on the evolution theory and the effects of genetic mutations. When data mining is the only tool to figure out important mechanisms in a less developed scientific discipline, comparison can reveal very important underlying principles. Taking the 1989 September 21st 7.6-magnitude earthquake in Taiwan as the example, comparison of the damages between the epic center and the central mountain ridge with that between the epic center and the sea shows much more severe damages on the mountain side. It indicates that the earthquake waves may have been reflected from the mountains thus to enhance the destruction. In plasma physics, similarities may be found, for example, in the Kelvin-Helmholtz instability, two stream instability, and trapped particle instability, etc., allowing deeper appreciation of the phenomena as such. Comparison is particularly useful when listening to a seminar. Compare the speaker's new knowledge with your familiar concepts and accept and digest as you follow along. This is the chance for synergy and the way to make your head think harder and spin faster during a talk and learn effectively as always.

Homework 7.1: In a group meeting, a student presented the following slide. Figure out the error in the presentation.

$$\vec{\vec{I}} \equiv \sum_{\sigma} n_\sigma q_\sigma \vec{v}_\sigma \vec{v}_\sigma$$

$$\vec{\vec{\Lambda}} \equiv \sum_{\sigma} n_\sigma q_\sigma (\vec{v}_\sigma - \vec{U})(\vec{v}_\sigma - \vec{U}) = \vec{\vec{I}} + Q\vec{U}\vec{U} - 2\vec{J}\vec{U}$$

$$Q \equiv \sum_{\sigma} n_\sigma q_\sigma,\ \vec{J} \equiv \sum_{\sigma} n_\sigma q_\sigma \vec{v}_\sigma,\ \rho\vec{U} \equiv \sum_{\sigma} n_\sigma m_\sigma \vec{v}_\sigma$$

Version control is an essential discipline in software development. Its purpose is to ensure that the path to successful implementation is not compromised. Comparison between old and new codes would reveal the mistakes made and computer bugs quickly fixed.

7.2 Basics in Numerical Methods

Many programming concepts are helpful in developing a numerical solution and in communicating the written program with coworkers. In particular, there are the basics in **Object Oriented Programming (OOP)**, namely, data abstraction, encapsulation, inheritance, modularity and polymorphism. Great programming sense can be developed from criteria used for evaluating software, though subjective and qualitative, consist of robustness, usability, scalability, portability and versatility. They are important for developing a better code. A word of caution however is, before indulging oneself with heavy computing, always use brain power before computer power, and always use one CPU before multi-CPU.

There are three important issues regarding the numerical methods: the **stability**, the **accuracy**, and the **speed**. The first question to ask when implementing a new algorithm is whether it is stable. Since plasma physics is often concerned with physical instabilities, it is not uncommon to mistake a numerical instability for the physical one. Beyond the stability, the numerical accuracy is critical for the correct description of the physical phenomenon. Once a numerical solution is timely found, do pay attention to the **numerical convergence** and the **sensitivity study** to avoid surprises. The real physical solution has to be the one when the grid cell size goes to zero ($\Delta X \rightarrow 0$), and the trend of numerical convergence has to confirm that. The parametric dependency of the solution has to be consistent with the physical law, and the sensitivity study may reveal the pitfalls of hidden singularities, unphysical initializations, and immature algorithms. Moreover, speed is crucial for the success of computing when a problem is pushing the computing limit. Nonlinear many-body problems tend to be worse than N^2 scaling, where N is the number of particles. Many N^2-scaling algorithms are often difficult to pursue for large N. The improvement to *NlogN* scaling is most desirable. Parallel computing, suffering from the communication bottlenecks, requires parallel algorithms to address this issue. But if a problem is not efficiently solved by the sequential program, it is often futile to solve it by brute force in parallelism. The best numerical algorithm perhaps is beyond the mathematical framework. It relies on logics and intuition to attain the best efficiency.

7.2.1 *Courant-Friedrichs-Lewy Condition and the Numerical Stability*

When solving a partial differential equation, the time step $\Delta\mathrm{T}$ greater than certain value for the given grid size $\Delta\mathrm{X}$ in space will result in a numerical instability. This is the **Courant–Friedrichs–Lewy (CFL) condition** that can be viewed as the "light cone" constraint. By defining $C \equiv \Delta\mathrm{X}/\Delta\mathrm{T}$ as the 'light speed' of the numerical system, this CFL condition can be stated as $C \geq \mathrm{V}$, or $\Delta\mathrm{T} \leq \Delta\mathrm{X}/\mathrm{V}$, where V is the physical speed in the system. This restricts the information propagation to be less than the grid size in each time step so that the grid data is not leaped over, but updated along the way.

Consider a simple oscillation function with a single harmonic $f(\vec{r},t) = f_0(\vec{r})\exp(-i\omega t)$ that is governed by

$$\frac{\partial f}{\partial t} = -i\omega f. \tag{7.3}$$

We can express Eq. (7.1) in finite difference scheme as

$$\frac{f^{n+1/2} - f^{n-1/2}}{\Delta t} = -i\omega f^n, \tag{7.4}$$

where $f^n = f(\vec{r}, n\Delta t)$. If Δt is small enough, a factor q can be defined,

$$q \equiv \frac{f^{n+1/2}}{f^n} = \frac{f^n}{f^{n-1/2}}.$$

Substituting Eq. (7.3) into Eq. (7.2), we have

$$q^2 + i\omega\Delta t q - 1 = 0, \tag{7.5}$$

and the solution is

$$q = -\frac{i\omega\Delta t}{2} \pm \sqrt{1 - \left(\frac{\omega\Delta t}{2}\right)^2}. \tag{7.6}$$

Utilizing the expression of the oscillation function, we obtain $f^n = f_0 \exp(-i\omega n\Delta t)$ as well as

$$q = \frac{f^{n+1/2}}{f^n} = \exp\left(-\frac{1}{2}i\omega\Delta t\right). \tag{7.7}$$

To maintain the numerical stability, it requires that $|q| \leq 1$, translated to $\omega\Delta t/2 \leq 1$ according to Eq. (7.5). Therefore, assuming $\omega = 2\pi/T$, we can rewrite it as $\Delta t \leq T/\pi$.

Now consider the wave equation

$$\nabla^2 f + \frac{\omega^2}{c^2} f = 0, \tag{7.8}$$

where the plane wave solution can be expressed as $f(\vec{r}, t) = f_0 \exp[i(\vec{k} \cdot \vec{r} - \omega t)]$. Applying the finite difference scheme, we express the second order partial derivative as

$$\frac{\partial^2 f}{\partial^2 x} \approx \frac{f(x + \Delta x) - 2f(x) + f(x - \Delta x)}{(\Delta x)^2}, \tag{7.9}$$

to give

$$\frac{\partial^2 f}{\partial^2 x} \approx \frac{\exp(ik_x\Delta x) - 2 + \exp(-ik_x\Delta x)}{(\Delta x)^2} f = -\frac{\sin^2\left(\frac{1}{2}k_x\Delta x\right)}{\left(\frac{1}{2}\Delta x\right)^2} f. \tag{7.10}$$

Utilizing Eq. (7.10), we can rewrite Eq. (7.8) as

$$\frac{\sin^2\left(\frac{1}{2}k_x\Delta x\right)}{\left(\frac{1}{2}\Delta x\right)^2} + \frac{\sin^2\left(\frac{1}{2}k_y\Delta y\right)}{\left(\frac{1}{2}\Delta y\right)^2} + \frac{\sin^2\left(\frac{1}{2}k_z\Delta z\right)}{\left(\frac{1}{2}\Delta z\right)^2} - \frac{\omega^2}{c^2} = 0,$$

or

$$\left(\frac{1}{2}c\Delta t\right)^2 \left[\frac{\sin^2\left(\frac{1}{2}k_x\Delta x\right)}{\left(\frac{1}{2}\Delta x\right)^2} + \frac{\sin^2\left(\frac{1}{2}k_y\Delta y\right)}{\left(\frac{1}{2}\Delta y\right)^2} + \frac{\sin^2\left(\frac{1}{2}k_z\Delta z\right)}{\left(\frac{1}{2}\Delta z\right)^2}\right] = \left(\frac{1}{2}\omega\Delta t\right)^2 \leq 1. \tag{7.11}$$

For arbitrary k_x, k_y and k_z, Eq. (7.11) is valid if $(c\Delta t)^2[(\Delta x)^{-2} + (\Delta y)^{-2} + (\Delta z)^{-2}] \leq 1$, which means

$$c\Delta t \leq \frac{1}{\sqrt{(\Delta x)^{-2} + (\Delta y)^{-2} + (\Delta z)^{-2}}}. \tag{7.12}$$

Equation (7.12) is known as Courant–Friedrichs–Lewy condition. Setting $\Delta x = \Delta y = \Delta z = \delta$ for three-dimensional situation, we have $c\Delta t \leq \delta/\sqrt{3}$, $c\Delta t \leq \delta/\sqrt{2}$ in 2-D, and $c\Delta t \leq \Delta x$ in 1-D.

In devising the finite difference scheme, it is noteworthy that the **explicit scheme** calculating the future state variable with the current one can be less stable and requires smaller Δt as a result. The **implicit scheme** on the other hand, solving the equation by both the current and future state variables, can be more time consuming and difficult to implement, but in general more stable and allowing larger Δt for time advancement.

7.2.2 *Simpson Rule and the Numerical Accuracy*

After a numerical solution is found timely, keep in mind that the numerical convergence and sensitivity study needs be checked to avoid surprises. Asymptology will always be helpful by comparing the limits. The numerical accuracy cannot be compromised before finalizing a numerical investigation. To begin with, the finite difference scheme may, in general, truncate at higher order and compute at smaller grid size. Chapter 2 shows the differences in dealing with the particle trajectories with the Rugge-Kutta second order and fourth order and with the particle characteristics. When singularity is involved, great care needs be exercised. An example is the evaluation of the plasma dispersion function (cf. Sec. 8.2.2). In most undertakings, however, it may be no more than a good exercise of the asymptotic expansion. The tradeoff could be the increased computational cost, which is not a serious issue given today's computational resources.

As applied in the previous Chapters, the Simpson rule for numerical integration serves well to gain the numerical accuracy. It has the following expansion scheme for a definitive integral,

$$\int_a^b f(x)dx \approx \frac{h}{3}[f(x_0) + 4f(x_1) + 2f(x_2) + 4f(x_3) \\ +2f(x_4) + \cdots + 4f(x_{n-1}) + f(x_n)] + O(h^5). \quad (7.13)$$

This basically is derived from the integration of a segment by,

$$\int_{x_0}^{x_2} f(x)dx \approx \frac{h}{3}[f(x_0) + 4f(x_1) + f(x_2)], \quad (7.14)$$

where $2h = |x_2 - x_0|$. When summing up all the segments, the interface grids would get the coefficient 2, while the first and the last grid points remain with the unity coefficient. The $O(h^5)$ accuracy is shown in Homework 7.2, and the algorithm is implemented in the program below. For higher dimensions of integrals, expansions with use of special functions are often preferred for less computational cost.

Homework 7.2: Prove that the expansion in Eq. (7.13) leaves the remaining error in the order of $O(h^5)$.

```
function V=Simpson(a,N,dt)          %a: data, N: length of data, dt, data
V=a*0;                              interval
V(2)=(a(1)+a(2))*dt/2;              % return a vector of the same length as
V(3)=(a(1)+4*a(2)+a(3))*dt/3;       a.
if(N<=3) return; end;               % V(j) is the value of the complete
v=a(1);                             integral up to the j point.
n=fix(N/2);
if(mod(N,2)==0) M=n-1; else M=n;
end;
for i=1:M
  j=2*i;
  v=v+4*a(j);
  V(j+1)=(v+a(j+1))*dt/3;
  v=v+2*a(j+1);
  V(j)=(V(j+1)+V(j-1))/2;
end;
if(mod(N,2)==0)
  V(N)=(v-a(j+1))*dt/3+(a(N-
1)+a(N))/2*dt;
end;
```

Homework 7.3: Find the magnetic field $\vec{B} = \nabla\psi \times \hat{e}_z$ in a cylindrical plasma of unity radius $a = 1$. The magnetic flux is governed by

$$d^2\psi/dr^2 + (1/r)(d\psi/dr) + \alpha e^{\psi} = 0,$$

where the plasma current $j_z = \alpha e^{\psi}$. Make use of the fact that $\psi = \psi_0 - \ln[1 + \frac{1}{8}\alpha r^2 e^{\psi_0}]^2$ and the central value ψ_0 is governed by $\exp(\psi_0) = (4/\alpha)(1 \pm \sqrt{1 - \alpha/2})$ to compare the results from the integration by Simpson method and from direct differentiation of ψ.

7.2.3 *Fast Fourier Transform and the Computing Speed*

The fast Fourier transform (FFT) is a good example of applying the strategy of divide-and-conquer to implement an algorithm that can speed up the computation. Discrete Fourier transforming a data set of length N, $\{X_m; m = 0..N-1\}$, to N Fourier components $\{F_k; k = 0..N-1\}$, with the formulae $F_k = \sum_{m=0}^{N-1} W^{mk} X_m$ may be expressed as the sum of two discrete Fourier transforms, each of length $N/2$:

$$F_k = \sum_{m=0}^{N/2-1} e^{-2\pi ik(2m)/N} X_{2m} + \sum_{m=0}^{N/2-1} e^{-2\pi ik(2m+1)/N} X_{2m+1}$$

$$= \sum_{m=0}^{N/2-1} e^{-2\pi ikm/(N/2)} X_{2m} + W^k \sum_{m=0}^{N/2-1} e^{-2\pi ikm/(N/2)} X_{2m+1}$$

$$\equiv F_k^e + W^k F_k^o,$$

where $W \equiv e^{-2\pi i/N}$. This divide-and-conquer strategy not only allows recursive operation, but also reduces summations to evaluations at $N/2, N/4, N/8, N/16, \ldots,$

thus from order N processes to order $\log_2 N$ for individual Fourier component. Since there are N components to evaluate, the method converts an order N^2 computational effort to order $N \ln N$. Without this improvement, Fourier transform of data with large N components can quickly run beyond the computing power of even today's computer. The FFT was applied in Sec. 4.4.2 to reduce the numerical inaccuracy in the spatial derivatives.

7.3 Robust Algorithms

Programs in the production mode would require great numerical efficiency and computational speed. By contrast, in the research mode, numerical solutions to a specific problem may only be needed for a few runs with different parameters for a quick check on a conjecture. Therefore, the speed of computation is not of great concern, but the time to get the answer is. That means a robust algorithm, however slow; capable of solving the generic problems by plug and run would be most desirable so to alleviate the need for developing a new algorithm time and again.

7.3.1 *Eigenvalue Solver and Power Method*

The power method is an eigenvalue solver that runs relatively slow by taking many iterations to arrive at an acceptable accuracy. But its applicability in solving most of the eigenvalue problems makes it the choice for quick answers.

The generic eigenvalue problem can be cast into the matrix equation, $\overleftrightarrow{A} \cdot \vec{x} = \lambda \vec{x}$, where $\overleftrightarrow{A}$ is the matrix, $\vec{x}$ the eigenvector and λ the eigenvalue. Instead of solving the original equation, the power method solves the following: $\overleftrightarrow{A} - \tau \overleftrightarrow{I}) \cdot \vec{y} = \vec{b}$ where $\vec{b}$ is the initial guess of the eigenvector and τ is the initial guess of the eigenvalue. Since $\vec{b}$ as a random guess is supposed to have components of all eigenvectors in the system, viz., $\vec{b} = \sum_i \beta_i \vec{x}_i$, the same applies to $\vec{y} = \sum_i \alpha_i \vec{x}_i$. It is clear that $\sum_i \alpha_i (\lambda_i - \tau) \vec{x}_i = \sum_i \beta_i \vec{x}_i$. Therefore, we have $\alpha_i(\lambda_i - \tau) = \beta_i$ by applying the orthogonality condition of the eigenvectors. Accordingly, $\vec{y} = \sum_j (\lambda_j - \tau)^{-1} \beta_j \vec{x}_j$. It is clear that the eigenvalue λ_τ closest to the initial guess τ will amplify the contribution of $\vec{x}_\tau$ to $\vec{y}$. Thus, replacing $\vec{b}$ by $\vec{y}$, and solving the equation iteratively, the eigenvector $\vec{y}$ will continue to project itself onto the eigenvector $\vec{x}_\tau$. Moreover, since $(\overleftrightarrow{A} - \lambda^j \overleftrightarrow{I}) \cdot \vec{y} = \vec{x}^j$, that implies $\vec{y} = (\overleftrightarrow{A} - \lambda^j \overleftrightarrow{I})^{-1} \cdot \vec{x}^j$, and $\vec{y} \cdot \vec{x}^j = (\lambda^{j+1} - \lambda^j)^{-1} |\vec{x}^j|^2$. Here the superscript index j refers to the numeral of iteration. Both the eigenvalue and eigenvector can then be solved iteratively as governed by the following equations;

$$\tau^{j+1} = \tau^j + \frac{|\vec{x}^j|^2}{|\vec{x}^j \cdot \vec{y}|}, \tag{7.15}$$

$$\vec{b}^{j+1} = \frac{\vec{y}}{|\vec{y}|}. \tag{7.16}$$

The following program implements the power method.

```
function [lambda,y]=Power(A,lambda,b)
lambda0=-10; iter=0;
n=length(A);
while(abs(lambda-lambda0)>1.0e-6&&iter<100)
  iter=iter+1;
  lambda0=lambda;
  N=A+diag(lambda*ones(1,n));
  y=N\b;
  lambda=lambda-(b'*b)/(b'*y),
  b=y/norm(y);
end;
```

%Solve the equation $A \cdot x = \lambda x$ by the POWER method. Here, A is the matrix of concer, lambda is the initial guess of the eigenvlaue, and b is the initial guess of the eigenvector. Returned are the eigenvalue and the eigenvecotor of the system that is closest to the initial guess.

Homework 7.4 demonstrates its application to solve the eigenvalue equation that in fact has the Bessel function solution.

Homework 7.4: Solve the following eigenvalue equation with use of the power method,

$$\frac{d^2\psi}{dr^2} + \frac{1}{r}\frac{d\psi}{dr} + \lambda\psi = 0,$$

that satisfies $\psi = 0$ at $r = 1$ and $d\psi/dr = 0$ at $r = 0$.

7.3.2 *Boundary Value Problems and Shooting Method*

Given a differential equation with the edge values specified, the shooting method, while requires many trial and error to satisfy the boundary conditions, is a plug-and-run algorithm that will get you the solution without hesitation. The solution is shot out from one edge. If the derivative thereof is not given, that is usually the case for the Dirichlet boundary condition, a trial value is initiated and subsequently corrected from the deviation at the end point. The accuracy in defining the slope would naturally reflect the accuracy in the solution.

In the Homework 6.4, the following equation: $\varepsilon y'' - (2x + 1)y' + y = 0$, with the values at the end points given: $y(0) = 1$ and $y(1) = 0$, where $\varepsilon \ll 1$, was solved by the asymptotic matching procedure. The composite solution is given by:

$$y = \sqrt{2x+1} - \sqrt{3}e^{3(x-1)/\varepsilon}.$$

The numerical analysis was done by cherry-picking the right slope. Here a numerical automaton is provided that finds the slope of high accuracy at around 100 iterations to reach an error at less than one part in a million.

```
function ShootingMethod
N=1000; dx=1/N;
x=(0:N)*dx; y=1+x*0; z=y;
format long; z(1)=0.75;
 epsilon=0.100;
for ip=1:1000
   for i=1:N
     y(i+1)=y(i)+z(i)*dx;
     z(i+1)=z(i)+((2*x(i)+1)*z(i)-
(y(i+1)+y(i))/2)*dx/epsilon;
    end;
 D=y(N)-epsilon*(z(N+1)-z(N))/dx+3*z(N),
if(abs(D)<1.0e-6) ip, break; end;
z(1)=z(1)-0.1^(ip/10)*sign(D);
end;
z(1),
Y=sqrt(2*x+1)-sqrt(3)*exp(3*(x-1)/epsilon);
plot(x,y,'g*',x,Y,'r -'); xlabel('x'); ylabel('y');
title('Shooting Method vs Multiple Length
Scale');
```

%Solve the equation by the shooting method and compare with the analytical solution.
%z(1)=0.924911458738780
% ip=123

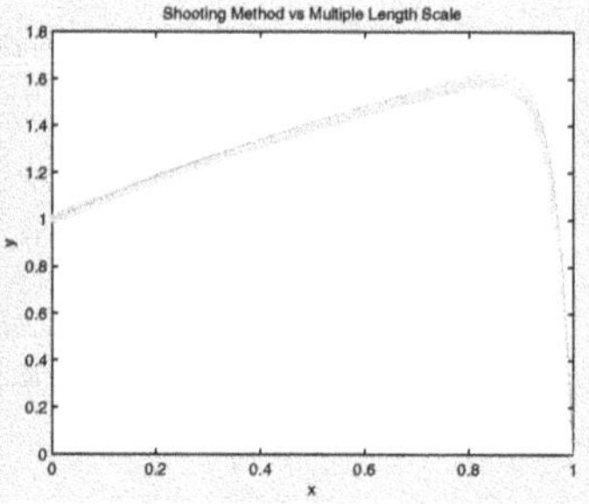

Turing to the more elaborate nonlinear equation, we here examine, by both the analytical and the shooting method, the **Bratu's equation**, which happens to be a limiting case for MHD equilibrium of minimum energy state (cf. Sec. 10.4),

$$\frac{d^2y}{dx^2} + \Lambda e^y = 0. \tag{7.17}$$

This nonlinear equation has the "Hamiltonian" as the constant of the motion,

$$H = \frac{1}{2}\left(\frac{dy}{dx}\right)^2 + \Lambda e^y. \tag{7.18}$$

```
function BratuSolubility
y0=0:0.001:10;
f=exp(-y0/2).*atanh(exp(-y0/2).*sqrt(exp(y0)-1));
LAMBDA=max(f)^2*2,
plot(y0,f.^2*2)
xlabel('y0'); ylabel('\lambda');
title('Solubility Condition for Bratu''s Equation');
PSIc=y0; LAMBDA=f.^2*2; N=length(y0);
save Bratu1D PSIc LAMBDA N;
```

```
LAMBDA =
 0.8785
```

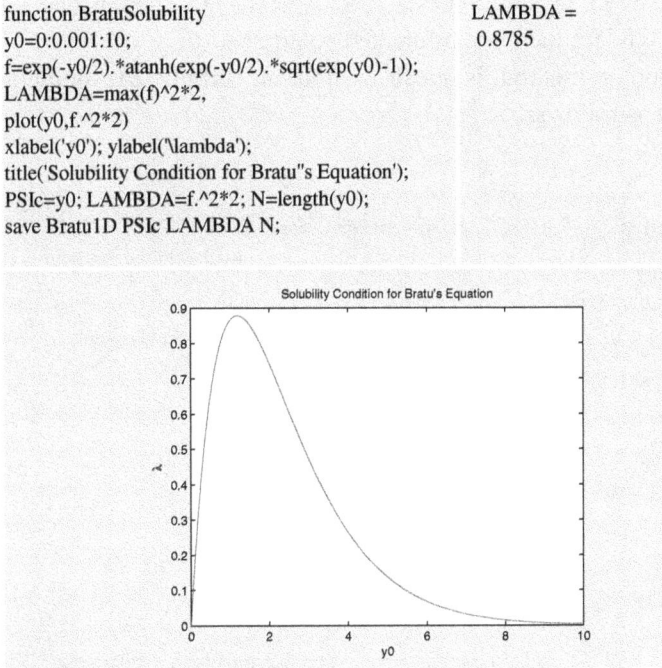

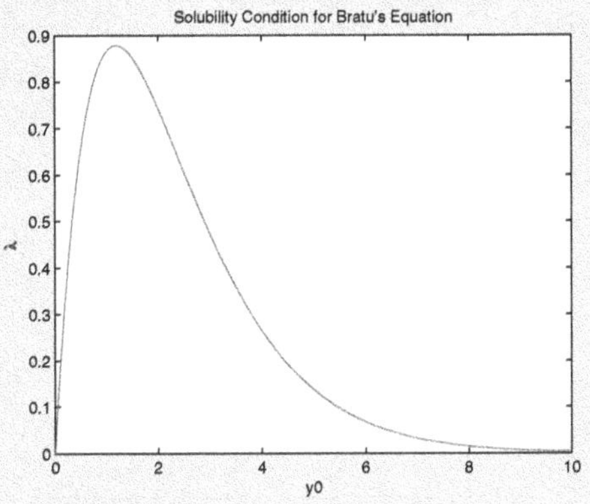

Applying Eq. (7.18) reduces Eq. (7.17) to the first order differential equation,

$$\frac{dy}{dx} = \pm\sqrt{2}\sqrt{H - \Lambda e^{y}}, \tag{7.19}$$

which can be integrated to give the following solution,

$$\int_0^y \frac{dy'}{\sqrt{e^{y_0} - e^{y'}}} = \pm\sqrt{2\Lambda}(x-1) = 2e^{-\frac{1}{2}y_0} \times \left(\tanh^{-1}(e^{-\frac{1}{2}y_0}\sqrt{e^{y_0}-1}) - \tanh^{-1}(e^{-\frac{1}{2}y_0}\sqrt{e^{y_0}-e^{y}})\right), \tag{7.20}$$

where we have taken the boundary conditions as $dy/dx|_{x=0} = y|_{x=1} = 0$ to give

$$H = \Lambda e^{y_0}, \quad H - \Lambda = \frac{1}{2}\left(\frac{dy}{dx}\right)^2\bigg|_{x=1}, \tag{7.21}$$

and assign $y(x=0) = y_0$. By substituting $x = 0$ and $y = y_0$ into Eq. (7.20), it gives a solubility condition, as numerically analyzed below,

$$\sqrt{\frac{1}{2}\Lambda} = e^{-\frac{1}{2}y_0}\tanh^{-1}\left(e^{-\frac{1}{2}y_0}\sqrt{e^{y_0}-1}\right) < 0.6627.$$

When $\Lambda > 0.8785$, no solution exists. At $\Lambda = 0.8785$, there is one solution of y_0, and for $\Lambda < 0.8785$, there are bifurcated solutions.

The shooting method is ideal to find the bifurcated solutions, and it is implemented as follows.

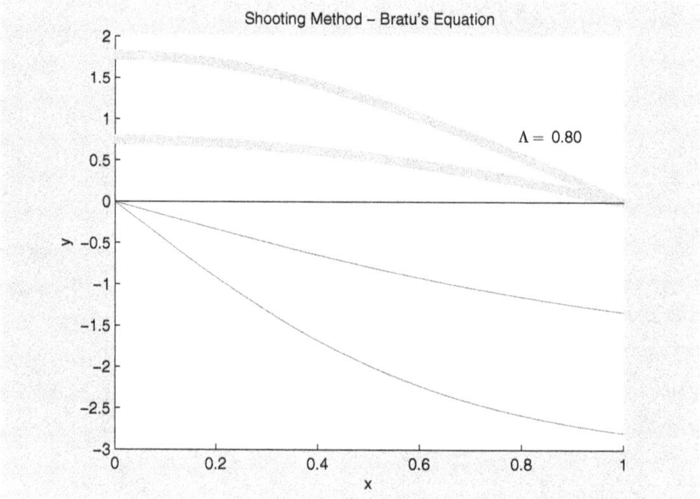


```
function Bratu(LAMBDA)
close all; clc;
N=1000; dx=1/N;
x=(0:N)*dx; y=1+x*0; z=y;
figure(1); hold on;
for ip=1:2
z(1)=0; y0=ip; y(1)=y0;
for iter=1:1000
for i=1:N
  y(i+1)=y(i)+z(i)*dx;
  z(i+1)=z(i)-
LAMBDA*exp((y(i+1)+y(i))/2)*dx;
end;
D=y(N+1);
if(abs(D)<1.0e-6) iter, D, y(1), break; end;
y(1)=y(1)+(-1)^ip*0.1^(iter/10)*sign(D);
end;
plot(x,y,'g*',x,z,'r-',x,x*0,'k-');
end;
title('Shooting Method - Bratu''s Equation');
xlabel('x'); ylabel('y');
text(0.8,0.8,'$\Lambda$=','interpreter','latex');
str=sprintf('%5.2f',LAMBDA);
text(0.85,0.8,str); hold off;
```

% The green lines are the y values of the bifurcated solutions, and the red lines are their derivatives.

y0=0.7450,
y0=1.7720,

Homework 7.5: Solve numerically the **Bratu's equation** in cylindrical coordinates,

$$d^2y/dr^2 + (dy/dr/r) + \Lambda e^y = 0$$

subject to the boundary conditions: $dy/dr|_{r=0} = y|_{r=1} = 0$. Work out the analytical theory and the condition for the bifurcation, and compare with the numerical solutions.

7.3.3 *Probabilistic Algorithm*

The probabilistic algorithm is versatile and capable of solving varieties of problems. It applies the random numbers to take a statistical ensemble that reflects the truth of the governing principle. Varieties in probabilistic algorithm include **Monte Carlo**, **Simulated Annealing**, **Genetic Algorithm**, and **Evaluation and Search**. It may serve as the artificial dynamics to simulate the many body systems such as soft matter, traffic event, rainfall projection, neutron transport, etc. The Quantum Monte Carlo method can solve the nonlinear many-body Schrodinger equation in principle, although it is limited by the computing power in practice. The magic of nondeterminism by tossing the coin provides the probability of finding a satisfying solution that would otherwise take a polynomial timescale to accomplish. Thus, probabilistic algorithm is often the choice to tackle the NP complete problem, such as the traveling salesman's problem, and is the technology behind the GPS route recalculating.

We demonstrate the **Monte Carlo (MC) method** by designing a code to evaluate the value of π. Fetch a large number of random points inside a square of unit length. If the distance from the origin is shorter than unity, then count that as a hit, otherwise a miss. The value of π is then given by the formula: π ~4*hits/total points. The code is as follows:

```
function PI=mcPI(n)
tic,
rand('state',sum(100*clock ));
p=rand(2,n);
r=sqrt(sum(p.^2,1));
hits=(r<1);
PI=4*sum(hits)/n
toc
```

```
% Type mcPI(64000000)
to get 4 digits of
accuracy.
PI = 3.1415
Elapsed time is
5.089915 seconds
```

In the variational principle, integrals are evaluated and parameters optimized to find the minimum energy, maximum entropy, or least action state, for examples. MC can be robust in evaluating these integrals to allow a variational Monte Carlo algorithm to work. The following program utilizes the function RANDN to generate

N random points of normal distribution. By summing up these points with proper weighting factor according to the integrand, the integral is obtained. Note that N up to 10^8 can easily be fetched, and the error $\sim \sqrt{1/N} < 0.1\%$ is possible. The following homework example demonstrates the MC method. The accuracy becomes an issue when the value p gets too large, however.

Homework 7.6: Evaluate the following integrals with use of the Monte Carlo method:

a. $I_1 = \int_0^\infty e^{-x^2/2} x^p dx,$ b. $I_2 = \int_0^\infty e^{-x^2/2} \log(1 + x^p) dx,$

c. $I_3 = \int_0^\infty e^{-x^2/2} \cos(px) dx,$ d. $I_4 = \int_0^\infty e^{-x} x^p dx,$

e. $I_5 = \int_0^\infty e^{-x} \log(1 + x^p) dx,$ f. $I_6 = \int_0^\infty e^{-x} \cos(px) dx.$

Earlier work on the diffusion coefficient (cf. Sec. 1.1.5) mimics the Markovian process which applies MC to simulate the Brownian motion. The algorithm of simulated annealing (SA) to mimic the thermal annealing process often works better than just simple MC. The necessary components of a simulated annealing program are: (1) an "energy" value to evaluate system state, (2) a "temperature" to access the excited states, (3) a Metropolis selection criterion to permit or forbid the transition, (4) a step size to allow phase space travel, (5) penalties to shape the preferred path of annealing, (6) a cooling recipe to reduce the temperature and the step size. The advantages of SA are: follow physical laws, allow better physical intuition, alter the phase space topology by flexible penalty functions, overcome local barriers to access other minima, and possess intrinsic parallelism. The disadvantages are: likely blocked from reaching global minimum, slow diffusive process, and no clear carpet sweeping mechanism.

The genetic algorithm (GA) is inspired by Evolution Biology and adopts the techniques such as inheritance, mutation, crossover and selection. Therefore, it is also termed as the Evolutionary Computation. It has the advantage over SA in that it could avoid the energy barrier SA could not jump over by eliminating the inferior genes. This is *elimination* and *substitution* at work. To implement a genetic algorithm, several components are necessary: (1) a genetic representation of solutions to the problem, (2) a way to create an initial population of solutions, (3) an evaluation function to rate fitness, (4) genetic operators to alter and to create the genetic composition of children. A population of individuals are created initially, and ranked according to the fitness of their genetic traits. The ranking of every individual is used to create the next generation of children, and the process continues for many generations until a satisfied individual (solution) is found.

While there are many genetic operators to balance inheritance with evolution, and there are many more man-made than natural selection rules to ensure 'talented' superstar to be produced, it is still hard to guarantee the global minimum to be found from the probabilistic algorithm. Moreover, to find an optimized solution to a problem, the methodology unfortunately has to go through heavy optimization itself. For example, mating best talents could easily create a pool of similar off springs of super genes and could lose the diversity in genetic makeup that may be necessary to eventually find the ultimate solution. Too little mutation also renders much slower evolution process. But the most serious shortcoming of the probabilistic algorithm is its intrinsic diffusive nature of finding the solution. Knowledge to perform evaluation and search however imperfect may as a rule speed up the computation by its determinism. This may be termed as the "**gene therapy**" for evolutionary computation with the aid of gene expression knowledge.

7.4 Advanced Numerical Topics

Numerical algorithm will continue to challenge us so to find solutions that are analytically intractable and humanly impossible. Moreover, time and money in running the real experiments can be saved by the computer simulation. The blind spots in experiment can be exposed by the diagnostics in the numerical experiment. Consequently, there is no shortage of advanced numerical topics to address. Here we present some promising topics that are very much of current interest and under active studies. To develop the ever improving programming skill, the evaluation criteria for quality software such as robustness, scalability, portability, versatility, can serve as good guidance for self learning.

7.4.1 *Computer Simulation*

Computer simulation can be an important tool for knowledge creation, and can help understand a phenomenon that is difficult to visualize in laboratory experiment. We here examine the **catenary** (cf. Sec. 8.8.1) by simulation. The mechanics of rigid body is often analytically solvable. For soft matter, the story is quite different. A linear chain, with two ends hinged at equal heights, relaxes to its ultimate configuration under the influence of the gravitational force. Mechanically according to Newton's law, the final equilibrium state must have vanishing net force: each hinge upholds with $F_{hinge} = \frac{1}{2}g\sigma l$ while the chain is pulled down by $F_{chain} = g\sigma l$, and vanishing net torque: $\tau = (\frac{1}{2}b)F_{hinge} = \int_0^{b/2} g\sigma x ds$, that also guarantees the chain to be symmetrical with respect to its center. The additional constraint is the length of the chain: $l = \int_l ds = \int_{-b/2}^{b/2} \sqrt{1 + z'^2} dx$. The detailed force balance is needed to determine the final equilibrium. There is however the bottom point of the chain suffers the force imbalance since the gravitational force is pulling down, yet the tensions, as the symmetry dictates, are pointing horizontally. The simulation utilizes a chain of particles connected by the spring coils at the spring constant k, determined to model properly the relaxation process. As illustrated in the simulation, a unique solution emerges, which is plotted as shown in green circles. The spring constant k is set at 5 while the gravitation is chosen at 0.05. This also indicates that the tension inside the chain is considerable. While the simulation can be an important tool, the underlying physics can be revealed as the minimum energy state by the variational principle (cf. Sec. 8.8.1). It has the simple solution $z/b = \cosh[(x - a)/b]$. Let $a = 0$ and make a shift of y coordinate, $z = z\min - b + b\cosh(x/b)$, which is plotted in red line to compare with the simulation result.

Homework 7.7: Utilize the same simulation model for the relaxed chain, but release the chain at one end. Follow its relaxation to the final configuration. Find the tension in the chain as compared to the gravitational force.

Homework 7.8: The Equivalence Principle: Consider a simple pendulum hanging in an elevator. The elevator moves at a constant acceleration a. Show by simulation that the period of the pendulum is $\sqrt{(g \pm a)/l}$, where the sign depends on whether the acceleration is downward or upward.

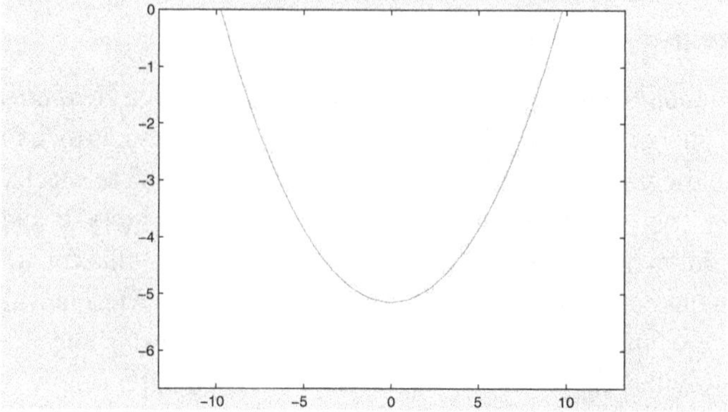

```
function RelaxedChain(b) %b is even                    % RelaxedChain(20)
global x y N k g;                                      % b is the width of the
if(mod(b,2)==1) b=fix(b)+1; end;                       ends.
N=b+1; k=5; g=0.05;
dx=b/N;
x=-b/2+dx/2:dx:b/2-dx/2;
y=-b/4*cos(pi*x/b)/2;
figure(1); iframe=0;
filename='RelaxedChain.gif';
for i=0:100*b
if(mod(i,b)==0)
    plot(x,y,'Og-'); axis([-2*b/3,2*b/3,-b/3,0]);
    iframe=iframe+1;
    title('Relaxed Chain');
    M(iframe)=getframe;
end;
[x,y]=displacement;
end;                                                   % Red line is the
                                                       minimum energy
save CHAIN x y b k g N;                                configuration. The
hold on;                                               spring constant k is set
X=-1:0.01:1;                                           at 5 while the
Y=min(y)-b/2+b/2*cosh(X);                              gravitational force is
plot(b/2*X,Y,'r-');                                    chosen as 0.05. This
title('Relaxed Chain');                                also indicates that the
M(iframe)=getframe;                                    tension inside the chain
hold off;                                              is in general quite
                                                       significant.
function [X,Y]=displacement
global x y N k g;
dt=0.1; f=x*0; fx=f; fy=f; X=f; Y=f; r=f;
r(2:N-1)=sqrt((x(2:N-1)-x(1:N-2)).^2+(y(2:N-1)-
y(1:N-2)).^2);
f(2:N-1)=-k*(r(2:N-1)-1);
fx(2:N-1)=f(2:N-1).*(x(2:N-1)-x(1:N-2))./r(2:N-1);
fy(2:N-1)=f(2:N-1).*(y(2:N-1)-y(1:N-2))./r(2:N-1);
r(2:N-1)=sqrt((x(2:N-1)-x(3:N)).^2+(y(2:N-1)-
y(3:N)).^2);
f(2:N-1)=-k*(r(2:N-1)-1);
fx(2:N-1)=fx(2:N-1)+f(2:N-1).*(x(2:N-1)-
x(3:N))./r(2:N-1);
fy(2:N-1)=fy(2:N-1)+f(2:N-1).*(y(2:N-1)-
y(3:N))./r(2:N-1);
X(2:N-1)=x(2:N-1)+fx(2:N-1)*dt;
Y(2:N-1)=y(2:N-1)+fy(2:N-1)*dt-g*dt;
X(1)=x(1); Y(1)=y(1); X(N)=x(N); Y(N)=y(N);
```

Particle simulation is an important tool for plasma physics study. The **particle in cell** (**PIC**) model is popular for its easiness in evaluating the self consistent electric field by loading charge densities to the nearby grids. Here, we simulate test particles under the cyclotron wave heating in the tokamak confined plasma.

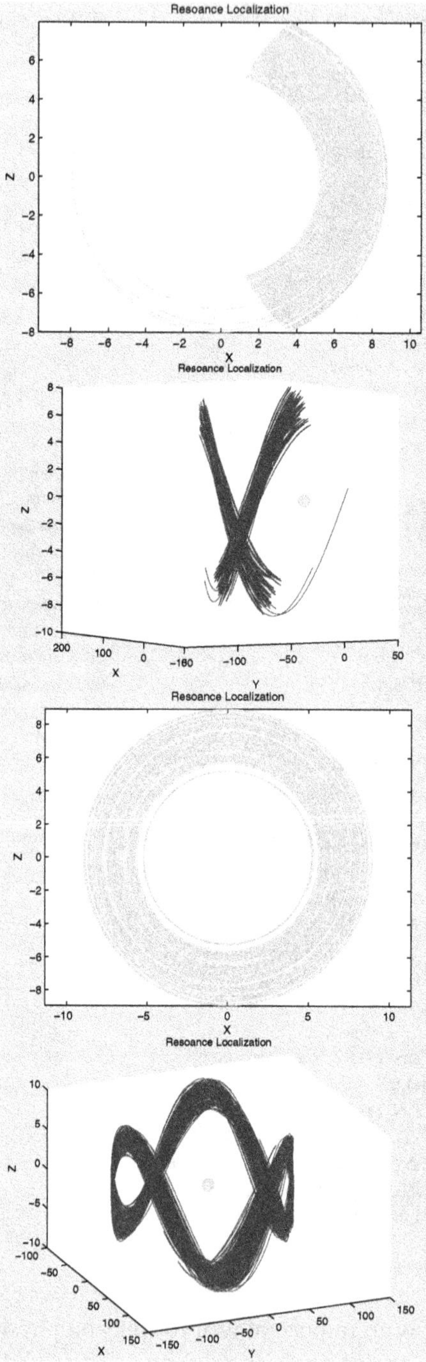
Resoance Localization
Z
X
Resoance Localization
Z
X
Y
Resoance Localization
Z
X
Resoance Localization
Z
X
Y

```
function ResonanceLocalization(RERUN)
close all; clc;
global bx by bz N MU E;
global X Y Z R V;
global R0 a0 B0 b0 Np;
global Nt Nc dt OMEGA0;
Init(RERUN),
tic; j=0; eField=0.01;
b=getBfield(R(:,1),R(:,2),R(:,3));
B=sqrt(DOT(b,b));
eb=b./repmat(B,1,3);
gradB=getGradB(R(:,1),R(:,2),R(:,3));
Vp=DOT(V,eb);% V|| magnitude
Vpara=repmat(Vp,1,3).*eb;  % V|| vector
E=DOT(V,V)/2;
E0=sum(E);
MU=(E-Vp.^2/2)./B;
Vperp=repmat(MU./B/1.76e7,1,3).*CROSS(eb,gradB);
R=R+(Vpara+Vperp)*dt/2;
sign=+1+0*(1:Np)';
for i=1:Nt;
  t=dt*i;
  b=getBfield(R(:,1),R(:,2),R(:,3));
  B=sqrt(DOT(b,b));
  Vp=sum(Vpara,2);
 for ip=1:Np
   if(abs(B(ip)-B0)<Vp(ip)/R0/1.76e7)
       dMU(ip)=rand(1)*dt*OMEGA0*1.0e16*eField;
       MU(ip)=MU(ip)+dMU(ip);
   end;
 end;
  eb=b./repmat(B,1,3);
  gradB=getGradB(R(:,1),R(:,2),R(:,3));
  PASS=E-MU.*B;
  Vb=zeros(Np,1); Vpb=MU;
 for ip=1:Np
  if(PASS(ip)<0)
    str=sprintf('Found turning point %e',B(ip)),
    dV(ip)=-MU(ip)*dot(eb(ip,:),gradB(ip,:))*dt;
    if(abs(Vp(ip))>abs(dV(ip)))
      s=sprintf('Error at turning point %e %e',Vp(ip),dV(ip)),
    end;
    Vb(ip)=dV(ip);
    sign(ip)=sign(ip)*(-1);
  else
    Vb(ip)=sign(ip)*sqrt(abs(E(ip)-MU(ip)*B(ip)))*sqrt(2);
    Vpb(ip)=Vpb(ip)+Vb(ip)^2/B(ip)/1.76e7;
  end;
 end;
```

```
    Vperp=repmat(Vpb./B,1,3).*CROSS(eb,gradB)/1.76e7;
    Vpara=repmat(Vb,1,3).*eb;
    V=Vperp+Vpara;
    R=R+V*dt;%    R=rk4(dt,R,sign);
    clear Vb Vpb;

    if(mod(i,Nc)==0&&abs(Nt-i)<=1.0e6)
       j=j+1;
       X(:,j)=R(:,1); Y(:,j)=R(:,2); Z(:,j)=R(:,3);
    end;
    E=DOT(Vpara,Vpara)/2+MU.*B;
end;
dE=abs(sum(E)-E0)./E0,
toc;
PrintRL;
```

```
function Init(RERUN)
global X Y Z R V;
global R0 a0 B0 b0 Np;
global Nt Nc dt OMEGA0;
R0=100; a0=10; c=3.0e10; Np=128; %Np is the number of particles
B0=1000; b0=300; OMEGA0=1.76e7*B0;
RandStream.setDefaultStream(RandStream('mt19937ar','seed',sum(100*clock)));
R=[R0+0.5*a0+a0/3*rand(Np,1), a0/3*rand(Np,1), a0/3*rand(Np,1)],
V=0.01*a0*OMEGA0*rand(Np,3)/sqrt(3),
if(RERUN) load RL R V Np;
else save RL R V Np; end;
Nt=2.0e6; Nc=1.0e3; dt=0.025/OMEGA0;
% X=(1:Nt/Nc)*0; Y=X; Z=X;
X=zeros(Np,1.0e3); Y=X; Z=X;
time=datestr(fix(clock),31),
```

```
function PrintRL
global X Y Z R V;
global R0 a0 B0 b0 Np;
figure(1); hold on;
for i=1:Np
plot3(X(i,:),Y(i,:),Z(i,:));
end;
plot3(0,0,0,'.g','MarkerSize',36); title('Resoance Localization');
xlabel('X'); ylabel('Y'); zlabel('Z'); hold off;
z=Z; R=sqrt(X.^2+Y.^2); x=R-R0;
figure(2);
r=sqrt(z.^2+x.^2); r0=mean(r); r0=mean(r0);
theta=0:0.1:2*pi;
plot(r0*cos(theta),r0*sin(theta),'y.'); hold on;
for i=1:Np
   plot(x(i,:),z(i,:),'g-');
end;
axis equal; xlabel('X'); ylabel('Z'); title('Resoance Localization');
dr=max(var(r)), time=datestr(fix(clock),31),
```

```
function gradB=getGradB(x,y,z)
global R0 a0 B0 b0 Np;
R2=x.^2+y.^2; R=sqrt(R2);
b=getBfield(x,y,z);
B=sqrt(sum(b.^2,2));
dB=-B0^2./B.*R0^2./R2.^2;
gradB=[dB.*x, dB.*y, dB*0];
ro=sqrt(z.^2+(R-R0).^2);
gradB=gradB+repmat(b0^2/a0^2./B.*(ro<a0),1,3).*[(R-R0).*x./R, (R-
R0).*y./R, z];
gradB=gradB-repmat(b0^2.*a0^2./B./ro.^4.*(ro>=a0),1,3).*[(R-R0).*x./R,
(R-R0).*y./R, z];
```

```
function b=getBfield(x,y,z)
global R0 a0 B0 b0 Np;
R2=x.^2+y.^2; R=sqrt(R2);
bx=-B0*R0*y./R2; by=B0*R0*x./R2; bz=0*bx;
ro=sqrt(z.^2+(R-R0).^2);
b=b0*ro./a0.*(ro<a0)+b0*a0./(ro+eps).*(ro>=a0);
bz=bz+b.*(R-R0)./(ro+eps);
bx=bx+b.*(-z)./(ro+eps).*x./R;
by=by+b.*(-z)./(ro+eps).*y./R;
b=[bx, by, bz];
```

The program of TokamakParticleTrajectory.m in Homework 3.8 is modified to advance the particle trajectory with the matrix operation by taking advantage of the intrinsic parallelism of the particle simulation. A heating term is also added to model the cyclotron heating. The matrix operation does improve the speed of computation greatly (cf. Homework 2.9), and paves the way for the Message Passing Interface (MPI) programming at later time. It can be examined in this code when the number of particles is doubled the time to complete the calculation is only increased by a fraction. The cyclotron heating increases the perpendicular energies of the resonant particles. As a result, the particles tend to become trapped particles with their banana tips localized near the cyclotron resonance layer, as shown in the first two figures below. The next two figures are the particle trajectories without wave heating from the same initial condition.

It is noteworthy that the above treatment may be generalized to combine the trajectories in toroidal geometry with the collision model in TPD (cf. Sec. 1.1.5) to simulate the neoclassical transport in tokamaks.

Homework 7.9: Generalize the inner $(\cdot)$ and cross $(\times)$ products of vectors to multiple particles in matrix operation.

7.4.2 *Finite Difference Time Domain*

Finite Difference Time Domain (**FDTD**), often refers to the numerical algorithms in the computational electrodynamics, a popular acronym, was coined by the researchers in the said field. The Fourier (frequency) domain deals with, in principle, the steady state situation, free from any numerical instability that might occur in the time domain. However, the frequency domain does not describe one-shot and transient physical phenomena. Moreover, interactions of multiple frequencies will hardly be accounted for in the Fourier domain.

Besides mindful of the Courant-Friedrichs-Lewy (CFL) condition when dealing with the FDTD, do recognize that the equations of temporal evolution are hyperbolic instead of elliptic, and the numerical instabilities abound, not to mention the numerical accuracy that could be lost in the long run. In this regard, the leap frog, the **velocity Verlet**, the **Rugge-Kutta** and the **Crank-Nicholson** schemes are algorithms worthy of attention. A straightforward Euler's method to solve the simple equation: $dy/dt = f(y,t)$ by $y^{n+1} = y^n + dt \cdot f(y^n, t^n)$ is often not accurate enough. It is an explicit scheme by taking the integrand from the previous time step that can also be less stable. An implicit scheme, for example, by taking the y value in f as the mean of y^{n+1} and y^n, while often is more stable, will need extensive algebra to express y^{n+1} in terms of y^n, that can often become intractable. The Rugge-Kutta method is a type of predictor-corrector scheme (cf. Sec. 3.1.3), that proceed by extrapolating a polynomial fit to the derivative from the previous points to the new point (the predictor step), then using this to interpolate the derivative (the corrector step).

We now solve the mode conversion problem in Sec. 6.2.3 by the FDTD,

$$\frac{\partial^2 \xi}{\partial t^2} - V_T^2 \nabla^2 \xi + \omega_p^2 \xi + \frac{\partial \omega_p^2}{\partial z} \delta E_z = 0, \tag{6.50}$$

$$\frac{\partial^2 \varsigma}{\partial t^2} - c^2 \nabla^2 \varsigma + \omega_p^2 \varsigma + \frac{\partial \omega_p^2}{\partial z} \delta E_x = 0, \tag{6.51}$$

where $\xi \equiv 4\pi \delta ne = (\partial \delta E_z / \partial z + i k_x \delta E_x)$ and $\varsigma \equiv \hat{e}_y \cdot \nabla \times \delta \vec{E} = \partial \delta E_x / \partial z - i k_x \delta E_z$ represent the longitudinal and the transverse components of the electric fields, respectively. The electric fields are governed by the following equations,

$$\frac{\partial^2 E_z}{\partial z^2} - k_x^2 E_z = \frac{\partial \xi}{\partial z} - i k_x \varsigma, \tag{7.21}$$

$$\frac{\partial^2 E_x}{\partial z^2} - k_x^2 E_x = \frac{\partial \varsigma}{\partial z} + i k_x \xi, \tag{7.22}$$

They can be solved by the Green's function method.

Homework 7.10: Find the Green's function solution in the $x - y$ plane for the 2d Poisson equation given the periodic solution in the x-direction.

The flow velocity is found from $\partial\delta\vec{v}/\partial t = -V_T^2\nabla\delta n/n_0 + e\delta\vec{E}/m$, the momentum equation, and the magnetic field from $\partial\delta B/\partial t = -c\nabla\times\delta\vec{E}$, the Faraday's law. Taking curl on the momentum equation gives,

$$n_0 m\frac{\partial\vec{\Omega}}{\partial t} = n_0 e\nabla\times\delta\vec{E} = -n_0 e\frac{\partial\delta B}{c\partial t}, \tag{7.23}$$

we find $\vec{\Omega} + e\delta\vec{B}/mc = const.$ is a constant of the motion, tantamount to the conservation of the canonical momentum $\vec{P} = \vec{p} + e\vec{A}/c$, where $\vec{\Omega} \equiv \nabla\times\delta\vec{v}$ is the vorticity. It is interesting to note that the flow associated with the EM wave in the plasma is intrinsically vortex flow, while the flow associated with the ES wave is compressible flow.

The density profile is prescribed with a vacuum region at the edge. A wave packet is launched in the vacuum from the low density side to reach the mode conversion layer of the resonance-cutoff pair. Four plots of the EM and ES wave properties are recorded in the movie clip. The time integration adopts the leap frog scheme. A 0th order outgoing wave boundary condition is imposed that has the wave travel to the outer grid point at the edge. While it does not absorb wave energy in its entirety, it serves to prevent the major wave energy from bouncing back. In the high density side where the plasma has a cut off, the reflection boundary is imposed by nullifying the ς value at the outer grid point,

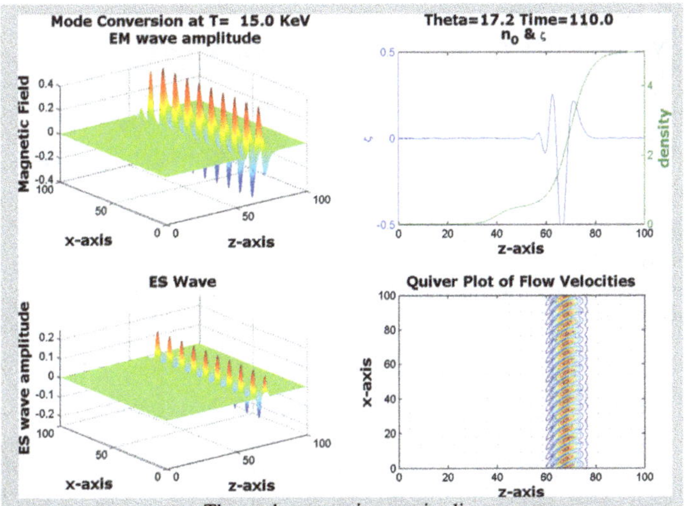

The mode conversion movie clip.

```
function FDTDmc
close all; clear all;
global A WavePacketWidth  omega0 nt2 nt3 T Vth c L N Lx dx;
global kx x dt dtx Lz dz z dtz nH nL zc Lc n0 dn0 ZZ;
init; datestr(now), tic;
Zt=zeros(N,N);Z=Zt; Ex=Zt; Ez=Zt; B=Zt; Vx=Zt; Vz=Zt;
X=Z;Xt=Zt; Tth=Vth^2;
k0=sqrt(1-kx^2),
theta=atan(kx/k0)*180/2/pi,
j=0; tf=nt3*dt; tend=nt2*dt; it=0;
fig=figure('position', [50 50 1000 650],'color','white');
msg=sprintf('wave packet on @t=0'),
for t=0:dt:tf
  ZS0=Z(:,1);
  it=it+1;
  if(it<fix(nt2))
     ZS=getZ(t); Z(:,1)=getZ(t-dt);
  else
     if(it<1.5*nt2) ZS=0*x;  Z(:,1)=ZS; end;
     if(it==fix(nt2))  msg=sprintf('wave packet off @t=%4.1f',t), end;
  end
  Zt=Zt+dtx*([Z(N,:);Z(1:N-1,:)]-2*Z+[Z(2:N,:);Z(1,:)]);
  Zt=Zt+dtz*([ZS,Z(:,1:N-1)]-2*Z+[Z(:,2:N),0*(1:N)']);
  Zt=Zt-dt*(Z.*repmat(n0,N,1)-Ex.*repmat(dn0,N,1));
  if(it>=fix(2*nt2)) ZS=(0*ZS0+2*Z(:,1))/2;
     if(it==fix(2*nt2)) msg=sprintf('outgoing wave boundary condition on
@t=%4.1f',t), end;
  end;
  Z=Z+dt*Zt;
  Xt=Xt+Tth*dtx*([X(N,:);X(1:N-1,:)]-2*X+[X(2:N,:);X(1,:)]);
  Xt=Xt+Tth*dtz*([0*(1:N)',X(:,1:N-1)]-2*X+[X(:,2:N),0*(1:N)']);
  Xt=Xt-dt*(X.*repmat(n0,N,1)+Ez.*repmat(dn0,N,1));
  X=X+dt*Xt; X=X.*(repmat(n0,N,1)>0.1);
  [Ex,Ez]=getE(X,Z,ZS);
  Vx=Vx-Tth*dt*([X(2:N,:);X(1,:)]-[X(N,:);X(1:N-1,:)])/2/dx-dt*Ex;
  Vz=Vz-Tth*dt*([X(:,2:N),0*(1:N)']-[0*(1:N)',X(:,1:N-1)])/2/dz-dt*Ez;
  Vx=Vx.*(repmat(n0,N,1)>0.1);  Vz=Vz.*(repmat(n0,N,1)>0.1);
  B=B-Z*dt; B(:,1)=0;
  if mod(it,20)==0
    j=j+1; Smax=max(ZS);
    PlotGraf(X,Z,Ex,Ez,Vx,Vz,B,theta,t); % PlotGraf(X,Z,Ex,Ez,t)
    M(j)=getframe(gcf);
   end
end
movie(M,1)
movie2avi(M,'FDTDmc.avi','Compression','None','FPS',15,'quality',100)
datestr(now), toc,
```

```
function [Ex,Ez]=getE(X,Z,ZS)
global A WavePacketWidth omega0 nt2 nt3 T Vth c L N Lx dx;
global kx x dt dtx Lz dz z dtz nH nL zc Lc n0 dn0 ZZ;
[Sx,Sz]=source(X,Z,ZS,N,dx,dz);
Ex=zeros(N,N);Ez=Ex;
ex=exp(-i*kx*x')*Sx*dx/Lx;
ez=exp(-i*kx*x')*Sz*dx/Lx;
   % Calculating Ex and Ez component
ex=ex*exp(-kx*abs(ZZ-ZZ'))/kx*dz/2;
ez=ez*exp(-kx*abs(ZZ-ZZ'))/kx*dz/2;
Ex=Ex+exp(i*kx*x)*ex;    %Green function solution
Ez=Ez+exp(i*kx*x)*ez;

ex=exp(i*kx*x')*Sx*dx/Lx;
ez=exp(i*kx*x')*Sz*dx/Lx;
   % Calculating Ex and Ez component
ex=ex*exp(-kx*abs(ZZ-ZZ'))/kx*dz/2;
ez=ez*exp(-kx*abs(ZZ-ZZ'))/kx*dz/2;
Ex=Ex+exp(-i*kx*x)*ex;%Green function
Ez=Ez+exp(-i*kx*x)*ez;
```

```
function init
global A WavePacketWidth omega0 nt2 nt3 T Vth c L N Lx dx;
global kx x dt dtx Lz dz z dtz nH nL zc Lc n0 dn0 ZZ;
A=0.5; dOmega=0.25; dk=0.25; omega0=1;
dt=0.05;
WavePacketWi dth=2*pi*2;%wave packet starts at t=0;
nt2=4*WavePacketWidth/dt;%wave packet ends
nt3=220/dt; %program ends
T=15; %keV
Vth=sqrt(T/500);
c=1; N=1000; L=9*2*pi; %L=12*2*pi;
Lx=100;dx=Lx/N; kx=L/Lx,
x=(1:N)'*dx;
dtx=dt/dx/dx; dtz=dt/dz/dz;
Lz=100; dz=Lz/N; z=(1:N)*dz;
ZZ=repmat(z,N,1);
[n0,dn0]=getDensity(Lz,z,dz,N);
```

```
function [n0,dn0]=getDensity(Lz,z,dz,N)
nH=4.5; nL=0.5;
zH=0.7*Lz; zL=0.4*Lz;
LH=8; LL=6;
n0=nH*(1+tanh((z-zH)/LH))/2+nL*(1+tanh((z-zL)/LL))/2;
dn0=diff(n0); dn0=[0,dn0];
```

```
function [Sx,Sz]=source(X,Z,ZS,N,dx,dz)
Sx=-([Z(:,2:N),0*(1:N)']-[ZS,Z(:,1:N-1)])/dz/2-getDx(X,N,dx);
Sz=getDx(Z,N,dx)-getDz(X,N,dx);
```

Homework 7.11: Design a function routine to evaluate the error in the Green's function solution in the FDTD program.

Homework 7.12: Design a wave packet to be launched from the vacuum into the plasma in the $x-z$ plane of a two dimensional geometry, where the x direction is periodic.

7.4.3 *Computational Fluid Dynamics*

Computational Fluid Dynamics (CFD) is an important research area due to its wide applications in aviation design, hydraulics application, device performance, manufacturing process, biomedical study, and not the least, the weather forecast especially for the extreme weather and global warming. The focus of CFD has been to solve the Navier Stokes equation, a special case of Eq. (4.18).

The equations of concern are hyperbolic in nature and require special attention to ensure the numerical stability. In the example of KHI in Chap. 4, it is recognized that numerical accuracy in evaluating the spatial derivatives is crucial and uncompromising, or numerical instability can develop in the long time scale, especially when the compressibility is unconstrained. The convection by flow can be better managed by the advection scheme, namely, solving the equation, $\partial\xi/\partial t + \vec{V}\cdot\nabla\xi = 0$, in terms of

$$\xi(\vec{X},t) = \xi\left(\vec{X} - \int_0^t \vec{V}(X(\tau),\tau)d\tau\right). \tag{7.24}$$

The advection of field variable is solved by taking its value at the prior position before it arrives at the grid point, thus constraining the variables in space and ensuring its stability as well as accuracy.

Decomposing a vector field into the longitudinal and the transverse components by the Helmholtz-Hodge Decomposition,

$$\vec{V} = \nabla\Phi + \nabla\times\vec{A}, \tag{4.47}$$

is helpful not only in the numerical algorithm, but also in exploring the underlying physics. For the flow velocities, there are the vortex flow versus the compressible flow, and for the electric fields, there are the EM wave versus the ES wave.

We here take a simple model to examine the earth climate. We want to solve the continuity equation,

$$\frac{\partial\rho}{\partial t} + V_\theta\frac{\partial}{\partial\theta}\rho + \frac{V_\varphi}{\sin\theta}\frac{\partial}{\partial\phi}\rho = 0, \tag{7.25}$$

the momentum equation,

$$\frac{\partial V_\theta}{\partial t} + V_\theta \frac{\partial}{\partial \theta}(V_\theta) + \frac{V_\phi}{\sin\theta}\frac{\partial}{\partial \phi}(V_\theta) = -\frac{1}{\rho}\frac{\partial}{\partial \theta}\left(\frac{1}{3}\rho T\right), \tag{7.26a}$$

$$\frac{\partial V_\phi}{\partial t} + V_\theta \frac{\partial}{\partial \theta}(V_\phi) + \frac{V_\phi}{\sin\theta}\frac{\partial}{\partial \phi}(V_\phi) = -\frac{1}{\rho \sin\theta}\frac{\partial}{\partial \phi}\left(\frac{1}{3}\rho T\right), \tag{7.26b}$$

and the energy equation which is simplified by neglecting the heat flux and taking the pressure $\vec{P} = \frac{1}{3}\rho T \vec{I}$ to give

$$\frac{\partial T}{\partial t} + \frac{4}{3}V_\theta \frac{\partial}{\partial \theta}T + \frac{4}{3}V_\phi \frac{\partial}{\partial \phi}T = -\frac{1}{3}\frac{T}{\rho}\left(V_\theta \frac{\partial}{\partial \theta}\rho + \frac{V_\phi}{\sin\theta}\frac{\partial}{\partial \phi}\rho\right), \tag{7.27}$$

where T is the temperature. Taking a thin surface layer approximation, and setting the length unit by the radius of earth $a = 6353\,\text{km}$ and the velocity unit by the thermal speed $V_{th} = 343.2\,\text{m/s}$, we choose the average density $\langle \rho_0 \rangle = 1$. Thus, the time unit is 5.14 hours which is the time needed for the sound wave to travel across the earth radius. The temperature is normalized to the room temperature. We consider the azimuthally symmetric equilibrium and fix the temperature at 36°C at equator where the latitude is at $\theta = 0$, and at the north and south poles -27°C, $\theta = \pm\pi/2$, respectively, we find $\Gamma_0 = \rho_0 V_0 = const.$, $W_0 = \Gamma_0 V_0 + \rho_0 T_0 = const.$ If we assume no equilibrium flow, then we end with $P_0 = \rho_0 T_0 = const.$ We choose $T_0 = (63\cos\theta - 27)$°C so that $T_0 \le 0$°C at $\theta \ge 64.6°$ which more or less coincides with the arctic circle. Normalizing to the room temperature 293°K, we have $T_0 = (63\cos\theta + 246)/293$. Assuming $\langle \rho_0 \rangle = 1$, we find $\rho_0 = 1/T_0/3.2327$. The problem is modeled in a spherical thin layer. Therefore, ρ, $\vec{V}$, and T are function of θ and ϕ only.

When the temperature goes up, the pressure goes up unless the density goes down so to reduce the pressure. That means the air would tend to flow away from the hot region where it would have less chance to have moisture air to arrive. This simple observation may correlate well with the draught as the continent gets hot. The flow tends to cross the globe in a few days, implying the weather pattern would generally last that long before it changes.

The external force cannot pump energy into the pressure energy directly, but accelerates the flow velocity. The flow velocity can be damped into the thermal energy, thus the pressure, by the internal friction force $(-\vec{F}^{\text{int}} \cdot \vec{v})$, while the pressure energy can be converted to the flow energy through the pressure gradient $(-\nabla P)$. This has the implication that higher atmospheric temperature can result in increased flow energy that may break the thermal barrier. A point is better appreciated when the polar vortex causes extreme icy storm, or tropical heat wave penetrates into the high latitude region. Any flow velocity field may result in the working force on the flow itself by the gradient of velocity energy ($-\frac{1}{2}\nabla\vec{V}^2$), and will also generate the vortex flow through the $(\nabla \times \vec{V}) \times \vec{V}$ component in the $\vec{V} \cdot \nabla\vec{V}$ term provided that the flow velocity is sufficiently large so that the second order terms in $\vec{V}$ are significant. A velocity shear gradient or counter streaming flow may cause the vortex flow, and small vortexes may merge to form larger vortex. The fact that the different energies in the fluid dynamics can be transformed from one form to the other suggests that human energy print can be very crucial for the weather changes. The warming up in an island due to human activities, for example, will make the pressure gradient pointing inward while the pressured flow would go the opposite so to reduce the density and balance the pressure. As a result, moisture from the surrounding ocean may have less chance to move toward the island and thus the occasional light rains will occur less often.

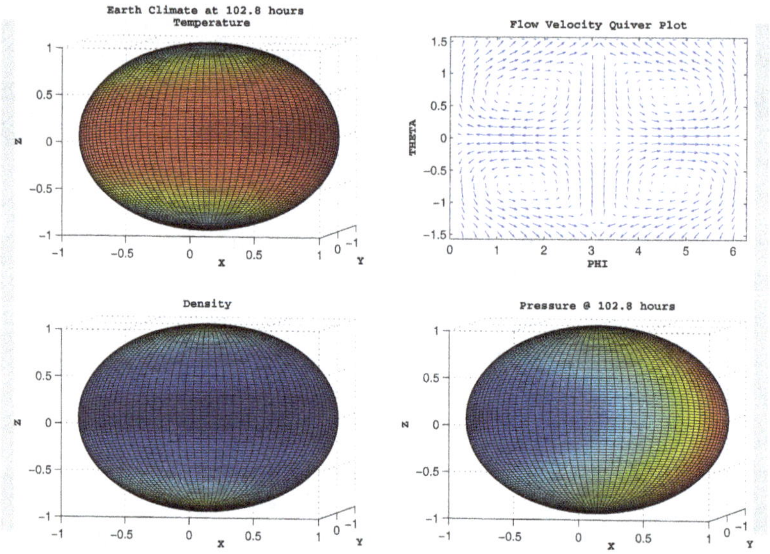

Homework 7.13: The ∇^2 operator is a key component in the viscous term or the Poisson's equation. To set it up in the 2d problem and beyond is a bit tricky. Write a function to load up the matrix for the Poisson's operator in the 2d geometry.

7.4.4 *Parallelism*

The vector operation can speed up the calculation greatly than a scalar operation with the do/while/for loop. It makes the coding more compact, easier to understand

and simpler to debug. It also provides the beneficial guidance to rewrite the code for MPI parallel programming when parallelism is needed for computational speed. If the operation can be treated by vector operation, it possesses the parallelism.

The **Message Passing Interface (MPI)** is the *de facto* standard for implementing programs on multiple processors for **distributed memory model**. If all CPUs share the full memory, it is in fact easier to implement a **shared-memory model** with use of OpenMP since the communication bottleneck is alleviated. MPI is popularly defined in C and Fortran languages for communication among different CPUs. As the PC world is hedging its way into the scientific computing arena, MPI may prove to be well worth the effort since PC clusters are rather cost-effective. The trend of multiple-cores per CPU may however, also make OpenMP indispensable. Mixing these two together are getting more attention recently.

MatlabMPI implements a subset of MPI and allows any Matlab program to be run on a parallel computer. The key innovation of MatlabMPI is that it implements the widely used MPI "look and feel" on top of standard Matlab file i/o, resulting in a "pure" Matlab implementation.

OpenMP is based upon the existence of multiple threads in the shared memory programming paradigm.

- **FORK:** the master thread then creates a ***team*** of parallel threads.
- The statements in the program that are enclosed by the parallel region construct are then executed in parallel among the various team threads.
- **JOIN:** When the team threads complete the statements in the parallel region construct, they synchronize and terminate, leaving only the master thread.

Software will be vital to knowledge creation in any discipline. Plasma fusion study has been in the forefront of computational physics for many decades since the start of Magnetic Fusion Energy program. Massively parallel computing is essential in this regard, and the current software technology relies on MPI and OpenMP or their derivatives. The graphic card process unit (GPU) has demonstrated its parallel computing power and will be a significant player in the parallel computing paradigm. Its language is also multiple-thread based.

Further Reading

Check into current progress and development of GPU in the following link: http://www.gpgpu.org/.

The notion of divide and conquer was documented in the Art of War (孫子兵法) written by **Sun Tzu** in the 6th century BC, a military treatise of war strategies. Among them, "The greatest conqueror is not fighting hundred wars and winning hundred wars, but defeating enemies without engaging in one." His philosophy was, in fact, not to adore war but to avoid war. The following sayings from the Chinese philosophers can be rather intriguing besides relevant to the methodology of problem solving.

格物為參	Investigate the physical phenomena as a reference;
窮理致知	Exhaust all the arguments to gain knowledge;
斷疑為悟	Alleviate any doubt to become enlightened.

Human history never shied away from elimination — even until today, many drone attacks are in the making. Before civilization, elimination has different names: murder, assassination, massacre, genocide, war, indicative of its effectiveness and inhumanity.

A recommended reading for algorithm is "Numerical Recipes — the Art of Scientific Computing" by Press, Teukolsky, Vetterling, and Flannery, Cambridge University Press. For plasma simulation, check into *Plasma Physics via Computer Simulations* by Birdsall and Langdon (1991); For MHD equations and algorithms in solving elliptical and hyperbolic equations, *Computational Methods in Plasma Physics* by Jardin (2010).

The book "Genetic Algorithms and Engineering Optimization" by Mitsuo Gen and Runwei Cheng, published by John Wiley and Sons, Inc, (2000) offers a thorough review on the genetic algorithm.

Since the perpendicular heating in cyclotron waves tends to pile up the resonant particles toward the low magnetic field side with the banana tips localized to the resonant surface, a poloidal electric field would result that can affect the ion transport Hsu (1984).

It is challenging to numerically solve bifurcated equations. The shooting method can be very robust to find the bifurcated solutions in the one-dimensional geometry, but difficult to generalize to the two-dimensional problem. Many other numerical algorithms (cf. Cliffe *et al.* 2011, Guevel 2011), such as the analytical continuation, and the direct iteration algorithm, often produce without hesitation one solution (the strong, the dominant), but not the other (the weak, the subdominant). A successful algorithm in solving the bifurcated problems has to alleviate the mode competition by the strong solution to find the weak. The direct iteration method, for example, starts from the neighborhood of the multiple eigensolutions of the corresponding linear operator, in this case $\Delta *$. The well educated guess is that the lowest two eigenstates may correspond to the bifurcated solutions, or at least lead to them. Variants of this method may look for two different solutions at the beginning on two different equations that only approach the same after extensive iterations.

A successful method in solving the MHD bifurcation utilizes the physical characteristics to search for the bifurcated solutions. It identifies the edge poloidal magnetic field as the key physical variable to be evaluated from two independent functions: One is related to the LHS of the equation and the other the RHS. Further discussion on this numerical method is presented in Sec. 11.4.2.

We have seen a few examples utilizing the physical characteristics to device an algorithm. Quite often applying the physical law can be more convenient and easier to solve a mathematical equation than the mathematical law would. The following is another good example. It applies the Monte Carlo method to solve the Schrödinger equation.

[Diffusive Quantum Monte Carlo Method] The DQMC originated from one of the best cited PRLs [Ceperley and Alder, 1980]. It may serve as a great learning example of the Monte Carlo method.

Take a single particle with mass m in one dimension system as example, the time-dependent Schrödinger equation is written as

$$i\hbar\frac{\partial}{\partial t}\psi(x,t) = H\psi(x,t) = \left[-\frac{\hbar^2}{2m}\frac{\partial^2}{\partial x^2} + V(x)\right]\psi(x,t). \tag{7.A.1}$$

The eigenfunction of the Schrödinger equation is the superposition of the basis of $\phi_n(x)$

$$\psi(x,t) = \sum_{n=0}^{\infty} c_n\phi_n(x)e^{-iE_nt/\hbar}. \tag{7.A.2}$$

The basis functions $\phi_n(x)$ are assumed orthonormal and real, i.e.,

$$\int_{-\infty}^{\infty} dx\phi_n(x)\phi_m(x) = \delta_{nm}. \tag{7.A.3}$$

Here, we shift the energy scale by a reference energy E_R, and replace the potential energy and the eigenenergy by $V(x) \to V(x) - E_R$ and $E_n \to E_n - E_R$. Thus, the wave function becomes $\psi(x,t) = \sum_{n=0}^{\infty} c_n\phi_n(x)e^{-i(E_n-E_R)t/\hbar}$.

In order to solve the Schrödinger equation numerically in Euclidean space, we take a wick rotation of real time to be imaginary time by a new variable $\tau \equiv it$. The time-dependent Schrödinger equation is transformed to

$$\hbar\frac{\partial}{\partial \tau}\psi(x,\tau) = \left[\frac{\hbar^2}{2m}\frac{\partial^2}{\partial x^2} - V(x) + E_R\right]\psi(x,\tau), \tag{7.A.4}$$

and the eigenfunction becomes

$$\psi(x,\tau) = \sum_{n=0}^{\infty} c_n\phi_n(x)e^{-(E_n-E_R)\tau/\hbar}. \tag{7.A.5}$$

In the imaginary time, the wave function at $\tau \to \infty$ takes the asymptotic behavior of

(1) $E_R > E_0 \Rightarrow \lim_{\tau\to\infty}\psi(x,\tau) \to \infty$;
(2) $E_R < E_0 \Rightarrow \lim_{\tau\to\infty}\psi(x,\tau) \to 0$;
(3) $E_R = E_0 \Rightarrow \lim_{\tau\to\infty}\psi(x,\tau) = c_0\phi_0$.

The above is the basis for the DQMC. The wave function of the system converges for $E_R = E_0$ as time goes to infinity, and yields the ground state for the system of interest.

The path integral formalism is applied to integrate the imaginary time-dependent Schrödinger equation. The wave function can be written as

$$\psi(x,\tau) = \int_{-\infty}^{\infty} dx_0 K(x,\tau|x_0,0)\psi(x_0,0),$$

where $K(x,\tau|x_0,0)$ is the propagator which gives the probability for a particle to move from a place to another in a time step.

$$K(x,\tau|x_0,0) = \lim_{N\to\infty}\int_{-\infty}^{\infty} dx_1 \ldots \int_{-\infty}^{\infty} dx_{N-1}\left(\frac{m}{h\Delta\tau}\right)^{\frac{N}{2}} \times \exp\left\{-\frac{\Delta\tau}{\hbar}\sum_{j=1}^{N}\left[\frac{m}{2\Delta\tau^2}(x_j - x_{j-1})^2 + V(x_j) - E_R\right]\right\} \tag{7.A.6}$$

Here, $\Delta\tau$ is the time step. Thus the wave function could be written as

$$\psi(x,\tau) = \lim_{N\to\infty}\left(\prod_{j=0}^{N-1}\int_{-\infty}^{\infty} dx_j\right)\prod_{k=1}^{N} W(x_k)P(x_k,x_{k-1})\psi(x_0,0),$$

where

$$P(x_k,x_{k-1}) \equiv \left(\frac{m}{h\Delta\tau}\right)^{\frac{1}{2}} \exp\left\{-\sum_{j=1}^{N}\left[\frac{m}{2\hbar\Delta\tau}(x_j - x_{j-1})^2\right]\right\} \tag{7.A.7}$$

related to the kinetic energy term of the Hamiltonian gives the probability density to generate particles from Gaussian distribution randomly. The weight function

$$W(x_k) \equiv \exp\left[-\frac{V(x_k)-E_R}{\hbar}\Delta\tau\right], \tag{7.A.8}$$

depending on the potential energy and the reference energy, controls the annihilation of creation of particles depending on whether they have more energy or less than E_R.

Standard Monte Carlo integration of N dimension system is

$$I = \int_{-\infty}^{\infty} dx_1 \ldots \int_{-\infty}^{\infty} dx_N f(x_1,\ldots x_N)p(x_1,\ldots x_N), \tag{7.A.9}$$

where N is the interval number, $p(x_1,\ldots x_N)$ is the probability density, and $f(x_1,\ldots x_N)$ is the function what we integrate. By definition, to evaluate $\psi(x,\tau)$ with (7.A.7) and (7.A.8)

$$p(x_1,\ldots x_N) = \prod_{k=1}^{N} P(x_k,x_{k-1}) \tag{7.A.10}$$

$$f(x_1,\ldots x_N) = \prod_{k=1}^{N} W(x_k) * \psi(x_0,0). \tag{7.A.11}$$

The variance $\sqrt{\hbar\Delta\tau/m}$ depends on the function $P(x_k,x_{k-1})$ of equation (7.A.7). In numerical scheme, the coordinate of the imaginary particle is defined as $x_k^{(i)}$ which means the position of the ith particle at kth time step. Based on the propagator, the position at the next time step is written as

$$x_k^{(i)} = x_{k-1}^{(i)} + \sqrt{\frac{\hbar\Delta\tau}{m}}\rho_k^{(i)}, \tag{7.A.12}$$

where $\rho_k^{(i)}$ is a Gaussian random number with zero mean and unit variance.

The following code with use of the diffusive quantum Monte Carlo method solves for the ground state of the hydrogen atom. Since the statistical error depends on the number of walkers deployed, the program quadruples the walker population in each rerun. This helps achieving the better accuracy at shorter time. Moreover, several separate runs may be made and the best result selected in each stage. When the expectation energy equals the averaged reference energy within the digits of accuracy allowed by the statistical error, namely, $1/\sqrt{N}$, the final result is achieved.

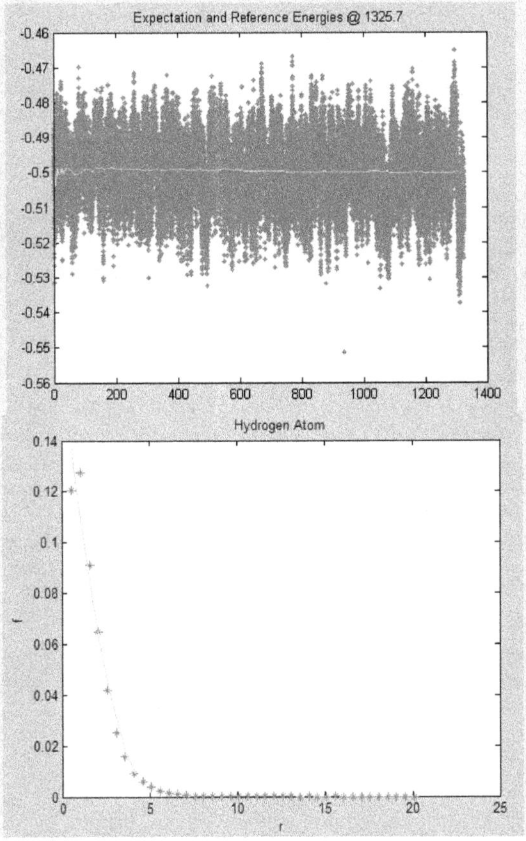
Expectation and Reference Energies @ 1325.7
-0.46
-0.47
-0.48
-0.49
-0.5
-0.51
-0.52
-0.53
-0.54
-0.55
-0.56
0
200
400
600
800
1000
1200
1400
Hydrogen Atom
0.14
0.12
0.1
0.08
0.06
0.04
0.02
0
f
0
5
10
15
20
25
r

```
function dqmcH(restart) % Hydrogen Atom
global bohr bohrR Zeff;
close all; clc;
tau=0.0; tauEnd=1000; Zeff=1;
bohr=27.2114; bohrR=0.529177;
count = 0;
N0=4096;
Nc=N0; % Creating a shell distribution
x = ones(3,2*N0)/sqrt(3)/5;
copies = ones(3,Nc)/sqrt(3)/5;
newN0 = Nc;
lstN0 = N0;
avgRefEnergy=0.0;
dTau = 0.05;
rand('state',sum(100*clock));
if(restart)
  load H.mat N0 Nc newN0 dTau count avgRefEnergy refE x;
  x=[x,x+randn(3,newN0)*sqrt(dTau)];
  x=[x,x+randn(3,newN0*2)*sqrt(dTau)];
  N0=N0*4; newN0=newN0*4; tauEnd=tauEnd/2, dTau,
  avgRefEnergy=avgRefEnergy/count, count=1;
else
  V=v(x(:,1:newN0));
  refEnergy=sum(V)/newN0;
end;
count0=count;
refE=[];
yEnergy = [];
xTime =[];
figure(1);
tic, datestr(now),
while(tau/2.<tauEnd)
  count=count+1;
  x=x(:,1:newN0) + randn(3,newN0).*sqrt(dTau);
  V=v(x);
  Vavg = sum(V)/newN0;
  if(count<=10) Vavg= -Zeff^2/2; refEnergy=-Zeff^2/2;
  else refEnergy = Vavg+ ...
    (1-newN0/N0)/max(dTau,10/sqrt(N0))/4; end;
  avgRefEnergy=avgRefEnergy+refEnergy;
  if(mod(count,500)==0)
    format short; disp([count,N0,newN0]);
    format long; disp([avgRefEnergy/count, refEnergy]),
    plot(xTime,refE,'r.',xTime,yEnergy,'g -');
   str=sprintf('Expectation and Reference Energies
@%7.1f',tau);
    title(str);
    getframe;
    if(mod(count,2000)==0) dTau=dTau/1.0125,
```

% run dqmcH(0) the first time
% then dqumcH(1)
39000 16384 16438
-0.500134641635311
-0.516874620579175
Energy =
-13.609469558063122 (eV)
refEnergy =
-0.500138516091112
ReferenceEnergy =
-13.609469216761687 (eV)

% Each rerun quadruples the walker population to improve the accuracy by a factor of 2. It also restarts the average reference energy to allow effective relaxation of the expected energy.

% Walk the walkers

% evaluate the reference energy

% Red dots represent the reference energy, and the green line the expectation energy, which in general falls in the middle of the band of the reference energy.

```
        save H.mat N0 Nc newN0 dTau count avgRefEnergy refE
x;
      end;                                                    %save data every so often
    end;
    yEnergy= [yEnergy,avgRefEnergy/count];
    xTime = [xTime, tau];
    refE=[refE,refEnergy];

    % Branch
    lstN0=newN0;
    nCpy = 0;
    lstN0=newN0;
    W=w(V,refEnergy,dTau);
    [mW,nW]=size(W);
    R=rand(mW,nW);
    if(mod(count,2)==0||count<20000) child=2; else child=3; end;
    m=min(floor(W+R),child);
    nCpy=sum(m);                                              % create 2 or 3 children
    copies=zeros(3,nCpy);
    iCpy=0;
    for i=1:lstN0,
      if(m(i)>0)
        for j=1:m(i)
         iCpy=iCpy+1;
         copies(:,iCpy)=x(:,i);                               % save the original population
        end;                                                  and add the replicated children
      end;
    end;
    newN0 = nCpy;
    x= copies(:,1:nCpy);                                      % update the new population
    tau = tau+dTau;
end
toc, datestr(now),
Energy=yEnergy(count-count0)*bohr,
refEnergy=sum(refE)/length(refE),
ReferenceEnergy=refEnergy*bohr,
figure(2);
plotR(x(1:3,1:newN0));                                        % plot the wave function
title('Hydrogen Atom');
figure(3);
plotWave(x);                                                  % plot the 3d electron cloud
save H.mat N0 Nc newN0 dTau count avgRefEnergy refE x;
```

`function V= v(x)`
`global bohr bohrR Zeff;`
`x1=x(1,:);`
`y1=x(2,:);`
`z1=x(3,:);`
`r=sqrt(x1.^2+y1.^2+z1.^2);`
`V=-Zeff./r;`
	% evaluate the potential energy

`function W = w(V,refEnergy,dTau)` `W = exp(-(V-refEnergy)*dTau);`	% evaluate the creation/annihilation probability

`function plotWave(x)`
`x1=x(1,:);`
`y1=x(2,:);`
`z1=x(3,:);`
`plot3(x1,y1,z1,'g*');`
`title('Electron Wave Function');`
`xlabel('X'); ylabel('Y'); zlabel('Z');`
	% plot the 3d electron cloud

`function plotR(f)`
`nmax=40;`
`r=sqrt(f(1,:).^2+f(2,:).^2+f(3,:).^2);`
`N=length(r);`
`xmin=min(r);`
`xmax=max(r);  xmax=20;`
`r=r-xmin;`
`dx=xmax/nmax;`
`F=fix(r*nmax/xmax)+1;`
`x=1:nmax; x=x*dx; x=x+xmin;`
`y=x*0;`
`for i=1:N`
`    index=min(F(i),nmax);`
`    if(index<=0) index=1; end;`
`    y(index)=y(index)+1;`
`end;`
`SUM=sum(y/dx);`
`f=y./x.^2/SUM;`
`normalization=sum(f.*x.^2*dx);`
`f=f/normalization;`
`normalization=sum(f.*x.^2*dx);`
`plot(x,f,'r*');`
`title('Distribution function');`
`ylabel('f');`
`xlabel('r');`
	% plot the 1d electron wave function

`hold on;`
`M=2;`
`P=polyfit(x(1:nmax/4),f(1:nmax/4),M);`
`y=0*x(1:nmax/4);`
`for i=1:M`
`y=(y+P(i)).*x(1:nmax/4);`
`end;`
`y=y+P(M+1);`
`plot(x(1:nmax/4),y,'g-');`
`hold off;`
	% A polynomial fit is plotted for comparison

Homework Hints

Homework 7.1: In a group meeting, a student presented a slide as follows:

$$\overleftrightarrow{I} \equiv \sum_\sigma n_\sigma q_\sigma \vec{v}_\sigma \vec{v}_\sigma$$

$$\overleftrightarrow{\Lambda} \equiv \sum_\sigma n_\sigma q_\sigma (\vec{v}_\sigma - \vec{U})(\vec{v}_\sigma - \vec{U}) = \overleftrightarrow{I} + Q\vec{U}\vec{U} - 2\vec{J}\vec{U}$$

$$Q \equiv \sum_\sigma n_\sigma q_\sigma,\ \vec{J} \equiv \sum_\sigma n_\sigma q_\sigma \vec{v}_\sigma,\ \rho\vec{U} \equiv \sum_\sigma n_\sigma m_\sigma \vec{v}_\sigma$$

Figure out the error in the presentation.

Compare the symmetry, it is clear that $2\vec{J}\vec{U} \rightarrow \vec{J}\vec{U} + \vec{U}\vec{J}$ is the correct expression as dyadic is not commutative among its components.

Homework 7.3: Find the magnetic field $\vec{B} = \nabla\psi \times \hat{e}_z$ in a cylindrical plasma of unity radius $a = 1$. The magnetic flux is governed by $d^2\psi/dr^2 + (1/r)(d\psi/dr) + \alpha e^\psi = 0$, where the plasma current $j_z = \alpha e^\psi$. Make use of the fact that $\psi = \psi_0 - \ln[1 + \frac{1}{8}\alpha r^2 e^{\psi_0}]^2$ and the central value ψ_0 is governed by $\exp(\psi_0) = (4/\alpha)(1 \pm \sqrt{1 - \alpha/2})$ to compare the results from the integration by Simpson method and from direct differentiation of ψ.

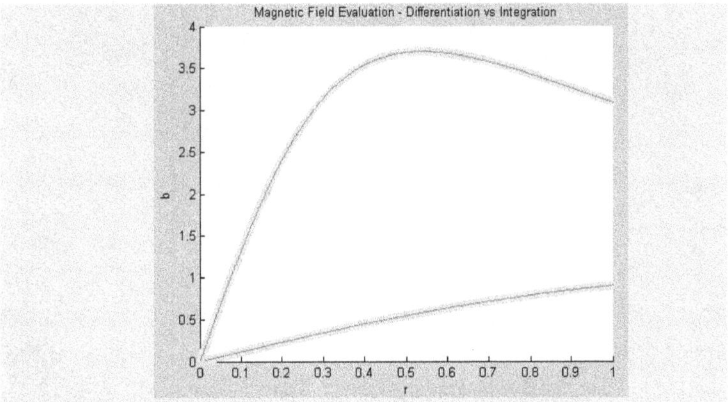

```
function b=getB
n=5000; dr=1/n;
r=(dr:dr:1)';
alpha=1.4;
figure(1); hold on;
p=2*log(4/alpha*(1+[-1,1]*sqrt(1-alpha/2))),
for icase=1:2
  psi0=p(icase);
  psi=psi0-2*log(1+alpha/8*r.^2*exp(psi0));
  bd=-[0;diff(psi)]/dr;
  bs=-Simpson(alpha*r.*exp(psi),n,dr)./r;
  plot(r,bd,'g*',r,bs,'r-');
end;
title('Magnetic Field Evaluation - Differentiation vs
Integration');
xlabel('r'); ylabel('b');
hold off;
```

% The red lines are solutions from the Simpson method; the green lines are from the differentiation method.

Homework 7.5: Solve numerically the ***Bratu's equation*** in cylindrical coordinates,

$$d^2y/dr^2 + (dy/dr/r) + \Lambda e^y = 0$$

subject to the boundary conditions: $dy/dr|_{r=0} = y|_{r=1} = 0$. Work out the analytical theory and the condition for the bifurcation, and compare with the numerical solutions.

Changing the variables as in the following: $\rho \equiv r^2$ and $z^2 \equiv e^{-y}$, and utilizing:

$$\frac{d^2}{dr^2} + \frac{1}{r}\frac{d}{dr} = 4\rho\frac{d^2}{d\rho^2} + 4\frac{d}{d\rho}, \quad \text{and} \quad 2\frac{dz}{d\rho} = -z\frac{dy}{d\rho}, \tag{7.5.1}$$

we find

$$\frac{d^2z}{d\rho^2} = -\frac{1}{2}\frac{dz}{d\rho}\frac{dy}{d\rho} - \frac{1}{2}z\frac{d^2y}{d\rho^2} = \frac{1}{z}\left(\frac{dz}{d\rho}\right)^2 - \frac{1}{2}z\frac{d^2y}{d\rho^2}, \tag{7.5.2}$$

and

$$\rho\frac{d^2y}{d\rho^2} = -\frac{2\rho}{z}\frac{d^2z}{d\rho^2} + \frac{2\rho}{z^2}\left(\frac{dz}{d\rho}\right)^2. \tag{7.5.3}$$

Therefore,

$$-\frac{2\rho}{z}\frac{d^2z}{d\rho^2} + \frac{2\rho}{z^2}\left(\frac{dz}{d\rho}\right)^2 - \frac{2}{z}\frac{dz}{d\rho} + \frac{\Lambda}{4z^2} = 0. \tag{7.5.4}$$

This nonlinear ordinary differential equation has nontrivial solution if the first term vanishes, i.e., $d^2z/d\rho^2 = 0$. Assuming $z = c_0 + c_1\rho$, we find

$$\frac{\Lambda}{8} - (c_0 + c_1\rho)c_1 + \rho c_1^2 = 0. \tag{7.5.6}$$

```
function Bratu7p5(LAMBDA)
close all; clc;
N=1000; dx=1/N;
x=(0.5:N)*dx; y=1+x*0; z=y;
Y0=(1+sqrt(1-LAMBDA/2)*[-
1,1])/LAMBDA*4;
Y0=2*log(Y0);
figure(1); hold on;
for ip=1:2
z(1)=0; y0=ip; y(1)=y0;
for iter=1:1000
for i=1:N-1
   y(i+1)=y(i)+z(i)*dx;
   z(i+1)=z(i)-z(i)./x(i)*dx;
   z(i+1)=z(i+1)-
LAMBDA*exp((y(i+1)+y(i))/2)*dx;
end;
D=y(N);
if(abs(D)<1.0e-6) iter, D, y(1), break; end;
y(1)=y(1)+(-1)^ip*0.1^(iter/10)*sign(D);
end;
w=Y0(ip)-
2*log(1+LAMBDA/8*exp(Y0(ip))*x.^2);
plot(x,y,'g*',x,w,'r-',x,x*0,'k-');
end;
title('Shooting Method - Bratu''s Equation');
xlabel('x'); ylabel('y');
text(0.8,1.5,'$\Lambda$=','interpreter','latex');
str=sprintf('%5.2f',LAMBDA);
text(0.85,1.5,str); hold off;
```

% The green lines are the solutions from the shooting method, and red lines are from the analytical solutions.

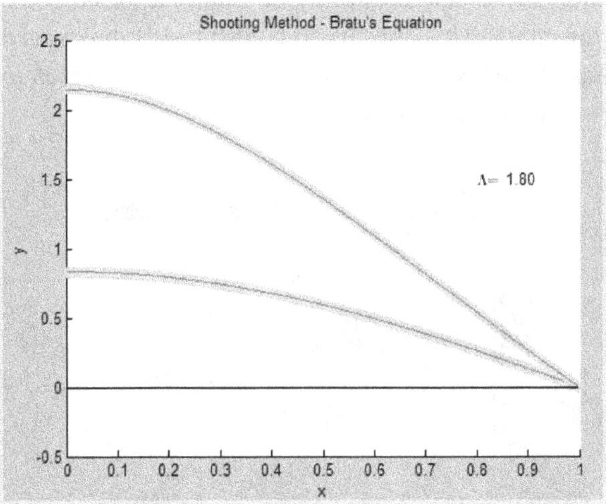

The solubility condition is $c_0 c_1 = \Lambda/8$. Changing back to the original variables and applying the boundary conditions, we arrive at $c_0 = \exp(-y_0/2)$ and $c_1 = \Lambda \exp(y_0/2)/8$. Therefore,

$$y = y_0 - \log\left(1 + \frac{1}{8}\Lambda e^{y_0} r^2\right)^2. \tag{7.5.7}$$

Taking the value $y = 0$ at $r = 1$, we find $e^{y_0/2} = 1 + \frac{1}{8}\Lambda e^{y_0}$, which can then be solved to give

$$e^{y_0/2} = 1 + \frac{1}{8}\Lambda e^{y_0} = \frac{1 \pm \sqrt{1 - \frac{1}{2}\Lambda}}{\frac{1}{4}\Lambda}. \tag{7.5.8}$$

The solubility condition is therefore $\Lambda \leq 2$. While there is no analytical proof that the solution of z is limited to no higher than the linear term of ρ, the shooting method, kind of carpet sweeping through the parameter space of Λ does appear to confirm that the solutions given by (7.5.7) are complete.

Homework 7.7: Utilize the same simulation model for the relaxed chain, but release the chain at one end. Follow its relaxation to the final configuration. Find the tension in the chain as compared to the gravitational force.

The analytical solutions often find the tension either too small or unbounded. As the simulation shows that the spring constant has to be 100 times g to maintain the length of the chain. This appears to confirm well the experimentally measured tension is at least 25 times the chain's weight. (*Classical Mechanics*, S. T. Thornton and J. B. Marion, Brooks/Cole, 2008, p. 335)

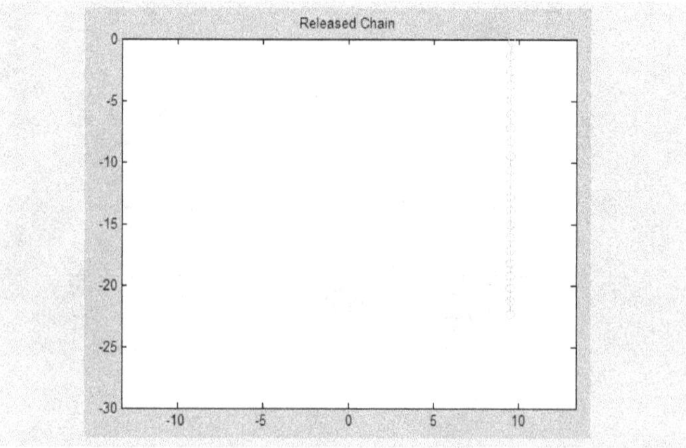

```
function ReleasedChain
global x y N k g;
load CHAIN x y b k g N;
figure(1); j=0;
for i=0:1500*b
    if(mod(i,100)==0)
      plot(x,y,'Og-'); axis([-2*b/3,2*b/3,-b*3/2,0]); j=j+1;
      M(j)=getframe;
    end;
    [x,y]=displacement;
end;
title('Released Chain');

function [X,Y]=displacement
global x y N k g;
dt=0.1; f=x*0; fx=f; fy=f; X=f; Y=f; r=f;
r(2:N-1)=sqrt((x(2:N-1)-x(1:N-2)).^2+(y(2:N-1)-y(1:N-2)).^2);
f(2:N-1)=-k*(r(2:N-1)-1);
fx(2:N-1)=f(2:N-1).*(x(2:N-1)-x(1:N-2))./r(2:N-1);
fy(2:N-1)=f(2:N-1).*(y(2:N-1)-y(1:N-2))./r(2:N-1);
r(2:N-1)=sqrt((x(2:N-1)-x(3:N)).^2+(y(2:N-1)-y(3:N)).^2);
f(2:N-1)=-k*(r(2:N-1)-1);
fx(2:N-1)=fx(2:N-1)+f(2:N-1).*(x(2:N-1)-x(3:N))./r(2:N-1);
fy(2:N-1)=fy(2:N-1)+f(2:N-1).*(y(2:N-1)-y(3:N))./r(2:N-1);
X(2:N-1)=x(2:N-1)+fx(2:N-1)*dt;
Y(2:N-1)=y(2:N-1)+fy(2:N-1)*dt-g*dt;
r(1)=sqrt((x(1)-x(2))^2+(y(1)-y(2))^2);
f(1)=-k*(r(1)-1);
fx(1)=fx(1)+f(1)*(x(1)-x(2))/r(1);
fy(1)=fy(1)+f(1)*(y(1)-y(2))/r(1);
X(1)=x(1)+fx(1)*dt;
Y(1)=y(1)+fy(1)*dt-g*dt;
X(N)=x(N); Y(N)=y(N);
```

Homework 7.9: Generalize the inner ($\cdot$) and cross ($\times$) products of vectors to multiple particles in matrix operation.

```
function C=CROSS(A,B)
C=A*0;
C(:,1)=A(:,2).*B(:,3)-A(:,3).*B(:,2);
C(:,2)=A(:,3).*B(:,1)-A(:,1).*B(:,3);
C(:,3)=A(:,1).*B(:,2)-A(:,2).*B(:,1);

function C=DOT(A,B)
C=sum(A.*B,2);
```

Homework 7.11: Design a function routine to evaluate the error in the Green's function solution in the FDTD program.

```
function CheckConsistency(X,Z,Ex,Ez,N,dx,dz)
   divE=getDIV(Ex,Ez,N,dx,dz);
errX=X-divE;
   curlE=getCURL(Ex,Ez,N,dx,dz);
errZ=Z-curlE;
maxXerr=max(max(abs(errX),[],1),[],2),
maxX=max(max(abs(X),[],1),[],2),
maxZerr=max(max(abs(errZ),[],1),[],2),
maxZ=max(max(abs(Z),[],1),[],2),
```

Homework 7.13: The ∇^2 operator is a key component in the viscous term or the Poisson's equation. To set it up in the 2d problem and beyond is a bit tricky. Write a function to load up the matrix for the Poisson's operator in the 2d geometry.

A field variable ϕ in the 2d geometry can be represented as ϕ_{ij} where the index i refers to ith grid in the x coordinate, and index j the jth grid in the y coordinate. In the matrix operation, we need to line up the set of ϕ_{ij}'s as a vector. Thus, we may keep

$$\Phi = [\phi_{11}, \phi_{12}, \ldots, \phi_{1N}, \phi_{21}, \phi_{22}, \ldots, \phi_{2N}, \ldots, \phi_{M1}, \phi_{M2}, \ldots, \phi_{MN},].$$

To translate into the matrix form, namely$M\Phi$, for the partial differential operator

$$\begin{aligned}\nabla^2\phi(x, y) &= \frac{\partial^2\phi(x, y)}{\partial x^2} + \frac{\partial^2\phi(x, y)}{\partial y^2} \\ &\rightarrow \frac{\phi_{i+1,j} - 2\phi_{i,j} + \phi_{i-1,j}}{\Delta x^2} + \frac{\phi_{i,j+1} - 2\phi_{i,j} + \phi_{i,j-1}}{\Delta y^2},\end{aligned}$$

we may construct submatrixes A and B, each of size $N \times N$ in such a way:

$$A = \begin{pmatrix} \frac{-2}{\Delta x^2} + \frac{-2}{\Delta y^2} & \frac{1}{\Delta y^2} & 0 & . & . & 0 \\ \frac{1}{\Delta y^2} & \frac{-2}{\Delta x^2} + \frac{-2}{\Delta y^2} & \frac{1}{\Delta y^2} & . & . & 0 \\ 0 & . & . & . & . & . \\ . & . & . & . & . & . \\ . & . & 0 & \frac{1}{\Delta y^2} & \frac{-2}{\Delta x^2} + \frac{-2}{\Delta y^2} & \frac{1}{\Delta y^2} \\ 0 & . & . & 0 & \frac{1}{\Delta y^2} & \frac{-2}{\Delta x^2} + \frac{-2}{\Delta y^2} \end{pmatrix}$$

and B is a diagonal matrix with the element value equal to $1/\Delta x^2$. We may therefore load up the matrix W as in the following:

$$
\begin{array}{c}
x=1, x=2, \ldots, x=M \\
W = \begin{array}{c} x=1 \\ x=2 \\ . \\ . \\ . \\ x=M \end{array}
\begin{pmatrix}
A & B & 0. & . & . & 0 \\
B & A & B & 0 & . & 0 \\
0 & B & A & B & . & 0 \\
. & . & . & . & . & . \\
0 & . & . & B & A & B \\
0 & . & . & 0 & B & A
\end{pmatrix}
\begin{pmatrix}
\phi(1,1:N) \\
\phi(2,1:N) \\
. \\
. \\
. \\
. \\
. \\
. \\
\phi(M,1:N)
\end{pmatrix}.
\end{array}
$$

Note that the boundary conditions have to be included in the final matrix equation by modifying M or in the driving terms.

```
function A=setA(N,dx2,dy2)
A=sparse(N);
A=A*0+diag(-2/dx2-2/dy2+0*(1:N));
A=A+diag(1/dy2+0*(1:N-1),1)+diag(1/dy2+0*(1:N-1),-1);

function B=setB(N,dx2)
B=sparse(N);
B=B*0+diag(1/dx2+0*(1:N));

function W=SetW(M,N,dx2,dy2)
W=sparse(M*N);
A=setA(N,dx2,dy2);
B=setB(N,dx2);
for i=1:M
   W(1+(i-1)*N:i*N,1+(i-1)*N:i*N)=A;
   if(i~=M) W(1+(i-1)*N:i*N,1+i*N:(i+1)*N)=B; end;
   if(i~=1) W(1+(i-1)*N:i*N,1+(i-2)*N:(i-1)*N)=B; end;
end;
M(1+(M-1)*N:M*N,1:N)=B;
M(1:N,1+(M-1)*N:M*N)=B;
```

Chapter Eight

Kinetic and Statistical Plasma Physics

"Wit beyond measure is man's greatest treasure."

— Harry Potter

In the N-body phase space description, the distribution function can be written as

$$\frac{d}{dt}F_N = \left(\frac{\partial}{\partial t} + H_N\right) F_N$$
$$= \left\{\frac{\partial}{\partial t} + \sum_i \vec{v}_i \cdot \nabla_i + \sum_i \vec{a}_i \cdot \frac{\partial}{\partial \vec{v}_i}\right\} F_N(\vec{r}_1, \ldots \vec{r}_N; \vec{v}_1, \ldots \vec{v}_N; t) = 0, \tag{8.1}$$

that follows the N-particle propagator H_N with the full kinematics to be tackled. This enables the study of anything beyond the local variables in the configuration space as suffice in the fluid model. While in theory the classical physics is retained in this equation, it is impossible to solve exactly, and the underlying principles are difficult to extract accordingly. On the other hand, the deterministic description of many body system given the initial and boundary values can be ambiguous or even misleading at times, since minute deviations in the initial values may yield drastic difference in the temporal behavior. Nonetheless, the kinetic description can be convenient for the statistical treatment that may give, time and again, more reliable predictions than otherwise.

The kinetic treatment often is cumbersome as it is algebra intense. We may opt to simplification by, for examples, the **Vlasov equation** or the **Fokker-Planck equation**, or treating one species with the kinetic description but others the fluid model. The reduced description, however, can also run into difficulties. The millennium problem of the Navier Stokes equations attests to the fact that in some circumstances the equations could even be ill-posed, and resorting to the direct simulation may be inevitable. The computational cost, however, can soon be out of reach in the direct simulation for moderate N even with the parallel computing. Alternatively, we need to go back to the first principles, lest we forget what have been neglected to have thus caused the ill behavior in the theoretical model. The scientific reasoning and physical intuition is indispensible and not to be replaced by the logical processes of arithmetic methodology alone. Nowhere is perhaps more true than the kinetic treatment with lengthy algebra, or we would not get very far in our understanding of the physics.

8.1 The Phase Space Description

A simplified kinetic theory is the one-body distribution function by summing up the contributions from particles in the phase space,

$$F(\vec{x},\vec{v},t) \equiv \sum_i \delta(\vec{x}-\vec{x}_i(t))\delta(\vec{v}-\vec{v}_i(t)). \tag{8.2}$$

Its time derivative is given by

$$\frac{\partial F(\vec{x},\vec{v},t)}{\partial t} = \sum_i \left[-\frac{\partial \vec{x}_i}{\partial t}\cdot\frac{\partial}{\partial \vec{x}}\delta(\vec{x}-\vec{x}_i(t))\delta(\vec{v}-\vec{v}_i(t)) - \frac{\partial \vec{v}_i}{\partial t}\cdot\frac{\partial}{\partial \vec{v}}\delta(\vec{v}-\vec{v}_i(t))\delta(\vec{x}-\vec{x}_i(t)) \right]. \tag{8.3}$$

The first term on the RHS can be simplified to $-\nabla\cdot\sum_i \vec{v}_i\delta(\vec{x}-\vec{x}_i(t))\delta(\vec{v}-\vec{v}_i(t)) = -\vec{v}\cdot\nabla F(\vec{x},\vec{v},t)$. The second term has the acceleration $\partial\vec{v}_i/\partial t = \vec{a}_{ext}+\sum_{j\neq i}\vec{a}_{ij}(t)$ due to fields externally imposed or generated internally by the particles. Evaluating the internal fields, in principle, requires **the pair correlation** that identifies the probability density function of jth particle at one phase point given the ith particle at the other phase point. This brings in the two-body distribution function that will be discussed in Sec. 8.6. Therefore, Eq. (8.3) can be rewritten as

$$\frac{\partial F}{\partial t} + \vec{v}\cdot\frac{\partial F}{\partial \vec{x}} + \vec{a}_{ext}\cdot\frac{\partial F}{\partial \vec{v}} + \frac{\partial}{\partial \vec{v}}\cdot\int d^3\vec{v}'d^3\vec{x}'\frac{q(\vec{x}-\vec{x}')}{m|\vec{x}-\vec{x}|'^3}F_2(\vec{x},\vec{v},\vec{x}',\vec{v}',t) = 0 \tag{8.3$'$}$$

and proceeds, as in the **BBGKY hierarchy**, beyond what will not terminate without an approximation. We will, however, treat only the one-body distribution function by following its propagator for now, equivalent to making $F_2(\vec{x},\vec{v},\vec{x}',\vec{v}',t) \approx F(\vec{x},\vec{v},t)F(\vec{x}',\vec{v}',t)$, and rewrite Eq. (8.3) as

$$\frac{\partial F}{\partial t} + \vec{v}\cdot\frac{\partial F}{\partial \vec{x}} + \frac{q}{m}\left(\vec{E}+\frac{\vec{v}}{c}\times\vec{B}\right)\cdot\frac{\partial F}{\partial \vec{v}} = 0. \tag{8.4}$$

This equation retains *the particle discreteness* of Eq. (8.2) in *the initial value* to render the kinetic description beyond just the smooth distribution function.

8.1.1 *The Fokker-Planck Equation*

Consider the smooth function of the phase space distribution

$$f(\vec{x},\vec{v},t) \equiv \frac{1}{n_0}\left\langle \sum_i \delta(\vec{x}-\vec{x}_i(t))\delta(\vec{v}-\vec{v}_i(t)) \right\rangle, \tag{8.5}$$

where the uniform density is factored out so that its phase space integration is normalized to unity, and it is ensemble averaged, denoted by the bracket $\langle\rangle$, over the fast time and short length scales. The fluctuating distribution function is defined as

$$n_0\delta f(\vec{x},\vec{v},t) \equiv F(\vec{x},\vec{v},t) - n_0 f(\vec{x},\vec{v},t), \tag{8.6}$$

that well preserves the particle discreteness. Thus, taking the ensemble average of Eq. (8.4), we find

$$\frac{\partial f}{\partial t} + \vec{v}\cdot\nabla f + \frac{q}{m}\left(\vec{E} + \frac{\vec{v}}{c}\times\vec{B}\right)\cdot\frac{\partial}{\partial\vec{v}}f$$
$$= -\frac{q}{m}\left\langle\left(\delta\vec{E}(\vec{x},t) + \frac{\vec{v}}{c}\times\delta\vec{B}(\vec{x},t)\right)\cdot\frac{\partial}{\partial\vec{v}}\delta f(\vec{x},\vec{v},t)\right\rangle, \tag{8.7}$$

where $\vec{E}$ and $\vec{B}$ are the ensemble averaged electric and magnetic fields on the equilibrium scales, and $\delta\vec{E}$ and $\delta\vec{B}$ are the fluctuating electric and magnetic fields due to the particle discreteness. The fluctuating distribution is governed by

$$\frac{\partial\delta f}{\partial t} + \vec{v}\cdot\nabla\delta f + \frac{q}{m}\left(\vec{E} + \frac{\vec{v}}{c}\times\vec{B}\right)\cdot\frac{\partial}{\partial\vec{v}}\delta f$$
$$= -\frac{q}{m}\left(\delta\vec{E} + \frac{\vec{v}}{c}\times\delta\vec{B}\right)\cdot\frac{\partial}{\partial\vec{v}}(f+\delta f) + \frac{q}{m}\left\langle\left(\delta\vec{E} + \frac{\vec{v}}{c}\times\delta\vec{B}\right)\cdot\frac{\partial}{\partial\vec{v}}\delta f\right\rangle.$$

The second order terms on the right hand side will be neglected when considering not so turbulent plasma. We will also neglect the magnetic fluctuations since magnetic field does no work on the particles and its force term $q\vec{v}\times\delta\vec{B}/mc\cdot\partial f/\partial\vec{v}$ vanishes in its entirety for the isotropic distribution function. Thus,

$$\frac{\partial\delta f}{\partial t} + \vec{v}\cdot\nabla\delta f + \frac{q}{m}\left(\vec{E} + \frac{\vec{v}}{c}\times\vec{B}\right)\cdot\frac{\partial}{\partial\vec{v}}\delta f = -\frac{q}{m}\delta\vec{E}\cdot\frac{\partial}{\partial\vec{v}}f. \tag{8.8}$$

The LHS of Eq. (8.8) follows the phase space trajectory governed by,

$$\frac{d\vec{x}(\tau)}{d\tau} = \vec{v}(\tau), \tag{8.9a}$$

$$\frac{d\vec{v}(\tau)}{d\tau} = \frac{q}{m}\left(\vec{E}(\vec{x}(\tau),\tau) + \frac{\vec{v}(\tau)}{c}\times B(\vec{x}(\tau),\tau)\right), \tag{8.9b}$$

and the particle arrives at the phase point $(\vec{x},\vec{v})$ at $\tau = t$. The solution to Eq. (8.8) can be put formally as $\delta f = -\int_{-\infty}^{t} d\tau\delta\vec{E}(\vec{x}(\tau),\tau)\cdot\partial f/\partial\vec{v}(\tau)$. Substituting δf into

Eq. (8.7), we arrive at the Fokker-Planck equation,

$$\frac{\partial f}{\partial t} + \vec{v}\cdot\nabla f + \frac{q}{m}\left(\vec{E} + \frac{\vec{v}}{c}\times\vec{B}\right)\cdot\frac{\partial}{\partial\vec{v}} f = -\frac{q}{m}\left\langle \delta\vec{E}\cdot\frac{\partial}{\partial\vec{v}}\delta f\right\rangle$$

$$= \frac{q^2}{m^2}\left\langle \delta\vec{E}(\vec{x},t)\cdot\frac{\partial}{\partial\vec{v}}\int_{-\infty}^{t} d\tau\,\delta\vec{E}(\vec{x}(\tau),\tau)\right\rangle\cdot\frac{\partial f}{\partial\vec{v}}. \quad (8.10)$$

By defining the velocity diffusion coefficient as $\overleftrightarrow{D} \equiv (q^2/m^2)\lim_{t\to\infty}\langle\delta\vec{E}(\vec{x},t)$ $\int_{-\infty}^{t} d\tau\delta\vec{E}(\vec{x}(\tau),\tau)\rangle$, where the second electric field term $\delta\vec{E}(\vec{x}(\tau),\tau)$ in $\overleftrightarrow{D}$ is evaluated along the particle trajectory governed by Eq. (8.9), the collision operator for the species a, can be rewritten as

$$C(f) \equiv \frac{\partial}{\partial\vec{v}}\cdot\overleftrightarrow{D}\cdot\frac{\partial}{\partial\vec{v}} f = \frac{\partial}{\partial\vec{v}}\frac{\partial}{\partial\vec{v}} : \overleftrightarrow{D} f - \frac{\partial}{\partial\vec{v}}\cdot(\vec{\mu} f)$$

$$= -\frac{\partial}{\partial\vec{v}}\cdot\left(\vec{\mu} f - \frac{\partial}{\partial\vec{v}}\cdot(\overleftrightarrow{D} f)\right), \quad (8.11)$$

where $\vec{\mu} \equiv \nabla_{\vec{v}}\cdot\overleftrightarrow{D}$ is the friction coefficient.

A commonly adopted Fokker-Planck operator is for plasma in the absence of the externally imposed electric or magnetic fields. It retains only the pitch angle scattering in the collision process. The diffusion coefficient can be generated by the potential function $g = \int d^3\vec{v}' f(\vec{v}')|\vec{v} - \vec{v}'|$, and is governed by

$$\overleftrightarrow{D} = \nu\nabla_{\vec{v}}\nabla_{\vec{v}} g = \nu\int d^3\vec{v}' f(\vec{v}')\left(\frac{\overleftrightarrow{I}}{|\vec{v}-\vec{v}'|} - \frac{(\vec{v}-\vec{v}')(\vec{v}-\vec{v}')}{|\vec{v}-\vec{v}'|^3}\right). \quad (8.12)$$

The friction coefficient is then given by,

$$\vec{\mu} = \nabla_{\vec{v}}\cdot\overleftrightarrow{D} = -2\nu\int d^3\vec{v}' f(\vec{v}')\frac{\vec{v}-\vec{v}'}{|\vec{v}-\vec{v}'|^3} = 2\nu\nabla_{\vec{v}}\int d^3\vec{v}'\frac{f(\vec{v}')}{|\vec{v}-\vec{v}'|}. \quad (8.13)$$

Here $\nu = (4\pi n_0 q^4/m^2)\log\Lambda$, where Λ is the ratio of the Debye length to the closest encounter $r_s \equiv e^2/k_B T$. Equations (8.12) and (8.13) will be derived in Sec. 8.3. Note that the following identities have been applied:

$$\nabla r = \hat{e}_r, \quad \nabla\vec{r} = \overleftrightarrow{I}, \quad \nabla\nabla r = \frac{\overleftrightarrow{I}}{r} - \frac{\vec{r}\vec{r}}{r^3}, \quad \nabla\frac{1}{r} = \frac{\vec{r}}{r^3}, \quad \nabla\cdot\left(\frac{\vec{r}}{r^3}\right) = 4\pi\delta(\vec{r}). \quad (8.14)$$

Homework 8.1: Show that the diffusion and friction coefficients of Eqs. (8.12) and (8.13) in the collisional term of Eq. (8.11) conserve the number of particles, the momentum and the energy.

8.2 Vlasov Equation and Landau Damping

Neglecting collisions in Eq. (8.10) gives the Vlasov equation:

$$\frac{\partial f}{\partial t} + \vec{v} \cdot \nabla f + \frac{q}{m}\left(\vec{E} + \frac{\vec{v}}{c} \times \vec{B}\right) \cdot \frac{\partial}{\partial \vec{v}} f = 0, \tag{8.15}$$

which may be regarded as the continuity equation in the phase space for a single-species plasma fluid. The Vlasov equation supports the kinetic plasma waves with particle-wave resonance effect that is absent in the MHD or two-fluid model.

Consider Vlasov equation in the simpler situation without external magnetic field. The linearized Vlasov equation is given by

$$\frac{\partial \delta f}{\partial t} + \vec{v} \cdot \nabla \delta f + \frac{q}{m} \delta \vec{E} \cdot \frac{\partial}{\partial \vec{v}} f_0 = 0. \tag{8.16}$$

Taking the Fourier transform in space, we arrive at

$$\frac{\partial}{\partial t} \delta f_k(\vec{v}, t) + i \vec{k} \cdot \vec{v} \delta f_k(\vec{v}, t) + \frac{q}{m} \delta \vec{E}_k(t) \cdot \frac{\partial}{\partial \vec{v}} f_0 = 0. \tag{8.17}$$

Expressing the electric field in terms of the electrostatic potential, $\delta \vec{E}_k(t) = -i\vec{k}\delta\phi_k$, and taking the Laplace transform of Eq. (8.17) in time gives

$$(s + i\vec{k} \cdot \vec{v})\delta \bar{f}_k - \delta f_k(0) = \frac{q}{m} i \delta\bar{\phi}_k \vec{k} \cdot \frac{\partial f_0}{\partial \vec{v}}. \tag{8.18}$$

The **Laplace transform** is defined by

$$\bar{f}(s) = \int_0^\infty e^{-st} f(t) dt, \; f(t) = \frac{1}{2\pi i} \int_{\sigma - i\infty}^{\sigma + i\infty} ds \bar{f}(s) e^{st}, \tag{8.19}$$

where a bar over the variables represents its Laplace transformed quantity. The Poisson's equation can be cast into

$$k^2 \delta\bar{\phi}_k = 4\pi n_0 q \int d^3\vec{v} \delta \bar{f}_k = \frac{4\pi n_0 q^2}{m} i \delta\bar{\phi}_k \int \frac{d^3\vec{v}}{s + i\vec{k} \cdot \vec{v}} \vec{k} \cdot \frac{\partial f_0}{\partial \vec{v}}$$
$$+ 4\pi n_0 q \int \frac{d^3\vec{v}}{s + i\vec{k} \cdot \vec{v}} \delta f_k(0).$$

Thus,

$$\delta\bar{\phi}_k \varepsilon(k, s) = \frac{4\pi n_0 q}{k^2} \int \frac{d^3\vec{v}}{s + i\vec{k} \cdot \vec{v}} \delta f_k(0). \tag{8.20}$$

The dielectric function is given by

$$\varepsilon(k, s) = 1 - \frac{\omega_p^2}{k^2} \int dv \frac{1}{v - is/k} \frac{\partial f_0}{\partial v}, \tag{8.21}$$

where the integration over the perpendicular velocity to the wave vector $\vec{k}$ has been carried out. In the low frequency limit ($s \to 0$), the real part of the dielectric function can be cast into

$$\varepsilon(k, 0) = 1 - \frac{\omega_p^2}{k^2} \int_{-\infty}^{\infty} dv \frac{1}{v} \frac{\partial f_0}{\partial v} = 1 + \frac{1}{k^2 \lambda_D^2}, \tag{8.22}$$

that recovers the Debye screening effect.

8.2.1 *The Landau Damping*

There are several choices to deal with the resonance pole in the velocity integration as given by the **Plemelj formula**

$$\frac{1}{v - v_0} = P \frac{1}{v - v_0} \pm i\pi \delta(v - v_0), \tag{8.23}$$

where P represents the principal value. Without the imaginary part, the dielectric function would support an oscillatory but constant-amplitude wave, but the additional plus or minus part of the residue would yield a damping or growing wave. Landau had the right interpretation of the physical phenomenon in that as long as the wave has been excited, the wave will be damped due to the fact that there are more particles travelling at slower speed, and less particles at faster speed than the wave. The net effect is the wave will give energy to the particles. Given the equilibrium distribution function as Maxwellian, $f_0 = \exp(-v^2/v_T^2)/\sqrt{\pi}/v_T$, we may evaluate the dielectric function as follows,

$$\varepsilon(k, -i\omega) = 1 - \frac{\omega_p^2}{k^2} \int_{-\infty}^{\infty} dv P \frac{1}{v - \omega/k} \frac{\partial f_0}{\partial v} + i\pi \frac{\omega_p^2}{k^2} \frac{\partial f_0}{\partial v}\bigg|_{v=\omega/k}, \tag{8.21a}$$

$$\approx 1 + \frac{\omega_p^2}{k^2} \int_{-\infty}^{\infty} dv \frac{k}{\omega} \frac{\partial f_0}{\partial v} \left(1 + \frac{kv}{\omega} + \frac{k^2 v^2}{\omega^2} + \frac{k^3 v^3}{\omega^3} + \cdots\right)$$

$$+ i\pi \frac{\omega_p^2}{k^2} \frac{\partial f_0}{\partial v}\bigg|_{v=\omega/k}$$

$$= 1 - \frac{\omega_p^2}{\omega^2} - \frac{3}{2} \frac{\omega_p^2}{\omega^4} k^2 v_T^2 - 2i \frac{\omega_p^2 \omega}{k^3 v_T^3} \sqrt{\pi} e^{-\omega^2/k^2 V_T^2}. \tag{8.24}$$

Note that the following integral identities have been used:

$$\frac{1}{\sqrt{\pi}}\int_{-\infty}^{\infty} dx \begin{pmatrix} 1 \\ x^2 \\ x^4 \end{pmatrix} \exp(-x^2) = \begin{pmatrix} 1 \\ \frac{1}{2} \\ \frac{3}{4} \end{pmatrix}. \tag{8.25}$$

This gives the real frequency

$$\omega^2 = \omega_p^2 + \frac{3}{2}k^2 v_T^2,$$

that is consistent with Eq. (5.4b). The additional kinetic effect gives rise to a damping rate

$$\gamma \approx -\frac{\sqrt{\pi}\,\omega_p^4}{k^3 v_T^3} e^{-\omega^2/k^2 v_T^2}. \tag{8.26}$$

As prescribed by the Landau damping, the positive sign in Plemelj formula has been chosen to reflect the energy-losing wave in the wave-particle resonance pole. For a non-Maxwellian distribution function with a positive slope in the resonance region, the wave will grow as the particle population is inverted in that more particles are moving faster than the wave.

Homework 8.2: Show that in the long wave length limit, the dispersion function from the Vlasov equation reduces to the following:

$$\varepsilon(\vec{k}, -i\omega) = 1 - \frac{\omega_p^2}{\omega^2} - \frac{\omega_p^4}{\omega^4}k^2\lambda_D^2 \approx 1 - \frac{\omega_p^2}{\omega^2}(1 + k^2\lambda_D^2)$$

that is consistent with Eq. (5.4a).

8.2.2 *The Plasma Dispersion Function*

The kinetic effect is manifested in the **Plasma Dispersion Function,**

$$Z(\varsigma) = \frac{1}{\sqrt{\pi}}\int_{-\infty}^{\infty} dx \frac{\exp(-x^2)}{x - \varsigma}, \tag{8.27}$$

for a Maxwellian plasma. There is no closed form for this integral, but the asymptotic expansions may give relatively accurate description for real ς. For $\varsigma \ll 1$,

$$Z(\varsigma) \approx -2\varsigma e^{-\varsigma^2}\left(1 + \frac{1}{3}\varsigma^2 + \frac{1}{10}\varsigma^4\right) + \cdots + i\sqrt{\pi}e^{-\varsigma^2}, \tag{8.28a}$$

and for $\varsigma \gg 1$,

$$Z(\varsigma) \approx -\frac{1}{\varsigma}\left(1 + \frac{1}{2}\frac{1}{\varsigma^2} + \frac{3}{4}\frac{1}{\varsigma^4} + \cdots\right) + i\sqrt{\pi}e^{-\varsigma^2}. \tag{8.28b}$$

The following program evaluates $Z(\varsigma)$ by the Simpson method (cf. Sec. 7.2.2) for its principal value. The algorithm takes care of the resonance singularity to ensure numerical accuracy by separating the integral into three parts,

$$Z(\varsigma) = \frac{1}{\sqrt{\pi}}\int_{-\infty}^{\zeta-\frac{1}{2}} dx\frac{\exp(-x^2)}{x-\varsigma} + \frac{1}{\sqrt{\pi}}\int_{\zeta+\frac{1}{2}}^{\infty} dx\frac{\exp(-x^2)}{x-\varsigma}$$
$$+\frac{1}{\sqrt{\pi}}\int_{\zeta-\frac{1}{2}}^{\zeta+\frac{1}{2}} dx\frac{\exp(-x^2)}{x-\varsigma}.$$

The last piece is evaluated by the following to ensure the numerical accuracy:

$$\int_{-\frac{1}{2}}^{\frac{1}{2}} \frac{ds}{s}e^{-(s+\zeta)^2} = e^{-(s+\zeta)^2}\ln|s|\Big|_{-\frac{1}{2}}^{\frac{1}{2}} + 2\int_{-\frac{1}{2}}^{\frac{1}{2}} (s\ln|s|)e^{-(s+\zeta)^2}ds$$
$$+2\zeta(s\ln|s|-s)e^{-(s+\zeta)^2}\Big|_{-\frac{1}{2}}^{\frac{1}{2}}$$
$$+4\zeta\int_{-\frac{1}{2}}^{\frac{1}{2}} (s\ln|s|-s)e^{-(s+\zeta)^2}(s+\zeta)ds$$

Homework 8.3: Prove the asymptotic expansions (8.24) for the plasma dispersion function

$$Z(\varsigma) = \frac{1}{\sqrt{\pi}}\int_{-\infty}^{\infty} dx\frac{\exp(-x^2)}{x-\varsigma},$$

and compare with the numerical results given in the program below.

```
function P=PlasmaDispersionFunction(zeta)
zn=zeta-1/2;
zp=zeta+1/2;
N=10000; dz=0.01;
x=zp+(0:N-1)*dz;
f=exp(-x.^2)./(x-zeta);
A=Simpson(f,N,dz);
x=(-N+1:0)*dz+zn;
f=exp(-x.^2)./(x-zeta);
B=Simpson(f,N,dz);
C=log(0.5)*(exp(-(zeta+0.5)^2)-exp(-(zeta-0.5)^2));
C=C+2*zeta*(0.5*log(0.5)-0.5)*exp(-(zeta+0.5)^2);
C=C-2*zeta*(-0.5*log(0.5)+0.5)*exp(-(zeta-0.5)^2);
x=-0.5+1/2/N:1/N:0.5;
f=x.*log(abs(x)).*exp(-(x+zeta).^2);
D=Simpson(f,N,1/N);
f=(x.*log(abs(x))-x).*(x+zeta).*exp(-(x+zeta).^2);
E=Simpson(f,N,1/N);
P=(A(N)+B(N)+C+D(N)*2+E(N)*4*zeta)/sqrt(pi);
```

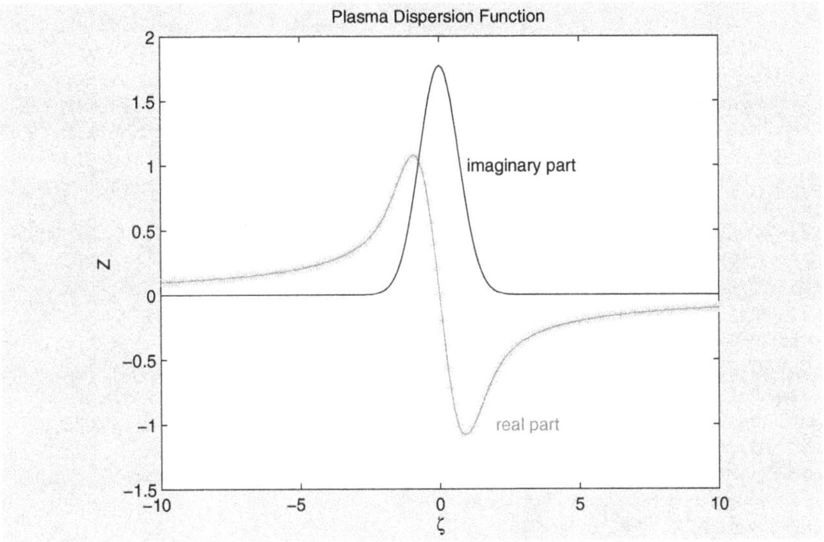

8.3 The Electric Field Autocorrelation

A **correlation function** describes the statistical dependency between random variables at two different points in space or time. The correlation of fluctuations is a good measure of the extent, or the order, to which their effects would affect the macroscopic properties. An autocorrelation refers to the correlation for the same variable. We are to find the two-time autocorrelation of the electric field defined as $\overleftrightarrow{S}(t,\tau) \equiv \langle \delta \vec{E}(\vec{x},t)\delta \vec{E}(\vec{x}(\tau),\tau)\rangle$ that has an essential role in the transport coefficients, and is the main component in the Fokker-Planck collision operator. The kinetic theory is particularly useful in this regard since the particle discreteness is retained in the formulation.

By considering plasmas without the external magnetic and electric fields, the unperturbed particle trajectory is governed by $d\vec{v}/dt = 0$ and $d\vec{x}/dt = \vec{v}$, with the condition that the particle arrives $\vec{x}$ at $\tau = t$. Therefore, $\vec{x}(\tau) = \vec{x} + (\tau - t)\ \vec{v}$. Expressing the electric field in its Fourier integral, $\delta \vec{E}(\vec{x}(\tau),\tau) = \int d^3\vec{k}\delta\vec{E}_k\ (\tau)e^{i\vec{k}\cdot\vec{x}+i\vec{k}\cdot\vec{v}(\tau-t)}$, and taking an ensemble average over the space gives

$$\overleftrightarrow{S}(t,\tau) = \left(\frac{2\pi}{L}\right)^3 \int d^3\vec{k}\langle \delta\vec{E}_{-k}(t)\delta\vec{E}_k(\tau)e^{i\vec{k}\cdot\vec{v}(\tau-t)}\rangle, \tag{8.29}$$

```
function PlotPlasmaDispersionFunction
zeta=-10:0.1:10;
P=zeta*0;
for i=1:length(zeta)
  P(i)=PlasmaDispersionFunction(zeta(i));
end;
nu=sqrt(pi)*exp(-zeta.^2);
Xc=1.4;
z=-Xc:0.01:Xc;
f=-2*z.*(1+z.^2/3+1/10*z.^4).*exp(-z.^2);
x=Xc:0.01:10;
g=-1./x-1./x.^3/2-3/4./x.^5;
y=-10:0.01:-Xc;
h=-1./y-1./y.^3/2-3/4./y.^5;
plot(zeta,P,'ro-',zeta,nu,'k*-',z,f,'g.-',x,g,'g.-',y,h,'g.-');
title('Plasma Dispersion Function','fontsize',16);
xlabel('\zeta','fontsize',16);
ylabel('Z','fontsize',16);
text(1,1,'imaginary part','color','k','fontsize',16);
text(2,-1,'real part','color','r','fontsize',16);
s=sprintf('asymptotic \n formulae');
text(-8,0.8,s,'color','g','fontsize',16);
```

% Plasma dispersion function for Maxwellian distribution
% The code applys the Simpson rule for integration and makes the asymptotic expansion near the sigularity to ensure numerical accuracy.
% The code can easily be generalized to other distribution functions.

% It plots the real and imaginary parts of the plasma dispersion function anc the asymptotic formulae also included for comparison.

where we have made use of the identity,

$$\left(\frac{1}{2\pi}\right)^3 \int d^3\vec{x}\, e^{i\vec{k}_1\cdot\vec{x}+i\vec{k}_2\cdot\vec{x}} = \delta(\vec{k}_1+\vec{k}_2). \qquad (8.30)$$

The remaining ensemble average is taken over the velocity of a Maxwellian distribution,

$$f_M(\vec{v}) = \left(\frac{1}{\sqrt{\pi}\,V_T}\right)^3 e^{-v^2/V_T^2}.$$

Making use of the following identity,

$$\int_{-\infty}^{\infty} dv\, \frac{1}{\sqrt{\pi}\,V_T} e^{-v^2/V_T^2} e^{ikv\tau} = e^{-\frac{1}{4}k^2 V_T^2\tau^2}, \qquad (8.31)$$

where the thermal spread in the particle velocity distribution causes the phase mixing and a reduced temporal correlation by the factor $\exp(-\frac{1}{4}k_2 V_T^2(\tau-t)^2)$, we find

$$\vec{S}(t,\tau) = \left(\frac{2\pi}{L}\right)^3 \int d^3\vec{k}\,\langle\delta\vec{E}_{-k}(t)\delta\vec{E}_k(\tau)\rangle e^{-\frac{1}{4}k^2 V_T^2(\tau-t)^2}. \qquad (8.29a)$$

To find the electric field $\delta\vec{E}_k(\tau)$, we keep the particle discreetness in the initial condition, $\delta f(\vec{x},\vec{v},0) = n_0^{-1}\sum_j \delta(\vec{x}-\vec{x}_{j0})\delta(\vec{v}-\vec{v}_{j0}) - f(\vec{v})$, that is Fourier transformed to

$$\delta f_k(\vec{v}, 0) = \frac{1}{(2\pi)^3 n_0} \sum_j e^{-i\vec{k}\cdot\vec{x}_{j0}} \delta(\vec{v} - \vec{v}_{j0}) - f(v)\delta(\vec{k}). \tag{8.32}$$

Substituting Eq. (8.32) into Eq. (8.18) gives

$$\delta \bar{f}_k = \frac{1}{(2\pi)^3 n_0} \sum_j e^{-\vec{k}\cdot\vec{x}_{j0}} \frac{\delta(\vec{v} - \vec{v}_{j0})}{s + i\vec{k}\cdot\vec{v}} - f\frac{\delta(\vec{k})}{s} + \frac{iq}{m}\frac{\delta\bar{\phi}_k}{s + i\vec{k}\cdot\vec{v}}\vec{k}\cdot\frac{\partial f}{\partial \vec{v}}, \tag{8.33}$$

and applying the Poisson's equation of Eq. (8.20) gives,

$$\varepsilon(\vec{k}, s)\delta\bar{\phi}_k = \frac{q}{2\pi^2 k^2} \sum_j \frac{e^{-i\vec{k}\cdot\vec{x}_{i0}}}{s + i\vec{k}\cdot\vec{v}_{j0}} - 4\pi n_0 \frac{\delta(\vec{k})}{s}. \tag{8.34}$$

Multiplying this equation by $-i\vec{k}$ gives the equation for the electric displacement,

$$\delta\vec{D}_k = \varepsilon(\vec{k}, s)\delta\bar{E}_k = \frac{-i\vec{k}q}{2\pi^2 k^2} \sum_j e^{-i\vec{k}\cdot\vec{x}_{j0}} \frac{1}{s + i\vec{k}\cdot\vec{v}_{j0}}. \tag{8.35}$$

The RHS is the free charge density (the bare particle effect along the unperturbed particle trajectory) resulted from the particle discreteness in the initial condition. The dielectric function, which is simplify the plasma dispersion function of Eq. (8.21), effectively dresses up the bare particle response for the electric field $\delta\bar{E}_k$ to the initial fluctuating density that persists by free streaming through the correlation time.

Taking the inverse Laplace transform in the long time limit as the following,

$$\frac{1}{2\pi i} \int_{\sigma - i\infty}^{\sigma + i\infty} \frac{e^{st}}{\varepsilon(\vec{k}, s)} \frac{ds}{s + i\omega} \xrightarrow{t\to\infty} \frac{e^{-i\omega t}}{\varepsilon(\vec{k}, -i\omega)}, \tag{8.36}$$

by keeping only the residue of the pole at $-i\omega$, but neglecting the contribution from the poles at the zeros of the dielectric function since they lead to damping effect, unrelated to the particle discreteness of concern here, we have

$$\delta\vec{E}_k(t) \xrightarrow{t\to\infty} \sum_j \frac{-iq\vec{k}}{2\pi^2 k^2} \frac{e^{-i\vec{k}\cdot\vec{x}_{j0} - i\vec{k}\cdot\vec{v}_{j0}t}}{\varepsilon(\vec{k}, -i\vec{k}\cdot\vec{v}_{j0})}. \tag{8.37}$$

Only terms referring to the same particle survives the ensemble average,

$$\langle \delta\vec{E}_{-k}(t)\delta\vec{E}_k(\tau)\rangle = \frac{q^2\vec{k}\vec{k}}{4\pi^4 k^4} \left\langle \sum_j \frac{e^{i\vec{k}\cdot\vec{v}_{j0}(t-\tau)}}{|\varepsilon(\vec{k}, -i\vec{k}\cdot\vec{v}_{j0})|^2} \right\rangle. \tag{8.38}$$

The dielectric function effectively dresses up the bare particle response to the fluctuating electric field by $\varepsilon(k, -ik_{\|}v_{j0\|}) \approx 1 + 1/k^2\lambda_D^2$, the Debye screening.

The two-time auto-correlation of the electric field in Eq. (8.32) now takes the simple form,

$$\overset{=}{S}(t,\tau) = \int d^3\vec{k}\,\frac{2n_0q^2\vec{k}\vec{k}}{\pi(k^2+k_D^2)^2}e^{-\frac{1}{2}k^2V_T^2(\tau-t)^2} \xrightarrow{|t-\tau|\to\infty} \frac{4}{3}\sqrt{\pi}\frac{n_0q^2}{\lambda_D}\overset{=}{I}. \quad (8.39)$$

The particle thermal motion has an effective decorrelation time on the order of $(kV_T)^{-1}$. The longer wavelength modes enjoy much longer correlation time. Equation (8.39) is integrable and gives a time asymptotic value $|\delta\vec{E}|^2 = 4\sqrt{\pi}n_0q^2/\lambda_D = \frac{1}{\sqrt{\pi}}\varepsilon_p n_0 k_B T$, where $\varepsilon_p = (n_0\lambda_D^3)^{-1}$ is the plasma parameter. Note that the electric field auto-correlation $\overset{=}{S}(t,\tau) = \overset{=}{S}(|t-\tau|)$ is real and function of the time difference only.

8.3.1 *The Diffusion Coefficient in the Fokker-Planck Equation*

We will now evaluate the diffusion coefficient from the electric field autocorrelation, which is rewritten as

$$\begin{aligned}
&\left\langle \delta\vec{E}(\vec{x},t)\int_{-\infty}^{t} d\tau\,\delta\vec{E}(\vec{x}(\tau),\tau)\right\rangle \\
&= \left\langle \int_{-\infty}^{\infty} e^{i\vec{k}\cdot\vec{x}}\delta\vec{E}_k(t)d^3\vec{k}\int_{-\infty}^{t}d\tau\int_{-\infty}^{\infty}d^3\vec{k}'\,\delta\vec{E}_{k'}(\tau)e^{i\vec{k}'\cdot(\vec{x}+\Delta\vec{x}(\tau))}\right\rangle \\
&= \left(\frac{2\pi}{L}\right)^3\left\langle \int_{-\infty}^{\infty} d^3\vec{k}\delta\vec{E}_k(t)\int_{-\infty}^{t} d\tau\delta\vec{E}_{-k}(\tau)e^{-i\vec{k}\cdot\Delta\vec{x}(\tau)}\right\rangle \\
&= \frac{2n_0q^2}{\pi}\left\langle \int_{-\infty}^{\infty} d^3\vec{k}\frac{\vec{k}\vec{k}}{k^4|\varepsilon(\vec{k},-i\vec{k}\cdot\vec{v}_{j0})|^2}\int_{-\infty}^{0} d\tau e^{i\vec{k}\cdot(\vec{v}_{j0}-\vec{v})\tau}\right\rangle \\
&= 2n_0q^2\int d^3\vec{v}' f'(\vec{v}')\int d^3\vec{k}\frac{\vec{k}\vec{k}}{k^4|\varepsilon(\vec{k},-i\vec{k}\cdot\vec{v}')|^2}\delta(\vec{k}\cdot(\vec{v}'-\vec{v})), \quad (8.40)
\end{aligned}$$

where we have applied Eq. (8.30) and the following identity,

$$\int_{-\infty}^{0} d\tau e^{i\vec{k}\cdot(\vec{v}'-\vec{v})\tau} = \pi\delta(\vec{k}\cdot(\vec{v}'-\vec{v})). \quad (8.41)$$

We find the diffusion coefficient,

$$\overset{=}{D} = \frac{2n_0q^4}{m^2}\int d^3\vec{v}' f(\vec{v}')\int d^3\vec{k}\frac{\vec{k}\vec{k}}{k^4|\varepsilon(\vec{k},-i\vec{k}\cdot\vec{v}')|^2}\delta(\vec{k}\cdot(\vec{v}'-\vec{v})). \quad (8.42)$$

Applying Eq. (8.22), $\varepsilon(k,0) = 1 + 1/k^2\lambda_D^2$, we integrate over the k space in Eq. (8.42) to get

$$\overleftrightarrow{D} = \frac{2n_0q^4}{m^2}\int d^3\vec{v}'f(\vec{v}')\frac{\overleftrightarrow{I} - \widehat{e}_{||}\widehat{e}_{||}}{|\vec{v} - \vec{v}'|}\int_0^\infty k_\perp dk_\perp \frac{2\pi k_\perp^2}{(k_\perp^2 + k_D^2)^2}$$

$$= \frac{4\pi n_0 q^4}{m^2}\int d^3\vec{v}'f(\vec{v}')\frac{\overleftrightarrow{I} - \widehat{e}_{||}\widehat{e}_{||}}{|\vec{v} - \vec{v}'|}\left[\log\left(1 + \frac{1}{k_D^2 r_s^2}\right) + \frac{1}{1 + k_D^2 r_s^2}\right]$$

where we have applied

$$\int d^3\vec{k}\frac{\vec{k}\vec{k}\delta(\vec{k}\cdot(\vec{v}' - \vec{v}))}{(k^2 + k_D^2)^2} = \int 2\pi k_\perp dk_\perp dk_{||}\frac{\vec{k}\vec{k}}{(k^2 + k_D^2)^2}\frac{\delta(k_{||})}{|\vec{v}' - \vec{v}|}$$

$$= \frac{\overleftrightarrow{I} - \widehat{e}_{||}\widehat{e}_{||}}{|\vec{v} - \vec{v}'|}\int_0^\infty k_\perp dk_\perp \frac{2\pi k_\perp^2}{(k_\perp^2 + k_D^2)^2}, \tag{8.43}$$

by defining $\widehat{e}_{||} \equiv (\vec{v} - \vec{v}')/|\vec{v} - \vec{v}'|$. Equation (8.43) is divergent at $k_\perp \to \infty$, corresponding to the closest encounter between particles, which we thus introduce as a cutoff. We then arrive at Eq. (8.12),

$$\overleftrightarrow{D} \approx \frac{4\pi n_0 q^4}{m^2}\int d^3\vec{v}'f(\vec{v}')\frac{\overleftrightarrow{I} - \widehat{e}_{||}\widehat{e}_{||}}{|\vec{v} - \vec{v}'|}\log\left(\frac{\lambda_D}{r_s}\right) = \nu\nabla_{\vec{v}}\nabla_{\vec{v}}g, \tag{8.12}$$

where $g = \int d^3\vec{v}'f(\vec{v}')|\vec{v} - \vec{v}'|$, $\nu = (4\pi n_0 q^4/m^2)\log\Lambda$. Here $\Lambda \equiv \lambda_D/r_s$ is the ratio of the Debye length λ_D to the length of the closest encounter r_s.

It is noteworthy that while the dielectric function is included in deriving the collisional effect, it only provides the Debye screening in the Fokker-Planck equation. The collective spectrum of fluctuations can in fact cause strong transport effect through the ExB drift by the low-frequency long-wavelength modes and the cyclotron resonance by the cyclotron harmonics. The Fokker-Planck equation, an unmagnetized proposition, can be very much limited in making valid prediction for the plasma transport phenomena.

Homework 8.4: Show that the H-theorem holds for the Fokker-Planck collision operator and the Maxwellian distribution yields the maximum entropy state to make the collision effect disappear.

8.3.2 *The Balescu-Lenard Equation*

An improvement upon the Fokker-Planck equation is to retain the dielectric effect. For that, we derive the collisional operator by the first form, namely $C(f) = -(q/m)(\partial/\partial\vec{v}) \cdot \langle\delta\vec{E}\delta f\rangle$ in Eq. (8.10). Equation (8.35) gives the time asymptotic distribution function as follows,

$$\delta\bar{f}_k(\vec{v},t) \xrightarrow[t\to\infty]{} \sum_j \frac{e^{-\vec{k}\cdot\vec{x}_{j0}-i\vec{k}\cdot\vec{v}_{j0}t}\delta(\vec{v}-\vec{v}_{j0})}{(2\pi)^3 n_0} - f\delta(\vec{k})$$
$$+\frac{q^2\vec{k}\cdot\frac{\partial f}{\partial\vec{v}}}{2\pi^2k^2m}\sum_j \frac{e^{-i\vec{k}\cdot\vec{x}_{j0}}}{\vec{k}\cdot(\vec{v}_{j0}-\vec{v})}\left(\frac{e^{-i\vec{k}\cdot\vec{v}t}}{\varepsilon(\vec{k},-i\vec{k}\cdot\vec{v})} - \frac{e^{-i\vec{k}\cdot\vec{v}_{j0}t}}{\varepsilon(\vec{k},-i\vec{k}\cdot\vec{v}_{j0})}\right)$$

We also need Eq. (8.37), namely,

$$\delta\vec{E}_k(t) \xrightarrow[t\to\infty]{} \sum_j \frac{-iq\vec{k}}{2\pi^2k^2}\frac{e^{-i\vec{k}\cdot\vec{x}_{j0}-i\vec{k}\cdot\vec{v}_{j0}t}}{\varepsilon(\vec{k},-i\vec{k}\cdot\vec{v}_{j0})}, \tag{8.37}$$

to evaluate $\langle\delta\vec{E}\delta f\rangle = \int d^3\vec{k}\langle\delta\vec{E}_{-k}\delta f_k\rangle$. Only the same particle has the net contribution. After taking the ensemble average over the smooth distribution function, we have,

$$\langle\delta\vec{E}\delta f\rangle = \int \frac{iq\vec{k}}{2\pi^2k^2}\frac{d^3\vec{k}}{\varepsilon(-\vec{k},i\vec{k}\cdot\vec{v})}\left(f(\vec{v}) + \frac{4\pi n_0q^2}{k^2m}\int d^3\vec{v}'\frac{f(\vec{v}')}{\vec{k}\cdot(\vec{v}'-\vec{v})}\right.$$
$$\left.\times\left[\frac{e^{i\vec{k}\cdot(\vec{v}'-\vec{v})t}}{\varepsilon(\vec{k},-i\vec{k}\cdot\vec{v})} - \frac{1}{\varepsilon(\vec{k},-i\vec{k}\cdot\vec{v}')}\right]\vec{k}\cdot\frac{\partial f}{\partial\vec{v}}\right)$$
$$= \int \frac{iq\vec{k}}{2\pi^2k^2}d^3\vec{k}\left(f(\vec{v})\frac{1}{\varepsilon(-\vec{k},i\vec{k}\cdot\vec{v})} + \frac{\omega_p^2}{k^2}\int d^3\vec{v}'\right.$$
$$\left.\times\frac{f(\vec{v}')}{|\varepsilon(\vec{k},-i\vec{k}\cdot\vec{v}')|^2}\frac{e^{i\vec{k}\cdot(\vec{v}'-\vec{v})t}-1}{\vec{k}\cdot(\vec{v}'-\vec{v})}\vec{k}\cdot\frac{\partial f}{\partial\vec{v}}\right)$$
$$= \int \frac{q\vec{k}}{2\pi^2k^2}d^3\vec{k}\left(f(\vec{v})\frac{i\varepsilon*(-\vec{k},i\vec{k}\cdot\vec{v})}{|\varepsilon(-\vec{k},i\vec{k}\cdot\vec{v})|^2} - \pi\frac{\omega_p^2}{k^2}\int d^3\vec{v}'\right.$$
$$\left.\times\frac{f(\vec{v}')\delta(\vec{k}\cdot(\vec{v}'-\vec{v}))}{|\varepsilon(\vec{k},-i\vec{k}\cdot\vec{v}')|^2}\vec{k}\cdot\frac{\partial f}{\partial\vec{v}}\right) \tag{8.44}$$

where we have set $\vec{v} = \vec{v}'$ in the dielectric function of the second term due to the singular nature of the pole and the oscillatory $e^{i\vec{k}\cdot(\vec{v}'-\vec{v})t}$ otherwise. We have also taken

$$(1 - e^{-i\vec{k}\cdot(\vec{v}'-\vec{v})t})/(\vec{k}\cdot(\vec{v}'-\vec{v}))$$
$$\xrightarrow{t\to\infty} \pi i\delta(\vec{k}\cdot(\vec{v}'-\vec{v})) = (\pi i/k)\delta(\widehat{k}\cdot(\vec{v}'-\vec{v})).$$

Only the imaginary part of the dielectric function $\varepsilon*(-\vec{k}, i\vec{k}\cdot\vec{v})$ in the first term survives the k integral that amounts to a factor $-\pi i(\omega_p^2/k^2) \quad \int d^3\vec{v}'\delta(\widehat{k}\cdot(\vec{v}'-\vec{v}))\widehat{k}\cdot\partial f(\vec{v}')/\partial\vec{v}'$. Defining $F(\widehat{k}\cdot\vec{v}) \equiv \int d^3\vec{v}' f(\vec{v}')\delta(\widehat{k}\cdot(\vec{v}'-\vec{v}))$, we end with

$$C(f) = \frac{\partial}{\partial\vec{v}}\cdot\int\frac{2n_0q^4}{m^2k^4}\frac{\vec{k}\vec{k}d^3\vec{k}}{|\varepsilon(\vec{k}, -i\vec{k}\cdot\vec{v})|^2}$$
$$\cdot\left(F(\widehat{k}\cdot\vec{v})\frac{\partial}{\partial\vec{v}}f(\vec{v}) - f(\vec{v})\frac{\partial}{\partial\vec{v}}F(\widehat{k}\cdot\vec{v})\right). \qquad (8.45)$$

Resorting to the low frequency dielectric response, namely, $\varepsilon(k, 0) = 1 + 1/k^2\lambda_D^2$, equivalent to the Debye screening, we get a symmetrical Fokker Planck collision operator,

$$C(f) = \frac{\nu}{2}\frac{\partial}{\partial\vec{v}}\cdot\int d^3\vec{v}'\frac{\vec{\vec{I}} - \widehat{e}_{||}\widehat{e}_{||}}{|\vec{v}-\vec{v}'|}\cdot\left(f(\vec{v}')\frac{\partial}{\partial\vec{v}}f(\vec{v}) - f(\vec{v})\frac{\partial}{\partial\vec{v}'}f(\vec{v}')\right). \qquad (8.45a)$$

Practically, we need the collisions among different species. That can be obtained by spelling out the correct momentum exchange effect,

$$C(f_a) = \frac{2\pi n_0q_a^2q_b^2}{m_a}\log\Lambda\frac{\partial}{\partial\vec{v}}\cdot\int d^3\vec{v}'\frac{\vec{\vec{I}} - \widehat{e}_{||}\widehat{e}_{||}}{|\vec{v}-\vec{v}'|}$$
$$\cdot\left(f_b(\vec{v}')\frac{\partial}{m_a\partial\vec{v}}f_a(\vec{v}) - f_a(\vec{v})\frac{\partial}{m_b\partial\vec{v}'}f_b(\vec{v}')\right) \qquad (8.45b)$$

While the dielectric effect may have been retained in Eq. (8.45), it is derived without the external magnetic field which is a necessary component in the magnetic fusion confinement. In the next sections, the magnetized plasma is examined, the magnetic field on particle trajectory is included and the dielectric response and the electric field fluctuation are derived.

8.4 The Dielectric Function of Magnetized Plasma

Linearizing and Fourier transforming Eq. (8.15) in space, we arrive at

$$\frac{\partial}{\partial t}\delta f_k(\vec{v},t) + i\vec{k}\cdot\vec{v}\delta f_k(\vec{v},t)$$
$$+\vec{v}\times\vec{\Omega}\cdot\frac{\partial}{\partial\vec{v}}\delta f_k(\vec{v},t) + \frac{q}{m}\delta\vec{E}_k(t)\cdot\frac{\partial}{\partial\vec{v}}f_0 = 0. \tag{8.46}$$

Utilizing $\vec{k}\cdot\vec{v} = k_\parallel v_\parallel + k_\perp v_\perp \cos\varphi$ and $\vec{v}\times\vec{\Omega}\cdot\partial/\partial\vec{v} = -\Omega\partial/\partial\varphi$, expressing the electric field in terms the electrostatic potential, $\delta\vec{E}_k(t) = -i\vec{k}\delta\phi_k$, and taking the Laplace transform of Eq. (8.46) gives

$$(s + ik_\parallel v_\parallel + ik_\perp v_\perp\cos\varphi)\delta\bar{f}_k - \delta f_k(0) - \Omega\frac{\partial}{\partial\varphi}\delta\bar{f}_k$$
$$= -\frac{2qi}{mv_T^2}(k_\parallel v_\parallel + k_\perp v_\perp\cos\varphi)\delta\bar{\phi}_k f_0, \tag{8.47}$$

where the 3d Maxwellian distribution $f_0 = \exp(-v^2/v_T^2)/(\sqrt{\pi}v_T)^3$ is assumed, and a bar over the variables represents its Laplace transformed quantity. Equation (8.47) can be cast into

$$\frac{\delta f_k(0)}{\Omega}e^{-G} + \frac{\partial}{\partial\varphi}(e^{-G}\delta\bar{f}_k)$$
$$= \frac{2iq}{mV_T^2}\frac{k_\parallel v_\parallel + k_\perp v_\perp\cos\varphi}{\Omega}e^{-G}\delta\bar{\phi}_k f_0. \tag{8.48}$$

Here we have defined $G(\varphi) \equiv (s\varphi + ik_\parallel v_\parallel\varphi + ik_\perp v_\perp\sin\varphi)/\Omega$. Thus,

$$\delta\bar{f}_k e^{-G} = \int^{\varphi} d\varphi' e^{-G(\varphi')}\left[-\frac{\delta f_k(0)}{\Omega} + \frac{2iq}{mV_T^2}\frac{k_\parallel v_\parallel + k_\perp\bar{v}_\perp\cos\varphi'}{\Omega}\delta\bar{\phi}_k f_0\right]. \tag{8.49}$$

The following identities are applied to simplify Eq. (8.49),

$$e^{ix\sin\varphi} = \sum_{n=-\infty}^{\infty} J_n(x)e^{in\varphi}, \tag{8.50}$$

$$\frac{1}{2}[J_{n+1}(x) + J_{n-1}(x)] = \frac{n}{x}J_n(x), \tag{8.51}$$

$$\int_0^\infty e^{-p^2x^2}J_n(\alpha x)J_n(\beta x)x dx = \frac{1}{2p^2}\exp\left(-\frac{\alpha^2+\beta^2}{4p^2}\right)I_n\left(\frac{\alpha\beta}{2p^2}\right),$$
$$[p,\alpha,\beta > 0]. \tag{8.52}$$

Thus,

$$\delta \bar{f}_k e^{-G} = -\frac{2iq\delta\bar{\phi}_k}{mV_T^2} f_0 \sum_n (k_\parallel \rho_\parallel + n) J_n(k_\perp \rho_\perp) \frac{e^{-\varphi(s/\Omega + ik_\parallel \rho_\parallel + in)}}{(s/\Omega + ik_\parallel \rho_\parallel + in)} - \int^{\varphi} d\varphi' e^{-G(\varphi')} \frac{\delta f_k(0)}{\Omega} \tag{8.49a}$$

The Poisson's equation can be cast into

$$\frac{k^2 \delta\bar{\phi}_k}{4\pi n_0 q} = -\frac{2iq\delta\bar{\phi}_k}{mV_T^2} \int d^3\vec{v} f_0 \sum_n J_n^2(k_\perp \rho_\perp) \frac{k_\parallel \rho_\parallel + n}{s/\Omega + ik_\parallel \rho_\parallel + in} - \int d^3\vec{v} e \int^{\varphi} d\varphi' e^{G(\varphi) - G(\varphi')} \frac{\delta f_k(0)}{\Omega} \tag{8.53}$$

Therefore,

$$\delta\bar{\phi}_k \varepsilon(k, s) = \frac{4\pi n_0 q}{k^2} \int d^3\vec{v} e^{G(\varphi)} \int^{\varphi} d\varphi' e^{-G(\varphi')} \frac{\delta f_k(0)}{\Omega} \ , \tag{8.54}$$

where the dielectric function is given by defining $\lambda \equiv k_\perp^2 V_T^2 / 2\Omega^2$,

$$\varepsilon(k, s) = 1 + \frac{2i\omega_p^2}{k^2 V_T^2} \sum_m e^{-\lambda} I_m(\lambda) \int_{-\infty}^{\infty} \frac{dv_\parallel}{\sqrt{\pi} V_T} e^{-v_\parallel^2 / V_T^2} \frac{k_\parallel v_\parallel + m\Omega}{s + ik_\parallel v_\parallel + im\Omega}. \tag{8.55}$$

The electrostatic cyclotron wave is retained in Eq. (8.55), together with higher cyclotron harmonics. They are termed as the **Bernstein waves** (Bernstein 1958). These modes are due to the kinetic effect of particle gyration that modulates the electric field into higher harmonics. The Bernstein waves are Landau damped if $k_\parallel$ is nonzero, and in the same token, they can be excited if there are inverted population in the distribution function where the resonance condition meets.

Homework 8.5: Show that in the long wavelength limit, the dispersion function from the Vlasov equation reduces to the following:

$$\varepsilon(\vec{k}, -i\omega) = \ 1 - 2\frac{\omega_p^2}{k^2} \frac{k_\perp^2}{\omega^2 - \Omega^2} \frac{\omega^2}{\Omega^2} - \frac{\omega_p^2}{k^2} \frac{k_\parallel^2}{\omega^2}.$$

8.5 The Electric Field Autocorrelation in Magnetized Plasma

We are to find the electric field autocorrelation $\vec{\vec{S}}(t, \tau) \equiv \langle \delta\vec{E}(\vec{x}, t) \delta\vec{E}(\vec{x}(\tau), \tau) \rangle$ in magnetized plasma under a uniform external magnetic field $\vec{B}$ but no equilibrium electric field. The particle trajectory governed by $d\vec{v}/dt = \vec{v} \times \vec{\Omega}$ and $d\vec{x}/dt = \vec{v}$, where $\vec{\Omega} = qB\hat{e}_z/mc$ is the gyrofreqeuncy, can be solved with use of the constant of the motion, namely, the canonical momentum, $\vec{v}(\tau) - \vec{x}(\tau) \times \vec{\Omega} = \vec{v} - \vec{x} \times \vec{\Omega}$. The initial conditions are set at $\tau = t$ for the velocity to equal to $\vec{v}$, $\vec{v}(\tau) = \vec{v}_\perp \cos\Omega(\tau - t) + (\vec{v} \times \hat{e}_z)\sin\Omega(\tau - t) + v_{||}\hat{e}_z$, and the position to arrive at $\vec{x}$, $\vec{x}(\tau) = \vec{x} + (\vec{v} - \vec{v}(\tau)) \times \vec{\Omega} + v_{||}(\tau - t)\hat{e}_z$. Expressing the electric field in terms of the Fourier representation,

$$\delta\vec{E}(\vec{x}(\tau), \tau) = \int d^3\vec{k}\delta\vec{E}_k(\tau)e^{i\vec{k}\cdot\vec{x}+i\vec{k}\cdot\hat{e}_z\times[\vec{v}(\tau)-\vec{v}]/\Omega+iv_z k_z(\tau-t)}, \tag{8.56}$$

and taking an ensemble average over the space uniformity gives the autocorrelation of the electric fields

$$\vec{\vec{S}}(t, \tau) = \left(\frac{2\pi}{L}\right)^3 \int d^3\vec{k}\langle\delta\vec{E}_{-k}(t)\delta\vec{E}_k(\tau)e^{i\vec{k}\cdot\hat{e}_z\times[\vec{v}(\tau)-\vec{v}]/\Omega+iv_z k_z(\tau-t)}\rangle. \tag{8.57}$$

Taking the ensemble average over the Maxwellian distribution and utilizing Eq. (8.31), we find

$$\vec{\vec{S}}(t, \tau) = \left(\frac{2\pi}{L}\right)^3 \int d^3\vec{k}\langle\delta\vec{E}_{-k}(t)\delta\vec{E}_k(\tau)\rangle \int_{-\infty}^{\infty} dv_x \times \int_{-\infty}^{\infty} dv_y \frac{e^{-(v_x^2+v_y^2)/V_T^2}}{\pi V_T^2} e^{i\vec{k}\cdot\hat{e}_z\times[\vec{v}(\tau)-\vec{v}]/\Omega} e^{-\frac{1}{4}k_z^2 V_T^2(\tau-t)^2} \tag{8.58}$$

Expanding $\vec{k}$ and $\vec{v}(\tau)$ in terms of their components,

$$\begin{aligned}\vec{k}\cdot\hat{e}_z \times (\vec{v}(\tau) - v) = {} & v_y(k_y \sin\Omega(\tau - t) - k_x \cos\Omega(\tau - t) + k_x) \\ & + v_x(k_x \sin\Omega(\tau - t) + k_y \cos\Omega(\tau - t) - k_y),\end{aligned}$$

and integrating over v_x and v_y by applying Eq. (8.31), we have

$$\vec{\vec{S}}(t, \tau) = \left(\frac{2\pi}{L}\right)^3 \int d^3\vec{k}\langle\delta\vec{E}_k(\tau)\delta\vec{E}_{-k}(t)\rangle e^{-\lambda[1-\cos\Omega(\tau-t)]} e^{-\frac{1}{4}k_{||}^2 V_T^2(\tau-t)^2}, \tag{8.59}$$

where $\lambda = k_\perp^2 V_T^2/2\Omega^2 = k_\perp^2\rho^2$, L the plasma size, $V_T \equiv \sqrt{2k_B T/m}$ the thermal velocity.

By keeping the particle discreetness in the initial condition as given by Eq. (8.32), and multiplying the electric potential of Eq. (8.54) by $-i\vec{k}$ to nullify the $\delta(\vec{k})$ term and gives the electric field,

$$\varepsilon(\vec{k},s)\delta\vec{E}_k = \frac{-i\vec{k}q}{2\pi^2k^2}\int d^3\vec{v}e^{G(\varphi)}\int^{\varphi}\frac{d\varphi'}{\Omega}e^{-G(\varphi')}\sum_j e^{-i\vec{k}\cdot\vec{x}_{j0}}\delta(\vec{v}-\vec{v}_{j0}). \tag{8.60}$$

Taking the φ' integration over $e^{-G(\varphi')}$, where $G(\varphi) \equiv (s\varphi + ik_\parallel v_\parallel \varphi + ik_\perp v_\perp \sin\varphi)/\Omega$ was defined, we have,

$$\varepsilon(\vec{k},s)\delta\vec{E}_k = \frac{-i\vec{k}q}{2\pi^2k^2}\int d^3\vec{v}\sum_{n'} J_{n'}(k_\perp\rho_\perp)e^{\varphi\left(\frac{s}{\Omega}+ik_\parallel\rho_\parallel+in'\right)}\sum_n J_n(k_\perp\rho_\perp)$$
$$\times\sum_j e^{-i\vec{k}\cdot\vec{x}_{j0}}\frac{e^{-\varphi\left(\frac{s}{\Omega}+ik_\parallel\rho_\parallel+in\right)}}{s+ik_\parallel v_\parallel+in\Omega}\delta(\vec{v}-\vec{v}_{j0})$$

After the integration over the velocity space, we end with

$$\varepsilon(\vec{k},s)\delta\vec{E}_k = \frac{-i\vec{k}q}{2\pi^2k^2}\sum_{n,j} e^{-i\vec{k}\cdot\vec{x}_{j0}}J_n^2\left(\frac{k_\perp v_{\perp j0}}{\Omega}\right)\frac{1}{s+ik_\parallel v_{\parallel j0}+in\Omega}. \tag{8.61}$$

Similar to the unmagnetized case of Eq. (8.35), the RHS is the bare particle effect along the unperturbed gyrating particle trajectory resulted from the particle discreteness in the initial condition. The dielectric function, which is the dispersion function of Eq. (8.55), dresses up the bare particle response for the electric field $\delta\bar{E}_k$ to the initial fluctuating density that persists by free streaming through the correlation time.

The inverse Laplace transform is taken in the long time limit as in Eq. (8.36), that keeps only the residue of the pole at $-ik_\parallel v_{\parallel j0} - in\Omega$, but neglects the contribution from the poles at the zeros of the dielectric function. The latter corresponds to the electrostatic Bernstein waves of the equilibrium plasma, and is unrelated to the particle discreteness of concern here. Thus,

$$\delta\vec{E}_k(t) \xrightarrow{t\to\infty} \sum_{j,n}\frac{-iq\vec{k}}{2\pi^2k^2}J_n^2\left(\frac{k_\perp v_{j0\perp}}{\Omega}\right)\frac{e^{-i\vec{k}\cdot\vec{x}_{j0}-ik_\parallel v_{j0\parallel}t-in\Omega t}}{\varepsilon(\vec{k},-ik_\parallel v_{j0\parallel}-in\Omega)}, \tag{8.62}$$

The dielectric function serves as the dress-up screening of the individual particles that would otherwise be "bare". Only terms referring to the same particle survives the ensemble average in the two-time auto-correlation of the electric field,

$$\langle\delta\vec{E}_{-k}(t)\delta\vec{E}_k(\tau)\rangle$$

$$
= \frac{q^2\vec{k}\vec{k}}{4\pi^4 k^4}\left\langle \sum_{n.n',j} J_n^2\left(\frac{k_\perp v_{j0\perp}}{\Omega}\right) J_{n'}^2\left(\frac{k_\perp v_{j0\perp}}{\Omega}\right)\right.
$$

$$
\left. \times \frac{e^{-i\Omega(nt-n'\tau)}}{\varepsilon(\vec{k}, -ik_{\|}v_{j0\|} - in\Omega)\varepsilon(-\vec{k}, ik_{\|}v_{j0\|} + in'\Omega)} \right\rangle
$$

Thus,

$$
\overline{\overline{S}}(t,\tau) = \int d^3\vec{k} \sum_{n,n',l} \overline{\overline{S}}_{nn'l} e^{il\Omega(t-\tau) - i\Omega(nt-n'\tau) - \frac{1}{2}k_{\|}^2 v_T^2(\tau - t)^2}, \tag{8.57a}
$$

where

$$
\overline{\overline{S}}_{nn'l} = \left\langle \frac{2n_0 q^2 \vec{k}\vec{k}}{\pi k^4 \varepsilon_n \varepsilon_{n'}^*} J_n^2\left(\frac{k_\perp v_{0\perp}}{\Omega}\right) J_{n'}^2\left(\frac{k_\perp v_{0\perp}}{\Omega}\right)\right\rangle I_l(\lambda) e^{-\lambda} \quad , \tag{8.57b}
$$

Homework 8.6: Show that the electric field auto-correlation (8.57) $\overline{\overline{S}}(t,\tau) = \overline{\overline{S}}(|t-\tau|)$ is real and function of the time difference only.

Here, $\varepsilon_n \equiv \varepsilon(\vec{k}, -ik_{\|}v_{0\|} - in\Omega)$ is subscripted with its major cyclotron harmonic, the particle index is dropped without ambiguity, the ensemble average is taken over the velocity $\vec{v}_0$, and the identity $e^{z\cos\varphi} = \sum_{l=-\infty}^{\infty} I_l(z)e^{il\varphi}$ has been applied, where I_l is the modified Bessel function of the first kind. Since the integer indexes n, n', l, are summing up from $-\infty$ to ∞, it can be shown that the electric field auto-correlation $\overline{\overline{S}}(t,\tau) = \overline{\overline{S}}(|t-\tau|)$ is an even function of the time difference only.

Homework 8.7: Work out the collision operator $C(f) = -(q/m)(\partial/\partial\vec{v}) \cdot \langle\delta\vec{E}\delta f\rangle$ for the magnetized plasma.

8.6 The Pair Correlation

The pair correlation is defined as the *conditional probability density* of finding another particle at $\vec{r}$, given a particle at the coordinate origin. To quantify this effect, a two-body distribution function needs be solved, which is defined as in the following:

$$
F_2(\vec{x}_1, \vec{v}_1, \vec{x}_2, \vec{v}_2, t)
$$

$$
\equiv \sum_{i,j} \delta(\vec{x}_1 - \vec{x}_i(t))\delta(\vec{v}_1 - \vec{v}_i(t))\delta(\vec{x}_2 - \vec{x}_j(t))\delta(\vec{v}_2 - \vec{v}_j(t)). \tag{8.63}
$$

Taking the time derivative on F_2 results in the time evolution equation,

$$\frac{\partial F_2}{\partial t} + \vec{v}_1 \cdot \frac{\partial F_2}{\partial \vec{x}_1} + \vec{a}_1 \cdot \frac{\partial F_2}{\partial \vec{v}_1} + \vec{v}_2 \cdot \frac{\partial F_2}{\partial \vec{x}_2} + \vec{a}_2 \cdot \frac{\partial F_2}{\partial \vec{v}2} - \frac{\partial}{\partial \vec{v}_1}$$
$$\cdot \int d\Gamma_3 q \frac{\partial \varphi_{13}}{\partial \vec{x}_1} F_3 - \frac{\partial}{\partial \vec{v}_2} \cdot \int d\Gamma_3 q \frac{\partial \varphi_{23}}{\partial \vec{x}_2} F_3 = 0 \tag{8.64}$$

where $\varphi_{ij} \equiv q/|\vec{x}_i - \vec{x}_j|$ is the interaction potential, and $d\Gamma_i \equiv d^3\vec{x}_i d^3\vec{v}_i$ the phase volume element. We will focus onto the unmagnetized plasma and limit the particle interaction as predominantly electrostatic. We may assume the higher order correlations are generally weaker than its lower order distribution functions, and write the two body distribution as

$$F_2(1,2) = F_1(1)F_1(2) + P(1,2), \tag{8.65}$$

where $P(1,2)$ is the two body correlation function. Moreover, the three body distribution function is expanded as in the following:

$$\begin{aligned} F_3(1,2,3) = F_1(1)F_1(2)F_1(3) + P(1,2)F_1(3) + P(2,3)F_1(1) \\ + P(3,1)F_1(2) + T(1,2,3). \end{aligned} \tag{8.66}$$

We will neglect the ternary correlation $T(1,2,3)$, and assume that the pair correlation can be expressed as $P(1,2) = F_1(1)F_1(2)p(r_{12})$, and take the one body distribution as Maxwellian distribution, $F_1(j) = \exp(-v_j^2/V_T^2)/\pi^{3/2}V_T^3$. By utilizing Eq. (8.3′) to eliminate the evolution of the one body distribution function, Eq. (8.64) is simplified for the stationary state to

$$\vec{v}_1 \cdot \frac{\partial p(1,2)}{\partial \vec{x}_1} + \vec{v}_2 \cdot \frac{\partial p(1,2)}{\partial \vec{x}_2} - \frac{q}{m}\frac{\partial \varphi_{12}}{\partial \vec{x}_1} \cdot \frac{\partial \ln F_1(1)}{\partial \vec{v}_1} - \frac{q}{m}\frac{\partial \varphi_{21}}{\partial \vec{x}_2} \cdot \frac{\partial \ln F_1(2)}{\partial \vec{v}_2}$$
$$- \int d\vec{r}_3 n_0 \frac{q}{m} \left(\frac{\partial \ln F_1(1)}{\partial \vec{v}_1} \cdot \frac{\partial \varphi_{13}}{\partial \vec{x}_1} (p(1,2) + p(2,3)) \right.$$
$$\left. + \frac{\partial \ln F_1(2)}{\partial \vec{v}_2} \cdot \frac{\partial \varphi_{23}}{\partial \vec{x}_2} (p(1,2) + p(3,1)) \right) = 0, \tag{8.67}$$

or,

$$(\vec{v}_1 - \vec{v}_2) \cdot \left(\frac{\partial p(1,2)}{\partial \vec{x}_1} + \frac{2q}{mV_T^2}\frac{\partial \varphi_{12}}{\partial \vec{x}_1} \right)$$
$$+ \frac{2n_0 q}{mV_T^2} \int d\vec{x}_3 \left(\vec{v}_1 \cdot \frac{\partial \varphi_{13}}{\partial \vec{x}_1} p(2,3) + \vec{v}_2 \cdot \frac{\partial \varphi_{23}}{\partial \vec{x}_2} p(3,1) \right) = 0, \tag{8.67a}$$

by utilizing $\int d\vec{r}_3 \partial\varphi_{13}/\partial\vec{x}_1 = 0 = \int d\vec{r}_3 \partial\varphi_{23}/\partial\vec{x}_2$. Making a Galilean transformation to stay in the moving frame of particle 2 so that $\vec{v}_2' = 0$, $\vec{v}_1' = \vec{v}_1 - \vec{v}_2$, and

$\vec{x}'_{ij} = \vec{x}_{ij}$, we reduce the equation to the following

$$\vec{v}'_1 \cdot \left(\frac{\partial p(1,2)}{\partial \vec{x}'_1} + \frac{2q}{mV_T^2} \frac{\partial \varphi_{12}}{\partial \vec{x}'_1} + \frac{2q}{mV_T^2} \cdot \int d\vec{x}'_3 \frac{\partial \varphi_{13}}{\partial \vec{x}'_1} p(2,3) \right) = 0. \tag{8.68}$$

For arbitrary $\vec{v}'_1$, we have

$$\frac{\partial p(1,2)}{\partial \vec{x}'_1} + \frac{2q}{mV_T^2} \frac{\partial \varphi_{12}}{\partial \vec{x}'_1} + \frac{2q}{mV_T^2} \cdot \int d\vec{x}'_3 \frac{\partial \varphi_{13}}{\partial \vec{x}'_1} p(2,3) = 0. \tag{8.68a}$$

Take $\nabla'_1 \cdot$ on the above equation, and omit the prime on the variables, we end with

$$\nabla_1^2 \left(p(1,2) + \frac{2q^2}{mV_T^2} \frac{1}{r_{12}} \right) = \frac{8\pi n_0 q^2}{mV_T^2} \int d\vec{x}_3 \delta(\vec{r}_{13}) p(2,3) = \frac{1}{\lambda_D^2} p(2,1), \tag{8.69}$$

that is simplified to

$$\nabla_1^2 p(1,2) - \frac{4\pi q^2}{k_B T} \delta(\vec{r}_{12}) == \frac{1}{\lambda_D^2} p(1,2). \tag{8.69a}$$

The particular solution to this equation is given by

$$p(1,2) = -\frac{q^2}{k_B T} \frac{e^{-r_{12}/\lambda_D}}{r_{12}}, \tag{8.70}$$

Note that we have applied the identity: $\nabla_1^2 (1/r_{12}) = -4\pi \delta(r_{12})$.

Homework 8.8: The Debye screened Poisson's equation is given by

$$\lambda_D^2 \nabla^2 \Phi - \Phi = -4\pi \lambda_D^2 \rho(\vec{r}).$$

Express Φ in the integral form for an arbitrary source of $\rho(\vec{r})$.

8.7 Application of the Fokker-Planck Equation

The classical transport effect can be found by solving the linearized Fokker-Planck equation,

$$\vec{v} \cdot \left[-2 \frac{q_\sigma}{m_\sigma} \frac{\vec{E}}{V_T^2} f_{\sigma 0} + \frac{\nabla n_0}{n_0} f_{\sigma 0} + \frac{\nabla V_T}{V_T} \left(\frac{2v^2}{V_T^2} - 3 \right) f_{\sigma 0} \right] + \frac{q_\sigma}{m_\sigma} \frac{\vec{v}}{c} \times \vec{B} \cdot \frac{\partial}{\partial \vec{v}} f_{\sigma 1}$$

$$= C(f_{\sigma 0}, f_{\sigma' 1}) + C(f_{\sigma 1}, f_{\sigma' 0}) = \nu_{\sigma\sigma'} \frac{\partial}{\partial \vec{v}} \cdot \int d^3\vec{v}' \frac{\vec{\vec{I}} - \widehat{e}_{||}\widehat{e}_{||}}{|\vec{v} - \vec{v}'|} \cdot$$

$$\times \left(f_{\sigma' 1}(\vec{v}') \frac{\partial f_{\sigma 0}(\vec{v})}{m_\sigma \partial \vec{v}} - f_{\sigma 0}(\vec{v}) \frac{\partial f_{\sigma' 1}(\vec{v}')}{m_{\sigma'} \partial \vec{v}'} \right.$$

$$\left. + f_{\sigma' 0}(\vec{v}') \frac{\partial f_{\sigma 1}(\vec{v})}{m_\sigma \partial \vec{v}} - f_{\sigma 1}(\vec{v}) \frac{\partial f_{\sigma' 0}(\vec{v}')}{m_{\sigma'} \partial \vec{v}'} \right), \tag{8.71}$$

where the density gradient is also included, and $\nu_{\sigma\sigma'} = (2\pi n_0 q_\sigma^2 q_{\sigma'}^2 / m_\sigma) \log \Lambda$. This equation represents the mathematical framework for the classical transport. Note that the equilibrium is satisfied when the distribution function is Maxwellian $f_M = (\sqrt{\pi} V_T)^{-3} \exp(-v^2/V_T^2)$.

8.7.1 *Particle Diffusion*

We assume the ion velocity is negligible compared to the electron velocity so that $\vec{v}_e - \vec{v}_i \approx \vec{v}_e$ and $f_{i1}(\vec{v}')$ is insignificant in the collision process. Taking $f_{e1} = g(\vec{v}_\perp) f_M$ to look for the particular solution as driven by the free energy due to the particle density gradient, we have,

$$\vec{v} \cdot \frac{\nabla n_0}{n_0} - \Omega_e \frac{\partial}{\partial \phi} g = \frac{\nu_c}{m_e v^3} \left(\frac{\partial}{\partial \mu} (1 - \mu^2) \frac{\partial}{\partial \mu} + \frac{1}{1 - \mu^2} \frac{\partial^2}{\partial \phi^2} \right) g, \tag{8.72}$$

where $\mu \equiv \cos\theta$, and θ and ϕ are the spherical angles of $\vec{v}$.

Homework 8.9: Show that by taking $f_{e1} = g(\vec{v}_\perp) f_M$ and neglecting the ion velocity, the linearized collision operator can be simplified to,

$$C(f_{e0}, f_{i1}) + C(f_{e1}, f_{i0}) \approx \frac{\nu_c}{m_e v^3} f_M \left(\frac{\partial}{\partial \mu} (1 - \mu^2) \frac{\partial}{\partial \mu} + \frac{1}{1 - \mu^2} \frac{\partial^2}{\partial \phi^2} \right) g,$$

where $\nu_c = (2\pi n_0 q_e^2 q_i^2 / m_e) \log \Lambda$.

Assuming that $g = \sum_m \alpha_m e^{im\phi}$, we can find the solution to Eq. (8.72) is given by

$$g = \frac{v}{\Omega_e} \sqrt{1 - \mu^2} \frac{\partial n_0 / \partial x}{n_0} \sin\phi + \frac{\nu_c}{m_e v^2 \Omega_e^2}$$

$$\times \left(\frac{\partial}{\partial \mu} (1 - \mu^2) \frac{\partial}{\partial \mu} - \frac{1}{1 - \mu^2} \right) \sqrt{1 - \mu^2} \frac{\partial n_0 / \partial x}{n_0} \cos\phi. \tag{8.73}$$

Homework 8.10: Assuming that $g = \sum_m a_m e^{im\phi}$, show that the solution to Eq. (8.72) is given by Eq. (8.73).

The first term in g does not contribute to the particle flux which can be expressed as,

$$\Gamma_x = n_0 \int d^3\vec{v}\, v_x f_{e1} = \int dv_{\|} dv_{\perp} d\phi f_M v_{\perp}^2 \cos^2\phi \frac{\nu_c}{m_e v^2 \Omega_e^2}$$
$$\times \left(\frac{\partial}{\partial\mu}(1-\mu^2)\frac{\partial}{\partial\mu} - \frac{1}{1-\mu^2}\right)\sqrt{1-\mu^2}\frac{\partial n_0}{\partial x}$$
$$= \int v dv f_M \int_{-1}^{1} \frac{d\mu}{2}\sqrt{1-\mu^2}\frac{\pi\nu_c}{m_e\Omega_e^2}$$
$$\times \left(\frac{\partial}{\partial\mu}(1-\mu^2)\frac{\partial}{\partial\mu} - \frac{1}{1-\mu^2}\right)\sqrt{1-\mu^2}\frac{\partial n_0}{\partial x}. \quad (8.74)$$

Therefore, the diffusion coefficient is extracted from Eq. (8.74) as $D = \Gamma_x/(\partial n_0/\partial x)$, and is given by

$$D = \frac{2}{3\sqrt{\pi}V_T}\frac{\nu_c}{m_e\Omega_e^2} = \frac{4\nu_c}{3\sqrt{\pi}m_e V_T^3}\frac{2m_e V_T^2}{m_e\Omega_e^2} = \nu_{ei}\frac{k_B T_e}{m_e\Omega_e^2} = \nu_{ei}\rho_e^2. \quad (8.75)$$

Homework 8.11: Derive the diffusion coefficient by directly solving the following simple collision model equation

$$\vec{v}\cdot\frac{\nabla n_0}{n_0} f_{e0} - \Omega_e\frac{\partial}{\partial\phi} f_{e1} = -\nu_{ei} f_{e1}.$$

8.7.2 *Spitzer Resistivity*

The response function to an externally imposed dc electric field can be found through the following:

$$-2\vec{v}\cdot\frac{e}{m_e}\frac{\vec{E}}{V_T^2} f_{e0} - \Omega_e\frac{\partial}{\partial\phi} f_{e1} = C(f_{e0}, f_{i1}) + C(f_{e1}, f_{i0}). \quad (8.76)$$

Consider the cold ions, we then simplify the collision operator,

$$-2v_z\frac{e}{m_e}\frac{E_z}{V_T^2} f_{e0} - \Omega_e\frac{\partial}{\partial\phi} f_{e1}(\vec{v}) \approx \nu_c\frac{\partial}{\partial\vec{v}}\cdot\frac{v^2\overline{\overline{I}} - \vec{v}\vec{v}}{|\vec{v}|^3}\cdot\frac{\partial f_{e1}(\vec{v})}{m_e\partial\vec{v}}. \quad (8.77)$$

The resultant electron perturbed distribution is expected to be azimuthally symmetric so that we may assume $f_{e1} = g f_M$. Therefore,

$$-2v\mu e\frac{E_z}{V_T^2} = \frac{\nu_c}{m_e v^3}\left(\frac{\partial}{\partial\mu}(1-\mu^2)\frac{\partial}{\partial\mu}\right)g. \tag{8.78}$$

The solution is given by

$$g = -\frac{m_e}{\nu_c}v^4 e\frac{E_z}{V_T^2}\left(2\mu + \log\left(\frac{1+\mu}{1-\mu}\right)\right). \tag{8.79}$$

That leads to the current response,

$$j_{\|} = \int d^3\vec{v}\, f_{e1}\vec{v}e = 11.28 n_0\frac{m_e}{\nu_c}V_T^3 e^2 E_z, \tag{8.80}$$

and the conductivity is simply $j_{\|}/E_{\|}$.

Other transport coefficients can be found through the Fokker Planck model. For example, the thermal conductivities are given by $\kappa_{\|} = 3.2nk_B T_e\tau_e/m_e$ along the field lines, and $\kappa_{\perp} = 4.7nk_B T_e/m_e\tau_e\Omega_{ce}^2$ the perpendicular.

8.8 The Statistical Ensemble

Ensemble average has been applied in this chapter to obtain the electric field autocorrelation. The concept of statistical ensemble goes beyond the expectation value. It can also extract statistically significant features in a complex system. The deterministic description of many body system given the initial and boundary values can be ambiguous or even misleading at times, since minute deviations in the initial or boundary values may yield drastic differences in the temporal behavior. A linear analysis, for example, is to set up the equilibrium to study its stability. If the equilibrium is not the minimum energy state, it will make a transition to lower its energy. Therefore, it is unstable, but the physical process is just a simple relaxation. The linear stability analysis would not reveal that until the nonlinear dynamics is fully explored. On the other hand, by comparing all the states in a statistical ensemble, the lowest energy state thereof will have the statistical significance to ensure its properties as the prevailing representation. Moreover, even if a minimum energy state is unstable to certain perturbations, it will not and cannot make further transformation for the given constraints. Electron in the atomic orbital is a good example. Under the constant centripetal force, it would not radiate EM waves, or the material would collapse.

The statistical treatment has the advantage of revealing general and common traits than detailed and rare properties, and may give, time and again, more reliable predictions than otherwise. It often leads to better global understanding free from the mechanical-and-deterministic constraints that can be too stringent and too many. The microcanonical ensemble would place the system to its minimum energy state, while the canonical ensemble allows certain probabilities for higher energy states thus the fluctuations.

8.8.1 *The Catenary*

A linear chain (cf. Sec. 7.4.1), hinged at its two ends at equal heights, would relax to its ultimate configuration under the influence of the gravitational force. The U-shape curve is termed the **catenary** that often appears in the design of arches and bridges. The problem can be solved by the Newtonian mechanics by balancing the tensile force. The only trouble is that at the bottom of the curve where $\theta = 0$ there is no tension in the $\hat{e}_z$ direction to balance the gravitational force.

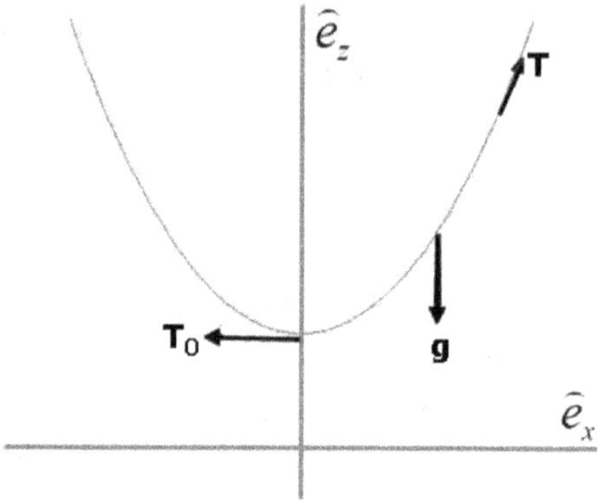

However, applying the variational principle to the potential energy given by $\varepsilon = g\rho \int_{-h}^{h} z ds = 2g\rho \int_0^h z\sqrt{1+x'^2}dz$ yields the following equation,

$$\begin{aligned}\delta\varepsilon &= 2g\rho \int_0^h \frac{zx'\delta x'}{\sqrt{1+x'^2}}dz = 2g\rho \int_0^h \frac{zx'}{\sqrt{1+x'^2}}\delta\frac{dx}{dz}dz \\ &= 2g\rho \int_0^h \frac{zx'}{\sqrt{1+x'^2}}\frac{d\delta x}{dz}dz = 2g\rho\frac{zx'}{\sqrt{1+x'^2}}\delta x|_0^h \\ &\quad -2g\rho \int_0^h \frac{d}{dz}\left(\frac{zx'}{\sqrt{1+x'^2}}\right)\delta x dz. \end{aligned} \tag{8.81}$$

where an integration by parts was taken. Given the arbitrary virtual displacement δx, we seek for the extreme energy state that requires $\delta\varepsilon = 0$ so that

$$\frac{zx'}{\sqrt{1+x'^2}} = b = const, \tag{8.82}$$

holds true. It has the simple solution

$$\frac{z}{b} = \cosh\left(\frac{x-a}{b}\right). \tag{8.83}$$

This concludes that the catenary solution is the minimum energy state and is stable to any perturbation since given the prescribed constraints there is no other lower energy state to relax to.

Homework 8.12: Instead of taking z but x as the independent variable, minimize the energy $\varepsilon = g\sigma \int_{-b/2}^{b/2} z\sqrt{1+z'^2}dx$. Show that the Euler's equation is given by

$$\sqrt{1+z'^2} = \frac{d}{dx}\left(\frac{zz'}{\sqrt{1+z'^2}}\right).$$

Find its solution and compare with the solution of Eq. (8.83).

Question arises as whether an 'open system' driven by an external environment can relax to its minimum energy state as the microcanonical ensemble in statistical physics often refers to an isolated system. The process of energy exchange is necessary for the chain to lower its energy (cf. Sec. 7.4.1 for computer simulation), but ceases to do so at the end of the relaxation, as the chain cannot go below its lowest energy state.

8.8.2 *The Taylor State*

The Taylor state minimizes the magnetic energy $\varepsilon_B = \frac{1}{8\pi}\int d\tau(B_z^2 + B_\theta^2)$ of the reversed field pinch subject to the constraint of the constancy of magnetic helicity, $H_B = \int d\tau \vec{A}\cdot\vec{B}$. The justification of this proposition lies in the fact that the long wavelength modes tend to decay slower compared to the short wavelength modes. As $\vec{B} = \nabla \times \vec{A}$ makes the magnetic field weighted more toward the short wavelength spectrum than the vector potential $\vec{A}$ given the same spectrum. As a result, the magnetic helicity could hold its constancy while the magnetic energy relaxes. By applying the variational calculus

$$\begin{aligned} 4\pi\,\delta\varepsilon_B + \frac{1}{2}\lambda\delta H_B &= \int d\tau\delta\vec{B}\cdot\vec{B} + \frac{1}{2}\lambda\int d\tau\delta\vec{A}\cdot\vec{B} + \frac{1}{2}\lambda\int d\tau\vec{A}\cdot\delta\vec{B} \\ &= \int d\tau\nabla\times\delta\vec{A}\cdot\vec{B} + \frac{1}{2}\lambda\int d\tau\delta\vec{A}\cdot\nabla\times\vec{A} \end{aligned}$$

$$+\frac{1}{2}\lambda\int d\tau\vec{A}\cdot\nabla\times\delta\vec{A}$$

$$=\int d\tau\delta\vec{A}\cdot(\nabla\times\vec{B}+\nabla\times\vec{A}), \tag{8.84}$$

the minimum energy state is governed by

$$\nabla\times\vec{B}=\lambda\vec{B}, \tag{8.85}$$

where λ is the Lagrangian multiplier.

Since the current $\vec{J}=\nabla\times\vec{B}=\lambda\vec{B}$ gives $\vec{J}\times\vec{B}=0$, it is a force free state (cf. Homework 4.3). In the cylindrical geometry of infinite length in the $\widehat{e}_z$-direction and no θ variation, we find

$$\frac{1}{r}\frac{\partial}{\partial r}rB_\theta+\lambda B_z=0,\quad -\frac{\partial}{\partial r}B_z+\lambda B_\theta=0. \tag{8.86}$$

```
function Taylor( lambda )
%lambda=4;
N=1000;
dr=1/N;
r=dr:dr:1;
b0=1;
bz=b0*besselj(0, lambda *r);
bt=b0*besselj(1, lambda *r);
plot(r,bz,'g-',r,bt,'r-',r,r*0,'k-');
title('Taylor State');
xlabel('r');
ylabel('B');
text(0.4,0.7,'B_\theta','color','r');
text(0.6,0.6,'B_z','color','g');
```

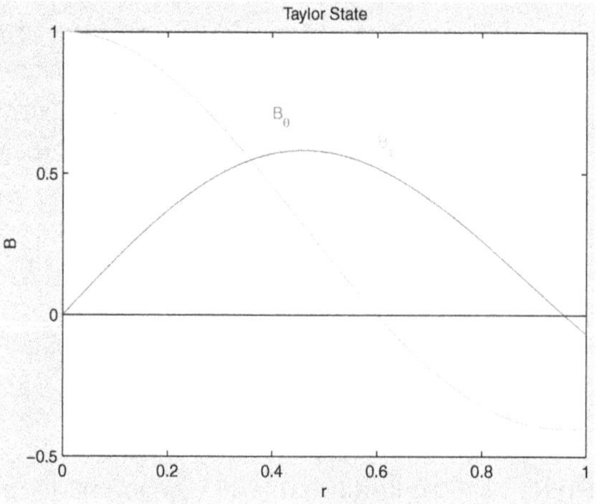

Eliminating B_z in favor of B_θ results in

$$\frac{\partial}{\partial r}\left(\frac{1}{r}\frac{\partial}{\partial r}rB_\theta\right)+\lambda^2B_\theta=\frac{\partial^2}{\partial r^2}B_\theta+\frac{1}{r}\frac{\partial}{\partial r}B_\theta-\frac{1}{r^2}B_\theta+\lambda^2B_\theta=0, \tag{8.87}$$

and the solutions are given by $B_\theta=b_0J_1(\lambda r)$and $B_z=b_0J_0(\lambda r)$. The magnetic fields can be reversed field at the plasma edge.

8.8.3 *The Canonical Profile*

The Canonical Profile minimizes both the magnetic and pressure energies of the tokamak plasma subject to the constraint of the constancy of the total current. After the variational calculus, it results in the following governing equation,

$$\Delta*\psi=-c_Fe^{\lambda\psi}-R^2c_Pe^{\frac{5}{4}\lambda\psi}=-\frac{4\pi}{c}RJ_\phi. \tag{8.88}$$

that exhibits the profile consistency, bifurcated solutions and the loss of equilibrium condition. The profile consistency is referred to the fact that a unique current profile holds true for the given safety factors at the center and at the edge. Both the pressure and the magnetic energies follow their canonical profiles at different temperatures since the toroidal geometry appears to prohibit the full equilibration among these two energies. Further discussions on the Tokamak equilibrium profiles can be found in the final two chapters.

8.9 The Temperature Measure

The temperature is a measure of the variation in the internal energy U with respect to the variation in the entropy $S = -\sum_i p_i \ln p_i$, i.e., $k_B T \equiv (\partial S/\partial U)^{-1}$, where, p_i is the probability to find the system in the ith microstate. The canonical ensemble assumes that the energy exchange between the system and its environment leads to an equilibrium but allowing good probabilities at different energy levels and maximizing the entropy of the system for the given expected energy. Moreover, the most probable state is given by the Gibbs distribution $P(\langle U\rangle = E) \propto \exp(-\beta E)$ that is consistent with Boltzmann's H-theorem and the Boltzmann distribution. The Gibbs distribution goes beyond the notion of the canonical ensemble, and implies that any probability distribution having the Markovian property of statistical independence will follow **the central limit theorem** with a variance.

Homework 8.13: The Central Limit Theorem: Any sum of independent and identically-distributed random variables tends to yield a normal distribution with a finite variance. Define $X = \langle x_i\rangle \equiv \frac{1}{N}\sum_{i=1}^{N} x_i$, where x_i follows a probability distribution $p(x_i)$ of any kind, and $N \gg 1$. Then X follows the distribution $P(X) = \frac{1}{\sqrt{2\pi}\sigma}\exp\{-(X-\mu)^2/2\sigma^2\}$, where $\mu = \langle X\rangle$ is the mean and $\sigma = \sqrt{\langle X^2\rangle - \langle X\rangle^2}$ the variance. Devise a numerical scheme as follows: Let $\frac{1}{2} \geq R_n \geq -\frac{1}{2}$ be a random number, and take a spatial step of $\delta x_n = f R_n$ with $1 > f > 0$. Repeat the process for a large number of steps and compare the final distribution of X with $P(X)$.

Homework 8.14: Given an isotropic Maxwellian distribution, show that

(a) the root mean square velocity is given by $V_{rms} = (3k_B T/m)^{1/2}$,
(b) the most probable speed is $V_{mps} = (2k_B T/m)^{1/2}$.
(c) The energy fluctuation is defined as $\sigma_E \equiv \{\langle E^2\rangle - \langle E\rangle^2\}^{1/2}/\langle E\rangle$, derive σ_E.

8.9.1 *The Negative Temperature State*

When the entropy of the system, or the number of configurations given the same energy content, is diminishing as the energy is increasing, the system possesses the **negative temperature** state. A good example in plasma physics is the 2d guiding center plasma moves only with the ExB drift. Accordingly, the internal energy is given by $U = \frac{1}{2}\sum_{i\neq j} q_i q_j \ln|\vec{r}_i - \vec{r}_j|$. The entropy can be evaluated from the density distribution of the electron and ion charges by $S = -\sum_{\sigma=i,e} n_\sigma \ln n_\sigma$, where n_i and n_e are the ion and electron density, respectively. When the energy increases, the charges of the same species have to get into closer proximity so that the entropy decreases as the available number of microstates is reduced.

The energy density is given by $\varepsilon_\sigma(\vec{x}) = n_\sigma(\vec{x})q_\sigma \int d\tau' g(|\vec{x} - \vec{x}'|)n(\vec{x}') = n_\sigma(\vec{x})q_\sigma\Phi(\vec{x})$, where $n(\vec{x}) = n_i(\vec{x}) - n_e(\vec{x})$. Assuming the Gibb's distribution for the individual charge,

$$n_\sigma(\vec{x}) = n_0 e^{-\beta\Phi(\vec{x}')q_\sigma}, \tag{8.89}$$

and taking the spatial derivative on the density, we have

$$\frac{\partial n_s(\vec{x})}{\partial \vec{x}} = -\beta n_s \varsigma_s \frac{\partial \Phi(\vec{x})}{\partial \vec{x}} \quad . \tag{8.90}$$

This implies that $\partial(\ln n_e(\vec{x}) + \ln n_i(\vec{x}))/\partial\vec{x} = 0$. Therefore, $n_e(\vec{x})n_i(\vec{x}) = const.$ The Poisson's equation is given by $\nabla^2\Phi + 4\pi n_0\varsigma(\exp(-\beta\varsigma\Phi) + \exp(\beta\varsigma\Phi)) = 0$. Defining $\Psi \equiv \varsigma\Phi$, and the length scale $\bar{x} \equiv \sqrt{8\pi n_0 \varsigma^2}x$, we arrive at the **Sinh-Poisson equation,**

$$\bar{\nabla}^2\Psi + \sinh\beta\Psi = 0. \tag{8.91}$$

The system energy is given by

$$E = \frac{1}{8\pi}\int d\tau|\nabla\Psi|^2. \tag{8.92}$$

The positive temperature states tend to have smaller energy content than the negative temperature states.

8.9.2 *The Magnetic Temperature*

To explore the magnetic fluctuation driven by the plasma thermal energy, the concept of a **magnetic temperature** is appealing. The magnetic perturbations tend to be weaker than the electrostatic fluctuations because when they are excited, both the electric and magnetic components must be present due to the Faraday's law, $\nabla \times \vec{E} = -\partial\vec{B}/\partial t/c$. Therefore, we may treat the plasma thermal energy as the energy reservoir, which provides the energy exchange with the magnetic fluctuations, and maintains the magnetic temperature as a result.

If we take the Gibb's distribution so that $P = N\exp(-\beta_M \varepsilon_M)$, where ε_M is the local magnetic energy. We may find the entropy from $S = -\int d\tau \int d\Gamma P \ln P$ where $d\Gamma = d\varepsilon_M$is the phase volume element, and $\int d\Gamma P = \int_0^\infty d\varepsilon_M P = 1$. The internal magnetic energy may be evaluated as $U = \int d\tau \int d\Gamma P \varepsilon_M$. It is straightforward to show that $\beta_M = \partial S/\partial U$.

> **Homework 8.15:** Show that $\beta_M = \partial S/\partial U$, if the system follows the Gibb's distribution, $P = N\exp(-\beta_M \varepsilon_M)$.

Consider a simple system with plasma current $J_z(r)$flowing along the $\hat{e}_z$ direction in a cylindrical plasma of radius a. The entropy can be expressed as $S = -\int d\tau (J_z/J_0)\ln(J_z/J_0)$, where J_0 is a normalization constant taken as the volume averaged current density such that $\int d\tau J_\phi / \int d\tau J_0 = 1$. The magnetic field is then given by $B_\theta(r) = (4\pi/rc)\int_0^r J_z(r')r'dr'$, and the magnetic energy $U = \frac{1}{2}\int_0^a r'dr' B_\theta^2(r')$. We may thus find the magnetic temperature from $\beta_M = \partial S/\partial U$ by varying the current profile. This example may be generalized to a more realistic situation.

The magnetic fluctuations may in principle be found by following the calculation of the electrostatic fluctuations (cf Secs. 8.3 and 8.5). The particle discreteness drives the fluctuating current in analogy to Eq. (8.61) whose right hand side has the charge density due to particle discreteness, while the left hand side has the dielectric response due to the collective wave characteristics. Equations (5.61) and (5.62) generalized to include the discrete particle effect would yield the electric field as in the following:

$$\vec{\vec{M}} \cdot \delta\vec{E} = \begin{pmatrix} S - n_z^2 & -iD & n_x n_z \\ iD & S - n^2 & 0 \\ n_x n_z & 0 & P - n_x^2 \end{pmatrix} \cdot \begin{pmatrix} \delta E_x \\ \delta E_y \\ \delta E_z \end{pmatrix} = \frac{4\pi i}{\omega}\delta\vec{j}(\omega, \vec{k}), \quad (8.93)$$

where the fluctuating current $\delta\vec{j}(\vec{x}, t) = \sum_j e\vec{w}_j(t)\delta(\vec{x} - \vec{x}_j(t)) - \vec{J}(\vec{x})$ arises from summing up the contributions from individual particles less the ensemble averaged current $\vec{J}(\vec{x}) = \langle \sum_j e\vec{w}_j(t)\delta(\vec{x} - \vec{x}_j(t))\rangle$. The particle trajectories can be treated by the unperturbed orbitals. By utilizing the Faraday's law $\delta\vec{B} = \vec{n} \times \delta\vec{E}$, the magnetic field can be found and so does its autocorrelation. Here $\vec{\vec{M}}$ is taken from the dielectric response of the cold plasma (cf. Eq. (5.62)). It may be generalized to the hot plasma response as in Eq. (5.69), or even the kinetic description and beyond such as in the relativistic regime.

Further Reading

Kinetic theory can be learned from Montgomery's book on "Theory of the Unmagnetized Plasma" in depth and without the overburdened algebra.

Krommes (2002) presented a systematic description of the statistical theories of plasma turbulence.

Ion species are generally considered to be more consistent with the neoclassical theory. A review article by Hirshman and Sigmar (1981) is a good reference. Viscosity has been identified as an important mechanism for momentum/rotation confinement in tokamak. It is almost impossible to keep track of the numerous works on this topic. A glimpse into the latest work, check into Shaing *et al.* 2013. Application of Fokker Planck equation to electron dynamics in fluctuating magnetic and ambipolar electric fields can be found in Harvey *et al.* (1981).

Chu (1982) applied the concept of magnetic temperature to analyze the experimental data. The most probable states in magnetohydrodynamics were proposed in Montgomery *et al.* (1979) to maximize the entropy.

Montgomery and Joyce (1974) first revealed the interesting negative temperature state for the 2d guiding center plasma model that was envisioned by Taylor (1971) to justify the Bohm diffusion (1949).

Homework Hints

Homework 8.1: Show that the diffusion and friction of Eqs. (8.12) and (8.13) in the collisional term of Eq. (8.11) conserve the number of particles, the momentum and the energy.

The collisional operator can be recognized to conserve the number of particles since

$$\dot{N} \equiv \int C(f) d^3\vec{v} = \int d^3\vec{v}\frac{\partial}{\partial\vec{v}} \cdot \left(\overline{\overline{D}} \cdot \frac{\partial}{\partial\vec{v}} f\right) = \left(\widehat{n} \cdot \overline{\overline{D}} \cdot \frac{\partial}{\partial\vec{v}} f\right)\Bigg|_{\vec{v}\to\infty} = 0,$$

when evaluating at the outer boundary of the velocity space, where $\widehat{n}$ is the outward unit vector.

The momentum integral can be rewritten as,

$$\frac{1}{m}\dot{\vec{M}} \equiv \int \vec{v} C(f) d^3\vec{v} = \int d^3\vec{v}\frac{\partial}{\partial\vec{v}} \cdot \left((\overline{\overline{D}} \cdot \frac{\partial}{\partial\vec{v}} f)\vec{v}\right) - \int d^3\vec{v}\,\overline{\overline{D}} \cdot \frac{\partial}{\partial\vec{v}} f(\vec{v})$$

$$= -\int d^3\vec{v}\,\overline{\overline{D}} \cdot \frac{\partial}{\partial\vec{v}} f(\vec{v}) = -\nu \int d^3\vec{v} \int d^3\vec{v}'(\nabla_v\nabla_v|\vec{v} - \vec{v}'|) f(\vec{v}') \cdot \frac{\partial}{\partial\vec{v}} f(\vec{v})$$

$$= \nu \int d^3\vec{v}(\nabla_v\nabla_v \int f(\vec{v}') d^3\vec{v}' \nabla_v|\vec{v} - \vec{v}'|) f(\vec{v})$$

$$= -\nu \int d^3\vec{v}(\nabla_v \nabla_v \int f(\vec{v}')d^3\vec{v}'\nabla_{v'}|\vec{v} - \vec{v}'|)f(\vec{v})$$

$$= \nu \int d^3\vec{v} \int d^3\vec{v}' \left(\nabla_v \nabla_v |\vec{v} - \vec{v}'|\right) f(\vec{v}) \frac{\partial}{\partial \vec{v}'} f(\vec{v}')$$

$$= \frac{\nu}{2} \int d^3\vec{v} \int d^3\vec{v}' \left(\nabla_v \nabla_v |\vec{v} - \vec{v}'|\right) \left(f(\vec{v}) \frac{\partial}{\partial \vec{v}'} f(\vec{v}') - f(\vec{v}') \frac{\partial}{\partial \vec{v}} f(\vec{v}) \right)$$

Thus, we may switch $\vec{v}$ and $\vec{v}'$ and find that $\dot{\vec{M}} = -\dot{\vec{M}}$ and the integral does vanish. Note that $\nabla_v |\vec{v} - \vec{v}'| = -\nabla_{v'} |\vec{v} - \vec{v}'|$, and $\nabla_v \nabla_v |\vec{v} - \vec{v}'| = \nabla_{v'} \nabla_{v'} |\vec{v} - \vec{v}'|$.

Similar to the momentum conservation, it is easier to prove the energy conservation by the symmetrical structure in the collision operator, which can be rewritten as (cf. Eq. (8.45a)),

$$C(f) \rightarrow \frac{\nu}{2} \frac{\partial}{\partial \vec{v}} \cdot \int d^3\vec{v}' \frac{\vec{\vec{I}} - \hat{e}_{\|}\hat{e}_{\|}}{|\vec{v} - \vec{v}'|} \left(f(\vec{v}') \frac{\partial}{\partial \vec{v}} f(\vec{v}) - f(\vec{v}) \frac{\partial}{\partial \vec{v}'} f(\vec{v}') \right)$$

The energy integral $\dot{E} \equiv \frac{1}{2} \int m\vec{v}^2 C(f) d^3\vec{v}$ gives,

$$\dot{E} = m\frac{\nu}{2} \int d^3\vec{v}\vec{v} \cdot \int d^3\vec{v}' \frac{\vec{\vec{I}} - \hat{e}_{\|}\hat{e}_{\|}}{|\vec{v} - \vec{v}'|} \left(f(\vec{v}') \frac{\partial}{\partial \vec{v}} f(\vec{v}) - f(\vec{v}) \frac{\partial}{\partial \vec{v}'} f(\vec{v}') \right)$$

Thus, we may switch $\vec{v}$ and $\vec{v}'$ to find that

$$\dot{E} = m\frac{\nu}{2} \int d^3\vec{v}'\vec{v}' \cdot \int d^3\vec{v} \frac{\vec{\vec{I}} - \hat{e}_{\|}\hat{e}_{\|}}{|\vec{v} - \vec{v}'|} \left(f(\vec{v}) \frac{\partial}{\partial \vec{v}'} f(\vec{v}') - f(\vec{v}') \frac{\partial}{\partial \vec{v}} f(\vec{v}) \right)$$

$$= m\frac{\nu}{4} \int d^3\vec{v} \int d^3\vec{v}(\vec{v} - \vec{v}') \cdot \frac{\vec{\vec{I}} - \hat{e}_{\|}\hat{e}_{\|}}{|\vec{v} - \vec{v}'|} \left(f(\vec{v}) \frac{\partial}{\partial \vec{v}'} f(\vec{v}') - f(\vec{v}') \frac{\partial}{\partial \vec{v}} f(\vec{v}) \right)$$

$$= m\frac{\nu}{4} \int d^3\vec{v} \int d^3\vec{v}\hat{e}_{\|} \cdot (\vec{\vec{I}} - \hat{e}_{\|}\hat{e}_{\|}) \left(f(\vec{v}) \frac{\partial}{\partial \vec{v}'} f(\vec{v}') - f(\vec{v}') \frac{\partial}{\partial \vec{v}} f(\vec{v}) \right) = 0$$

Homework 8.3: Find the asymptotic expansions for the plasma dispersion function

$$Z(\varsigma) = \frac{1}{\sqrt{\pi}} \int_{-\infty}^{\infty} dx \frac{\exp(-x^2)}{x - \varsigma},$$

and compare with the numerical results given in the program above.

The plasma dispersion function has the asymptotic expansions for $|\varsigma| \ll 1$,

$$Z(\varsigma) = \frac{e^{-\zeta^2}}{\sqrt{\pi}} \int_{-\infty}^{\infty} ds \frac{e^{-s^2-2s\zeta}}{s}$$
$$\approx \frac{-2}{\sqrt{\pi}} \zeta e^{-\zeta^2} \int_{-\infty}^{\infty} ds \left(1 + \frac{2}{3} s^2 \zeta^2 + \frac{2}{15} s^4 \zeta^4 \ldots\right) e^{-s^2}$$
$$\approx -2\zeta e^{-\zeta^2} \left(1 + \frac{1}{3}\zeta^2 + \frac{1}{10}\zeta^4\right), \tag{8.3.1}$$

and for $|\varsigma| \gg 1$,

$$Z(\varsigma) \approx -\frac{1}{\sqrt{\pi}} \frac{1}{\zeta} \int_{-\infty}^{\infty} dx e^{-x^2} \left(1 - \frac{x}{\zeta} + \frac{x^2}{\zeta^2} - \frac{x^3}{\zeta^3} + \frac{x^4}{\zeta^4} - \cdots\right)$$
$$\approx -\frac{1}{\zeta} \left(1 + \frac{1}{2} \frac{1}{\zeta^2} + \frac{3}{4} \frac{1}{\zeta^4} + \cdots\right). \tag{8.3.2}$$

These formulae are calculated at the divide of 1.5. It is interesting to note that outside the region $2 > |\varsigma| > 1$, the consistency with the numerical result is rather acceptable.

> **Homework 8.5:** Show that in the long wavelength limit, the dispersion function from the Vlasov equation reduces to the following:
>
> $$\varepsilon(\vec{k}, -i\omega) = 1 - 2\frac{\omega_p^2}{k^2} \frac{k_\perp^2}{\omega^2 - \Omega^2} \frac{\omega^2}{\Omega^2} - \frac{\omega_p^2}{k^2} \frac{k_\parallel^2}{\omega^2}.$$

$$I_m(\lambda) \approx \left(\frac{1}{2}\lambda\right)^m / \Gamma(m+1), \quad I_{-m}(\lambda) = I_m(\lambda)$$

$$\varepsilon(k, -i\omega) = 1 + \frac{2\omega_p^2}{k^2 V_T^2} \sum_m e^{-\lambda} I_m(\lambda) \int_{-\infty}^{\infty} \frac{dv_\parallel}{\sqrt{\pi} V_T} e^{-v_\parallel^2/V_T^2} \frac{k_\parallel v_\parallel + m\Omega - \omega + \omega}{k_\parallel v_\parallel + m\Omega - \omega}$$
$$\approx 1 + \frac{1}{k^2 \lambda_D^2} \left(1 + e^{-\lambda} I_0(\lambda) \int_{-\infty}^{\infty} \frac{ds}{\sqrt{\pi}} e^{-s^2} \frac{\omega/k_\parallel V_T}{s - \omega/k_\parallel V_T}\right.$$
$$\left. + e^{-\lambda} I_1(\lambda) \int_{-\infty}^{\infty} \frac{ds}{\sqrt{\pi}} e^{-s^2} \left(\frac{\omega}{\Omega - \omega} + \frac{\omega}{-\Omega - \omega}\right)\right)$$
$$\approx 1 + \frac{1}{k^2 \lambda_D^2} - \frac{1}{k^2 \lambda_D^2} (1 - \lambda) \left(1 + \frac{k_\parallel^2 V_T^2}{2\omega^2} + \frac{3}{4} \frac{k_\parallel^4 v_T^4}{\omega^4}\right)$$
$$- \frac{1}{k^2 \lambda_D^2} e^{-\lambda} I_1(\lambda) \frac{2\omega^2}{\omega^2 - \Omega^2} \approx 1 - 2\frac{\omega_p^2}{k^2} \frac{k_\perp^2}{\omega^2 - \Omega^2} \frac{\omega^2}{\Omega^2} - \frac{\omega_p^2}{k^2} \frac{k_\parallel^2}{\omega^2}$$

$$\lambda \equiv k_\perp^2 V_T^2 / 2\Omega^2, \quad \lambda_D = V_T / \sqrt{2} \omega_p.$$

```
function ComparePlasmaDispersionFunction
zeta=-10:0.1:10;
P=zeta*0;
for i=1:length(zeta);
  P(i)=PlasmaDispersionFunction(zeta(i));
end;
Xc=1.4;
z=-Xc:0.01:Xc;
f=-2*z.*(1+z.^2/3+1/10*z.^4).*exp(-z.^2);
x=Xc:0.01:10;
g=-1./x-1./x.^3/2-3/4./x.^5;
y=-10:0.01:-Xc;
h=-1./y-1./y.^3/2-3/4./y.^5;
plot(zeta,P,'g*',zeta,P,'r -',zeta,zeta*0,'k-',z,f,'b-',x,g,'b-',y,h,'b-');
title('Plasma Dispersion Function');
xlabel('\zeta'); ylabel('Z');
text(-1,1.3,'asymptotic expansion','color','b');
text(0,0.5,'-2\zeta(1+\zeta^2/3+\zeta^4/10)exp(-\zeta^2)','color','b');
text(2.5,-0.5,'-1/\zeta(1+1/\zeta^2/2+3/\zeta^4/4)','color','b');
```

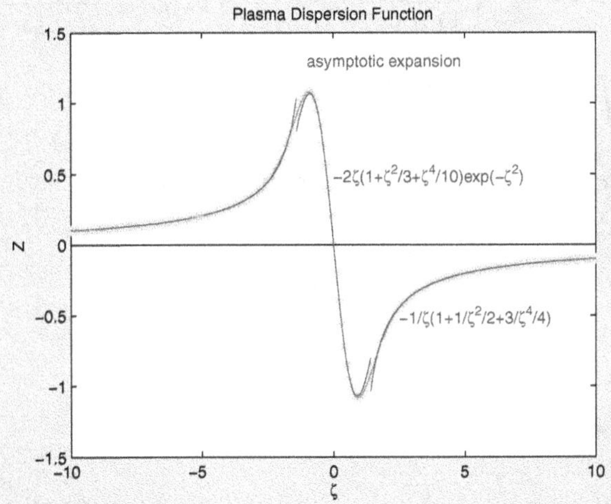

Homework 8.7: Work out the collision operator $C(f) = -(q/m)(\partial/\partial\vec{v}) \cdot \langle\delta\vec{E}\delta f\rangle$ for the magnetized plasma.

We will assume the equilibrium distribution function is azimuthally symmetric,

$$\frac{\partial}{\partial t}\delta f_k(\vec{v},t) + i\vec{k}\cdot\vec{v}\delta f_k(\vec{v},t) - \Omega\frac{\partial}{\partial\varphi}\delta f_k(\vec{v},t)$$

$$-\frac{q}{m}\delta\phi_k(t)i\left(k_{\|}\frac{\partial}{\partial v_{\|}}f + k_{\perp}v_{\perp}\cos\varphi\frac{\partial f}{\partial v_{\perp}^2/2}\right) = 0. \tag{8.46}$$

Taking the Laplace transform in t gives,

$$\frac{s+ik_{\|}v_{\|}+ik_{\perp}v_{\perp}\cos\varphi}{\Omega}\delta f_k(\vec{v},t)-\frac{\partial}{\partial\varphi}\delta f_k(\vec{v},t)$$
$$-\frac{q}{m}\delta\phi_k(t)i\left(\frac{k_{\|}}{\Omega}\frac{\partial}{\partial v_{\|}}f+\frac{k_{\perp}v_{\perp}}{\Omega}\cos\varphi\frac{\partial f}{\partial v_{\perp}^2/2}\right)=\frac{\delta f_k(0)}{\Omega} \tag{8.7.1}$$

Replacing with $G(\varphi)=(s\varphi+ik_{\|}v_{\|}\varphi+ik_{\perp}v_{\perp}\sin\varphi)/\Omega$ gives,

$$e^{G(\varphi)}\frac{\partial}{\partial\varphi}e^{-G(\varphi)}\delta f_k(\vec{v},t)=-\frac{iq}{m}\delta\phi_k(t)\left(\frac{k_{\|}}{\Omega}\frac{\partial}{\partial v_{\|}}f+\frac{k_{\perp}v_{\perp}}{\Omega}\cos\varphi\frac{\partial f}{\partial v_{\perp}^2/2}\right)-\frac{\delta f_k(0)}{\Omega}. \tag{8.7.2}$$

After the φ integration, we arrive at the Eq. (8.49a), without assuming the Maxwellian distribution for the equilibrium,

$$\delta\bar{f}_k e^{-G}=-\frac{iq\delta\bar{\phi}_k}{m}\sum_n\left(\frac{k_{\|}}{\Omega}\frac{\partial f}{\partial v_{\|}}+n\frac{\partial f}{\partial v_{\perp}^2/2}\right)J_n(k_{\perp}\rho_{\perp})\frac{e^{-\varphi(s/\Omega+ik_{\|}\rho_{\|}+in)}}{(s/\Omega+ik_{\|}\rho_{\|}+in)}$$
$$-\int^{\varphi}d\varphi' e^{-G(\varphi')}\frac{\delta f_k(0)}{\Omega} \tag{8.49a}$$

The following identities are applied to simplify Eq. (8.49),

$$e^{ix\sin\varphi}=\sum_{n=-\infty}^{\infty}J_n(x)e^{in\varphi}, \tag{8.50}$$

$$\frac{1}{2}[J_{n+1}(x)+J_{n-1}(x)]=\frac{n}{x}J_n(x), \tag{8.51}$$

The initial distribution is given by Eq. (8.32).

$$\delta f_k(\vec{v},0)=\frac{1}{(2\pi)^3 n_0}\sum_j e^{-i\vec{k}\cdot\vec{x}_{j0}}\delta(\vec{v}-\vec{v}_{j0})-f_0(v)\delta(\vec{k}). \tag{8.32}$$

With use of the expansion

$$e^{G(\varphi)}=e^{s\varphi/\Omega+ik_{\|}\rho_{\|}\varphi}\sum_{n=-\infty}^{\infty}J_n\left(\frac{k_{\perp}v_{\perp}}{\Omega}\right)e^{in\varphi},$$

in Eq. (8.49a), we have

$$\delta\bar{f}_k(\vec{v},s)=-\frac{iq\delta\bar{\phi}_k}{m}\sum_n\left(\frac{k_{\|}}{\Omega}\frac{\partial f}{\partial v_{\|}}+n\frac{\partial f}{\partial v_{\perp}^2/2}\right)J_nJ_{n'}\frac{e^{i\varphi(n'-n)}}{(s/\Omega+ik_{\|}\rho_{\|}+in)}$$
$$-\int^{\varphi}d\varphi' e^{-G(\varphi')}\frac{\delta f_k(0)}{\Omega} \tag{8.7.3}$$

The last term on the RHS gives,

$$\int^{\varphi} d\varphi' e^{G(\varphi)-G(\varphi')} \frac{\delta f_k(0)}{\Omega}$$

$$= e^{G(\varphi)} \int^{\varphi} \frac{d\varphi'}{\Omega} \sum_{n=-\infty}^{\infty} J_n \left(\frac{k_\perp v_\perp}{\Omega}\right) e^{-in\varphi'}$$

$$\times \left(\frac{e^{-s\varphi'/\Omega - ik_\parallel v_\parallel \varphi'/\Omega}}{(2\pi)^3 n_0} \sum_j e^{-i\vec{k}\cdot\vec{x}_{j0}} \delta(\vec{v} - \vec{v}_{j0}) - f_0(v)\delta(\vec{k}) \right)$$

$$= \sum_{n,n'=-\infty}^{\infty} J_n \left(\frac{k_\perp v_\perp}{\Omega}\right) J_{n'} \left(\frac{k_\perp v_\perp}{\Omega}\right) \sum_j e^{-i\vec{k}\cdot\vec{x}_{j0}} \frac{\delta(\vec{v} - \vec{v}_{j0})}{(2\pi)^3 n_0}$$

$$\times \left(\frac{e^{i(n-n')\varphi_{j0}}}{-s - ik_\parallel v_\parallel - in\Omega} \right) + \frac{1}{s} f(v)\delta(\vec{k})$$

$$\delta \bar{f}_k(\vec{v}, s)$$

$$= -\int^{\varphi} d\varphi' e^{-G(\varphi')} \frac{\delta f_k(0)}{\Omega} - \frac{iq}{\varepsilon(\vec{k}, s)} \frac{4\pi n_0 q}{mk^2} \int d^3\vec{v} e^{G(\varphi)}$$

$$\times \int^{\varphi} d\varphi' e^{-G(\varphi')} \frac{\delta f_k(0)}{\Omega} \sum_n \left(\frac{k_\parallel}{\Omega} \frac{\partial f}{\partial v_{||}} + n \frac{\partial f}{\partial v_\perp^2/2} \right) J_n J_{n'}$$

$$\times \frac{e^{i\varphi(n'-n)}}{(s/\Omega + ik_\parallel \rho_\parallel + in)} \tag{8.7.4}$$

$$\delta\bar{\phi}_k \varepsilon(k, s) = \frac{4\pi n_0 q}{k^2} \int d^3\vec{v} e^{G(\varphi)} \int^{\varphi} d\varphi' e^{-G(\varphi')} \frac{\delta f_k(0)}{\Omega}$$

$$\delta \bar{f}_k(\vec{v}, t) \xrightarrow[t\to\infty]{} \sum_j \sum_{n,n'=-\infty}^{\infty} J_n J_{n'} e^{i(n-n')\varphi_{j0}} \frac{e^{-i\vec{k}\cdot\vec{x}_{j0} - ik_{||} v_{j0} t - in\Omega t} \delta(\vec{v} - \vec{v}_{j0})}{(2\pi)^3 n_0}$$

$$- f\delta(\vec{k}) - \frac{iq\delta\bar{\phi}_k}{m} \sum_{n,n'} \left(\frac{k_\parallel}{\Omega} \frac{\partial f}{\partial v_{||}} + n \frac{\partial f}{\partial v_\perp^2/2} \right) J_n J_{n'} e^{i\varphi(n'-n)} e^{-in\Omega t - ik_{||} v_{||} t}$$

$$= \sum_j \sum_{n,n'=-\infty}^{\infty} J_n J_{n'} e^{i(n-n')\varphi_{j0}} \frac{e^{-i\vec{k}\cdot\vec{x}_{j0} - ik_{||} v_{j0} t - in\Omega t} \delta(\vec{v} - \vec{v}_{j0})}{(2\pi)^3 n_0}$$

$$- f\delta(\vec{k}) + \frac{q^2}{2\pi^2 k^2 m}$$

$$\sum_{j,n,n',l}\left(\frac{k_{\|}}{\Omega}\frac{\partial f}{\partial v_{\|}}+n\frac{\partial f}{\partial v_{\perp}^{2}/2}\right)\frac{J_nJ_{n'}e^{i\varphi(n'-n)}}{k_{\|}(\vec{v}_{j0}-\vec{v})_{\|}}J_l^2\frac{e^{-i\vec{k}\cdot\vec{x}_{j0}}}{k_{\|}\rho_{\|}+n-k_{\|}\rho_{j0\|}-l}$$
$$\times\left(\frac{e^{-il\Omega t-ik_{\|}v_{\|}t}}{\varepsilon(\vec{k},-ik_{\|}v_{\|}-il\Omega)}-\frac{e^{-il\Omega t-ik_{\|}v_{j0\|}t}}{\varepsilon(\vec{k},-ik_{\|}v_{j0\|}-il\Omega)}\right)$$

$$\delta\bar{f}_k(\vec{v},t)\xrightarrow[t\to\infty]{}\sum_j\frac{e^{-\vec{k}\cdot\vec{x}_{j0}-i\vec{k}\cdot\vec{v}_{j0}t}\delta(\vec{v}-\vec{v}_{j0})}{(2\pi)^3n_0}-f\delta(\vec{k})$$
$$+\frac{q^2\vec{k}\cdot\frac{\partial f}{\partial\vec{v}}}{2\pi^2k^2m}\sum_j\frac{e^{-i\vec{k}\cdot\vec{x}_{j0}}}{\vec{k}\cdot(\vec{v}_{j0}-\vec{v})}\left(\frac{e^{-i\vec{k}\cdot\vec{v}t}}{\varepsilon(\vec{k},-i\vec{k}\cdot\vec{v})}-\frac{e^{-i\vec{k}\cdot\vec{v}_{j0}t}}{\varepsilon(\vec{k},-i\vec{k}\cdot\vec{v}_{j0})}\right)$$

The dielectric function is given by

$$\varepsilon(k,s)=1+\frac{2i\omega_p^2}{k^2V_T^2}\sum_m e^{-\lambda}I_m(\lambda)\int_{-\infty}^{\infty}\frac{dv_{\|}}{\sqrt{\pi}V_T}e^{-v_{\|}^2/V_T^2}\frac{k_{\|}v_{\|}+m\Omega}{s+ik_{\|}v_{\|}+im\Omega}. \tag{5}$$

To evaluate $\langle\delta\vec{E}\delta f\rangle=\int d^3\vec{k}\langle i\vec{k}\delta\phi_{-k}(t)\delta f_k(\vec{v},t)\rangle$, only the same particle has the net contribution.

$$\delta\vec{E}_k(t)\xrightarrow[t\to\infty]{}\sum_{j,n}\frac{-iq\vec{k}}{2\pi^2k^2}J_n^2\left(\frac{k_{\perp}v_{j0\perp}}{\Omega}\right)\frac{e^{-i\vec{k}\cdot\vec{x}_{j0}-ik_{\|}v_{j0\|}t-in\Omega t}}{\varepsilon(\vec{k},-ik_{\|}v_{j0\|}-in\Omega)}. \tag{8.62}$$

Applying

$$\delta\bar{f}_k(\vec{v},t)=\sum_j\sum_{n,n'=-\infty}^{\infty}J_nJ_{n'}e^{i(n-n')\varphi_{j0}}\frac{e^{-i\vec{k}\cdot\vec{x}_{j0}-ik_{\|}v_{j0\|}t-in\Omega t}\delta(\vec{v}-\vec{v}_{j0})}{(2\pi)^3n_0}$$
$$-f\delta(\vec{k})+\frac{q^2}{2\pi^2k^2m}\sum_{j,n,n',l}\left(\frac{k_{\|}}{\Omega}\frac{\partial f}{\partial v_{\|}}+n\frac{\partial f}{\partial v_{\perp}^2/2}\right)$$
$$\times\frac{J_nJ_{n'}e^{i\varphi(n'-n)}J_l^2e^{-i\vec{k}\cdot\vec{x}_{j0}}}{k_{\|}\rho_{j0\|}+l-k_{\|}\rho_{\|}-n}$$
$$\times\left(\frac{e^{-il\Omega t-ik_{\|}v_{\|}t}}{\varepsilon(\vec{k},-ik_{\|}v_{\|}-il\Omega)}-\frac{e^{-il\Omega t-ik_{\|}v_{j0\|}t}}{\varepsilon(\vec{k},-ik_{\|}v_{j0\|}-il\Omega)}\right)$$

we have

$$\langle \delta \vec{E}_{-k}(t)\delta f_k(\vec{v},t)\rangle$$

$$= \Bigg\langle \sum_{j,l'} \frac{iq\vec{k}}{2\pi^2 k^2} J_{l'}^2\left(\frac{k_\perp v_{j0\perp}}{\Omega}\right) \frac{e^{ik_{\|} v_{j0\|}t - il'\Omega t}}{\varepsilon(-\vec{k}, ik_{\|} v_{j0\|} - il'\Omega)}$$

$$\times \Bigg\{ \sum_{n,n'} J_{n'} J_n e^{i(n'-n)\varphi_{j0}} \frac{e^{-ik_{\|} v_{j0\|}t - in\Omega t}\delta(\vec{v} - \vec{v}_{j0})}{(2\pi)^3 n_0}$$

$$+ \frac{q^2}{2\pi^2 k^2 m} \sum_{j,n,n',l} \left(\frac{k_{\|}}{\Omega}\frac{\partial f}{\partial v_{\|}} + n\frac{\partial f}{\partial v_\perp^2/2}\right) \frac{J_n J_{n'} e^{i\varphi(n'-n)} J_l^2}{k_{\|}\rho_{j0\|} + l - k_{\|}\rho_{\|} - n}$$

$$\times \left(\frac{e^{-il\Omega t - ik_{\|} v_{\|} t}}{\varepsilon(\vec{k}, -ik_{\|} v_{\|} - il\Omega)} - \frac{e^{-il\Omega t - ik_{\|} v_{j0\|} t}}{\varepsilon(\vec{k}, -ik_{\|} v_{j0\|} - il\Omega)} \right) \Bigg\} \Bigg\rangle$$

$$= \Bigg\langle \sum_{j,l'} \frac{iq\vec{k}}{2\pi^2 k^2} J_{l'}^2\left(\frac{k_\perp v_{j0\perp}}{\Omega}\right) \frac{e^{ik_{\|} v_{j0\|}t - il'\Omega t}}{\varepsilon(-\vec{k}, ik_{\|} v_{j0\|} - il'\Omega)}$$

$$\times \Bigg\{ \sum_{n} J_n^2 \frac{e^{-ik_{\|} v_{j0\|}t - in\Omega t}\delta(\vec{v} - \vec{v}_{j0})}{(2\pi)^3 n_0}$$

$$+ \frac{q^2}{2\pi^2 k^2 m} \sum_{j,n,l} \left(\frac{k_{\|}}{\Omega}\frac{\partial f}{\partial v_{\|}} + n\frac{\partial f}{\partial v_\perp^2/2}\right) \frac{J_n^2 J_l^2}{k_{\|}\rho_{j0\|} + l - k_{\|}\rho_{\|} - n}$$

$$\times \left(\frac{e^{-il\Omega t - ik_{\|} v_{\|} t}}{\varepsilon(\vec{k}, -ik_{\|} v_{\|} - il\Omega)} - \frac{e^{-il\Omega t - ik_{\|} v_{j0\|} t}}{\varepsilon(\vec{k}, -ik_{\|} v_{j0\|} - il\Omega)} \right) \Bigg\} \Bigg\rangle$$

$$= \sum_{n} \frac{iq\vec{k}}{2\pi^2 k^2} \Bigg\{ J_n^2\left(\frac{k_\perp v_\perp}{\Omega}\right) J_n^2\left(\frac{k_\perp v_\perp}{\Omega}\right) \frac{f(\vec{v})}{\varepsilon(-\vec{k}, ik_{\|} v_{\|} + in\Omega)}$$

$$+ \frac{4\pi n_0 q^2}{k^2 m} \int d^3\vec{v}' f(\vec{v}') \sum_{j,l} J_l^2\left(\frac{k_\perp v'_\perp}{\Omega}\right) \frac{1}{\varepsilon(-\vec{k}, ik_{\|} v'_{\|} + il\Omega)}$$

$$\times \Bigg\{ \left(\frac{k_{\|}}{\Omega}\frac{\partial f}{\partial v_{\|}} + n\frac{\partial f}{\partial v_\perp^2/2}\right) \frac{J_n^2 J_l^2}{k_{\|}\rho'_{\|} + l - k_{\|}\rho_{\|} - n}$$

$$\times \left(\frac{e^{ik_{\|}(v_{\|} - v'_{\|})t}}{\varepsilon(\vec{k}, -ik_{\|} v_{\|} - il\Omega)} - \frac{1}{\varepsilon(\vec{k}, -ik_{\|} v'_{\|} - il\Omega)} \right) \Bigg\}$$

$$\delta\vec{E}_k(t) \xrightarrow[t\to\infty]{} \sum_j \frac{-iq\vec{k}}{2\pi^2k^2}\frac{e^{-i\vec{k}\cdot\vec{x}_{j0}-i\vec{k}\cdot\vec{v}_{j0}t}}{\varepsilon(\vec{k}, -i\vec{k}\cdot\vec{v}_{j0})}, \tag{8.37}$$

$$\delta\bar{f}_k(\vec{v},t) \xrightarrow[t\to\infty]{} \sum_j \frac{e^{-\vec{k}\cdot\vec{x}_{j0}-i\vec{k}\cdot\vec{v}_{j0}t}\delta(\vec{v}-\vec{v}_{j0})}{(2\pi)^3n_0} - f\delta(\vec{k})$$

$$+\frac{q^2\vec{k}\cdot\frac{\partial f}{\partial\vec{v}}}{2\pi^2k^2m}\sum_j \frac{e^{-i\vec{k}\cdot\vec{x}_{j0}}}{\vec{k}\cdot(\vec{v}_{j0}-\vec{v})}\left(\frac{e^{-i\vec{k}\cdot\vec{v}t}}{\varepsilon(\vec{k},-i\vec{k}\cdot\vec{v})} - \frac{e^{-i\vec{k}\cdot\vec{v}_{j0}t}}{\varepsilon(\vec{k},-i\vec{k}\cdot\vec{v}_{j0})}\right)$$

$$\langle\delta\vec{E}\delta f\rangle = \int \frac{iq\vec{k}}{2\pi^2k^2}\frac{d^3\vec{k}}{\varepsilon(-\vec{k},i\vec{k}\cdot\vec{v})}$$

$$\times\left(f(\vec{v}) + \frac{4\pi n_0q^2}{k^2m}\int d^3\vec{v}'\frac{f(\vec{v}')}{\vec{k}\cdot(\vec{v}'-\vec{v})}\right.$$

$$\left.\times\left[\frac{e^{i\vec{k}\cdot(\vec{v}'-\vec{v})t}}{\varepsilon(\vec{k},-i\vec{k}\cdot\vec{v})} - \frac{1}{\varepsilon(\vec{k},-i\vec{k}\cdot\vec{v}')}\right]\vec{k}\cdot\frac{\partial f}{\partial\vec{v}}\right)$$

$$= \int \frac{q\vec{k}}{2\pi^2k^2}d^3\vec{k}\left(f(\vec{v})\frac{i\varepsilon*(-\vec{k},i\vec{k}\cdot\vec{v})}{|\varepsilon(-\vec{k},i\vec{k}\cdot\vec{v})|^2}\right.$$

$$\left.-\pi\frac{\omega_p^2}{k^2}\int d^3\vec{v}'\frac{f(\vec{v}')\delta(\vec{k}\cdot(\vec{v}'-\vec{v}))}{|\varepsilon(\vec{k},-i\vec{k}\cdot\vec{v}')|^2}\vec{k}\cdot\frac{\partial f}{\partial\vec{v}}\right) \tag{8.44}$$

where we have set $\vec{v} = \vec{v}'$ in the dielectric function of the second term due to the singular nature of the pole and the oscillatory $e^{i\vec{k}\cdot(\vec{v}'-\vec{v})t}$ otherwise. We have also taken

$$(1-e^{-i\vec{k}\cdot(\vec{v}'-\vec{v})t})/(\vec{k}\cdot(\vec{v}'-\vec{v})) \xrightarrow{t\to\infty} \pi i\delta(\vec{k}\cdot(\vec{v}'-\vec{v})) = (\pi i/k)\delta(\hat{k}\cdot(\vec{v}'-\vec{v})).$$

Only the imaginary part of the dielectric function $\varepsilon*(-\vec{k},i\vec{k}\cdot\vec{v})$ in the first term survives the k integral that amounts to a factor $-\pi i(\omega_p^2/k^2)\int d^3\vec{v}'\delta(\hat{k}\cdot(\vec{v}'-\vec{v}))\hat{k}\cdot\partial f(\vec{v}')/\partial\vec{v}'$. Defining $F(\hat{k}\cdot\vec{v}) \equiv \int d^3\vec{v}' f(\vec{v}')\delta(\hat{k}\cdot(\vec{v}'-\vec{v}))$, we end with

$$C(f) = \frac{\partial}{\partial\vec{v}}\cdot\int\frac{2n_0q^4}{m^2k^4}\frac{\vec{k}\vec{k}d^3\vec{k}}{|\varepsilon(\vec{k},-i\vec{k}\cdot\vec{v})|^2}$$

$$\cdot\left(F(\hat{k}\cdot\vec{v})\frac{\partial}{\partial\vec{v}}f(\vec{v}) - f(\vec{v})\frac{\partial}{\partial\vec{v}}F(\hat{k}\cdot\vec{v})\right) \tag{8.45}$$

Resorting to the low frequency dielectric response, namely, $\varepsilon(k,0) = 1 + 1/k^2\lambda_D^2$, with the Debye screening, we get a symmetrical Fokker Planck collision operator,

$$C(f) = \frac{\nu}{2}\frac{\partial}{\partial\vec{v}} \cdot \int d^3\vec{v}' \frac{\overline{\overline{I}} - \widehat{e}_{||}\widehat{e}_{||}}{|\vec{v} - \vec{v}'|} \cdot \left(f(\vec{v}')\frac{\partial}{\partial\vec{v}} f(\vec{v}) - f(\vec{v})\frac{\partial}{\partial\vec{v}'} f(\vec{v}') \right) \tag{8.45a}$$

Practically, we need the collisions among different species. That can be obtained by spelling out the correct momentum exchange effect,

$$C(f_a) = \frac{2\pi n_0 q_a^2 q_b^2}{m_a} \log\Lambda \frac{\partial}{\partial\vec{v}} \cdot \int d^3\vec{v}' \frac{\overline{\overline{I}} - \widehat{e}_{||}\widehat{e}_{||}}{|\vec{v} - \vec{v}'|} \cdot \left(f_b(\vec{v}')\frac{\partial}{m_a\partial\vec{v}} f_a(\vec{v}) - f_a(\vec{v})\frac{\partial}{m_b\partial\vec{v}'} f_b(\vec{v}') \right) \tag{8.45b}$$

Therefore,

$$\begin{aligned}
\overline{\overline{D}}_v &= \frac{q^2}{m^2}\left\langle \delta\vec{E}(\vec{x},t) \int_{-\infty}^{t} d\tau\, \delta\vec{E}(\vec{x}(\tau),\tau) \right\rangle \\
&= \frac{q^2}{m^2}\int d^3\vec{k} \sum_{nl} e^{-i\Omega(n-l)t} \left\langle \frac{2n_0 q^2 \vec{k}\vec{k}}{\pi k^4 \varepsilon_n \varepsilon_l^*} J_n^2\left(\frac{k_\perp v_{0\perp}}{\Omega}\right) J_l^2\left(\frac{k_\perp v_{0\perp}}{\Omega}\right) \right\rangle I_l(\lambda)e^{-\lambda} \\
&\xrightarrow[t\to\infty]{} \frac{q^2}{m^2}\int d^3\vec{k} \sum_{l} I_l(\lambda)e^{-\lambda} \left\langle \frac{2n_0 q^2 \vec{k}\vec{k}}{\pi k^4 |\varepsilon_l|^2} J_l^4\left(\frac{k_\perp v_{0\perp}}{\Omega}\right) \right\rangle \\
\vec{\mu}_v &= \nabla_{\vec{v}} \cdot \overline{\overline{D}}_v = -2\frac{\vec{v} - \vec{v}'}{|\vec{v} - \vec{v}|^3} \Rightarrow \nabla_{\vec{v}} \cdot \vec{\mu}_v = -8\pi\delta(\vec{v} - \vec{v}'),
\end{aligned}$$

Homework 8.9: Show that by taking $f_{e1} = g(\vec{v}_\perp)f_M$ and neglecting the ion velocity, the linearized collision operator can be simplified to,

$$C(f_{e0}, f_{i1}) + C(f_{e1}, f_{i0}) \approx \frac{\nu_c}{m_e v^3} f_M \left(\frac{\partial}{\partial\mu}(1-\mu^2)\frac{\partial}{\partial\mu} + \frac{1}{1-\mu^2}\frac{\partial^2}{\partial\phi^2} \right) g,$$

where $\nu_c = (2\pi n_0 q_e^2 q_i^2/m_e)\log\Lambda$.

since

$$C(f_{e1}, f_{i0}) \approx \nu_{ei}\frac{\partial}{\partial\vec{v}} \cdot \int d^3\vec{v}' \frac{\overline{\overline{I}} - \widehat{e}_{||}\widehat{e}_{||}}{|\vec{v} - \vec{v}'|} \cdot \left(f_{i'0}(\vec{v}')\frac{\partial f_{e1}(\vec{v})}{m_e\partial\vec{v}} - f_{e1}(\vec{v})\frac{\partial f_{i0}(\vec{v}')}{m_i\partial\vec{v}'} \right)$$

$$= \nu_{ei}\frac{\partial}{\partial \vec{v}} \cdot \int d^3\vec{v}'\frac{\vec{\vec{I}} - \hat{e}_{||}\hat{e}_{||}}{|\vec{v}|} \cdot \left(f_{i'0}(\vec{v}')\frac{\partial f_{e1}(\vec{v})}{m_e\partial \vec{v}}\right)$$

$$= \frac{\nu_{ei}}{m_e}\frac{\partial}{\partial \vec{v}} \cdot \frac{\vec{\vec{I}} - \hat{e}_{||}\hat{e}_{||}}{|\vec{v}|} \cdot \frac{\partial f_{e1}(\vec{v})}{\partial \vec{v}}$$

it leads to

$$C(f_{e0}, f_{i1}) + C(f_{e1}, f_{i0}) \approx \frac{\nu_c}{m_e}\frac{\partial}{\partial \vec{v}} \cdot \left(\frac{\vec{\vec{I}} - \hat{e}_{||}\hat{e}_{||}}{v} \cdot \frac{\partial f_{e1}(\vec{v})}{\partial \vec{v}}\right)$$

$$= \frac{\nu_c}{m_e} f_M \frac{\partial}{\partial \vec{v}} \cdot \left(\frac{\vec{\vec{I}} - \hat{e}_{||}\hat{e}_{||}}{v} \cdot \frac{\partial g}{\partial \vec{v}}\right)$$

$$= \frac{\nu_c}{m_e v^3} f_M \left(\frac{\partial}{\partial \mu}(1-\mu^2)\frac{\partial}{\partial \mu} + \frac{1}{1-\mu^2}\frac{\partial^2}{\partial \phi^2}\right) g.$$

Homework 8.11: Derive the diffusion coefficient by directly solving the following simple collision model equation

$$\vec{v} \cdot \frac{\nabla n_0}{n_0} f_{e0} - \Omega_e \frac{\partial}{\partial \phi} f_{e1} = -\nu_{ei} f_{e1}.$$

The model equation leads to

$$v_\perp \cos\phi \frac{\partial n_0/\partial x}{n_0} f_{e0} = \Omega_e \frac{\partial}{\partial \phi} f_{e1} - \nu_{ei} f_{e1} = \Omega_e e^{\nu_{ei}\phi/\Omega_e}\frac{\partial}{\partial \phi}(e^{-\nu_{ei}\phi/\Omega_e} f_{e1}). \tag{8.11.1}$$

Therefore,

$$f_{e1} = \frac{v_\perp}{\Omega_e} e^{\nu_{ei}\phi/\Omega_e} \int^{\phi} d\phi' e^{-\nu_{ei}\phi'/\Omega_e} \cos\phi' \frac{\partial n_0/\partial x}{n_0} f_{e0}$$

$$= \frac{v_\perp}{\Omega_e}\frac{\partial n_0/\partial x}{n_0} f_{e0}\frac{1}{2}\left(\frac{e^{i\phi}}{i - \frac{\nu_{ei}}{\Omega_e}} + \frac{e^{-i\phi}}{-i - \frac{\nu_{ei}}{\Omega_e}}\right)$$

$$= v_\perp \frac{\partial n_0/\partial x}{n_0} f_{e0} \frac{-\nu_{ei}\cos\phi + \Omega_e \sin\phi}{\Omega_e^2 + \nu_{ei}^2}. \tag{8.11.2}$$

The particle flux is then given by

$$\Gamma_x = n_0 \int d^3\vec{v}\, v_x f_{e1} = \int dv_{||} v_\perp dv_\perp d\phi v_\perp^2 \frac{\partial n_0}{\partial x} \cos\phi f_{e0} \frac{-\nu_{ei}\cos\phi + \Omega_e \sin\phi}{\Omega_e^2 + \nu_{ei}^2}$$

$$= -\int \pi d v_{\|} v_\perp d v_\perp v_\perp^2 \frac{\partial n_0}{\partial x} f_{e0} \frac{\nu_{ei}}{\Omega_e^2 + \nu_{ei}^2}$$

$$= -\frac{V_T^2}{2}\frac{\partial n_0}{\partial x}\frac{\nu_{ei}}{\Omega_e^2 + \nu_{ei}^2} \approx -\frac{V_T^2}{2}\frac{\partial n_0}{\partial x}\frac{\nu_{ei}}{\Omega_e^2}, \tag{8.11.3}$$

and the diffusion coefficient is

$$D = \frac{V_T^2}{2}\frac{\nu_{ei}}{\Omega_e^2} = \frac{V_T^2}{2}\frac{\nu_{ei}}{\Omega_e^2} = \frac{k_B T_e}{m_e}\frac{\nu_{ei}}{\Omega_e^2} = \nu_{ei}\rho_e^2, \quad \nu_{ei} = \frac{2\pi n_0 q_e^2 q_i^2}{m_e}\log\Lambda.$$

```
function CentralLimitTheorem
tic, clear all; close all; clc;
RandStream.setDefaultStream(RandStream('mt19937ar','seed',sum(1
00*clock)));
N=400000; Nt=10000;
xn=0*(1:N); X=xn;
for it=1:Nt
   f=0.1;
   Rn=rand(1,N)-0.5;
   xn=xn+f*Rn;
end;
X=xn; Width=9;
maxX=max(X),
X=Width*X/maxX;
mu=sum(X)/N,
sigma=sqrt(sum(X.^2)/N-mu^2),
iX=fix(abs(X)*10);
iXmin=min(iX);
iX=iX-iXmin+1;
iXmax=max(iX),
f=0*(1:iXmax);
f or i=1:N
   f(iX(i))=f(iX(i))+1;
end;
dx=0.1;
x=0:dx:Width;
save CLT x f N sigma;
figure(1);
plot(x,f(1:length(x))/N/2/dx,'g*',x,exp(-
x.^2/2/sigma^2)/sqrt(2*pi)/sigma,'r-');
title('The Central Limit Theorem');
toc,
```

Homework 8.13: The Central Limit Theorem: Any sum of independent and identically-distributed random variables tends to yield a normal distribution with a finite variance. Define $X = \langle x_i \rangle \equiv \frac{1}{N}\sum_{i=1}^{N} x_i$, where x_i follows a probability distribution $p(x_i)$ of any kind, and $N \gg 1$. Then X follows the distribution

$$P(X) = \frac{1}{\sqrt{2\pi}\,\sigma} \exp\{-(X-\mu)^2/2\sigma^2\},$$

where $\mu = \langle X \rangle$ is the mean and $\sigma = \sqrt{\langle X^2 \rangle - \langle X \rangle^2}$ the variance. Devise a numerical scheme as follows: Let $\frac{1}{2} \geq R_n \geq -\frac{1}{2}$ be a random number, and take a spatial step of $\delta x_n = f R_n$ with $1 > f > 0$. Repeat the process for a large number of steps and compare the final distribution of X with $P(X)$.

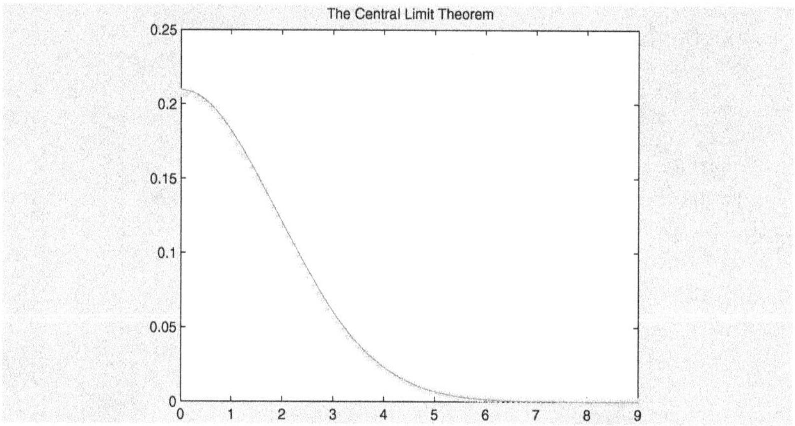

Homework 8.15: Show that $\beta_M = \partial S/\partial U$, if the system follows the Gibb's distribution, $P = N\exp(-\beta_M \varepsilon_M)$.

Given $P = N\exp(-\beta_M\varepsilon_M)$, we have $S = -\int d\tau \int d\Gamma P \ln P = \int d\tau \int d\Gamma P$ $(\beta_M \varepsilon_M - \ln N) = \beta_M U - \ln N$. Thus,

$$\partial S/\partial U = \beta_M - \partial \ln N/\partial U = \beta_M.$$

Since the normalization constant N is not function of the internal energy, $\partial N/\partial U = 0$.

Chapter Nine

Plasma Transport

"If scientific reasoning were limited to the logical processes of arithmetic, we should not get very far in our understanding of the physical world."

Vannevar Bush (1890–1974)

The importance of transport goes beyond the magnetic and inertia confinement in fusion application. The solar flare, for example, involves drastic activities of magnetic fields, energetic particles, and vortex flows to result in solar wind, solar prominence, sun spots, sun quake, and coronal mass ejection. Its impact on earth has been termed the space weather capable of affecting communication satellites and sensitive electronic devices, or even crippling the power grid. Its magnificent showing of aurora is often a northern delight.

Transport can be caused by collisions, large amplitude waves, turbulence, and major disruptions. Particle diffusion, $D = \langle (\Delta x)^2/\Delta t \rangle$, as described in Chap. 1 by taking the ensemble average of test particles subject to the pitch angle scattering, is the least loss mechanism for the confined plasma. Moreover, the binary collision of pitch angle scattering is an oversimplified collision model in plasma where the long range high coherence plasma fluctuations can be more effective in causing the transport. Plasma turbulence is also expected to enhance the transport which is often termed as the anomalous transport. Large scale MHD instabilities can eject substantial plasma, and lose its entirety if major disruption occurs. However the severe disruptions will be categorized as MHD activities for it is fast and furious. Transport would customarily refer to the loss mechanisms for confined plasma that maintains an equilibrium state.

The classical transport due to collisions is sensitive to the magnetic geometry and is further termed as the neoclassical transport. By contrast, the anomalous transport often refers to anything beyond the classical or neoclassical transport. However, even in thermal equilibrium plasma, transport may not be characterized by the classical scaling. The Fokker Planck collision model retains only the binary collisions, the least of the actions from the two-time electric-field autocorrelation. Its relevance or precision cannot always be presumed, as it lacks many important physical effects: The magnetic field, a zeroth order quantity in evaluating transport coefficients (cf. Sec. 8.7), is often ignored in the collision operator; Except the Debye screening in the dielectric function, particle response disregards the electric field or magnetic field fluctuation spectrum; Only the scattering in the velocity space, not the phase space, is included; The particle-wave interactions also are often neglected in its entirety.

This chapter will consider transport effects for the relatively stable plasma with two distinct space and time scales. A variable f whose ensemble averaged quantity is denoted as $\langle f\rangle$ with the bracket $\langle\rangle$, varies on the slow time and long length scale, while its fluctuation, denoted as Δf, on the fast time and short length scale. Moreover, $\langle f\rangle$ is much greater than Δf so that the plasma is at most weakly turbulent, and an asymptotic expansion is readily applicable.

9.1 Basics in Plasma Transport

Several related topics in transport are first discussed here.

9.1.1 *Kubo Formula*

The diffusion coefficient is often conveniently evaluated in terms of Kubo's formula, $D = \int_0^\infty d\tau \langle v(0)v(\tau)\rangle$. We may start from the definition of the diffusion coefficient given by $D = \langle(\Delta x)^2/\Delta t\rangle$ to find,

$$
\begin{aligned}
D &= \frac{1}{\Delta t}\int_0^{\Delta t} dt_1 \int_0^{\Delta t} dt_2 \langle v(t_1)v(t_2)\rangle \\
&= \frac{1}{2}\frac{1}{\Delta t}\int_{-\Delta t}^{\Delta t} d\tau \int_0^{\Delta t} dT \left\langle v\left(\frac{1}{2}(T-\tau)\right) v\left(\frac{1}{2}(T+\tau)\right)\right\rangle \\
&= \frac{1}{2}\frac{1}{\Delta t}\int_{-\Delta t}^{\Delta t} d\tau \int_0^{\Delta t} dT \langle v(0))v(\tau)\rangle \\
&= \frac{1}{2}\int_{-\Delta t}^{\Delta t} d\tau \langle v(0))v(\tau)\rangle = \int_0^{\Delta t} d\tau \langle v(0))v(\tau)\rangle.
\end{aligned}
\tag{9.1}
$$

By passing $\Delta t \to \infty$, it gives the Kubo formula. The basic assumption behind the formula is that the two time ensemble $\langle v(t_1)v(t_2)\rangle = \langle v(\tau)v(\tau + t_2 - t_1)\rangle = \langle v(\tau)v(0)\rangle = \langle v(-\tau)v(0)\rangle$ depends only on the time difference $\tau = t_2 - t_1$, irrespective of its mean $\frac{1}{2}T = \frac{1}{2}(t_2 + t_1)$ or the time order. To ensure the physical process of diffusion, irreversibility is intrinsically built in by truncating the correlation to low orders, and to ensure the integral over the infinite time domain exists, temporally damping autocorrelation is implicitly assumed, that rarely but surely requires a **renormalization** by taking into account the damping by diffusion itself if no other damping mechanism is at work.

9.1.2 *Einstein Relation*

The particle response to an arbitrary force $\vec{f}$ can be expressed as a flow $\vec{v} = \mu \vec{f}$ and a corresponding particle flux $\vec{j} = n\vec{v} = n\mu\vec{f}$, where μ is the **particle mobility**. For a conservative force field $\vec{f} = -\nabla\phi$, it would have the equilibrium particle distribution in accordance with the Boltzmann function $n = n_0 \exp(-\phi/k_B T)$. The particle flux under a fluctuating force f results in the **Markovian process** of diffusion flux given by $\vec{j} = D\nabla n = -nD\nabla\phi/k_B T = n\vec{f}D/k_B T$. If we equate this diffusion flux with the particle flux due to the flow, that would yield the **Einstein relation** $D = \mu k_B T$. Furthermore, since the electrical current $j_E = nqv$ generated under the electric field Ecan be expressed as $j_E = nq\mu E = \sigma E$ he electrical conductivity can be expressed as $\sigma = nq\mu$. Thus, we arrive also at $D = \sigma k_B T/nq$, another expression for Einstein relation.

9.1.3 *The Classical B^{-1} and Bohm B^{-2} Transport Scaling*

How the plasma transport is reduced by the strength of the external magnetic field is of great concern in studying magnetic confinement of fusion plasma. In the uniform magnetic field, a particle under collision will execute random walks across the field lines by the step size of gyroradius $\rho \equiv v_T/\Omega$, where v_T is its thermal velocity, $\Omega \equiv qB/mc$ the gyrofrequency. The diffusion rate defined by $D \equiv (\Delta x)^2/\Delta t$ is given by $\nu_c \rho^2$, with the rather favorable B^{-2} scaling law, where ν_c is the collision frequency. On the other hand, in a collisionless plasma, the low-frequency long-wavelength spectrum can cause particles the ExB drift. The wave coherence time can be so extremely long as to cause virtually free streaming across the field lines. Thus, the transport would be the only mechanism to limit the run of its own course and to result in a self correction by quenching the coherent transport through the diffusive damping. To quantify these statements, we may write down the diffusive damping time as $\tau_D \equiv 1/k_\perp^2 D$, where $k_\perp$ is the wave number perpendicular to the magnetic field. Therefore, the step size is $c\delta E\tau_D/B$, and the diffusion coefficient is $D \sim (\Delta x)^2/\tau_D \sim c^2\delta E^2/B^2 k_\perp^2 D \sim c\delta E/Bk_\perp$. It clearly yields for the diffusion a scaling law of B^{-1} for the 2d plasma. The thermal fluctuation is typically a small portion of the particle thermal energy. It is reduced by the plasma parameter $\varepsilon_p \equiv (n_0\lambda_D^3)^{-1} \ll 1$, and is given by $|\delta E|^2 \sim \frac{1}{\sqrt{\pi}}\varepsilon_p n_0 k_B T \sim 4\sqrt{\pi} n_0 q^2 \lambda_D^{-1}$ (cf. Eq. (8.39)), where n_0 is the plasma density, λ_D the Debye length, and T the plasma temperature. Taking $1/k_\perp \sim \lambda_D$ and substituting the electric field by the thermal energy, we have $D \sim (2cq\pi^{1/4}/B)\sqrt{\lambda_D n_0} \sim \sqrt{\varepsilon_p} ck_B T/qB/2\pi^{3/4}$, that has the B^{-1} scaling law, as Bohm *et al.* conjectured in 1949 from observing the lossy transport in plasma machines in the early days.

9.1.4 *The Hsu $B^{-3/2}$ Diffusion in 3d Plasma*

It is unrealistic to assume a 2d plasma with only ExB drifts and no parallel decoherence. A 3d collisionless plasma has a decoherence time characterized by the particle streaming along the field lines, namely, $\tau_c \sim 1/k_{||}v_T$. Replacing the diffusive damping time by the parallel decoherence time, the B^{-1} scaling disappears and the B^{-2} scaling prevails (Vahala 1974), while the magnitude is very much enhanced by the factor $(\lambda_D/\rho)^2$ (O'Neil 1985) due to the ExB drifts. It is interesting to note that the transport scaling tends to indicate neither B^{-1} nor B^{-2} in the magnetic confinement of stellarator or tokamak, but lies somewhat in between. In fact, there are cyclotron harmonics that can cause resonance diffusion in the velocity space leading to gyroradius enlargement and particle diffusion. The diffusion process arises from the electric field autocorrelation complete with two electric field components. This provides an effective diffusion mechanism not entirely due to the ExB drift, but the dc harmonic in one component and the cyclotron harmonic in the other. As the cyclotron harmonic is in tune with the particle gyration, it is effectively stationary as seen by the particles, but weakened by the finite Larmor radius (FLR) effect, i.e., $I_1(\lambda)e^{-\lambda} \sim \lambda \equiv \frac{1}{2}k_\perp^2\rho^2 \ll 1$ in the thermal fluctuation spectrum (cf. Eq. (8.57b)). If we order the parallel decoherence and the perpendicular diffusive damping on the same time scale, namely, $\Omega \gg k_\perp^2 D \sim k_{||}v_T \gg \nu_c$, which in fact is the correct ordering to obtain the greatest diffusion value, it results in a diffusion $D = \langle \Delta v \Delta \tau \Delta a \Delta \tau \rangle \sim (c\delta E_\perp/B)(k_\perp^2 D)^{-1}(k_\perp^2\rho^2 q\delta E_\perp/m)(k_{||}v_T)^{-1}$, that has the $B^{-3/2}$ scaling by applying $\delta E^2 \sim \frac{1}{\sqrt{\pi}}\varepsilon_p n_0 k_B T$, as predicted by Hsu *et al.* in 2013.

9.2 Transport Effects in Fluid Equations

The description of particle and fluid duality as formulated in Chap. 4 has the potential of resolving the deficiencies in the fluid model by recuperating particle discreteness. It can provide a paradigm that may save computational effort by mixing the fluid and particle simulations, and may alleviate pitfalls of nonexistence, nonuniqueness or unphysical solutions by including the missing physical effects in the fluid model. This advantage is demonstrated by deriving all the transport terms in the fluid equations in the following.

The fluid description as discussed in Chap. 4 is generalized here to include the transport effects that are caused by the plasma fluctuations due to particle discreteness. The formulation is valid for weak turbulence inasmuch as the fluctuating quantities are smaller compared to the equilibrium quantities and truncation to the second order in the fluctuating quantities is justified.

Consider the ith particle having velocity $\vec{w}_i(t) = \vec{v}_i(t) + \Delta\vec{w}_i(t)$ and displacement $\vec{x}_i(t) = \vec{\chi}_i(t) + \Delta\vec{x}_i(t)$that are separated into the ensemble averaged and the perturbed motion. The perturbed fluid response is driven by the fluctuating electric field $\Delta\vec{E}$ due to the particle discreteness such that $\langle\Delta\vec{E}\rangle = 0$, $\langle\Delta\vec{w}_i(t)\rangle = 0$, and $\langle\Delta\vec{x}_i(t)\rangle = 0$. The averaged particle acceleration is due to the external force, $\langle\vec{a}_i(t)\rangle \equiv \vec{F}(\vec{\chi}_i(t), t)/m$, while the fluctuating internal acceleration is given by $\Delta\vec{a}_i(t) = q\Delta E(\vec{x}_i(t), t)/m$.

The delta function can be expanded to the second order for small perturbations as in the following:

$$\begin{aligned}\delta(\vec{x} - \vec{\chi}_i - \Delta\vec{x}_i) &= \delta(\vec{x} - \vec{\chi}_i) - \Delta\vec{x}_i \cdot \nabla\delta(\vec{x} - \vec{\chi}_i)\\ &\quad + \frac{1}{2}\Delta\vec{x}_i\Delta\vec{x}_i : \nabla\nabla\delta(\vec{x} - \vec{\chi}_i) + \cdots .\end{aligned} \tag{9.2}$$

The density evolution with transport effects can then be expressed as

$$\begin{aligned}\frac{\partial n}{\partial t} &= \left\langle \sum_i \frac{\partial}{\partial t}\delta(\vec{x} - \vec{\chi}_i - \Delta\vec{x}_i)\right\rangle \approx -\nabla \cdot \left\langle \sum_i \vec{w}_i\delta(\vec{x} - \vec{\chi}_i)\right\rangle\\ &\quad + \frac{1}{2}\left\langle \sum_i (\Delta\vec{w}_i\Delta\vec{x}_i + \Delta\vec{x}_i\Delta\vec{w}_i) : \nabla\nabla\delta(\vec{x} - \vec{\chi}_i))\right\rangle\\ &\quad - \frac{1}{2}\left\langle \sum_i \Delta\vec{x}_i\Delta\vec{x}_i : \nabla\nabla\nabla \cdot \vec{w}_i\delta(\vec{x} - \vec{\chi}_i)\right\rangle .\end{aligned} \tag{9.3}$$

Therefore, by assuming that the fluctuating electric field is relatively uncorrelated with the particle trajectory as sampled by particles along the unperturbed trajectories, it gives

$$\frac{\partial}{\partial t}n = -\nabla \cdot n\vec{v} + \nabla\nabla : (\vec{\vec{D}}n) - \frac{1}{2}\nabla\nabla : \left(\nabla \cdot (n\vec{v}\vec{\vec{X}})\right), \tag{9.4}$$

with the diffusion coefficient, $\vec{\vec{D}} \equiv \frac{1}{2}\langle\Delta\vec{w}\Delta\vec{x} + \Delta\vec{x}\Delta\vec{w}\rangle$, and a **mixing tensor** $\vec{\vec{X}} \equiv \langle\Delta\vec{x}\Delta\vec{x}\rangle$ on the particle flux.

Homework 9.1: Take time derivative on Eq. (9.2) and sum up all particles to show that the continuity equation with transport effect is given by Eq. (9.4).

Similarly, taking time derivative on the particle flux

$$
\begin{aligned}
(\vec{v}_i + \Delta\vec{w}_i)\delta(\vec{x} - \vec{\chi}_i - \Delta\vec{x}_i) = {} & (\vec{v}_i + \Delta\vec{w}_i)\delta(\vec{x} - \vec{\chi}_i) \\
& - (\vec{v}_i + \Delta\vec{w}_i)\Delta\vec{x}_i \cdot \nabla\delta(\vec{x} - \vec{\chi}_i) \\
& + \frac{1}{2}\vec{v}_i\Delta\vec{x}_i\Delta\vec{x}_i : \nabla\nabla\delta(\vec{x} - \vec{\chi}_i) + \cdots,
\end{aligned}
$$

we have the following equation of momentum evolution,

$$
\begin{aligned}
\frac{\partial}{\partial t}\vec{\Gamma} = {} & \frac{\partial}{\partial t}\sum_i \vec{v}_i\delta(\vec{x} - \vec{\chi}_i) - \nabla \cdot \left\langle \sum_i \Delta\vec{x}_i\Delta\vec{w}_i\frac{\partial}{\partial t}\delta(\vec{x} - \vec{\chi}_i) \right\rangle \\
& - \nabla \cdot \left\langle \sum_i \Delta\vec{w}_i\Delta\vec{w}_i\delta(\vec{x} - \vec{\chi}_i) \right\rangle - \nabla \cdot \left\langle \sum_i \Delta\vec{x}_i\Delta\vec{a}_i\delta(\vec{x} - \vec{\chi}_i) \right\rangle \\
& + \frac{1}{2}\nabla\nabla : \left\langle \sum_i (\Delta\vec{w}_i\Delta\vec{x}_i + \Delta\vec{x}_i\Delta\vec{w}_i)\delta(\vec{x} - \vec{\chi}_i)\vec{v}_i \right\rangle \\
& + \frac{1}{2}\nabla\nabla : \left\langle \sum_i \Delta\vec{x}_i\Delta\vec{x}_i\frac{\partial}{\partial t}(\delta(\vec{x} - \vec{\chi}_i)\vec{v}_i) \right\rangle + \cdots
\end{aligned}
\tag{9.5}
$$

Therefore,

$$
\begin{aligned}
\frac{\partial}{\partial t}\vec{\Gamma} = {} & -\frac{1}{m}\nabla \cdot \overleftrightarrow{K} + \frac{\vec{F}}{m} + \nabla\nabla : (n\vec{v}\overleftrightarrow{D}^T) - \nabla \cdot ((\overleftrightarrow{W} + \overleftrightarrow{A}_x)n) + \nabla\nabla : (\overleftrightarrow{D}n\vec{v}) \\
& - \frac{1}{2}\frac{1}{m}\nabla \cdot (\nabla\nabla : (\overleftrightarrow{X}\overleftrightarrow{K})),
\end{aligned}
\tag{9.6}
$$

where we have applied Eq. (4.17) to replace the first term on the RHS of Eq. (9.5). In Eq. (9.6), $\overleftrightarrow{W} \equiv \langle \Delta\vec{w}\Delta\vec{w} \rangle$ is the **shear stress tensor**, $\overleftrightarrow{A}_x \equiv \langle \Delta\vec{x}\Delta\vec{a} \rangle$ is the **random work**, and $\overleftrightarrow{D}^T \equiv \langle \Delta\vec{x}\Delta\vec{w} \rangle$ is the incomplete diffusion tensor.

Taking the time derivative on the particle energy,

$$
\begin{aligned}
& \frac{1}{2}m(\vec{v}_i + \Delta\vec{w}_i)^2\delta(\vec{x} - \vec{\chi}_i - \Delta\vec{x}_i) \\
& \quad = \frac{1}{2}m\vec{v}_i^2\delta(\vec{x} - \vec{\chi}_i) + m\vec{v}_i \cdot \Delta\vec{w}_i\delta(\vec{x} - \vec{\chi}_i) + \frac{1}{2}m\Delta\vec{w}_i^2\delta(\vec{x} - \vec{\chi}_i) \\
& \qquad - \frac{1}{2}m\vec{v}_i^2\Delta\vec{x}_i \cdot \nabla\delta(\vec{x} - \vec{\chi}_i) - m\vec{v}_i \cdot \Delta\vec{w}_i\Delta\vec{x}_i \cdot \nabla\delta(\vec{x} - \vec{\chi}_i) \\
& \qquad + \frac{1}{4}m\vec{v}_i^2\Delta\vec{x}_i\Delta\vec{x}_i : \nabla\nabla\delta(\vec{x} - \vec{\chi}_i) + \cdots,
\end{aligned}
$$

we have the evolution of the energy equation,

$$\frac{\partial \in}{\partial t} = \frac{\partial}{\partial t}\left\langle \sum_i \frac{1}{2} m \vec{v}_i^2 \delta(\vec{x} - \vec{\chi}_i) \right\rangle + \left\langle \sum_i m \Delta\vec{w}_i \cdot \Delta\vec{a}_i \delta(\vec{x} - \vec{\chi}_i) \right\rangle$$
$$+ \frac{1}{2} \sum_i m \Delta\vec{w}_i^2 \frac{\partial}{\partial t} \delta(\vec{x} - \vec{\chi}_i) - \nabla \cdot \left\langle \sum_i m \Delta\vec{x}_i \Delta\vec{w}_i \frac{\partial}{\partial t}\left(\vec{v}_i \delta(\vec{x} - \vec{\chi}_i)\right) \right\rangle$$
$$- \left\langle \sum_i m \vec{v}_i \cdot (\Delta\vec{a}_i \Delta\vec{x}_i + \Delta\vec{w}_i \Delta\vec{w}_i) \cdot \nabla \delta(\vec{x} - \vec{\chi}_i) \right\rangle$$
$$+ \frac{1}{2} \nabla\nabla : \left\langle \sum_i (\Delta\vec{w}_i \Delta\vec{x}_i + \Delta\vec{x}_i \Delta\vec{w}_i) \frac{1}{2} m \vec{v}_i^2 \delta(\vec{x} - \vec{\chi}_i) \right\rangle$$
$$+ \frac{1}{2} \nabla\nabla : \left\langle \sum_i \Delta\vec{x}_i \Delta\vec{x}_i \frac{\partial}{\partial t}\left(\frac{1}{2} m \vec{v}_i^2 \delta(\vec{x} - \vec{\chi}_i)\right) \right\rangle. \tag{9.7}$$

Therefore,

$$\frac{\partial}{\partial t} \in (\vec{x}, t) = \vec{v} \cdot \vec{F} - \nabla \cdot (\vec{v} \in + \overset{\rightrightarrows}{P} \cdot \vec{v} + \vec{Q}) + trace(\overset{\rightrightarrows}{A}_w) nm$$
$$- \frac{1}{2} \nabla \cdot (trace(\overset{\rightrightarrows}{W}) nm\vec{v}) - \nabla \cdot (\overset{\rightrightarrows}{D}^T \cdot (\vec{F} - \nabla \cdot \overset{\rightrightarrows}{K}))$$
$$- \nabla \cdot ((\overset{\rightrightarrows}{A}_x + \overset{\rightrightarrows}{W}) \cdot nm\vec{v}) + \nabla\nabla : (\overset{\rightrightarrows}{D} \in) + \frac{1}{2} \nabla\nabla : (\overset{\rightrightarrows}{X} \vec{v} \cdot \vec{F})$$
$$- \frac{1}{2} \nabla \cdot \nabla\nabla : (\overset{\rightrightarrows}{X} (\vec{v} \in + \overset{\rightrightarrows}{P} \cdot \vec{v} + \vec{Q})), \tag{9.8}$$

where $\overset{\rightrightarrows}{A}_w \equiv \langle \Delta\vec{a} \Delta\vec{w} \rangle$ is the **velocity diffusion**, and Eq. (4.19) has been applied.

We now have the complete set of transport coefficients: the mixing tensor $\overset{\rightrightarrows}{X} \equiv \langle \Delta\vec{x} \Delta\vec{x} \rangle$, the diffusion coefficient $\overset{\rightrightarrows}{D} \equiv \frac{1}{2}\langle \Delta\vec{w} \Delta\vec{x} + \Delta\vec{x} \Delta\vec{w} \rangle$, the random work $\overset{\rightrightarrows}{A}_x \equiv \langle \Delta\vec{a} \Delta\vec{x} \rangle$, the shear stress tensor $\overset{\rightrightarrows}{W} \equiv \langle \Delta\vec{w} \Delta\vec{w} \rangle$, and the velocity diffusion $\overset{\rightrightarrows}{A}_w \equiv \langle \Delta\vec{a} \Delta\vec{w} \rangle$. The random work $\overset{\rightrightarrows}{A}_x$ is always accompanied by the shear stress tensor $\overset{\rightrightarrows}{W}$. Both the diffusion coefficient $\overset{\rightrightarrows}{D}$ and the mixing tensor $\overset{\rightrightarrows}{X}$ appear in every equation, and cause particle, momentum, and energy transport. The mixing tensor provides a **mixing length** given by $r_m = \sqrt{trace(\overset{\rightrightarrows}{X})}$ that makes the domain therein indistinguishable and renders the precision within meaningless.

The diffusion coefficient has a different appearance from the Kubo formula. However, taking $\Delta \vec{x}(0) = \int_{-t}^{0} dt' \Delta \vec{w}(t')$, we have

$$\vec{\vec{D}} = \frac{1}{2} \int_{-t}^{0} dt' \langle \Delta \vec{w}(0) \Delta \vec{w}(t') + \Delta \vec{w}(t') \Delta \vec{w}(0) \rangle$$

$$= \frac{1}{2} \int_{-t}^{0} d\tau \langle \Delta \vec{w}(0) \Delta \vec{w}(\tau) + \Delta \vec{w}(\tau) \Delta \vec{w}(0) \rangle \xrightarrow{t \to \infty} \int_{0}^{\infty} \langle \Delta \vec{w}(0) \Delta \vec{w}(\tau) \rangle d\tau.$$

By passing $\Delta t \to \infty$, it gives the Kubo formula. Therefore, they are equivalent provided that the velocity autocorrelation is dependent only on the time difference but independent of the mean time or time order. Note that the correct interpretation for the velocity in the Kubo formula is that it is the perturbed velocities due to the fluctuations.

Assuming that the transport coefficients are much weaker in the spatial variation than the equilibrium quantities, the fluid equations with transport effects are reduced to the following:

(1) the particle transport equation,

$$\frac{\partial}{\partial t} n = -\nabla \cdot (n\vec{v}) + \vec{\vec{D}} : \nabla\nabla n - \frac{1}{2} \vec{\vec{X}} : \nabla\nabla\nabla \cdot (n\vec{v}), \qquad (9.4a)$$

(2) the momentum transport equation,

$$\frac{\partial}{\partial t} n\vec{v} + \nabla \cdot (n\vec{v}\vec{v}) = -\frac{1}{m}\nabla \cdot \vec{\vec{P}} + \frac{1}{m}\vec{F} + (\vec{\vec{D}}^T + \vec{\vec{D}}) \cdot \nabla\nabla \cdot (n\vec{v})$$

$$- (\vec{\vec{A}}_x + \vec{\vec{W}}) \cdot \nabla n - \frac{1}{2}\frac{1}{m}\vec{\vec{X}} : \nabla\nabla\nabla \cdot \vec{\vec{K}}, \qquad (9.6a)$$

(3) the energy transport equation,

$$\frac{\partial}{\partial t} \in (\vec{x}, t) = \vec{v} \cdot \vec{F} - \nabla \cdot (\vec{v} \in + \vec{\vec{P}} \cdot \vec{v} + \vec{Q}) + trace(\vec{\vec{A}}_w)nm$$

$$- \frac{1}{2}mtrace(\vec{\vec{W}})\nabla \cdot (n\vec{v}) - (\vec{\vec{D}}^T \cdot \nabla) \cdot (\vec{F} - \nabla \cdot \vec{\vec{K}})$$

$$- (\vec{\vec{A}}_x + \vec{\vec{W}}) \cdot \nabla \cdot (nm\vec{v}) + \vec{\vec{D}} : \nabla\nabla \in$$

$$+ \frac{1}{2}\vec{\vec{X}} : \nabla\nabla(\vec{v} \cdot \vec{F} - \nabla \cdot (\vec{v} \in + \vec{\vec{P}} \cdot \vec{v} + \vec{Q})). \qquad (9.8a)$$

Contrast to the viscosity as commonly defined in the Navier-Stokes equation (cf. Eq. (1.2)), Eq. (9.6a) would entertain a few more transport effects .

9.3 The Transport Coefficients

The transport coefficients have been derived by *counting the particle population, momentum and energy in the configuration space*, instead of *the phase space* as the kinetic approach would. This simplification reduces the effort in deriving the transport coefficients from the microscopic mechanisms. To examine the transport among the different species and across the gradients we assume the ensemble averaged quantities to vary on the slow time and long length scales, and the fluctuating fields of fast time and short length scales.

Consider the equation of motion for a particle in the electromagnetic force,

$$\frac{d\vec{w}_i}{d\tau} = \frac{q}{m}\left(\delta\vec{E} + \frac{\vec{w}_i \times \vec{B}}{c}\right) = \frac{q}{m}\delta\vec{E} + \vec{w}_i \times \vec{\Omega}. \tag{9.9}$$

Here, the constraint is that the particle has to arrive at the phase point $(\vec{x}_i, \vec{w}_i)$ at $\tau = t$. We may define the time variable $\Im \equiv \tau - t$ so that the constraint for solving the particle trajectories is $\vec{x}_i(\Im) = \vec{x}_i$ and $\vec{w}_i(\Im) = \vec{w}_i$at $\Im = 0$. Imposing a uniform magnetic field but no electric field externally, and integrating Eq. (9.9), we have the evolution of the canonical momentum,

$$\vec{w}_i(\Im) - \vec{x}_i(\Im) \times \vec{\Omega} - \frac{q}{m}\int_{\Im}^{0} d\tau' \delta\vec{E}(\vec{x}_i(\tau'), \tau') = \vec{w}_i - \vec{x}_i \times \vec{\Omega}. \tag{9.10}$$

Separating the velocity and the displacement into the unperturbed as governed by $d\vec{v}_i/d\Im = \vec{v}_i \times \vec{\Omega}$, $d\vec{\chi}_i/d\Im = \vec{v}_i$, and that driven by the electric field fluctuations, $d\Delta\vec{w}_i/d\Im = \Delta\vec{w}_i \times \vec{\Omega} + q\delta\vec{E}/m$, and $\Delta\vec{w}_i(\Im) - \Delta\vec{x}_i(\Im) \times \vec{\Omega} = (q/m)\int_0^{\Im} d\tau \delta\vec{E}(\vec{x}_i(\tau), \tau)$, we find

$$\vec{v}_i(\Im) = \vec{w}_{i\perp}\cos\Omega\Im + (\vec{w}_i \times \widehat{e}_z)\sin\Omega\Im + w_{i\parallel}\widehat{e}_z, \tag{9.11a}$$

$$\vec{\chi}_i(\Im) = \vec{x}_i + \widehat{e}_z \times [\vec{v}_i(\Im) - \vec{w}_i]/\Omega + w_{i\parallel}\Im\widehat{e}_z, \tag{9.11b}$$

$$\begin{aligned}\Delta\vec{w}_i(\Im) = \frac{q}{m}\int_{\Im}^{0} dt[&\delta\vec{E}_\perp(\vec{\chi}_i(t), t)\cos\Omega(\Im - t) \\ &+ (\delta\vec{E}_\perp(\vec{\chi}_i(t), t) \times \widehat{e}_z)\sin\Omega(\Im - t) + \widehat{e}_z\delta E_\parallel(\vec{\chi}_i(t), t)],\end{aligned} \tag{9.11c}$$

$$\begin{aligned}\Delta\vec{x}_i(\Im) = \frac{q}{m}&\left[\int_{\Im}^{0} dt \int_{t}^{0} dt'\widehat{e}_z\delta E_\parallel(\vec{\chi}_i(t'), t') - \int_{\Im}^{0} dt\widehat{e}_z \times \delta\vec{E}_\perp(\vec{\chi}_i(t), t)/\Omega\right] \\ &+ \widehat{e}_z \times \Delta\vec{w}_i(\Im)/\Omega.\end{aligned} \tag{9.11d}$$

The electric field $\delta\vec{E}$ in the time integral is evaluated along the unperturbed particle trajectory. We may express the transport coefficients in terms of the autocorrelation of the electric field $\overleftrightarrow{S}(\tau_1, \tau_2) \equiv \langle \delta\vec{E}(\vec{x}(\tau_1), \tau_1)\delta\vec{E}(\vec{x}(\tau_2), \tau_2)\rangle$. For example, by making use of Eqs. (9.11c) and (9.11d), the diagonal terms of the cross-field diffusion coefficient is then expressed as $\overleftrightarrow{D}_\perp = (qc/2mB)\int_\Im^0 d\tau_1 \int_\Im^0 d\tau_2 S_{xx}(\tau_1, \tau_2)(\hat{e}_x\hat{e}_x + \hat{e}_y\hat{e}_y)(\sin\Omega(\tau_1 - \Im) + \sin\Omega(\tau_2 - \Im))$. It can be transformed to a more convenient form in terms of positive time scales by taking $t_1 \equiv \tau_1 - \Im$, $t_2 \equiv \tau_2 - \Im$, and $t \equiv -\Im$ to give

$$\overleftrightarrow{D}_\perp = \frac{qc}{2mB}\int_0^t dt_1 \int_0^t dt_2 S_{xx}(t_1, t_2)(\hat{e}_x\hat{e}_x + \hat{e}_y\hat{e}_y)(\sin\Omega t_1 + \sin\Omega t_2). \tag{9.12}$$

> **Homework 9.2:** Show that Eq. (9.11) satisfies the equations of motion, namely, $d\vec{v}_i/d\Im = \vec{v}_i \times \vec{\Omega}$, $d\vec{\chi}_i/d\Im = \vec{v}_i$, and $d\Delta\vec{w}_i/d\Im = \Delta\vec{w}_i \times \vec{\Omega} + q\delta\vec{E}/m$, and the boundary conditions: $\vec{x}_i(\Im) = \vec{x}_i$ and $\vec{w}_i(\Im) = \vec{w}_i$ at $\Im = 0$.

Defining the following matrixes:

$$\overleftrightarrow{I}_\perp \equiv \begin{pmatrix} 1 & 0 & 0 \\ 0 & 1 & 0 \\ 0 & 0 & 0 \end{pmatrix}, \quad \overleftrightarrow{I}_z \equiv \begin{pmatrix} 0 & 0 & 0 \\ 0 & 0 & 0 \\ 0 & 0 & 1 \end{pmatrix}, \quad \overleftrightarrow{I}_t \equiv \begin{pmatrix} 0 & -1 & 0 \\ 1 & 0 & 0 \\ 0 & 0 & 0 \end{pmatrix}, \tag{9.13}$$

where $\overleftrightarrow{I}_\perp$ is the 2d cross-field identity matrix, $\overleftrightarrow{I}_z$ is the $\hat{e}_z\hat{e}_z$ component, and $\overleftrightarrow{I}_t$ is the 2d antisymmetric off-diagonal matrix. We will assume that $\overleftrightarrow{S}$ is function of the time difference $|\tau_1 - \tau_2|$ only, and drop the odd terms from the orbital effect, namely the $\sin\Omega(\tau_1 - \tau_2)$ terms. We have the following:

(1) the velocity diffusion

$$\overleftrightarrow{A}_w = \langle \Delta\vec{a}\Delta\vec{w}\rangle = \frac{q^2}{m^2}\int_t^0 dt_1[S_{xx}(|t_1|)(\overleftrightarrow{I}_\perp \cos\Omega t_1 + \overleftrightarrow{I}_t \sin\Omega t_1) + S_{zz}(|t_1|)\overleftrightarrow{I}_z] \tag{9.14}$$

(2) the random work,

$$\overleftrightarrow{A}_x = \langle \Delta\vec{a}\Delta\vec{x}\rangle = \overleftrightarrow{I}_z\frac{q^2}{m^2}\int_0^t dt_2 \int_{t_2}^t dt_1 S_{zz}(|t_1|)$$
$$+\frac{qc}{mB}\int_0^\Im dt_1 S_{xx}(|t_1|)[\overleftrightarrow{I}_\perp \sin\Omega t_1 - \overleftrightarrow{I}_t(1 - \cos\Omega t_1)]. \tag{9.15}$$

(3) the shear stress tensor,

$$\overleftrightarrow{W} = \langle \Delta\vec{w}\Delta\vec{w}\rangle = \frac{q^2}{m^2}\int_t^0 dt_1 \int_t^0 dt_2[S_{xx}(|t_1 - t_2|)\overleftrightarrow{I}_\perp \cos\Omega(t_1 - t_2)$$
$$+\overleftrightarrow{I}_z S_{zz}(|t_1 - t_2|)]. \tag{9.16}$$

(4) the diffusion coefficient,

$$\overleftrightarrow{D} = \frac{1}{2}\langle \Delta\vec{w}\Delta\vec{x} + \Delta\vec{x}\Delta\vec{w}\rangle$$

$$D_{zz} = \frac{q^2}{m^2}\int_0^t dt_1 \int_0^t d\tau \int_0^\tau dt_2 S_{zz}(|t_1 - t_2|)$$

$$\overleftrightarrow{D}_\perp = \frac{qc}{2mB}\int_0^t dt_1 \int_0^t dt_2 S_{xx}(|t_1 - t_2|)$$

$$\times(\overleftrightarrow{I}_\perp(\sin\Omega t_2 + \sin\Omega t_1) + \overleftrightarrow{I}_t(\cos\Omega t_1 + \cos\Omega t_2 - 2\cos\Omega(t_1 - t_2)). \tag{9.17}$$

(5) the mixing tensor,

$$\overleftrightarrow{X} \equiv \langle \Delta\vec{x}\Delta\vec{x}\rangle,$$

$$\overleftrightarrow{X}_{zz} = \frac{q^2}{m^2}\int_0^{\Im} d\tau \int_0^\tau dt_1 \int_0^{\Im} d\tau' \int_0^{\tau'} dt_2 \overleftrightarrow{I}_z S_{zz}(|t_1 - t_2|),$$

$$\overleftrightarrow{X}_\perp = \frac{c^2}{B^2}\int_0^t dt_1 \int_0^t dt_2 S_{xx}(|t_1 - t_2|)$$

$$\times[\overleftrightarrow{I}_\perp(1 - \cos\Omega t_1 - \cos\Omega t_2 + \cos\Omega(t_1 - t_2))]. \tag{9.18}$$

Homework 9.3: Derive equations (9.14) to (9.18).

9.4 The Cross Field Transport

Recall that the magnetized plasma dielectric function is given by

$$\varepsilon(k,s) = 1 + \frac{2i\omega_p^2}{k^2V_T^2}\sum_m e^{-\lambda}I_m(\lambda)\int_{-\infty}^{\infty}\frac{dv_\parallel}{\sqrt{\pi}V_T}e^{-v_\parallel^2/V_T^2}\frac{k_\parallel v_\parallel + m\Omega}{s + ik_\parallel v_\parallel + im\Omega}. \tag{8.55}$$

The two-time auto-correlation of the electric field is then given by

$$\overleftrightarrow{S}(t,\tau) = \int d^3\vec{k}\sum_{n,n',l}\overleftrightarrow{S}_{nn'l}e^{il\Omega(t-\tau)-i\Omega(nt-n'\tau)-\frac{1}{2}k_\parallel^2 v_T^2(\tau-t)^2}, \tag{8.59}$$

where

$$\overleftrightarrow{S}_{nn'l} = \left\langle \frac{2n_0q^2\vec{k}\vec{k}}{\pi k^4\varepsilon_n\varepsilon_{n'}^*}J_n^2\left(\frac{k_\perp v_{0\perp}}{\Omega}\right)J_{n'}^2\left(\frac{k_\perp v_{0\perp}}{\Omega}\right)\right\rangle I_l(\lambda)e^{-\lambda}, \tag{8.57b}$$

The dc component in the strong magnetic field limit has $\vec{\vec{S}}_{000} \approx \langle 2n_0 q^2 \vec{k}\vec{k}/(\pi k^4 |\varepsilon_0|^2)\rangle$ and $\varepsilon_0 \approx 1 + 1/k^2\lambda_D^2$ to the leading order.

9.4.1 *The 2d Diffusion Coefficient*

We are now ready to evaluate the diffusion coefficient. It is noteworthy that by freezing $S_{xx}(t_1, t_2)$ in the cyclotron period, we find

$$D_\perp \equiv (qc/2mB) \int_0^t dt_1 \int_0^t dt_2 S_{xx}(t_1, t_2)(\sin \Omega t_1 + \sin \Omega t_2)$$

$$\approx \lim_{\tau \to \infty} (\frac{c}{B})^2 \int_0^t dt' \langle \delta E_x(\vec{x}, 0) \delta E_x(\vec{x}(t'), t')\rangle,$$

in agreement with the 2d guiding center plasma diffusion due to the ExB drift. Looking into the case of $k_\| = 0$, we recognize that the most important contribution to the diffusion comes from the dc component, since by taking $n = n' = l = 0$ and $\lambda \to 0$, the diffusion coefficient

$$D = \lim_{\tau \to \infty} \frac{qc}{mB} \int_0^\tau dt_1 \int_0^\tau dt_2 \int d^3\vec{k} \frac{n_0 q^2}{\pi k^2 |\varepsilon_0|^2} \sin \Omega t_1, \tag{9.19}$$

becomes unbounded in the t_2 integration. This **secularity** is due to the convection by the ExB drifts that is virtually free streaming. Thus, damping by the diffusive process itself must be taken into consideration, and a renormalization is called for. By adding the factor $\exp(-k^2 D t_2)$, it leads to

$$D = \frac{qc}{mB} \int d^3\vec{k} \frac{n_0 q^2}{\pi k^2 |\varepsilon_0|^2} \frac{1 - e^{-k^2 D\tau}}{k^2 D} \frac{1 - \cos \Omega\tau}{\Omega}$$

$$\xrightarrow{\tau \to \infty} \frac{c^2}{DB^2} \int d^3\vec{k} \frac{n_0 q^2}{\pi |k^2 + k_D^2|^2}. \tag{9.20}$$

Thus, the 2d diffusion coefficient is given by

$$D = \frac{c}{B} \sqrt{\int_0^\infty dk \frac{4k^2 n_0 q^2}{|k^2 + k_D^2|^2}} = \frac{qc}{B} \sqrt{\frac{\pi n_0}{k_D}} = \frac{1}{4\sqrt{\pi}} \frac{ck_B T}{qB\sqrt{n_0 \lambda_D^3}} \to \frac{1}{8\sqrt{\pi}} \sqrt{\varepsilon_p} \frac{V_T^2}{\Omega}. \tag{9.20a}$$

Note that the diffusion coefficient has the scaling law $D \sim T^{1/4} n_0^{1/4} B^{-1}$ that is weakly dependent on both the temperature and the density.

Homework 9.4: Take tokamaks of ITER grade, find the confinement time for the given B^{-1}diffusion coefficient of Eq. (9.20a).

9.4.2 *Particle-Fluid Duality and Simulation of 2d Diffusion*

Since the averaged and smoothed particle density can be described by the fluid equations, and the particle discreteness effect arises from a small fraction of the total density, it may therefore be possible to take advantage of the **particle-fluid duality** by employing a smaller number of particles to simulate the plasma without incurring serious error.

In principle, the fluid equations of motion, Eqs. (9.4a), (9.6a) and (9.8a) can be advanced once the discrete particle effects have been tracked. Following the result of $|\delta \vec{E}|^2 = 4\sqrt{\pi} n_0 q^2/\lambda_D = \frac{1}{\sqrt{\pi}} \varepsilon_p n_0 k_B T$ from Sec. 8.3, where $\varepsilon_p = (n_0 \lambda_D^3)^{-1}$ is the plasma parameter, it implies that $|\delta n/n_0|^2 = \sqrt{\pi}/n_0 \lambda_D^3$. This allows a normalization of the fluctuating density, and the values of $\Delta \vec{a}$, $\Delta \vec{w}$, and $\Delta \vec{x}$ can be evaluated. The corresponding ensemble averaged transport coefficients are taken on the fluid time scale and can be updated on the fly.

We will consider a simple model for a 2d plasma under uniform external magnetic field with the periodic boundary condition in both x and y directions. The fluid velocity is nullified, resulting in the constant fluid density $n_f = n_0 = const.$, that eliminates the fluid evolution in its entirety. By loading a smaller particle density n_s in the simulation, the fluctuating particle density n_x is evaluated. The particle discreteness measured by n_x/n_s is of $O(1/\sqrt{n_s})$, a very high number compared to $n_d/n_f \sim O(\varepsilon_p)$. This will have to be corrected. We take $n_d = n_f \sqrt{\pi^{1/2} \varepsilon_p}/(n_x/n_s)$. While this normalization may not be the best way to treat the particle discreteness since numerical discrepancy does occur, it allows us to find the magnetic field scaling law of the diffusion coefficient as in the following.

The fluctuating electric field is found from n_d by the Poisson solver utilizing the matrix inversion (cf function getM and invMgc2d) as listed below. The particles are then advanced by the ExB drift.

The background plasma is simulated with 4 grids in one Debye length, while the computational size of 20 Debye length long with the periodic boundary condition. A Debye square for simulation has 1600 particles or equivalently, $6.4e^4$ particles in the Debye sphere, which is a far-cry from the $O(10^{10})$ in the real machines. The fluctuation is adjusted in the simulation according to the plasma parameters in the ITER machine. Once particles are advanced, the particle density is recalculated on the grids. The discrete particle effect is again normalized, and the electric field recalculated to advance the particles. The process is repeated for sufficient long time until the diffusion coefficient reaches a plateau value when the diffusive damping makes the process saturate.

Since the particles are advanced with only the ExB drifts, there is no thermal velocity to refer to. However, it is important to ensure the test particle transport

```
function invMgc2d
L=20; % system dimension 20 Debye length
dx=0.25; % grid length
Nl=L/dx, dy=dx;
x=0:dx:L; y=x;
M=getM(Nl,dx,dy);
invM=inv(M);
save invMgc2d invM;

function M=getM(N,dx,dy)
A=sparse(N);
A=A+diag(-2/dx^2-2/dy^2+0*(1:N));
A=A+diag(1/dy^2+0*(1:N-
1),1)+diag(1/dy^2+0*(1:N-1),-1);
B=sparse(N);
B=B+diag(1/dx^2+0*(1:N));
M=sparse(N*N);
for i=1:N
   M(1+(i-1)*N:i*N,1+(i-1)*N:i*N)=A;
   if(i<N)
       M(1+(i-1)*N:i*N,1+i*N:(i+1)*N)=B;
       end;
   if(i>1)
       M(1+(i-1)*N:i*N,1+(i-2)*N:(i-1)*N)=B;
       end;
  end;
M(1:N,1+(N-1)*N:N*N)=B;
M(1+(N-1)*N:N*N,1:N)=B;
```

```
% Find the inversion matrix to solve the Poisson's equation.
% The submatrixes A and B are constructed to load up the Poisson operator, the M matrix.
% M is then inverted and the result is saved in the mat file invMgc2d.

% B makes the Poisson operator periodic.
```

is compared on the unbiased Initialization. The plasma is prepared with the 'thermalized distribution', which is obtained from the long time evolution of randomly loaded initial particle positions. The energy conservation is maintained at an error of 10% or so through the entire simulation. Transport is convective at the beginning of the simulation and the diffusion coefficient increases linearly with time. It takes the diffusive damping to make the diffusion process saturate at later time that depends on the magnitude of the diffusion coefficient itself. Therefore, the time duration for the simulation varies with different B values, and is adjusted accordingly to save computational time.

The diffusion coefficient is calculated for the external magnetic field ranging from 3 to 9 tesla. Its dependency on the magnetic field is shown in the accompanying figure to obey the 1/B scaling law. The magnitude however differs by more than an order of magnitude from that obtained in Eq. (9.20a). The issue may be easily rectified by the normalization coefficient if a better formula can be argued.

```
function gc2d
close all; clc; clear all;
global dt Nl L Np xp yp Xt Yt Vt
global Theta B Tc Xtest Ytest;
T=10000;
n0=1.0e14;
wpe=8.98e3*sqrt(n0),
vth=4.19e7*sqrt(T),
Debye=vth/wpe,
epsilon=1/n0/Debye^3,
L=20; dx=0.25;
Nd=1600; % particles per grid square
Nl=L/dx; dy=dx;
x=dx:dx:L; y=x;
X=repmat(x,Nl,1); Y=repmat(y',1,Nl);
RO=0*X+1; Diffusion=[]; bField=[];
% Initialize the plasma
Np=Nd*Nl*Nl;
runTime=[5,6,7,9,10,12.4,14]*1000;
for B=3:9 %Tesla
omega=1.76e7*1.0e4*B; B,
OMEGA=omega/wpe,
rand('state',sum(100*clock));
xp=rand(Np,1)*L; yp=rand(Np,1)*L;
load 2dgcS9 xp yp;
% initialize the test particles
Ntest=100;
Xt=L/2+L/8*(rand(Ntest,1)-0.5);
Yt=L/2+L/8*(rand(Ntest,1)-0.5);
X0=Xt; Y0=Yt;
dt=0.5; Tc=1000; Time=runtime(B-2);
time=[0]; distance=[0];
Xtest=[]; Ytest=[]; endTime=0;
ROp=getDensity(xp,yp,L,Nl,dx,dy);
ns=sum(sum(ROp,1),2)/Nl^2,
n3=sqrt(sum(sum((ROp/ns-1).^2,1),2))*Nl/(L)^3,
calibration=sqrt(epsilon*sqrt(pi))/n3,
nd=(ROp-ns)*calibration;
[Ex,Ey]=getEfield(nd,Nl,dx,dy);
EXt=0*(1:Ntest)'; EYt=0*(1:Ntest)';
xp0=xp; yp0=yp;
vxp=xp*0;  vyp=yp*0;
tic, datestr(now),
Exp=xp*0; Eyp=xp*0;
xDensity=(1:L)*0; yDensity=(1:L)*0;
tic,
```

```
% a 2d gc particle-fluid duality code to investigate the 1/B diffusion
% plasma temperature in eV
% particle density per cc
% plasma frequency
% thermal velocity
% Debye length
% the plasma parameter
%magnetic field in terms of plasma frequency
% system dimension 10 Debye length
% 1600 particles per unit grid

%uniform density normalized to n0

% load particles in the thermalized distribution

% initialize the test particles
```

```
for it=1:Time
   [Exp,Eyp]=getE(Ex,Ey,xp,yp,dx,dy,Nl,L);
   vxp=Eyp/OMEGA;
   vyp=-Exp/OMEGA;
   xp=xp+vxp*dt; yp=yp+vyp*dt;
   xp=Boxin(xp,L); yp=Boxin(yp,L);
   [EXt,EYt]=getE(Ex,Ey,Xt,Yt,dx,dy,Nl,L);
   VXt=EYt/OMEGA;
   VYt=-EXt/OMEGA;
   Xt=Xt+VXt*dt; Yt=Yt+VYt*dt;
   Xt=Boxin(Xt,L); Yt=Boxin(Yt,L);
   Xtest=[Xtest,Xt]; Ytest=[Ytest,Yt];
   if(mod(it,Tc/5)==0) it, toc, tic,
      eEnergy=sum(sum(Ex.^2+Ey.^2,1),2)/L^2,
      d=sqrt(sum((Xt-X0).^2+(Yt-Y0).^2))/Ntest,
      D=getD(Debye,vth);
      time=[time,endTime+it*dt]; distance=[distance,D];
      figure(1); axis equal; hold on;
      plot(X0,Y0,'rx',Xt,Yt,'g.');
      s=sprintf('Test Particles @ time= %d  B=%d Tesla',it,B);
      title(s);
      getframe;
      if(mod(it,Tc)==0)
         density=sum(nd,1); datestr(now);
         figure(2); hold off;
         plot(time/wpe,distance);
 s=sprintf('Diffusion as function of time for B=%d Tesla',B);
         title(s); xlabel('second'); ylabel('cm^2/sec');
         getframe; datestr(now);
save 2dgc dt Nl L Np xp yp Xt Yt B Tc Nd Xtest Ytest time distance;
         figure(3); surf(ROp);
      end;
   end;
   ROp=getDensity(xp,yp,L,Nl,dx,dy);
   nd=ns*(ROp/ns-1)*calibration;
   [Ex,Ey]=getEfield(nd,Nl,dx,dy);
end;
toc, datestr(now), hold off;
D=getD(Debye,vth);
bField=[bField,B], Diffusion=[Diffusion,D],
FileName=sprintf('2dgc%d.mat',B);
save(FileName,'dt','Nl','L','Np','xp','yp','Xt','Yt','B','Tc','Nd','Xtest','Ytest',
'time','distance');
end;
save dc bField Diffusion;
PlotDiffusion(bField,Diffusion);
```

```
% count
density
% fluid density
in simulation
% convert to
3d density
% calibrate
according to
the fluctuation-
dissipation
theorem
%uniform
density
normalized to
n0

% The electric
field is
normalized to
the time scale
of plasma
frequency and
the Debye
length scale.

B=3 tesla
it =      5000
eEnergy =  4.5448e-04
D =  1.7346e+02

B=4 tesla
it =      6000
eEnergy =  4.4722e-04
D =1.3561e+02

B=5 tesla
it =      7000
eEnergy =  4.4514e-04
D =    1.2016e+02

B=6 tesla
it =  9000
eEnergy =  4.5549e-04
```

```
function xp=Boxin(xp,L)
n=floor(xp/L);
xp=xp-n*L;

function [ex,ey]=getE(Ex,Ey,xp,yp,dx,dy,Nl,L)
Ixp=fix(xp/L*Nl); Iyp=fix(yp/L*Nl);
dxp=xp-Ixp*dx; dyp=yp-Iyp*dy;
Ixp=Ixp+1; Iyp=Iyp+1;
ex=0*xp; ey=ex;
for i=1:length(xp)
   ix=Ixp(i); iy=Iyp(i);
   ixp=ix+1; iyp=iy+1;
   if(ixp>Nl) ixp=1; end; if(iyp>Nl) iyp=1; end;
   dx=dxp(i); dy=dyp(i);
   ex(i)=Ex(ix,iy)*(1-dx)*(1-dy)+Ex(ixp,iy)*dx*(1-dy);
   ex(i)=ex(i)+Ex(ix,iyp)*dy*(1-dx)+ Ex(ixp,iyp)*dx*dy;
   ey(i)=Ey(ix,iy)*(1-dx)*(1-dx)+Ey(ixp,iy)*dx*(1-dy);
    ey(i)=ey(i)+Ey(ix,iyp)*dy*(1-dx)+ Ey(ixp,iyp)*dx*dy;
end;
clear Ixp Iyp dxp dyp ;

function [Ex,Ey]=getEfield(nd,N,dx,dy)
load invMgc2d invM;
PHI=invM*reshape(nd,N*N,1);
PHI=reshape(PHI,N,N);
Ex=getDx(PHI,N,dx);
Ey=getDy(PHI,N,dy);
clear invM PHI;
```

```
D = 93.5563

B=7 tesla
it =    10000
eEnergy =  4.4503e-04
D =  78.4726

B=8 tesla
it = 12400
eEnergy =  4.5091e-04
D =  75.8767.

B=9 tesla
it =    14000
eEnergy =  4.5033e-04
D =  65.3027
```

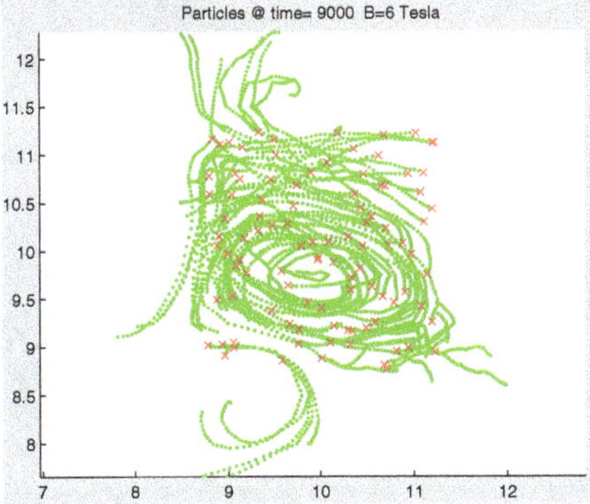

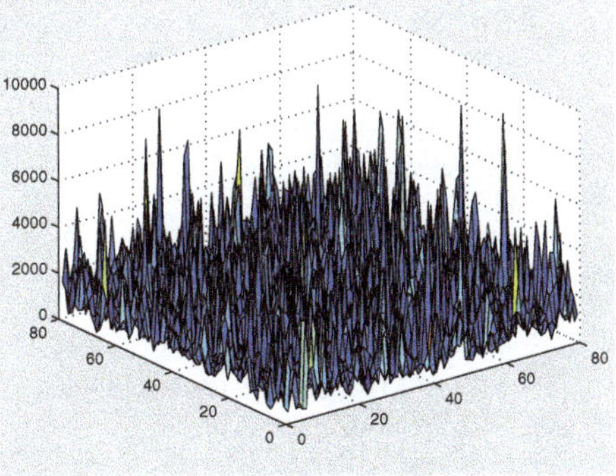

```
function ROp=getDensity(xp,yp,L,Nl,dx,dy)                    % test particle
ROp=zeros(Nl);                                                 trajectories
Ixp=fix(xp/L*Nl); Iyp=fix(yp/L*Nl);
dxp=xp-Ixp*dx; dyp=yp-Iyp*dy;
Ixp=Ixp+1; Iyp=Iyp+1;Np=length(xp);
for i=1:Np
   ix=Ixp(i); iy=Iyp(i);
   if(ix<0) ix=Nl; end; if(iy<0) iyp=Nl; end;
   ixp=ix+1; iyp=iy+1;
   if(ixp>Nl) ixp=1; end; if(iyp>Nl) iyp=1; end;
   ROp(ix,iy)=ROp(ix,iy)+(1-dxp(i))*(1-dyp(i));
   ROp(ixp,iyp)=ROp(ixp,iyp)+dxp(i)*dyp(i);
   ROp(ix,iyp)=ROp(ix,iyp)+(1-dxp(i))*dyp(i);
   ROp(ixp,iy)=ROp(ixp,iy)+dxp(i)*(1-dyp(i));
end;

function dX=getDx(x,N,dx)
dX=([x(2:N,:);x(1,:)]-[x(N,:);x(1:N-1,:)])/dx/2;
                                                             % background particle
function dY=getDy(y,N,dy)                                      distribution
dY=([y(:,2:N),y(:,1)]-[y(:,N),y(:,1:N-1)])/dy/2;

function d2X=getD2x(x,N,dx)
d2X=([x(2:N,:);x(1,:)]-2*x+[x(N,:);x(1:N-1,:)])/dx/dx;

function d2Y=getD2y(y,N,dy)
d2Y=([y(:,2:N),y(:,1)]-2*y+[y(:,N),y(:,1:N-1)])/dy/dy;

function Diffusion=getDiffusion
global dt Nl L Np xp yp Xt Yt OMEGA Tc Nd Xtest Ytest
T=10000;  n0=1.0e14;
wpe=8.98e3*sqrt(n0);  OMEGA,
omega=OMEGA*wpe/1.76e7/10^4,
vth=4.19e7*sqrt(T);
Debye=vth/wpe;
epsilon=1/n0/Debye^3;                                        % evaluate the diffusion
[m,n]=size(Xtest);                                             coefficient
x0=Xtest(:,1); y0=Ytest(:,1);
Diffusion=[]; Time=[];
for index=fix(n/2):n
   xend=Xtest(:,index);  yend=Xtest(:,index);
   dx=xend-x0;   dy=yend-y0;
   D=sum(dx.^2+dy.^2);
   D=D/m/index/dt/Tc;
   Diffusion=[Diffusion,D]; Time=[Time,index*dt];
end;
figure(2);
plot(Time,Diffusion);
D=sum(Diffusion)/length(Diffusion),
Diffusion=D*vth^2/wpe,
```

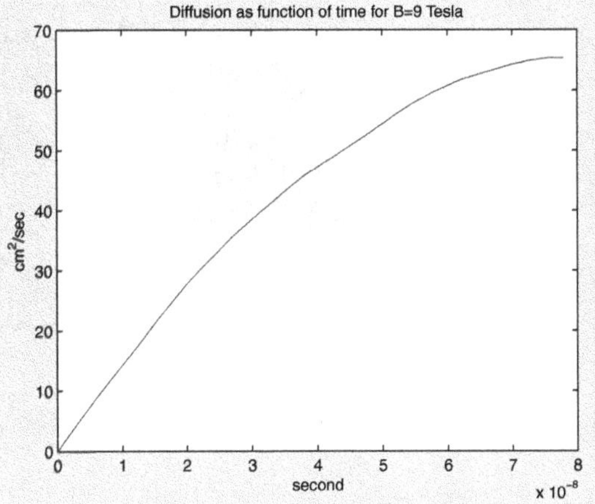


```
function PlotDiffusion(B,D)   % the  diffusion coefficient of 2d guiding center
plasma

N=length(B);
n=fix(length(N)/2+1);
b=B(1):0.1:B(N);
d=D(n)*B(n)./b;
plot(B,D,'g*',b,d,'r-');
xlabel('B (tesla)');
ylabel('D (cm^2/sec)');
title('2d gc diffusion');
text(6, 100,'1/B','color','r','fontsize',14);
```

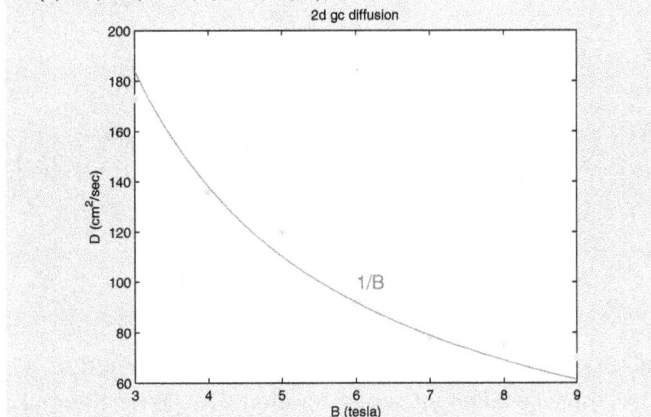

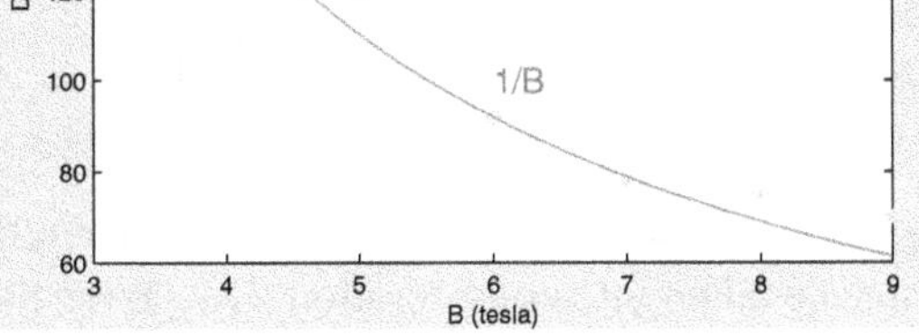

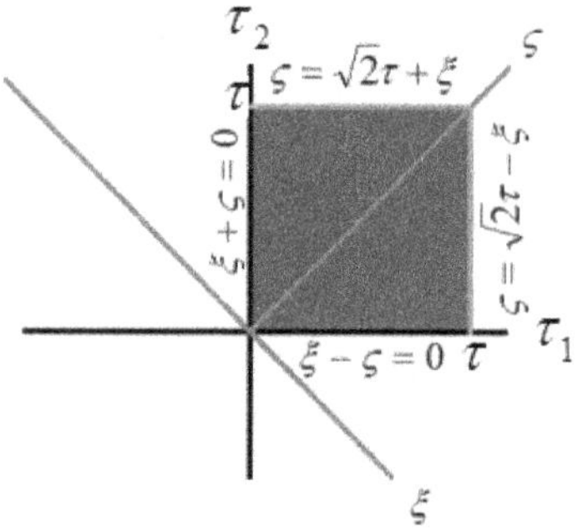

Fig. 9.1. Transformation from (t_1, t_2) to (ξ, ς).

9.4.3 *The 3d Diffusion Coefficient*

In the 3d case, while the major contribution still comes from the long wavelength modes, the cyclotron resonance can no longer be neglected. Exchanging t_1 and t_2 leaves D invariant as switching the indexes of n and n' and negating l makes it unchanged. Several observations are readily available. Only low integers of n, n', and l are important since the FLR effect in $S_{nn'l}$ makes them higher order in magnetic field dependence otherwise. Changing variables to $\xi = \Omega(t_1 - t_2)$ and $\varsigma = \Omega(t_1 + t_2)$, we define $T = \Omega\tau$ and convert the integration from $\Omega^2 \int_0^\tau dt_1 \int_0^\tau dt_2$ to $\int_{-\frac{T}{\sqrt{2}}}^{0} d\xi \int_{-\xi}^{\sqrt{2}T+\xi} d\varsigma + \int_0^{\frac{T}{\sqrt{2}}} d\xi \int_{\xi}^{\sqrt{2}T-\xi} d\varsigma$ (cf. Fig. 9.1).

Thus, setting $l = 0$, we have

$$D = \lim_{T\to\infty} \frac{2qc}{mB} \mathrm{Im} \int d^3\vec{k} \sum_{n,n'} \frac{S_{nn'0}}{\Omega^2} \times \left[\left(\int_{-\frac{T}{\sqrt{2}}}^{0} d\xi \int_{-\xi}^{\sqrt{2}T+\xi} d\varsigma + \int_0^{\frac{T}{\sqrt{2}}} d\xi \int_{\xi}^{\sqrt{2}T-\xi} d\varsigma \right) \times \left(e^{-i\frac{n+n'}{2}\xi - \frac{1}{2}k_{\|}^2\rho^2\xi^2} e^{-i\frac{n-n'}{2}\varsigma} e^{\frac{1}{2}i(\varsigma+\xi)}\right) \right], \tag{9.21}$$

that has a secularity in the ς integration when $n = n' + 1$. Therefore, damping by the diffusive process must be taken into consideration. Due to the thermal decorrelation $e^{-\frac{1}{2}k_{\|}^2\rho^2\xi^2}$, the major contribution comes from small values of ξ. We may neglect ξ relative to T, and $e^{-k_\perp^2 D\xi/\Omega}$ relative to $e^{-in'\xi - \frac{1}{2}k_{\|}^2\rho^2\xi^2}$ to arrive at

$$D \approx \lim_{\tau\to\infty} \frac{qc}{2mB} \int d^3\vec{k} \mathrm{Im} \sum_{n'} S_{n'+1,n',0} \int_{-\infty}^{\infty} e^{-in'\xi - \frac{1}{2}k_{\|}^2\rho^2\xi^2} \frac{d\xi}{\Omega} \frac{1}{k_\perp^2 D}. \tag{9.22}$$

With use of the identity, $\int_{-\infty}^{\infty} dx \exp(\pm iax - b^2x^2/2) = \sqrt{2\pi} e^{-a^2/2b^2}/b$, it is clear that the $n' = 0$ harmonic dominates. We find

$$D \approx \lim_{\tau\to\infty} \frac{qc}{mB}\sqrt{\frac{\pi}{2}} \int d^3\vec{k}\mathrm{Im}S_{100}\frac{1}{|k_{\parallel}|v_T}\frac{1}{k_{\perp}^2 D}, \tag{9.23}$$

where $S_{100} \approx \langle (n_0 q^2/\pi k^2 \varepsilon_1 \varepsilon_0^*)(k_\perp v_{0\perp}/\Omega)^2 \rangle$. Taking

$$\varepsilon_0 \approx 1 + k_D^2/k^2 + i\sqrt{\pi}(k_D^2/k^2)|v_{0\parallel}/v_T|e^{-|v_{0\parallel}/v_T|^2}$$

and $\varepsilon_1 \approx 1$ we have

$$\mathrm{Im}(1/\varepsilon_0^*) = Im\left(1 + \frac{k_D^2}{k^2} - i\frac{k_D^2}{2k^2}\right)^{-1} \approx \frac{k_D^2}{2k^2}\left(1 + \frac{k_D^2}{k^2}\right)^{-2}.$$

Hence, keeping only lowest order terms in the FLR expansion and using $\langle v_{0\perp}^2 \rangle = v_T^2$, one obtains

$$D \approx \frac{c^3}{B^3}\frac{n_0 q v_T m}{D}\alpha = \frac{c^{3/2}}{B^{3/2}}\sqrt{n_0 q m v_T \alpha} \to \frac{1}{4\pi}\sqrt{\alpha}\sqrt{\varepsilon_p}\frac{V_T^2}{\Omega}\sqrt{\frac{\omega_p}{\Omega}}, \tag{9.23a}$$

where

$$\alpha \equiv \frac{1}{2}\sqrt{\frac{\pi}{2}}\int_0^{\infty} x^3 dx \int_{1/\Xi}^{\infty} dz z^{-1}(x^2+z^2)^{-1}(x^2+z^2+1)^{-2} \approx \frac{1}{4}\sqrt{\frac{\pi}{2}}\log \Xi$$

with $\Xi = \frac{L}{\lambda_D}$. The diffusion coefficient has the scaling law $D \sim T^{1/4} n_0^{1/2} B^{-3/2}$ that is weakly dependent on both the temperature and the density.

Homework 9.5: Take tokamaks of ITER grade, find the confinement time for the given $B^{-3/2}$ diffusion coefficient of Eq. (9.23).

9.4.4 *The Cross-Field Transport Coefficients*

While equations (9.4a), (9.6a) and (9.8a) take up different transport terms, the transport coefficients along the field lines have similar dependency on the electric field autocorrelation as that in the Fokker Planck equation. Thus, they are not expected to depart from the classical picture.

It is rather involved to evaluate the integrals for the cross-field transport coefficients. A leading order estimate indicates that both the random work $A_x \sim \varepsilon_p V_T^2(\omega_p/\Omega)^2$, and the velocity diffusion $A_w \sim \varepsilon_p V_T^2 \omega_p(\omega_p/\Omega)$, are reduced by the plasma parameter. They have similar characteristics as the classical effect. The stress tensor, on the other hand, having similar secularity as the diffusion coefficient is found to be $W \sim D\Omega \sim \sqrt{\varepsilon_p \omega_p/\Omega} V_T^2$, that is enhanced from the classical effect of binary collisions, and can cause significant momentum and energy transport. The mixing tensor $X \sim D/\Omega \sim \sqrt{\varepsilon_p \omega_p/\Omega}\rho^2$, yields a mixing length that is Debye length times the factor $(\varepsilon_p)^{1/4}(\omega_p/\Omega)^{5/4}$.

Further Reading

It has been widely accepted that the electrical conductivity is classical. The neoclassical prediction of the bootstrap current due to the potato orbital (Shaing and Hazeltine 1997) has also been considered to exist despite the turbulent nature of the plasma fluctuations. The trigger from low to high (L-H) confinement transition in tokamaks remain elusive despite better insight into the current pedestal of bifurcated MHD equilibrium states (Hsu and Chu 1987, Toi *et al.* 1989), the radial electric field (Shaing and Crume 1989), or the zonal flow (Manz *et al.* 2012).

The viscosity in the pure electron plasma was found in the experiment to be many orders of magnitude greater than the classical collision theory due to the long range ExB drift (Kriesel and Driscoll 2001), while particle and energy transport was found at 10 times greater than the classical theory due to the long range interactions (Anderegg, 1997).

Along the classical collision model, Braginskii (1965) has the comprehensive documentation on the transport coefficients.

The Large Helical Device (LHD) group reported the confinement dependence on the magnetic field (Miyazawa *et al.* 2006). The energy confinement scaling of tokamaks may be found in Sauter and Martin (2000).

The 2d guiding center plasma was proposed by Taylor and McNamara (1971) to explain the scenario of Bohm diffusion, that leads to the negative temperature state (Montgomery and Joyce 1974) and vortex diffusion (Dawson *et al.* 1971). Theoretical prediction of $B^{-3/2}$ diffusion scaling law was proved in Hsu *et al.* (2013). The removal of the secularity by renormalization is intuitively sound, and mathematically simpler. The perturbation treatment of strong plasma turbulence by Dupree (1966) to remove the secularity by the ensemble average of the background waves leads to the resonance broadening by the diffusive effect. Cumulant expansion was derived by Weinstock (1969) by applying the statistical theory to the strong turbulence. Along similar line of attack, the turbulence at the tokamak edge was treated by Terry and Diamond (1985). The basic theory of renormalization group analysis of turbulence was investigated by Yakhot and Orszag (1986). Statistical theory of turbulent transport for non-Markovian effects was developed by Zagorodny and Weiland (1999).

The ambipolar electric field due to the fast escape of electrons may alter the neoclassical impurity transport (Hsu *et al.* 1981).

The neoclassical transport regularly adopts the Fokker Planck collision operator that is obtained by assuming no magnetic effect, unperturbed and unmagnetized particle trajectories, and is irrespective of fluctuation spectrum. The neoclassical transport coefficients are often obtained by assuming negligible MHD activities and no electrostatic or electromagnetic turbulences. Moreover, the adiabatic invariants such as the magnetic moment μ while holds true in the vacuum magnetic field, can break down easily given inhomogenous or nonstationary fields (cf. Sec. 3.2.1). Despite these deficiencies, neoclassical theory has been able to provide insightful understanding of transport phenomena in tokamak experiments.

Homework Hints

Homework 9.1: Take time derivative on Eq. (9.2) and sum up all particles to show that the continuity equation with transport effect is given by Eq. (9.4).

Given

$$\delta(\vec{x}-\vec{\chi}_i-\Delta\vec{x}_i)=\delta(\vec{x}-\vec{\chi}_i)-\Delta\vec{x}_i\cdot\nabla\delta(\vec{x}-\vec{\chi}_i)\\ +\frac{1}{2}\Delta\vec{x}_i\Delta\vec{x}_i:\nabla\nabla\delta(\vec{x}-\vec{\chi}_i)+\cdots,$$

the density evolution with transport effects can then be expressed as

$$\frac{\partial}{\partial t}n=\left\langle\sum_i\frac{\partial}{\partial t}\delta(\vec{x}-\vec{\chi}_i-\Delta\vec{x}_i)\right\rangle\\ =\left\langle\sum_i\frac{\partial}{\partial t}\left(\delta(\vec{x}-\vec{\chi}_i)+\frac{1}{2}\Delta\vec{x}_i\Delta\vec{x}_i:\nabla\nabla\delta(\vec{x}-\vec{\chi}_i)\right)\right\rangle.$$

Note that terms of the first order in the fluctuating quantities vanish after the ensemble average,

$$\left\langle\sum_i\frac{\partial\Delta\vec{x}_i}{\partial t}\cdot\nabla\delta(\vec{x}-\vec{\chi}_i)\right\rangle=\left\langle\sum_i\Delta\vec{x}_i\cdot\nabla\frac{\partial}{\partial t}\delta(\vec{x}-\vec{\chi}_i)\right\rangle=0. \tag{9.1.1}$$

Thus,

$$\frac{\partial}{\partial t}n=\left\langle\sum_i\left(-\frac{\partial}{\partial\vec{x}}\delta(\vec{x}-\vec{\chi}_i)\cdot\frac{\partial\vec{\chi}_i}{\partial t}+\frac{1}{2}\frac{\partial}{\partial t}\left(\Delta\vec{x}_i\Delta\vec{x}_i:\nabla\nabla\delta(\vec{x}-\vec{\chi}_i)\right)\right)\right\rangle\\ =-\nabla\cdot\left\langle\sum_i\vec{w}_i\delta(\vec{x}-\vec{\chi}_i)\right\rangle\\ +\frac{1}{2}\left\langle\sum_i\left(\Delta\vec{w}_i\Delta\vec{x}_i+\Delta\vec{x}_i\Delta\vec{w}_i\right):\nabla\nabla\delta(\vec{x}-\vec{\chi}_i)\right\rangle\\ +\frac{1}{2}(\Delta\vec{x}_i\Delta\vec{x}_i:\nabla\nabla\frac{\partial}{\partial t}\delta(\vec{x}-\vec{\chi}_i)). \tag{9.1.2}$$

Applying the relation, $\partial\delta(\vec{x}-\vec{\chi}_i)/\partial t=-\nabla\cdot(\vec{w}_i\delta(\vec{x}-\vec{\chi}_i))$ to the last term, we have

$$\frac{\partial}{\partial t}n=-\nabla\cdot\left\langle\sum_i\vec{w}_i\delta(\vec{x}-\vec{\chi}_i)\right\rangle\\ +\frac{1}{2}\left\langle\sum_i\left(\Delta\vec{w}_i\Delta\vec{x}_i+\Delta\vec{x}_i\Delta\vec{w}_i\right):\nabla\nabla\delta(\vec{x}-\vec{\chi}_i)\right\rangle\\ -\frac{1}{2}\left\langle\sum_i\Delta\vec{x}_i\Delta\vec{x}_i:\nabla\nabla\nabla\cdot\vec{w}_i\delta(\vec{x}-\vec{\chi}_i)\right\rangle. \tag{9.3}$$

Therefore,

$$\frac{\partial}{\partial t}n = -\nabla \cdot n\vec{v} + \nabla\nabla : (\overset{=}{D}n) - \frac{1}{2}\nabla\nabla : (\nabla \cdot (n\vec{v}\overset{=}{X})), \tag{9.4}$$

Homework 9.3: Derive equations (9.14) to (9.18).

Given

$$\Delta\vec{a}_i(\Im) = \frac{q}{m}\delta\vec{E}(\vec{\chi}_i(\Im), \Im),$$

$$\Delta\vec{w}_i(\Im) = \frac{q}{m}\int_{\Im}^{0} dt[\delta\vec{E}_\perp(\vec{\chi}_i(t), t)\cos\Omega(\Im - t)$$

$$+ (\delta\vec{E}_\perp(\vec{\chi}_i(t), t) \times \widehat{e}_z)\sin\Omega(\Im - t) + \widehat{e}_z\delta E_{||}(\vec{\chi}_i(t), t)],$$

$$\Delta\vec{x}_i(\Im) = \frac{q}{m}\left[\int_{\Im}^{0} dt \int_{t}^{0} dt'\widehat{e}_z\delta E_{||}(\vec{\chi}_i(t'), t') - \int_{\Im}^{0} dt\widehat{e}_z \times \delta\vec{E}_\perp(\vec{\chi}_i(t), t)/\Omega\right]$$

$$+\widehat{e}_z \times \Delta\vec{w}_i(\Im)/\Omega = \frac{q}{m}\int_{\Im}^{0} dt \int_{t}^{0} dt'\widehat{e}_z\delta E_{||}(\vec{\chi}_i(t'), t')$$

$$-\frac{c}{B}\int_{\Im}^{0} dt\widehat{e}_z \times \delta\vec{E}_\perp(\vec{\chi}_i(t), t)\,(1 - \cos\Omega(\Im - t))$$

$$+\frac{e}{B}\int_{\Im}^{0} dt\delta\vec{E}_\perp(\vec{\chi}_i(t), t)\sin\Omega(\Im - t),$$

we find

$$\overset{=}{A}_w = \langle\Delta\vec{a}(\Im)\Delta\vec{w}(\Im)\rangle$$

$$= \frac{q^2}{m^2}\int_{0}^{\Im} d\tau_1[\langle\delta\vec{E}_\perp(\vec{\chi}_i(\Im), \Im)\delta\vec{E}_\perp(\vec{\chi}_i(\tau_1), \tau_1)\rangle\cos\Omega(\Im - \tau_1)$$

$$+ \langle\delta\vec{E}_\perp(\vec{\chi}_i(\Im), \Im)\delta\vec{E}_\perp(\vec{\chi}_i(\tau_1), \tau_1)\rangle \times \widehat{e}_z\sin\Omega(\Im - \tau_1)\}$$

$$+ \widehat{e}_z\widehat{e}_z\delta E_{||}(\vec{\chi}_i(\Im), \Im)\delta E_{||}(\vec{\chi}_i(\tau_1), \tau_1)]$$

$$= \frac{q^2}{m^2}\int_{0}^{\Im} d\tau_1[S_{xx}(|\Im - \tau_1|)\{\overset{=}{I}_\perp\cos\Omega(\Im - \tau_1)$$

$$+ \overset{=}{I}_t\sin\Omega(\Im - \tau_1)\} + S_{zz}(|\Im - \tau_1|)\overset{=}{I}_z]$$

$$= \frac{q^2}{m^2}\int_{t}^{0} dt_1[S_{xx}(|t_1|)\{\overset{=}{I}_\perp\cos\Omega t_1 - \overset{=}{I}_t\sin\Omega t_1\} + S_{zz}(|t_1|)\overset{=}{I}_z], \tag{9.14}$$

where $t_1 \equiv \tau_1 - \Im$, $t \equiv -\Im$ is applied. We also assume $\langle\delta E_x(\vec{\chi}_i(\Im), \Im)\delta E_y(\vec{\chi}_i(t), t)\rangle = 0$.

The random work can be expressed as

$$
\begin{aligned}
\vec{\vec{A}}_x &= \langle \Delta\vec{a}\,\Delta\vec{x}\rangle \\
&= \frac{q^2}{m^2}\int_{\Im}^{0} dt \int_{t}^{0} dt' \widehat{e}_z\widehat{e}_z \langle \delta E_{||}(\vec{\chi}_i(\Im), \Im)\delta E_{||}(\vec{\chi}_i(t'), t')\rangle \\
&\quad - \frac{qc}{mB}\int_{\Im}^{0} dt \langle \delta\vec{E}_{\perp}(\vec{\chi}_i(\Im), \Im)\widehat{e}_z \times \delta\vec{E}_{\perp}(\vec{\chi}_i(t), t)\rangle (1 - \cos\Omega(\Im - t)) \\
&\quad + \frac{c}{B}\int_{\Im}^{0} dt \langle \delta\vec{E}_{\perp}(\vec{\chi}_i(\Im), \Im)\delta\vec{E}_{\perp}(\vec{\chi}_i(t), t)\rangle \sin\Omega(\Im - t) \\
&= \vec{\vec{I}}_z \frac{q^2}{m^2}\int_0^{\Im} d\tau_2 \int_0^{\tau_2} d\tau_1 S_{zz}(|\Im - \tau_1|) \\
&\quad + \frac{qc}{mB}\int_0^{\Im} d\tau_1 S_{xx}(|\Im - \tau_1|)[\vec{\vec{I}}_{\perp} \sin\Omega(\Im - \tau_1) \\
&\quad + \vec{\vec{I}}_t(1 - \cos\Omega(\Im - \tau_1))] = \vec{\vec{I}}_z \frac{q^2}{m^2}\int_0^{t} dt_2 \int_{t_2}^{t} dt_1 S_{zz}(|t_1|) \\
&\quad + \frac{qc}{mB}\int_0^{t} dt_1 S_{xx}(|t_1|)[\vec{\vec{I}}_{\perp} \sin\Omega t_1 - \vec{\vec{I}}_t(1 - \cos\Omega t_1)],
\end{aligned}
\tag{9.15}
$$

The shear stress tensor is given by,

$$
\begin{aligned}
\vec{\vec{W}} &= \langle \Delta\vec{w}\,\Delta\vec{w}\rangle = \frac{q^2}{m^2}\int_0^{\Im} d\tau_1 \int_0^{\Im} d\tau_2 \{[\delta\vec{E}_{\perp}(\vec{\chi}_i(\tau_1), \tau_1)\cos\Omega(\Im - \tau_1) \\
&\quad + (\delta\vec{E}_{\perp}(\vec{\chi}_i(\tau_1), \tau_1) \times \vec{e}_z)\sin\Omega(\Im - \tau_1)][\delta\vec{E}_{\perp}(\vec{\chi}_i(\tau_2), \tau_2)\cos\Omega(\Im - \tau_2) \\
&\quad + (\delta\vec{E}_{\perp}(\vec{\chi}_i(\tau_2), \tau_2) \times \vec{e}_z)\sin\Omega(\Im - \tau_2)] \\
&\quad + \widehat{e}_z\widehat{e}_z \langle \delta E_{||}(\vec{\chi}_i(\tau_2), \tau_2)]\delta E_{||}(\vec{\chi}_i(\tau_1), \tau_1)\rangle\} \\
&= \frac{q^2}{m^2}\int_0^{\Im} d\tau_1 \int_0^{\Im} d\tau_2 [S_{xx}(|\tau_1 - \tau_2|)\vec{\vec{I}}_{\perp} \cos\Omega(\tau_1 - \tau_2) \\
&\quad - S_{xx}(|\tau_1 - \tau_2|)\vec{\vec{I}}_{\perp} \sin\Omega(\tau_1 - \tau_2) + \vec{\vec{I}}_z S_{zz}(|\tau_1 - \tau_2|)] \\
&= \frac{q^2}{m^2}\int_0^{t} dt_1 \int_0^{t} dt_2 [S_{xx}(|t_1 - t_2|)\vec{\vec{I}}_{\perp} \cos\Omega(t_1 - t_2) + \vec{\vec{I}}_z S_{zz}(|t_1 - t_2|)].
\end{aligned}
\tag{9.16}
$$

The diffusion coefficient is,

$$
\begin{aligned}
\overset{=}{D} &\equiv \frac{1}{2}(\langle \Delta\vec{w}\Delta\vec{x}\rangle + \langle \Delta\vec{x}\Delta\vec{w}\rangle),\\
D_{zz} &= \frac{q^2}{m^2}\int_0^{\Im} d\tau_1 \int_0^{\Im} d\tau \int_0^{\tau} d\tau_2 S_{zz}(|\tau_1-\tau_2|)\\
\langle \Delta\vec{w}\Delta\vec{x}\rangle_\perp &= \frac{qc}{mB}\left\{\int_0^{\Im} d\tau_1[\delta\vec{E}_\perp(\tau_1)\cos\Omega(\Im-\tau_1)\right.\\
&\quad + (\delta\vec{E}_\perp(\tau_1)\times\widehat{e}_z)\sin\Omega(\Im-\tau_1)]\\
&\quad \times\left[\int_0^{\Im} d\tau_2(\delta\vec{E}_\perp(\tau_2)\times\widehat{e}_z)(1-\cos\Omega(\Im-\tau_2))\right.\\
&\quad \left.\left.+\int_0^{\Im} d\tau_2\delta\vec{E}_\perp(\tau_2)\sin\Omega(\Im-\tau_2)\right]\right\}\\
&= \frac{qc}{mB}\int_0^{\Im} d\tau_1\int_0^{\Im} d\tau_2 S_{xx}(|\tau_1-\tau_2|)\overset{=}{I}_\perp(\sin\Omega(\tau_1-\tau_2)\\
&\quad +\sin\Omega(\Im-\tau_1)) + \frac{qc}{mB}\int_0^{\Im} d\tau_1\int_0^{\Im} d\tau_2 S_{xx}(\widehat{e}_y\widehat{e}_x-\widehat{e}_x\widehat{e}_y)\\
&\quad \times(\cos\Omega(\Im-\tau_1)(1-\cos\Omega(\Im-\tau_2))\\
&\quad -\sin\Omega(\Im-\tau_1)\sin\Omega(\Im-\tau_2))\\
&= \frac{qc}{mB}\int_0^{\Im} d\tau_1\int_0^{\Im} d\tau_2 S_{xx}(|\tau_1-\tau_2|)(\overset{=}{I}_\perp(\sin\Omega(\tau_1-\tau_2)\\
&\quad +\sin\Omega(\Im-\tau_1)) + \overset{=}{I}_t(\cos\Omega(\Im-\tau_1)-\cos\Omega(\tau_1-\tau_2)))\\
\overset{=}{D}_\perp &= \frac{qc}{2mB}\int_0^{\Im} d\tau_1\int_0^{\Im} d\tau_2 S_{xx}(|\tau_1-\tau_2|)[\overset{=}{I}_\perp(\sin\Omega(\Im-\tau_2)\\
&\quad +\sin\Omega(\Im-\tau_1)) + \overset{=}{I}_t(\cos\Omega(\Im-\tau_1)\\
&\quad +\cos\Omega(\Im-\tau_2)-2\cos\Omega(\tau_1-\tau_2))]\\
&= \frac{qc}{2mB}\int_0^{\Im} dt_1\int_0^{\Im} dt_2 S_{xx}(|t_1-t_2|)(\overset{=}{I}_\perp(\sin\Omega t_2+\sin\Omega t_1)\\
&\quad +\overset{=}{I}_t(\cos\Omega t_1+\cos\Omega t_2-2\cos\Omega(t_1-t_2))).
\end{aligned}
\tag{9.17}
$$

Finally, the mixing tensor is given by

$$
\overset{=}{X} \equiv \langle \Delta\vec{x}\Delta\vec{x}\rangle
$$

$$\overset{\rightleftharpoons}{X}_{zz} = \frac{q^2}{m^2}\int_0^{\Im} d\tau \int_0^{\tau} d\tau_1 \int_0^{\Im} d\tau' \int_0^{\tau'} d\tau_2 \overset{\rightleftharpoons}{I}_z S_{zz}(|\tau_1 - \tau_2|)$$

$$\begin{aligned}
\overset{\rightleftharpoons}{X}_{\perp} &= \frac{c^2}{B^2}\int_0^{\Im} d\tau_1 \int_0^{\Im} d\tau_2 S_{xx}(\tau_1, \tau_2)[\overset{\rightleftharpoons}{I}_{\perp}(1 - \cos\Omega(\Im - \tau_1) \\
&\quad - \cos\Omega(\Im - \tau_2) + \cos\Omega(\tau_1 - \tau_2))] \\
&\quad + \frac{c^2}{2B^2}\int_0^{\Im} d\tau_1 \int_0^{\Im} d\tau_2 S_{xx}(\tau_1, \tau_2)\overset{\rightleftharpoons}{I}_t[\sin\Omega(\Im - \tau_1) \\
&\quad - \sin\Omega(\Im - \tau_2) - \sin\Omega(\tau_1 - \tau_2)] \\
&= \frac{c^2}{B^2}\int_0^{t} dt_1 \int_0^{t} dt_2 S_{xx}(|t_1 - t_2|)[\overset{\rightleftharpoons}{I}_{\perp}(1 - \cos\Omega t_1 \\
&\quad - \cos\Omega t_2 + \cos\Omega(t_1 - t_2))]
\end{aligned} \tag{9.18}$$

> **Homework 9.5:** Take tokamaks of ITER grade, find the confinement time for the given $B^{-3/2}$ diffusion coefficient of Eq. (9.23).

If we take $n_0 \sim 10^{14}/cc$, $T \sim 10KeV$, $B \sim 6Tesla$, we find $\lambda_D \sim 7.4e^{-3}cm$, $n_0\lambda_D^3 \sim 3e^7$, $V_{Te} \sim 5e^9 cm/\sec$, $\Omega_e \sim 1e^{12}$, $\omega_{pe} \sim 5.6e^{11}$, $\alpha \approx \frac{1}{4}\sqrt{\frac{\pi}{2}}\log\Xi \sim 6$ and $D \sim 7e^2 cm^2/sec$. For a minor radius of 1 meter, the confinement time is about 15 seconds.

Chapter Ten

Magnetohydrodynamics

"Histories make men wise; poets, witty; the mathematics, subtle;
natural philosophy, deep; moral, grave;
logic and rhetoric, able to contend."

Francis Bacon (1561–1626)
British Philosopher, Essayist, Statesman

The kinetic equations can often be too complicated to solve. On the other hand, the wave, transport, and fluid effects, may be understood by a few smoothed and averaged local variables in the configuration space to reveal the governing principles by simplification with clarification. Resorting to the MHD model thus has many advantages. By neglecting some detailed structures, the large-scale phenomena may be explored from the macroscopic quantities to establish the global understanding of (fusion) plasma, allow fundamentally new design concept, give better intuition for the experimental observables, and study major instabilities that severely limit the achievable plasma parameters.

10.1 The MHD Equilibrium Properties

Many assumptions go into the MHD description. Alternation of any of these assumptions would recuperate features neglected such as the two-fluid or kinetic effects, which may or may not be unimportant to the underlying mechanisms in question. One key assumption is the **charge neutrality** condition, $\rho_q = 0$, which would be well satisfied when the length scale is longer than the Debye length. The other key assumption is the low frequency activities that fall well below the electron and ion cyclotron frequencies to result in only the E $\times$ B drifts for the charged particles with negligible polarization drifts and FLR effects. Thus, MHD theory is particularly suitable for describing activities of long time and large length scales.

Considering a plasma of electron and single nucleon species in the one fluid description, we have from Chapter Four, the continuity equation given by Eq. (4.25),

$$\frac{\partial}{\partial t}\rho + \nabla \cdot (\rho \vec{U}) = 0. \tag{10.1}$$

The charge neutrality simplifies Eq. (4.26), to

$$\nabla \cdot \vec{J} = 0, \tag{10.2}$$

and abridges the momentum equation to

$$\rho\frac{\partial \vec{U}}{\partial t}+\rho\vec{U}\cdot\nabla\vec{U}=-\nabla\cdot\overleftrightarrow{P}+\frac{1}{c}\vec{J}\times\vec{B}. \tag{10.3}$$

The cross-field E × B drift gives

$$\vec{E}+\frac{1}{c}\vec{U}\times\vec{B}=0, \tag{10.4}$$

that makes the Faraday's law into

$$\frac{\partial}{\partial t}\vec{B}=\nabla\times(\vec{U}\times\vec{B}). \tag{10.5}$$

Once the magnetic field is found then the Ampere's law allows us to evaluate the current,

$$\vec{J}=\frac{c}{4\pi}\nabla\times\vec{B}, \tag{10.6}$$

which is consistent with Eq. (10.2) and without the current transport of Eq. (4.31). Only one more equation for the pressure, the MHD equations would be complete. The simple case of isotropic pressure is often assumed so that $\overleftrightarrow{P}=p\overleftrightarrow{I}$. We may take Eq. (4.33a) or the adiabatic law:

$$p\rho^{-\gamma}=\Theta=const. \tag{10.7}$$

that implies

$$\frac{d}{dt}\left(\frac{p}{\rho^{\gamma}}\right)=0. \tag{10.8}$$

An alternative equation to Eq. (10.8) can be cast as:

$$\frac{\partial p}{\partial t}+\vec{U}\cdot\nabla p=-\gamma p\nabla\cdot\vec{U}. \tag{10.9}$$

It is noteworthy that when additional particle, momentum, or energy sources are present, the pressure can be shaped through the energy equation accordingly. There is great freedom in the plasma equilibrium structure. The pressure needs not be isotropic, as readily tailored by Ohmic heating or otherwise.

10.1.1 *The MHD Equilibrium*

The steady state of MHD is described by:

$$\nabla\cdot(\rho\vec{U})=0, \tag{10.10}$$

$$\rho\vec{U}\cdot\nabla\vec{U} = -\nabla p + \frac{1}{c}\vec{J}\times\vec{B}, \tag{10.11}$$

$$\vec{J} = \frac{c}{4\pi}\nabla\times\vec{B}, \tag{10.12}$$

$$\nabla\times(\vec{U}\times\vec{B}) = 0, \tag{10.13}$$

Greater simplification can be made when the flow is absent, and the equations become

$$\frac{1}{4\pi}(\nabla\times\vec{B})\times\vec{B} = \nabla p. \tag{10.14}$$

$$\vec{J} = \frac{c}{4\pi}\nabla\times\vec{B}. \tag{10.15}$$

The equilibrium state has these equations governing its properties:

$$\begin{aligned}
\vec{B}\cdot\nabla p &= 0,\\
\vec{J}\cdot\nabla p &= 0,\\
\frac{1}{4\pi}\vec{B}\cdot\nabla\vec{B} &= \nabla\left(p + \frac{1}{8\pi}B^2\right),\\
\vec{B}\cdot\nabla\vec{J} &= \vec{J}\cdot\nabla\vec{B},\\
\nabla\cdot(\vec{B}\times\nabla p) &= 0,\\
\vec{B}\cdot\nabla(\vec{B}\cdot\vec{J}) &= \vec{J}\cdot\nabla(B^2).
\end{aligned} \tag{10.16}$$

Homework 10.1: Prove the equilibrium properties as listed in Eq. (10.16). The following vector identities are useful:

$$\nabla(\vec{A}\cdot\vec{B}) = \vec{A}\times(\nabla\times\vec{B}) + \vec{B}\times(\nabla\times\vec{A}) + (\vec{A}\cdot\nabla)\vec{B} + (\vec{B}\cdot\nabla)\vec{A},$$

$$\nabla\times(\vec{A}\times\vec{B}) = \vec{A}(\nabla\cdot\vec{B}) - \vec{B}(\nabla\cdot\vec{A}) + (\vec{B}\cdot\nabla)\vec{A} - (\vec{A}\cdot\nabla)\vec{B}.$$

10.1.2 *The Magnetic Surface*

The magnetic confinement as a rule applies the ingenuous construct of closed magnetic surfaces as protection layers to effectively lessen the plasma leakage. The magnetic field lines winding along on the magnetic surface are perpendicular to the norm of that surface as described by $\psi(\vec{r}) = const$. Therefore, $\vec{B}\cdot\nabla\psi = 0$. Since $\vec{B}\cdot\nabla p = 0$ from Eq. (10.16), it implies that the pressure is a function of the flux surfaces alone, i.e., $p = p(\psi)$. This is consistent with the fact that the pressure will quickly distribute along the field lines and equilibrate on the magnetic surfaces. The other two vectors lying on the flux surface, denoted as $\nabla\alpha$ and $\nabla\beta$, respectively, are perpendicular to the norm of the flux surface, namely $\nabla\psi$. The magnetic field can be expressed as $\vec{B} = a\nabla\alpha + b\nabla\beta = \nabla\psi \times \nabla\beta + b\nabla\beta$. The absence of magnetic monopole implies $\nabla \cdot \vec{B} = 0$ that requires $\nabla \cdot (b\nabla\beta) = 0 = b\nabla^2\beta + \nabla b \cdot \nabla\beta$. We now consider the confinement machine that possesses the toroidal symmetry, i.e., $\partial/\partial\phi = 0$. The vacuum magnetic field in the azimuthal $\widehat{e}_\phi$-direction can be attributed to an infinitely long current rod in the $\widehat{e}_z$-direction, and the plasma current is to flow on the flux surfaces. We may choose $\beta = \phi$. Therefore, $\nabla^2\phi = 0$, $\nabla b \cdot \nabla\phi = 0$ and $\vec{B} = \nabla\psi \times \nabla\phi + b(\psi, \alpha)\nabla\phi$. This readily gives the current

$$\frac{4\pi}{c}\vec{J} = \nabla \times \vec{B} = -\nabla^2\psi\nabla\phi - (\nabla\psi \cdot \nabla)\nabla\phi \\ + (\nabla\phi \cdot \nabla)\nabla\psi + \nabla b(\psi, \alpha) \times \nabla\phi$$

that results in

$$\frac{4\pi}{c}\vec{J} = -\nabla^2\psi\frac{\widehat{e}_\phi}{R} - \left(\frac{\partial\psi}{\partial R}\frac{\partial}{\partial R}\right)\left(\frac{\widehat{e}_\phi}{R}\right) + \frac{\widehat{e}_\phi}{R^2}\frac{\partial\psi}{\partial R} + \nabla b(\psi, \alpha) \times \nabla\phi$$

$$= -\Delta^*\psi\frac{\widehat{e}_\phi}{R} + \nabla b(\psi, \alpha) \times \frac{\widehat{e}_\phi}{R}. \tag{10.17}$$

Here,

$$\Delta^*\psi \equiv R\frac{\partial}{\partial R}\left(\frac{1}{R}\frac{\partial\psi}{\partial R}\right) + \frac{\partial^2\psi}{\partial Z^2} = R^2\nabla \cdot (R^{-2}\nabla\psi). \tag{10.18}$$

Thus,

$$\frac{4\pi}{c}\vec{J} \times \vec{B} = \left(-\Delta^*\psi\frac{\widehat{e}_\phi}{R} + \nabla b(\psi, \alpha) \times \nabla\phi\right) \times (\nabla\psi \times \nabla\phi + b(\psi, \alpha)\nabla\phi)$$

$$= -\Delta * \psi\nabla\psi|\nabla\phi|^2 - \nabla\phi\nabla\psi \cdot (\nabla b \times \nabla\phi) - b\nabla b|\nabla\phi|^2. \tag{10.19}$$

From $\frac{1}{c}\vec{J} \times \vec{B} = \nabla p = p'(\psi)\nabla\psi$, the $\nabla\phi$ component gives $\nabla\psi \cdot (\nabla b \times \nabla\phi) = 0$ which implies that $b = b(\psi)$ is a function of ψ alone. Thus, the magnetic field of toroidal axisymmetry is given by

$$\vec{B} = \nabla\psi \times \nabla\phi + b(\psi)\nabla\phi, \tag{10.20}$$

where $b(\psi)$ incurs the modification due to the poloidal plasma current.

We then find from the component of $\vec{J} \times \vec{B}$ along $\nabla\psi$, the governing equation for ψ,

$$\Delta^*\psi = -b(\psi)b'(\psi) - 4\pi R^2 p'(\psi). \tag{10.21}$$

This is the Grad-Sharfranov MHD equilibrium equation for the toroidally axisymmetric plasma. Inasmuch as both plasma pressure and toroidal magnetic field are homogeneous on the flux surfaces, their functional dependency on the flux surface is unspecific. The freedom in choosing $b(\psi)$ and $p(\psi)$ can be a shortcoming in the MHD equilibrium theory when making conclusions based on the convenient profiles to attest to the experimental data.

Homework 10.2: Derive the MHD equilibrium equation **for the force free toroidal state** with $\vec{J} = \lambda \vec{B}$ in the axisymmetric plasma.

10.1.3 *The Total Plasma Energy*

Multiplying the continuity equation with $\frac{1}{2}U^2$ and dotting the equation of motion with $\vec{U}$, and adding them together, we arrive at

$$\frac{\partial\left(\frac{1}{2}\rho\vec{U}^2\right)}{\partial t} + \nabla \cdot \left(\frac{1}{2}\rho\vec{U}^2\vec{U}\right) = -\vec{U} \cdot \nabla p + \frac{1}{c}\vec{U} \cdot \vec{J} \times \vec{B}. \tag{10.22}$$

Multiplying the Faraday's equation with $\vec{B}$ gives the following magnetic energy equation,

$$\frac{\partial}{\partial t}\left(\frac{1}{8\pi}\vec{B}^2\right) = \frac{1}{4\pi}\vec{B} \cdot \nabla \times (\vec{U} \times \vec{B}) = \frac{1}{c}\vec{U} \times \vec{B} \cdot \vec{J} + \frac{1}{4\pi}\nabla \cdot ((\vec{U} \times \vec{B}) \times \vec{B}). \tag{10.23}$$

The additional equation needed is that of the pressure energy,

$$\frac{1}{\gamma - 1}\left(\frac{\partial p}{\partial t} + \nabla \cdot (\vec{U}p)\right) = -p\nabla \cdot \vec{U}. \tag{10.24}$$

Adding the above three equations, we arrive at the conservation law of energy,

$$\frac{\partial}{\partial t}\left(\frac{1}{2}\rho U^2 + \frac{p}{\gamma - 1} + \frac{1}{8\pi}B^2\right) = -\nabla \cdot \left[\left(\frac{1}{2}\rho U^2 + \frac{\gamma p}{\gamma - 1}\right)\vec{U} + \frac{c}{4\pi}\vec{E} \times \vec{B}\right]. \tag{10.25}$$

By integrating Eq. (10.25) over the space as the surface terms nullify, it gives the total energy

$$W = \int d\tau \left(\frac{1}{8\pi}|\vec{B}|^2 + \frac{1}{\gamma - 1}p + \frac{1}{2}\rho U^2\right). \tag{10.26}$$

It is noteworthy that aside from the Poynting flux in the RHS of Eq. (10.25), the fluid energy flux is different from the energy density $\frac{1}{2}\rho U^2 + p/(\gamma - 1)$ times $\vec{U}$.

Homework 10.3: Derive Eq. (10.25) by adding the magnetic energy to Eq. (4.21).

10.1.4 *The Frozen Field Lines*

The magnetic flux $\Psi \equiv \int d\vec{a} \cdot \vec{B}$ defined as the magnetic field through any surface spanned by a given closed curve can be shown to be conserved in the ideal MHD. In fact, there is no net magnetic flux through the surface of an enclosed volume as it can easily be shown by $\Psi = \oint d\vec{a} \cdot \vec{B} = \int d\tau \nabla \cdot \vec{B} = 0$. Let us start with a resistive plasma so that $\vec{J} = \sigma \vec{E} = c\nabla \times \vec{B}/4\pi$. Taking curl of the current and replacing $\nabla \times \vec{E}$ by the Faraday's law, we arrive at

$$\frac{\partial \vec{B}}{\partial t} = -\frac{\eta c^2}{4\pi}\nabla \times (\nabla \times \vec{B}) = \frac{\eta c^2}{4\pi}\nabla^2 \vec{B} = D_M \nabla^2 \vec{B}, \tag{10.27}$$

```
function MagneticDiffusion      % Eq(10.28) is plotted accordingly.
clear all; close all; clc;      % t is the time, d is the diffusion coefficient.
d=0.01;
x=0:0.001:1.5;
T=[0.1,1,10];
COLOR=['g','b','r'];
figure(1); hold on;
for i=1:3
    t=T(i);
    B=exp(-
x.^2/4/d/t)/sqrt(4*pi*d*t);
    bE=sum(B.*B);
    B=B/sqrt(bE);
    plot(x,B,COLOR(i));
end;
str=sprintf('Magnetic Diffusion
at D=%5.3f',d);
title(str);
xlabel('space range');
ylabel('B field amplitude');
text(0.05,0.14,'t=0.1','color','g');
text(0.15,0.07,'t=1','color','b');
text(0.4,0.04,'t=10','color','r');
```

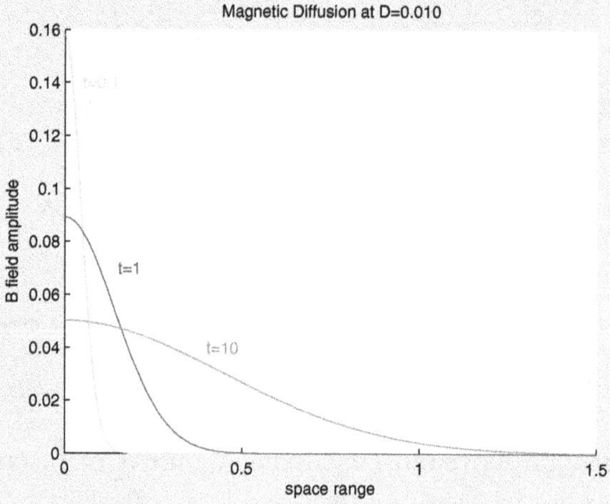

where the resistivity $\eta = 1/\sigma$, and D_M is the magnetic diffusivity. The solution to this equation, assuming constant D_M can be given as

$$\vec{B}(x,t) = \frac{\vec{b}_0}{\sqrt{4\pi D_M t}} e^{-x^2/4D_M t}. \tag{10.28}$$

It is clear that the magnetic field diffuses to a wider range if the plasma is resistive as also illustrated in the accompanying program.

From the Faraday's law, $\nabla \times \vec{E} = -(1/c)\partial \vec{B}/\partial t$. In the low frequency response of the plasma, both the electrons and ions are assumed to move together under the $\vec{E} \times \vec{B}$ drifts so that $\vec{E} + \vec{U} \times \vec{B}/c = 0$, the governing equation for the magnetic field in the ideal MHD can then be cast into

$$\frac{\partial \vec{B}}{\partial t} = \nabla \times (\vec{U} \times \vec{B}) = -\vec{U} \cdot \nabla \vec{B} + \vec{B} \cdot \nabla \vec{U} - \vec{B} \nabla \cdot \vec{U}, \quad (10.29)$$

that is

$$\frac{d\vec{B}}{dt} = \frac{\partial \vec{B}}{\partial t} + \vec{U} \cdot \nabla \vec{B} = \vec{B} \cdot \nabla \vec{U} - \vec{B} \nabla \cdot \vec{U}. \quad (10.29a)$$

If the flow is incompressible and is function of the flux surface only so that $\nabla \cdot \vec{U} = 0$ and $\vec{U} = \vec{U}(\psi)$, then $\vec{B} \cdot \nabla \vec{U} = 0$ and $d\vec{B}/dt = \partial \vec{B}/\partial t + \vec{U} \cdot \nabla \vec{B} = 0$. Therefore, the so called 'material derivative' of magnetic field $d\vec{B}/dt$ vanishes. That means the variation of the magnetic field moving along the flow is nullified, and the magnetic field is frozen into the flow.

Homework 10.4: Show that the vector representation of $\vec{V} = \alpha \nabla \beta$ is a subset of the Helmholtz-Hodge Decomposition that has both the transverse and the longitudinal components, but has null helicity, namely $\vec{V} \cdot \nabla \times \vec{V} = 0$. Show that the Clebsch decomposition given by $\vec{V} = \alpha \nabla \beta + \nabla \gamma$ allows for arbitrary helicity.

A class of vector potential described by the **Clebsch decomposition** $\vec{A} = \alpha \nabla \beta + \nabla \gamma$, that has $\nabla \alpha \times \nabla \beta \cdot \nabla \gamma \neq 0$ to ensure nonzero helicity, gives the magnetic field $\vec{B} = \nabla \alpha \times \nabla \beta$. Each gradient component defines an independent vector direction. Therefore, $\partial \vec{B}/\partial t = \nabla \times (\dot{\alpha} \nabla \beta + \alpha \nabla \dot{\beta})$ and $\nabla \times (\vec{U} \times \vec{B}) = \nabla \times \{\vec{U} \times (\nabla \alpha \times \nabla \beta)\} = \nabla \times \{(\nabla \alpha \vec{U} \cdot \nabla \beta - \nabla \beta \vec{U} \cdot \nabla \alpha)$. Thus, we have $\nabla \times (\dot{\alpha} \nabla \beta - \dot{\beta} \nabla \alpha + \nabla \beta \vec{U} \cdot \nabla \alpha - \nabla \alpha \vec{U} \cdot \nabla \beta) = 0$ that implies the quantity inside the parathesis is gradient of some function, i.e., $\dot{\alpha} \nabla \beta - \dot{\beta} \nabla \alpha + \nabla \beta \vec{U} \cdot \nabla \alpha - \nabla \alpha \vec{U} \cdot \nabla \beta = \nabla \Phi$. In terms of the material derivative of α and β, it gives

$$\frac{d\alpha}{dt} \nabla \beta - \frac{d\beta}{dt} \nabla \alpha = \nabla \Phi. \quad (10.30)$$

The physical meaning of Φ can be recognized as the longtitudinal potential that does not affect the magnetic field, we can therefore neglect it by choosing $\nabla \Phi = 0$. That leads us to the conclusion:

$$\frac{d\alpha}{dt} = \frac{d\beta}{dt} = 0. \quad (10.31)$$

This proves the magnetic field is frozen into the plasma flow in the ideal MHD model for the class of magnetic fields as represented by the Clebsch decomposition.

For the tokamak field, $\vec{B} = \nabla \psi \times \nabla \phi + b(\psi) \nabla \phi$, that may not be represented by the Clebsch decomposition, it is expected that other constraints are needed in order to assert that the field lines can be frozen into the flow.

10.2 The MHD Energy Principle

By defining a virtual displacement $\vec{\xi}$ from $\partial\vec{\xi}/\partial t = \vec{v}$, the continuity equation of Eq. (10.1) becomes

$$\delta\rho + \nabla \cdot (\rho\vec{\xi}) = 0, \tag{10.32}$$

and the Faraday's law becomes

$$\delta\vec{B} = \nabla \times (\vec{\xi} \times \vec{B}), \tag{10.33}$$

while the pressure equation becomes

$$\delta p + \vec{\xi} \cdot \nabla p = -\gamma p \nabla \cdot \vec{\xi}. \tag{10.34}$$

The equation of motion Eq. (10.3) can be rewritten as

$$\rho\frac{\partial^2\vec{\xi}}{\partial t^2} = \vec{F}(\vec{\xi}) = \nabla(\gamma p \nabla \cdot \vec{\xi} + \vec{\xi} \cdot \nabla p) + \frac{1}{c}(\vec{J} \times \delta\vec{B} + \delta\vec{J} \times \vec{B}). \tag{10.35}$$

The change in the plasma kinetic energy is given by the work done, $\frac{1}{2}\int d\tau \rho\vec{\xi} \cdot \partial^2\vec{\xi}/\partial t^2 = -\delta W = \frac{1}{2}\int d\tau \vec{F} \cdot \vec{\xi}$. Therefore,

$$\delta W = -\frac{1}{2}\int d\tau\vec{\xi} \cdot \left(\nabla(\gamma p \nabla \cdot \vec{\xi} + \vec{\xi} \cdot \nabla p) + \frac{1}{c}\vec{J} \times \delta\vec{B} + \frac{1}{4\pi}(\nabla \times \delta\vec{B}) \times \vec{B}\right). \tag{10.36}$$

With use of the integration by parts, the change in the potential energy can be cast into the plasma volume piece,

$$\delta W_p = \frac{1}{2}\int d\tau \left(\gamma p(\nabla \cdot \vec{\xi})^2 + \vec{\xi} \cdot \nabla p \nabla \cdot \vec{\xi} + \frac{1}{4\pi}|\delta\vec{B}|^2 + \frac{1}{c}\vec{J} \cdot (\vec{\xi} \times \delta\vec{B})\right), \tag{10.37}$$

and the plasma surface,

$$\delta W_s = -\frac{1}{2}\int d\vec{\sigma} \cdot (\gamma p\vec{\xi}\nabla \cdot \vec{\xi} + \vec{\xi}\vec{\xi} \cdot \nabla p + \frac{1}{4\pi}(\vec{\xi} \times \vec{B}) \times \delta\vec{B}). \tag{10.38}$$

If the surface is taken to coincide with the magnetic surface so that $\vec{B} \cdot d\vec{\sigma} = 0$, Eq. (10.38) is simplified to

$$\delta W_s = \frac{1}{2}\int d\vec{\sigma} \cdot \vec{\xi}\left(\delta p + \frac{1}{4\pi}\vec{B} \cdot \delta\vec{B}\right).$$

The first and the third term of δW_V are the stabilization energies due to the compressibility $\nabla \cdot \vec{\xi} \neq 0$ that represents the acoustic wave. The magnetic perturbations are corresponding to the slow and fast Alfvén waves. The pressure gradient in the second term and the plasma current in the last term of δW_V are the free energies that drive the interchange and the kink instabilities respectively. Taking the virtual displacement to have a growth rate γ so that $\vec{\xi} \propto \exp(\gamma t)$, we have

$$\frac{1}{2}\int d\tau\rho\vec{\xi}\cdot\partial^2\vec{\xi}/\partial t^2 = \frac{1}{2}\int d\tau\rho|\vec{\xi}|^2\gamma^2 = -\delta W \quad = \gamma^2\delta K.$$

Thus, $\delta W < 0$ results in an instability as the system picks up kinetic energy δK by falling off the potential cliff, while $\delta W > 0$ the system has to climb up the potential hill and is pushed back to thus maintain the stability.

10.2.1 *The Interchange Instability*

We now apply the δW analysis to the pressure driven instability. Consider the azimuthally symmetric magnetic configuration of either concave or convex in its flux surfaces, atypical in mirror machines. The plasma current is not of concern here and is assumed to be absent. The magnetic energy (cf. Eq. (10.28)) is given by

$$W_M = \frac{1}{8\pi}\int B^2 dV = \frac{1}{8\pi}\int B^2 A dl = \frac{1}{8\pi}\int \Phi^2 (dl/A),$$

where $\Phi \equiv BA$ is the magnetic flux, A is the area of the cross section and dl is the arc length. For the purpose of the discussion, we will refer to region I as the plasma, and region II as the vacuum region. The magnetic energy exchange is assumed to take place fast enough to conserve the magnetic flux so that $\Phi_I = \Phi_{II}$ during the process. Thus, swapping the magnetic fluxes given by $B_I A_I = B_{II} A_{II}$ leads to the energy difference given by

$$\delta W_M = \frac{1}{8\pi}\left[\left\{\Phi_I^2\int_{II}\frac{dl}{A} + \Phi_{II}^2\int_I\frac{dl}{A}\right\} - \left\{\Phi_I^2\int_I\frac{dl}{A} + \Phi_{II}^2\int_{II}\frac{dl}{A}\right\}\right] = 0. \tag{10.39}$$

Similarly, the pressure energy given by $W_p = pV/(\gamma - 1)$ (cf. Eq. (10.28)) leads to the energy difference given by

$$\delta W_p = \frac{1}{\gamma - 1}\left[\frac{p_I V_I^\gamma}{V_{II}^\gamma}V_{II} + \frac{p_{II} V_{II}^\gamma}{V_I^\gamma}V_I - p_I V_I - p_{II} V_{II}\right], \tag{10.40}$$

where $V = \int dlA = \Phi\int dl/B$. Since $p\rho^{-\gamma} = const$, or equivalently, $pV^\gamma = const$, we have $p_{II}^{new} = p_I V_I^\gamma V_{II}^{-\gamma}$ and $p_I^{new} = p_{II} V_{II}^\gamma V_I^{-\gamma}$ when the pressure energies are swapped. Assume $p_{II} = p_I + \delta p$, $V_{II} = V_I + \delta V$, and set $p_I = p$ and

$V_I = V$, we expand to the second order to get

$$\delta W_p = \frac{1}{\gamma - 1}\left[pV\left(1 - \frac{\delta V}{V}(\gamma - 1) + \frac{\delta V^2}{2V^2}(\gamma - 1)\gamma\right)\right.$$
$$\left. + p_{II}V_{II}\left(1 + \frac{\delta V}{V}(\gamma - 1) + \frac{\delta V^2}{2V^2}(\gamma - 1)(\gamma - 2)\right) - pV - p_{II}V_{II}\right]$$

Thus, $\delta W_p = (\gamma p_I \delta V/V + \delta p)\delta V$. A sufficient condition for stability, namely, $\delta W_p > 0$, is $\delta V \delta p > 0$. Thus, since the vacuum region has no plasma pressure, i.e., $p_{II} = 0$, we have $\delta p < 0$, and the stability condition is $\delta V < 0$, or equivalently, $\Phi\delta \int dl/B < 0$. Thus, if the plasma exchanges energy with the vacuum region of higher magnetic field then the plasma is stable. The plasma is unstable if the surrounding vacuum region has weaker magnetic field.

10.2.2 *The Surface Kink Instability*

We now apply the δW analysis to the current driven instability for an infinitely long current channel. This is interestingly related to the lightening phenomena, but is enlightening to fusion plasma study as well. The equilibrium current is assumed to be in the $\widehat{e}_z$ direction. We will consider the virtual displacement peaked at the plasma edge so that the volume δW is unimportant, and the instability can be featured as the surface kink.

By assuming the incompressible flow for the virtual displacement $\nabla \cdot \vec{\xi} = 0$, Eq. (10.38) is simplified to

$$\delta W_s = \frac{1}{2}\int d\vec{\sigma} \cdot \left(-\frac{1}{c}\vec{\xi}\vec{\xi} \cdot \vec{J} \times \vec{B} + \frac{1}{4\pi}\delta\vec{B} \times (\vec{\xi} \times \vec{B})\right),$$

where the equilibrium condition $\vec{J} \times \vec{B} = \nabla p$ has been applied. The perturbed magnetic field is given by

$$\delta\vec{B} = \vec{B} \cdot \nabla\vec{\xi} - \vec{\xi} \cdot \nabla\vec{B}, \tag{10.41}$$

which is expanded to give

$$\delta\vec{B} = B_\theta\left(\left(\frac{1}{r}\frac{\partial \xi_r}{\partial \theta} - \frac{\xi_\theta}{r}\right)\widehat{e}_r + \left(\frac{1}{r}\frac{\partial \xi_\theta}{\partial \theta} + \frac{\xi_r}{r}\right)\widehat{e}_\theta + \frac{1}{r}\frac{\partial \xi_z}{\partial \theta}\widehat{e}_z\right)$$
$$+ \left(\frac{B_\theta \xi_\theta}{r}\widehat{e}_r - \xi_r\frac{\partial B_\theta}{\partial r}\widehat{e}_\theta\right). \tag{10.41a}$$

Therefore,

$$\widehat{e}_r \cdot \delta\vec{B} \times (\vec{\xi} \times \vec{B}) = \xi_r(\vec{B} \cdot \delta\vec{B}) = \xi_r B_\theta\left(\frac{B_\theta}{r}\frac{\partial \xi_\theta}{\partial \theta} + \frac{B_\theta \xi_r}{r} - \xi_r\frac{\partial B_\theta}{\partial r}\right). \tag{10.42}$$

The pressure term has

$$\vec{B} \cdot \vec{J} \times \vec{\xi} = J_z[-\xi_\theta, \xi_r, 0] \cdot \vec{B} = \xi_r J_z B_\theta = \xi_r B_\theta \frac{c}{4\pi}(1/r)\partial r B_\theta/\partial r.$$

Thus,

$$\delta W_s = \frac{a}{8\pi} \int dz \oint d\theta \left[\xi_r B_\theta^2 \left(\frac{1}{r}\frac{\partial \xi_\theta}{\partial \theta} + \xi_r \frac{2}{r} \right) \right]_{r=a}. \tag{10.43}$$

Assuming $\xi_z = 0$ for simplicity so that

$$\frac{\partial \xi_\theta}{\partial \theta} + \frac{\partial r \xi_r}{\partial r} = 0, \tag{10.44}$$

we find

$$\delta W_s = \frac{aL}{8} \left[f B_\theta^2 \left(-\frac{\partial f}{\partial r} + \frac{f}{r} \right) \right]_{r=a} = \frac{L}{8} B_\theta^2 (r = a) \xi_0^2 (1 - \nu). \tag{10.45}$$

The last expression of Eq. (10.45) is obtained by choosing $f = \xi_0 (1+s)^\nu \exp(-\lambda s^2)$ where $s = (r/a) - 1$ and presumably $\lambda \gg 1$ so that the virtual displacement is fitting the definition of a surface mode, namely, peaked at the edge to alleviate the volume δW contribution, both in the plasma and in the vacuum regions. For the displacement with $\nu > 1$, the surface δW is sufficient to drive the kink instabilities, and the plasma current channel is unstable to the surface kinks of all $m > 0$ modes, irrespective of its axial mode numbers. The **sausage instability** of $m = 0$ has null instability cause in Eq. (10.43) and therefore cannot be driven unstable by the surface δW_s alone.

Homework 10.5: Examine the instability condition for the surface kinks of the current channel as described in Eq. (10.45) by including the plasma and vacuum δW.

10.2.3 *The Internal Kink Instability*

Now consider the volume δW for the stability of current channel. We will impose a conducting wall on the plasma surface so that the δW contribution, both on the plasma surface and in the vacuum region are negligible. Thus, we are examining the internal kinks only. Assuming the incompressible flow $\nabla \cdot \vec{\xi} = 0$, we have

$$\delta W_p = \frac{1}{2} \int d\tau \left(\frac{1}{4\pi} |\delta \vec{B}|^2 + \frac{1}{c} \vec{J} \cdot (\vec{\xi} \times \delta \vec{B}) \right), \tag{10.46}$$

and the perturbed magnetic field given by Eq. (10.41) is reduced to

$$\delta\vec{B} = \frac{B_\theta}{r}\frac{\partial\xi_r}{\partial\theta}\widehat{e}_r - \frac{\partial B_\theta\xi_r}{\partial r}\widehat{e}_\theta, \tag{10.41b}$$

where we have taken $\xi_z = 0$ for simplicity. Therefore,

$$|\delta\vec{B}|^2 = \left|\frac{B_\theta}{r}\frac{\partial\xi_r}{\partial\theta}\right|^2 + \left|\frac{\partial B_\theta\xi_r}{\partial r}\right|^2.$$

Applying $\vec{J}\times\vec{\xi} = J_z[-\xi_\theta, \xi_r, 0]$, we have

$$\vec{J}\times\vec{\xi}\cdot\delta\vec{B} = -\frac{c}{4\pi}\frac{1}{r}\frac{\partial B_\theta r}{\partial r}\left(B_\theta\xi_\theta\frac{\partial\xi_r}{r\partial\theta} + \xi_r\frac{\partial B_\theta\xi_r}{\partial r}\right).$$

Therefore,

$$\delta W_p = \frac{1}{8\pi}\int d\tau\left(\left|\frac{B_\theta}{r}\frac{\partial\xi_r}{\partial\theta}\right|^2 + \left|\frac{\partial B_\theta\xi_r}{\partial r}\right|^2 - \frac{1}{r}\frac{\partial B_\theta r}{\partial r}\left(B_\theta\xi_\theta\frac{\partial\xi_r}{r\partial\theta} + \xi_r\frac{\partial B_\theta\xi_r}{\partial r}\right)\right). \tag{10.47}$$

Taking $\xi_r = f(r)\cos(m\theta - kz)$ and $\xi_\theta = g(r)\sin(m\theta - kz)$ that gives from Eq. (10.44)

$$mg = -\frac{\partial rf}{\partial r}, \tag{10.48}$$

we have

$$\delta W_p = \frac{1}{16\pi}\int d\tau\left(\frac{B_\theta f}{r}\right)^2\left(m^2 - \left(1 + 2\frac{r\partial B_\theta}{B_\theta\partial r} + 2\frac{r\partial f}{f\partial r}\right) + \left(\frac{r\partial f}{f\partial r}\right)^2\right). \tag{10.49}$$

Thus, the instability condition is given by

$$\int d\tau\left(\frac{B_\theta f}{r}\right)^2\left(m^2 + (\frac{r\partial f}{f\partial r} - 1)^2\right) < 2\int d\tau\left(\frac{B_\theta f}{r}\right)^2\left(1 + \frac{r\partial B_\theta}{B_\theta\partial r}\right). \tag{10.50}$$

For the uniform current profile, $r\partial B_\theta/(B_\theta\partial r) = 1$. By choosing the displacement $f = \xi_0(r/a)$ so that $r\partial f/f\partial r = 1$, the instability condition can be cast into $m^2 < 4$. It is clear that the internal kink modes of $m = 1$ are unstable, while the $m = 2$ mode is marginally stable.

To find the $m = 0$ **internal sausage instability**, we have to examine Eq. (10.47), which is reduced to

$$\delta W_p = \frac{1}{8\pi}\int d\tau\left(\left|\frac{\partial B_\theta\xi_r}{\partial r}\right|^2 - \frac{1}{r}\frac{\partial B_\theta r}{\partial r}\xi_r\frac{\partial B_\theta\xi_r}{\partial r}\right). \tag{10.47a}$$

For the uniform current profile, and the displacement $f = \xi_0(r/a)$, the $m = 0$ is marginally stable. However, for $f = \xi_0(r/a)^\nu$ with $0 \le \nu < 1$ the sausage mode is unstable. Therefore, with the conducting wall surrounding the plasma current channel, only the $m = 0$ and $m = 1$ internal kinks are unstable.

Homework 10.6: Examine the peaked current profile to find the instability condition for the internal kinks as described by Eq. (10.50).

10.2.4 *The External Magnetic Field on the Kink Instability*

The free flowing current channel is vulnerable to kink instabilities as also evidenced from lightening striations and filamentations. An external magnetic field, providing stiffness to the plasma, in general, helps the plasma stability.

Imposing an external magnetic field in the axial direction, the MHD equilibrium is governed by

$$\partial p/\partial r = \frac{1}{c}(J_\theta B_z - J_z B_\theta) = -\frac{1}{4\pi}\left(B_z(\partial B_z/\partial r) + (B_\theta/r)(\partial r B_\theta/\partial r)\right).$$

Taking the incompressible flow $\nabla\cdot\vec{\xi} = 0$, we have from Eqs. (10.41a)

$$\delta\vec{B} = \left(\left(\vec{B}\cdot\nabla\xi_r - \frac{B_\theta\xi_\theta}{r}\right)\widehat{e}_r + \left(\vec{B}\cdot\nabla\xi_\theta + \frac{B_\theta\xi_r}{r}\right)\widehat{e}_\theta + \vec{B}\cdot\nabla\xi_z\widehat{e}_z\right)$$
$$+\left(\frac{B_\theta\xi_\theta}{r}\widehat{e}_r - \xi_r\frac{\partial\vec{B}}{\partial r}\right). \tag{10.41c}$$

Setting $\xi_z = 0$, $\xi_r = f(r)\cos(m\theta - kz)$, and $\xi_\theta = g(r)\sin(m\theta - kz)$, where $k = 2\pi n/L$, we find $\vec{B}\cdot\nabla\xi_\theta = (m - nq)(B_\theta/r)\xi_\theta$. Here we have defined the **safety factor**

$$q(r) \equiv \frac{2\pi r B_z}{L B_\theta}. \tag{10.51}$$

We want to examine the surface δW first. By taking

$$\delta\vec{B}\cdot\vec{B} = \left(B_\theta(\vec{B}\cdot\nabla\xi_\theta) + \xi_r\frac{B_\theta^2}{r} - \frac{1}{2}\xi_r\frac{\partial\vec{B}^2}{\partial r}\right)$$
$$= \left(B_\theta(\vec{B}\cdot\nabla\xi_\theta) + 2\xi_r\frac{B_\theta^2}{r} + 4\pi\,\xi_r\frac{\partial p}{\partial r}\right),$$

the surface δW is given by

$$\delta W_s = \frac{1}{2}\int d\vec{\sigma}\cdot\vec{\xi}\left(\delta p + \frac{1}{4\pi}\vec{B}\cdot\delta\vec{B}\right)$$
$$= \frac{a}{8\pi}\int dz\oint d\theta\xi_r\left[B_\theta(\vec{B}\cdot\nabla\xi_\theta) + 2\xi_r\frac{B_\theta^2}{r}\right]_{r=a}, \tag{10.52}$$

where we have utilized Eq. (10.9′) to replace δp by $-\xi_r \partial p/\partial r$. Moreover, choosing

$$f = \xi_0(1+s)^\nu \exp(-\lambda s^2) \quad \text{where } s = (r/a) - 1 \quad \text{with } \lambda \gg 1,$$

we find

$$\delta W_s = \frac{L}{8} B_\theta^2(r=a)\xi_0^2 \left(\left(1 + \frac{nq_a}{m}\right) - \left(1 - \frac{nq_a}{m}\right)\nu \right). \tag{10.52a}$$

We may draw a conclusion on the surface kinks. If the edge safety factor q_a is such that $m/n > q_a$, then the (m,n) surface kink mode will be unstable. On the other hand, if $q_a > m/n$, then the (m,n) surface kink mode will be stable. Since high m modes tend to benefit from the plasma and vacuum δW (cf. Homework 10.6), it is expected that only low (m,n) modes are serious instabilities to watch.

The magnetic field in Eq. (10.41d) can be simplified to

$$\delta\vec{B} = (\vec{B}\cdot\nabla\xi_r)\hat{e}_r + \left(\vec{B}\cdot\nabla\xi_\theta + \frac{B_\theta\xi_r}{r}\right)\hat{e}_\theta - \xi_r\frac{\partial\vec{B}}{\partial r}, \tag{10.41d}$$

that gives

$$|\delta\vec{B}|^2 = |\vec{B}\cdot\nabla\xi_r|^2 + \left|\vec{B}\cdot\nabla\xi_\theta + \frac{B_\theta\xi_r}{r} - \xi_r\frac{\partial B_\theta}{\partial r}\right|^2 + \left|\xi_r\frac{\partial B_z}{\partial r}\right|^2, \tag{10.41e}$$

and the δW contribution from the vacuum region is given by,

$$\delta W_v = \frac{1}{16\pi}\int d\tau \left(\frac{B_\theta f}{r}\right)^2 \times\left[m^2\left(1-\frac{nq}{m}\right)^2 + \left(1-\left(1-\frac{nq}{m}\right)\frac{\partial rf}{f\partial r} - \frac{r\partial B_\theta}{B_\theta\partial r}\right)^2 + \left(\frac{r\partial B_z}{B_\theta\partial r}\right)^2\right]. \tag{10.53}$$

Multiplying $\vec{J}\times\vec{\xi} = [-J_z\xi_\theta, J_z\xi_r, -J_\theta\xi_r]$ with $\delta\vec{B}$ of Eq. (10.41e), we have

$$\begin{aligned}\frac{4\pi}{c}\vec{J}\times\vec{\xi}\cdot\delta\vec{B} &= -\frac{1}{r}\frac{\partial B_\theta r}{\partial r}\xi_\theta(\vec{B}\cdot\nabla\xi_r) \\ &\quad + \frac{1}{r}\frac{\partial B_\theta r}{\partial r}\xi_r\left(\vec{B}\cdot\nabla\xi_\theta + \frac{B_\theta\xi_r}{r} - \frac{\partial B_\theta}{\partial r}\xi_r\right) - \left(\frac{\partial B_z}{\partial r}\xi_r\right)^2 \\ &= -\frac{1}{r}\frac{\partial B_\theta r}{\partial r}\left(\xi_\theta(\vec{B}\cdot\nabla\xi_r) - \xi_r(\vec{B}\cdot\nabla\xi_\theta) + \xi_r\left(\frac{\partial B_\theta}{\partial r}\xi_r - \frac{B_\theta\xi_r}{r}\right)\right) \\ &\quad - \left(\frac{\partial B_z}{\partial r}\xi_r\right)^2,\end{aligned}$$

$$\vec{J}\times\vec{\xi}\cdot\delta\vec{B} = -\frac{c}{4\pi}\frac{1}{r}\frac{\partial B_\theta r}{\partial r}\left(B_\theta\xi_\theta\frac{\partial\xi_r}{r\partial\theta} + \xi_r\frac{\partial B_\theta\xi_r}{\partial r}\right)$$

that reduces the plasma δW of Eq. (10.46) to

$$\delta W_p = \frac{1}{16\pi}\int d\tau \left(\frac{B_\theta f}{r}\right)^2 \left[(m-nq)^2 + \left(\left(1-\frac{nq}{m}\right)\frac{\partial rf}{f\partial r} - 2\right)^2 - 2\frac{\partial r B_\theta}{B_\theta \partial r}\right], \tag{10.54}$$

where Eq. (10.48) has been applied. Note that Eq. (10.54) recovers Eq. (10.49) when setting $n = 0$.

The stabilization effect in the middle term cannot be nullified if m/n is less than the central safety factor q_0 since the virtual displacement would become singular at the origin. Therefore, $q_0 > m/n$ provides the stability for (m, n) kink mode. By taking $m = nq$, it is clear that $\delta W_p = 0$ and the (m, n) mode is marginally stable for the uniform current profile.

The ohmic heating has the tendency to drive peaked current profile since plasma conductivity improves with temperature. It may therefore make the central $q_0 < 1$. If we assume a current profile to be $(1+\bar{r}^2)^{-2}$ so that the poloidal magnetic field is given by $2B_a\bar{r}/(1+\bar{r}^2)$, where $\bar{r} = r/a$ and $B_a = aB_0/q_aR_0$ is the poloidal magnetic field at the plasma edge. The safety factor is $q(r) = \frac{1}{2}q_a(1+\bar{r}^2)$. Thus, q_a, the edge q is twice the central q. We may choose the displacement $f = 1$ for $\bar{r} < r_s$ and $f = 0$ for $1 \geq \bar{r} \geq r_s$, where $r_s = \sqrt{1 - nq_0/m}$. For simplicity, we consider $n = 1$ only. The numerical result is shown in the accompanying program for $m = 1$ internal kink. It is unstable for $q_0 < 1$.

There appears to correlate well the sawtooth oscillation in the ohmic heated tokamak plasma and the internal kink. The peaked current profile drives the $m = 1$ internal kink which dumps the energy onto the outer region, the central q may then rise to 1 and stabilize the mode. Then the q profile is driven back to $q_0 < 1$ again, the process would repeat itself and present itself the fishbone structure. On the other hand, it is not difficult to imagine a major disruption scenoria if additional plasma instabilities are present.

Most magnetic confinement encloses the plasma with a conducting wall that helps the stability of surface kink modes. Moreover, the external magnetic field near the winding coil is stronger with a magnetic well providing stabilization effect through the last term in Eq. (10.53) of the vacuum δW. The surface kink may therefore be less threatening. The internal kink can also be stabilized if $q_0 > 1$. There are, however, nonideal MHD effects, notably the resistive MHD, that may result in the tearing modes localized near the rational surfaces. Additional free energies such as the α particles in burning plasma, the anisotropic pressure, the plasma flow are not included in the current model. They are capable of degrading the confinement through MHD instabilities or otherwise.

```
function Kink(m)                              % m=1 internal kink
syms r;
dWp=[]; n=1;
dq=0.01;  q0=1;
if(m==1) q0=0.8; elseif(m==2) q0=0.8; end;
for q=q0:dq:m
qr=q*(1+r^2); rs=sqrt(1-n*q/m),
Bp=2*r/(1+r^2);
dBp=diff(r*Bp,r);
F=Bp*dBp/r;
I=eval(int(F,r,0,rs));
G=(m-n*qr)^2*Bp^2/r;
J=eval(int(G,r,0,rs));
H=(m+n*qr)^2/r*Bp^2;
K=eval(int(H,r,0,rs));
inertia=eval(int(r,r,0,rs));
e=(-2*I+J+K)/(inertia+eps),
dWp=[dWp,e];
end;
plot(q0:dq:m,dWp),
title('\deltaW_p/\deltaK versus q_0');
xlabel('q_0'); ylabel('\deltaW_p/\deltaK ');
```

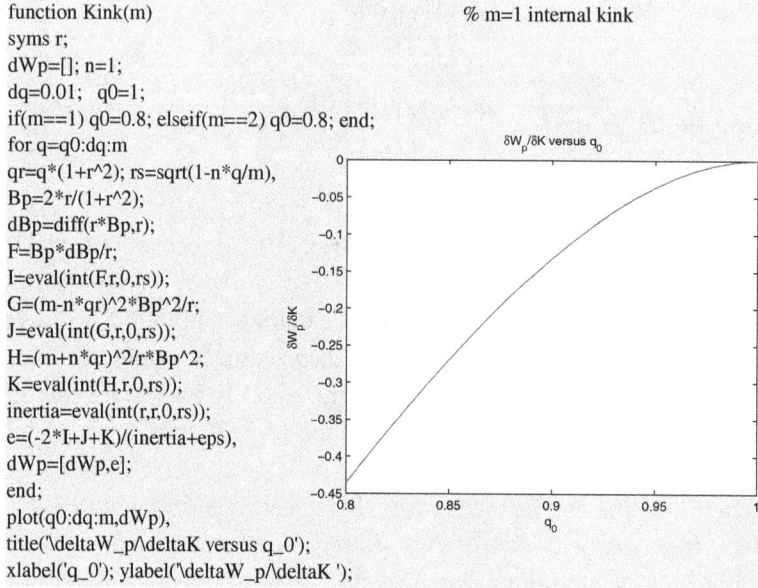

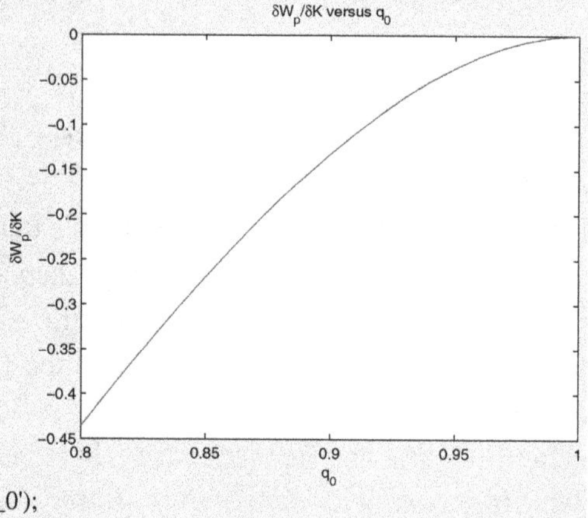

Homework 10.7: Investigate the $m = 1$ and $m = 2$ kink stability condition for the current profile $(1 + r^2/a^2)^{-2}$ assuming the central safety factor is $q_0 = 0.9$.

10.3 The Toroidal Magnetic Confinement

The plasma safety factor, q is important in toroidal magnetic confinement. It denotes the number of times a magnetic field line goes around a torus "the long way" (toroidally) for each time around 'the short way' (poloidally), i.e., $q \equiv \Delta\phi/\Delta\theta = \oint (d\theta/2\pi)(rB_T/RB_p)$, where r is the minor radius, R the major radius, and B_T and B_p are the toroidal and poloidal magnetic fields. The inverse of safety factor is the **rotational transform**, $2\pi/q \equiv \iota$ (iota). In addition to stiffening, by the axial magnetic field, the current channel to stabilize the kink instability, the safety factor provides the magnetic well to hold up higher plasma pressure.

10.3.1 *Magnetic and Velocity Well*

In the tokamak geometry (donut shape) with the magnetic field given by the toroidal magnetic field $\vec{B}_T = \hat{e}_\Phi B_0 R_0/R$ and the poloidal magnetic field $\vec{B}_p = \hat{e}_\theta b_0 r/a$ for $r < a$ due to a uniform toroidal current taken in the cylindrical approximation. The magnetic field lines trace a flux surface given by $Z^2 + (R - R_0)^2 = const.$ as shown in Homework 3.6. If the magnetic field is averaged over the flux surface, it gives

$$\langle \vec{B}^2 \rangle = B_0^2 \oint \frac{d\theta}{2\pi}\left(1+\frac{r}{R_0}\cos\theta\right)^{-2}, \tag{10.55}$$

a magnetic well that is effective in confining plasma pressure and to increase the plasma β as shown by the numerical result. The circular flux surface has equal access to the high and low field regions. Effort to shape the flux surface so to make field lines stay in the high field side more than the low field side, in principle, would increase the averaged magnetic well. The triangularity and elongation of the flux surface, for example, has been under study to optimize this benefit. The **D-shape plasma cross section** in ITER is a testament to this conclusion.

> **Homework 10.7:** Show that the leading oder of the averaged magnetic well as described by Eq. (10.55) in the limit of large aspect ratio , $r/R_0 \ll 1$, is $1.5(r/R_0)^2$.

By rearranging Eq. (10.11) to become

$$\nabla\left(p+\frac{1}{8\pi}B^2+\frac{1}{2}\rho U^2\right)=\frac{1}{4\pi}\vec{B}\cdot\nabla\vec{B}+\rho\vec{U}\times(\nabla\times\vec{U}), \tag{10.56}$$

```
function MagneticWell                % Eq(10.55) is integrated by the Simpson method.
B=[];
r=0.0:0.01:0.5;
theta=0.0:0.01:2*pi;
N=length(theta)
for i=1:length(r)
   f=1./(1+r(i)*cos(theta)).^2;
   b=Simpson(f,N,0.01)/2/pi;
   B=[B,b(N)];
end;
a=1.5;
b=1+a*r.^2;
plot(r,B,'g*',r,b,'r-');
title('Tokamak Magnetic Well');
xlabel('|R-R_0|/R_0');
ylabel('<B^2/B_0^2>');
text(0.4,1.2,'~1.5*r^2','color','r');
```

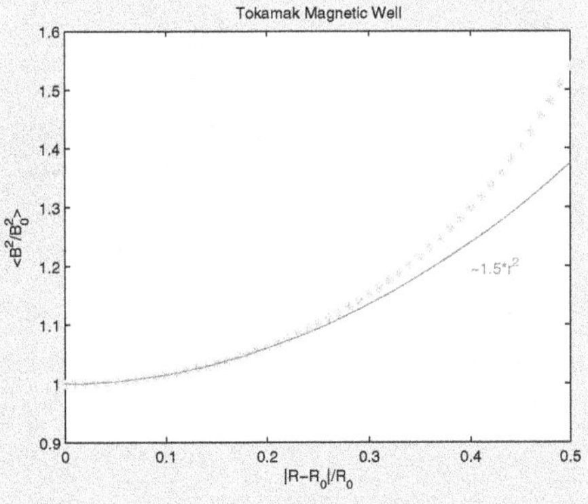

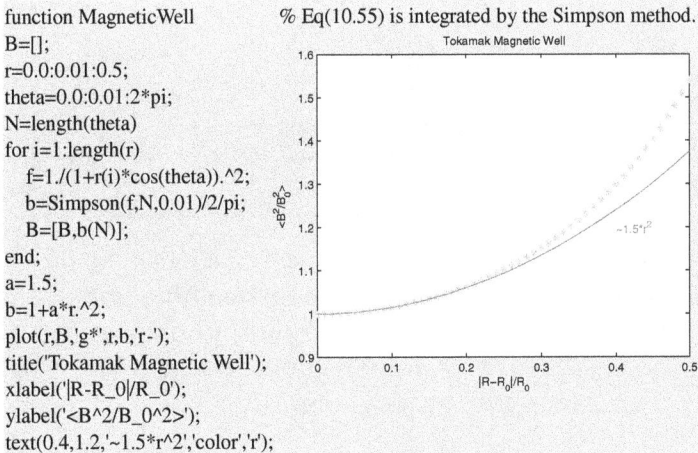

which shows the flow velocity is equivalent to the magnetic field and can therefore hold up additional plasma pressure if the flow profile is properly shaped.

10.3.2 *Toroidal Equilibrium*

The axisymmetric toroidal equilibrium can be described by the **Grad-Shafranov equation**,

$$\Delta^* \psi = -F(\psi)F'(\psi) - 4\pi R^2 p'(\psi). \tag{10.23}$$

Both $F(\psi)$ and $p(\psi)$ are often prescribed pertaining to the experimental measurements that lacks the theoretical foundation but sheds light for a more serious investigation. A simpler case is the force free toroidal equilibrium described by

$$\begin{aligned}
\frac{4\pi}{c}\vec{J} &= -\Delta^* \psi \nabla\phi + F'(\psi)\nabla\psi \times \nabla\phi = \lambda \vec{B}, \\
\Delta^* \psi &= -\lambda(\lambda\psi + F_0), \\
F' &= \lambda, \\
\vec{B} &= \nabla\psi \times \nabla\phi + (F_0 + \lambda\psi)\nabla\phi
\end{aligned}$$

that can be solved exactly but difficult to extract the underlying physics.

Homework 10.8: The toroidal MHD equilibrium can be described by

$$\Delta^* \psi = -\lambda^2 \psi - \lambda F_0$$

for the force free state. Solve the equation by the separation of variables.

Homework 10.9: Transform the differential operator Δ^* from coordinates of (R, Φ, Z) to (r, θ, z).

We will instead solve the force free toroidal equilibrium by the large-aspect-ratio expansion, i.e. $R \gg r$. By transforming from the (R, Φ, Z) coordinates to (r, θ, z) with $R = R_0 + r\cos\theta$, $Z = r\sin\theta$, it is straightforward to show that $\partial^2/\partial R^2 + \partial^2/\partial Z^2 = \partial^2/\partial r^2 + (1/r)\partial/\partial r + \partial^2/\partial z^2$. Applying $\partial/\partial R = \cos\theta(\partial/\partial r) - (\sin\theta/r)(\partial/\partial\theta)$, we are to solve

$$\begin{aligned}
&\left(\frac{\partial^2}{\partial r^2} + \frac{1}{r}\frac{\partial}{\partial r} + \frac{1}{r^2}\frac{\partial^2}{\partial\theta^2} - \frac{\cos\theta}{(R_0 + r\cos\theta)}\frac{\partial}{\partial r} + \frac{\sin\theta}{(R_0 + r\cos\theta)r}\frac{\partial}{\partial\theta}\right)\psi \\
&\quad = -\lambda^2\psi - \lambda F_0.
\end{aligned} \tag{10.57}$$

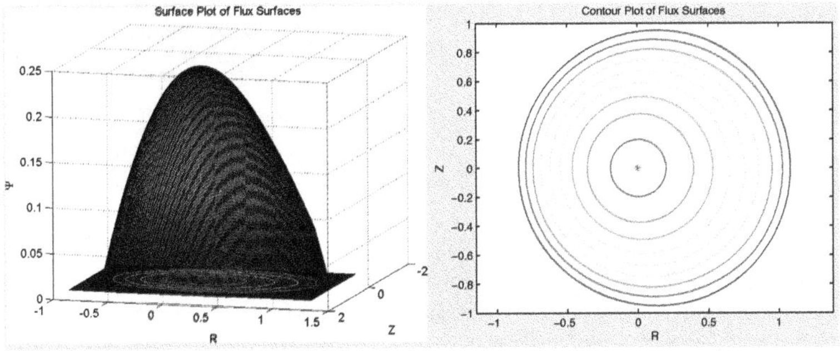

To lowest order, we have

$$\left(\frac{\partial^2}{\partial r^2}+\frac{1}{r}\frac{\partial}{\partial r}+\frac{1}{r^2}\frac{\partial^2}{\partial\theta^2}+\lambda^2\right)\psi^{(0)}=-\lambda F_0, \tag{10.58}$$

and the solution is $\psi^{(0)}=\psi_0 J_0(\lambda r)-F_0/\lambda$. To the next order, we have

$$\left(\frac{\partial^2}{\partial r^2}+\frac{1}{r}\frac{\partial}{\partial r}+\frac{1}{r^2}\frac{\partial^2}{\partial\theta^2}+\lambda^2\right)\psi^{(1)}=\frac{\cos\theta}{R_0}\frac{\partial}{\partial r}\psi^{(0)}. \tag{10.59}$$

The solution can be expressed as

$$\psi^{(1)}(r,\theta)=\frac{\psi_0\cos\theta}{\lambda R_0}\varphi(\lambda r). \tag{10.60}$$

Defining $\xi\equiv\lambda r$, we have

$$\left(\frac{\partial^2}{\partial\xi^2}+\frac{1}{\xi}\frac{\partial}{\partial\xi}-\frac{1}{\xi^2}+1\right)\varphi(\xi)=-J_1(\xi). \tag{10.61}$$

Its solution is

$$\varphi(\xi)=\int_0^{\xi}dt\,J_1(t)\,(Y_1(t)J_1(\xi)+Y_1(\xi)J_1(t))\,/\,(J_0(t)Y_1(t)-J_1(t)Y_0(t)).$$

Therefore,

$$\psi(r,\theta)=\psi_0\left[J_0(\lambda r)+\frac{\cos\theta}{\lambda R_0}\int_0^{\lambda r}dt\,J_1(t)\frac{Y_1(t)J_1(\lambda r)+Y_1(\lambda r)J_1(t)}{J_0(t)Y_1(t)-J_1(t)Y_0(t)}\right]-\frac{F_0}{\lambda}. \tag{10.62}$$

The numerical plot shows that the inner flux surfaces remain intact to the geometrical center, the outer flux surfaces shift outward.

% $F_0 = \psi_0 \lambda J_0(\lambda a)$ is chosen to make $\psi^{(0)} = 0$ at $r = a$. The outer flux surfaces are shifted by $\psi^{(1)}$ effect. F_0 has the physical meaning as the external current that produces the vacuum toroidal field.

```
function ForceFree
R=3; a=1;
dx=0.01;
x=-1:dx:1.2;
y=x;
N=length(x);
X=repmat(x,N,1);
Y=repmat(y',1,N);
lambda=1;
r=sqrt(X.^2+Y.^2);
theta=atan2(Y,X);
PSI=besselj(0,lambda*r)-besselj(0,lambda);
xi=lambda*x;
t=eps:dx:1+eps; t=t*lambda;Nt=length(t);
f=besselj(1,t).*bessely(1,t)./(besselj(0,t).*bessely(1,t)-besselj(1,t).*bessely(0,t));
g=besselj(1,t).*besselj(1,t)./(besselj(0,t).*bessely(1,t)-besselj(1,t).*bessely(0,t));
I=Simpson(f,Nt,dx*lambda);
J=Simpson(g,Nt,dx*lambda);
phi=I.*besselj(1,lambda*t)+J.*bessely(1,lambda*t);
PSI1=zeros(N);
for i=1:N
  for j=1:N
    s=r(i,j);
    is=fix(s/dx+1);
    if(is>Nt) is=Nt; elseif(is<1) is=1; end;
    PHI=phi(is);
    PSI1(i,j)=cos(theta(i,j))/R/lambda*phi(is);
  end;
end;
PSI=(PSI+PSI1);
PSI=PSI.*(PSI>0);

figure(1);
surfc(X,Y,PSI); xlabel('R'); ylabel('Z'); zlabel('\Psi');
view(-16,-12); title('Surface Plot of Flux Surfaces');
figure(2);
maxPSI=max(max(PSI,[],1),[],2);
dPSI=maxPSI/10;
[c,h]=contour(X,Y,PSI,[0.4:-0.025:0.01]); % clabel(c,h);
title('Contour Plot of Flux Surfaces');
axis([-1 1.2 -1 1]); xlabel('R'); ylabel('Z');
axis equal; hold on; plot(0,0,'r*'); hold off;
```

10.3.3 *Toroidal Alfvén Eignemode*

Due to the rotational transform, a wave in the toroidal geometry would experience periodicity in the poloidal angle as it propagates along the field lines given the toroidal axisymmetry. It is analogous to the Bloch electrons in the lattice, where the plane wave solution persists for the Schrodinger equation, but the **Bloch-Floquet theorem** gives an important phase factor that determines the energy band and band gap, the Bloch momentum and the current conduction.

We consider the MHD waves in the toroidal confinement with eigenfrequency ω, and the toroidal mode number n. Linearizing Eq. (10.3) gives the wave flow velocity driven by the $J \times B$ force whereas the pressure term is neglected,

$$-i\omega\rho_0\delta\vec{u} = \frac{1}{4\pi}(\nabla \times \delta\vec{b}) \times \vec{B} + \frac{1}{c}\vec{J} \times \delta\vec{b}. \tag{10.63}$$

Substituting $\delta\vec{u}$ into the linearized Faraday's law of Eq. (10.5), we have the eigenvalue equation,

$$-4\pi\omega^2\rho_0\delta\vec{b} = \nabla \times \left([(\nabla \times \delta\vec{b}) \times \vec{B} + (\nabla \times \vec{B}) \times \delta\vec{b}] \times \vec{B}\right). \tag{10.64}$$

Defining the eignevalue $\lambda \equiv \omega^2/\omega_A^2$, the Alfvén frequency $\omega_A \equiv V_A/R_0$, the Alfvén velocity $V_A \equiv \sqrt{B_0^2/4\pi\rho_0}$ and B_0 the magnetic field at the magnetic axis where $R = R_0, Z = 0$, we rewrite Eq. (10.63) as

$$\lambda\delta\vec{b} = \nabla \times (\vec{b} \times [(\nabla \times \delta\vec{b}) \times \vec{b} + (\nabla \times \vec{b}) \times \delta\vec{b}]), \tag{10.64a}$$

where we further define $\vec{b} \equiv \vec{B}/B_0$. We also define the unit vector $\hat{e}_B \equiv \vec{b}/b$. The structure of toroidal eigenmode obviously depends on the MHD equilibrium. We chose to neglect the pressure effect on the Alfvén wave, and will consider the force free toroidal equilibrium state (cf. Homework 10.2) for simplification by taking $\vec{J} = \sigma\vec{B}$, $\Delta^*\psi = -\sigma^2\psi$, and $\nabla \times \vec{b} = \mu\vec{b}$, where $\mu \equiv 4\pi\sigma/c$. Both compressional and shear Alfvén waves are described in Eq. (10.64a). Since the compressional Alfvén wave is energetically more difficult to excite (cf. Homework 5.6), we will consider the shear Alfvén wave only, and look for the wave solution that has $\delta\vec{b} \cdot \vec{b} \to 0$, $\delta\vec{j} \cdot \vec{b} \gg |\delta\vec{j} \times \vec{b}|$, and $k_{||} \gg k_{\perp}$. Equation (10.64a) is expanded to give,

$$\lambda\delta\vec{b} = \nabla \times \left(\frac{4\pi}{c}(\delta\vec{j}b^2 - \vec{b}\vec{b} \cdot \delta\vec{j}) + \mu\vec{b}\vec{b} \cdot \delta\vec{b} - \mu b^2\delta\vec{b}\right). \tag{10.64b}$$

Note that if we keep only the leading order in Eq. (10.64b) by neglecting derivatives on the equilibrium quantities in comparison to that on the wave quantities, we are left with the $\delta\vec{j}$ terms in Eq. (10.64b) that are further expanded to

$$\lambda\delta\vec{b}\frac{c}{4\pi} \approx b^2\nabla \times \delta\vec{j} - \nabla(\vec{b} \cdot \delta\vec{j}) \times \vec{b} \approx (\vec{b}\vec{b} \cdot \nabla \times \delta\vec{j} - (\vec{b} \cdot \nabla\delta\vec{j}) \times \vec{b}.$$

Here we have applied the following identities and retained only derivaties on the wave variables:

$$\nabla \times (f\vec{A}) = f\nabla \times \vec{A} + \nabla f \times \vec{A},$$

$$\nabla(\vec{A} \cdot \vec{B}) = \vec{A} \times (\nabla \times \vec{B}) + \vec{B} \times (\nabla \times \vec{A}) + \vec{A} \cdot \nabla\vec{B} + \vec{B} \cdot \nabla\vec{A}.$$

The term $\vec{b} \cdot \nabla \times \delta\vec{j}$ when expanded by local wave vectors in the WKB sense gives $\vec{b} \cdot \delta\vec{b}k^2 - \vec{b} \cdot \vec{k}\vec{k} \cdot \delta\vec{b}$, which has primarily the compressional $\vec{b} \cdot \delta\vec{b}$ and the longitudinal $\vec{k} \cdot \delta\vec{b}$ components, and is therefore ignored. Recall that the Alfvén waves are transverse waves. Therefore, utilizing

$$\nabla \times \delta\vec{b} = \frac{4\pi}{c}\delta\vec{j} \approx \frac{-in}{R}\delta b_Z \widehat{e}_R + \frac{in}{R}\delta b_R \widehat{e}_Z - \left(\frac{\partial \delta b_Z}{\partial R} - \frac{\partial \delta b_R}{\partial Z}\right)\widehat{e}_\phi. \tag{10.65}$$

and expanding $\lambda\delta\vec{b} \approx -\frac{4\pi}{c}(\vec{b} \cdot \nabla\delta\vec{j}) \times \vec{b}$ in the (R, Z, ϕ) directions, we have

$$\lambda\delta b_R = (\vec{b} \cdot \nabla)\left(\frac{in}{R}\delta b_R b_\phi + \left(\frac{\partial \delta b_z}{\partial R} - \frac{\partial \delta b_R}{\partial Z}\right) b_Z\right), \tag{10.66a}$$

$$\lambda\delta b_Z = (\vec{b} \cdot \nabla)\left(\frac{in}{R}\delta b_z b_\phi - \left(\frac{\partial \delta b_z}{\partial R} - \frac{\partial \delta b_R}{\partial Z}\right) b_R\right). \tag{10.66b}$$

With use of $\vec{b} \cdot \delta\vec{b} \approx b_R\delta b_R + b_Z\delta b_Z \approx 0$, we may replace $b_Z\partial\delta b_Z/\partial R \approx -b_R\partial\delta b_R/\partial R$ and $b_Z\partial\delta b_Z/\partial Z \approx -b_R\partial\delta b_R/\partial Z$ to get

$$\lambda\delta b_{R,Z} = -(\vec{b} \cdot \nabla)(\vec{b} \cdot \nabla)\delta b_{R,Z}, \tag{10.67}$$

which yields exactly the dispersion relation for the shear Alfvén wave. It, however, needs accuracy to order $\varepsilon \sim r/R \sim B_p/B_T \sim b_p \sim 1/m \sim 1/n \ll 1$ to have the toroidal effect.

Many other terms are of order ε in Eq. (10.64). They have to be checked for consistency to describe the shear Alfvén wave. A simpler model is analyzed here by taking from the dispersion function of Eq. (10.67), i.e.,

$$\frac{\omega^2}{\omega_A^2} = k_\parallel^2 R_0^2 = \left(\frac{n^2}{R^2}\frac{B_\varphi^2}{B_0^2} + \frac{m^2}{r^2}\frac{B_\theta^2}{B_0^2}\right) R_0^2 \approx n^2(1 - 2\varepsilon\cos\theta) + \frac{m^2}{q^2},$$

to yield the following eigenvalue eqution,

$$\lambda\Psi = n^2(1 - 2\varepsilon\cos\theta)\Psi - \frac{1}{q^2}\frac{\partial^2}{\partial\theta^2}\Psi. \tag{10.68}$$

This is a Mathieu's equation and its solutions are solved by the power method as in the accompanying code.

```
function Mathieu(n)                                              %Solve the equation by
a=2; R0=6.2; B0=5.3; I0=15e6; V=850; q=2;                        the POWER method and
epsilon=a/R0;                                                    compare with the
N=5000; dTheta=2*pi/N;                                           analytical solution.
theta=(dTheta:dTheta:2*pi)';
A=setM(n,N,epsilon,q,theta,dTheta);
b=cos(n*theta); lambda=1;
[lambda,J]=Power(A,lambda,N,dTheta,b);
plot(theta,J/J(1),'g.');
 title('Mathieu Function'); xlabel('\theta'); ylabel('\psi');

function [lambda,y]=Power(A,lambda,N,dTheta,b)
lambda0=-10; iter=0;
while(abs(lambda/lambda0-1)>1.0e-6&&iter<100)
   iter=iter+1;
   lambda0=lambda;
   M=A+diag(lambda*ones(N,1));
   y=M\b;
   lambda=lambda-(b'*b)/(b'*y),
   b=y/norm(y);
   D=norm(A*b+lambda*b),
end;
y=b; iter, lambda0,

function M=setM(n,N,epsilon,q,theta,dTheta)
M=sparse(N,N);
dt2=diag(ones(N-1,1),1)+diag(ones(N-1,1),-1)-
2*diag(ones(N,1));
dt2(1,N)=1; dt2(N,1)=1;
M=M+dt2/(dTheta*q)^2-n^2*(1-2*epsilon*diag(cos(theta)));
```

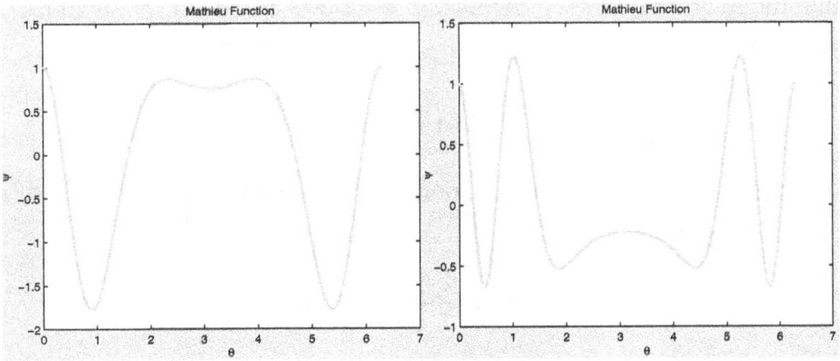

10.4 Stastistical MHD

The axisymmetric toroidal MHD equilibrium governed by the Grad-Shafranov equation,

$$\Delta^* \psi = -F(\psi)F'(\psi) - 4\pi R^2 p'(\psi) = -\frac{4\pi}{c} R J_\phi,$$

(cf. Eq. (10.23)), has countless degrees of freedom in the Hilbert space to construct $F(\psi)$ and $p(\psi)$ so to fit the experimental observables, notably the central and edge safety factors, the plasma beta, and more. While solutions thus obtained by choosing model equilibrium profiles would reveal important theoretical properties, the question arises as what this arbitrariness means in the fundamental physics.

10.4.1 *The Microcanonical Ensemble*

From the theoretical point of view, it is natural to ask whether the statistical physics may apply here in that the microcanonical ensemble would place the system to its minimum energy state, while the canonical ensemble allows certain probabilites for higher energy states thus the MHD fluctuations (cf. Sec. 7.7). The missing link is a constraint lest the system fall into vacuum. Higher than second order in field variables decaying faster than the energy itself are ruled out as proper constraints. A simple one in fact is readily available as we ask given the same total plasma current

$$\frac{4\pi}{c}I = \int dS\left(\frac{F(\psi)F'(\psi)}{R} + 4\pi Rp'(\psi)\right)$$
$$= \int \frac{d\tau}{2\pi R^2}\left(F(\psi)F'(\psi) + 4\pi R^2 p'(\psi)\right), \tag{10.69}$$

which state of $F(\psi)$ and $p(\psi)$ has the lowest total plasma energy. Here, J_ϕ is integrated over the minor cross section, and converted to the volume integration. Recall that the magnetic field is given by $\vec{B} = \nabla\psi \times \nabla\phi + F(\psi)\nabla\phi$ of Eq. (10.22), and the total plasma energy is from Eq. (10.28) without the flow energy,

$$W = \int \frac{d\tau}{8\pi R^2}\left(|\nabla\psi|^2 + |F(\psi)|^2 + \frac{8\pi}{\gamma - 1}R^2 P(\psi)\right). \tag{10.70}$$

We are to minimize the total plasma energy W subject to the constancy of the total plasma current I so that

$$\delta W + \frac{1}{\lambda}\frac{4\pi}{c}\delta I = 0, \tag{10.71}$$

by applying the variational principle for arbitray perturbations of $\delta\psi$, where λ^{-1} is the Lagrangian multiplier.

10.4.2 *The Canonical Profile*

Varying the poloidal magnetic energy gives

$$\delta W_p = \int \frac{d\tau}{4\pi R^2}\nabla\psi \cdot \nabla\delta\psi = \int \frac{d\tau}{4\pi}\nabla \cdot (\delta\psi R^{-2}\nabla\psi) - \int \frac{d\tau}{4\pi}\nabla \cdot (R^{-2}\nabla\psi)\delta\psi.$$

The first term can be converted to the surface integral and is inconsequential if the perturbations are null, $\delta\psi = 0$, on the plasma surface that limits the relaxation process to internal instabilities. Dropping the surface term, we get $\delta W_p = -\int \frac{d\tau}{4\pi R^2}\Delta*\psi\delta\psi$. Equation (10.71) without much more algebra results in the following equation,

$$-\Delta * \psi + FF' + \frac{3}{2}4\pi R^2 P' = 2\lambda^{-1}\left(\frac{\partial}{\partial\psi}FF' + 4\pi R^2\frac{\partial}{\partial\psi}p'\right). \tag{10.72}$$

Replacing $\Delta * \psi$ from the equilibrium equation, Eq. (10.23), we end with

$$FF' + \frac{5}{4}4\pi R^2 P' = \lambda^{-1}\left(\frac{\partial}{\partial\psi}FF' + 4\pi R^2\frac{\partial}{\partial\psi}p'\right), \tag{10.73}$$

which has to be satisfied for arbitrary R and ψ. Therefore, it demands that

$$FF' = \lambda^{-1}\frac{\partial}{\partial\psi}(FF'), \quad \frac{5}{4}p' = \lambda^{-1}\frac{\partial}{\partial\psi}p'. \tag{10.74}$$

Solving Eq. (10.74) to give $FF' = c_F\exp(\lambda\psi)$ and $p' = c_P\exp(\frac{5}{4}\lambda\psi)$, we then have the minimum energy state of canonical profiles,

$$\Delta * \psi = -c_F e^{\lambda\psi} - R^2 c_P e^{\frac{5}{4}\lambda\psi} = -c_F\left(e^{\lambda\psi} + \frac{R^2}{R_0^2}\beta_J e^{\frac{5}{4}\lambda\psi}\right) = -\frac{4\pi}{c}RJ_\phi. \tag{10.75}$$

Here $\beta_J \equiv C_p R_0^2/C_F$ is a measure of the fraction of toroidal current the pressure induced. Note that $p(\psi) = \frac{5}{4\lambda}c_P(\exp(\frac{5}{4}\lambda\psi) - 1) + p_0$, with p_0 the pressure at $\psi = 0$. Thus, β_J is proportional to the plasma β_p, but differs substantially if p_0 is significant. In the similar line of attack, one may consider, for example, maximizing $H = \int d\tau J_\phi \ln J_\phi$, the entropy of the system. It has also been shown by Montgomery *et al.* to have the canonical profile.

We first consider zero beta and simplified geometry. In Sec. 8.3.2, and Homework 8.5, solutions to Eq. (10.75) in the slab and cylindrical geometries were found for equilibria of $\beta_J = 0$. They reveal bifurcated solutions, profile consistency, and the loss of equilibrium condition that may be interpreted as a **major disruption** scenario. The results are summarized as in the following: By normalizing the length scale to the plasma radius a, the slab geometry has the governing equation, $d^2\psi/dx^2 + \Lambda e^\psi = 0$. Defining $\Lambda = C_F a^2\lambda$, we have the solution,

$$\begin{aligned}\int_0^\psi \frac{dy}{\sqrt{e^{\psi_c} - e^y}} &= \pm\sqrt{2\Lambda}(x-1)\\ &= 2e^{-\frac{1}{2}\psi_c}\left(\tanh^{-1}(\sqrt{1-e^{-\psi_c}}) - \tanh^{-1}(\sqrt{1-e^{\psi-\psi_c}})\right),\end{aligned} \tag{10.76}$$

where the boundary conditions are taken as $d\psi/dx|_{x=0} = \psi|_{x=1} = 0$. By substituting $x = 0$ where $\psi = \psi_c$ into Eq. (10.76), it gives a solubility condition, $\sqrt{\frac{1}{2}\Lambda} = e^{-\frac{1}{2}\psi_c} \tanh^{-1}(\sqrt{1 - e^{-\psi_c}}) < 0.6627$. When $\Lambda > 0.8785$, no solution exists. At $\Lambda = 0.8785$, there is one solution of ψ_c, and for $\Lambda < 0.8785$, there are bifurcated solutions.

> **Homework 10.10: MHD equilibrium — slab model:** Solve the magnetic flux function numerically as described by
>
> $$d^2\varphi/dx^2 = -c_F e^{\lambda\psi}$$
>
> with use of matrix inversion.

The cylindrical geometry has the governing equation, $d^2\psi/dr^2 + (1/r)d\psi/dr + \Lambda e^{\psi} = 0$, where the length is normalized. The solution is given by

$$\psi = \psi_c - \log\left(1 + \frac{1}{8}\Lambda e^{\psi_c} r^2\right)^2, \tag{10.77}$$

where we take the boundary condition $\psi(r = 0) = \psi_c$. Assigning the value $\psi = 0$ at $r = 1$, we find $e^{\psi_c/2} = 1 + \frac{1}{8}\Lambda e^{\psi_c}$, which can then be solved to give

$$e^{\frac{1}{2}\psi_c} = 1 + \frac{1}{8}\Lambda e^{\psi_c} = \frac{1 \pm \sqrt{1 - \frac{1}{2}\Lambda}}{\frac{1}{4}\Lambda}. \tag{10.78}$$

No equilibrium solution can exist when $\Lambda > 2$. The solubility condition is therefore $\Lambda \leq 2$. The toroidal current and the poloidal magnetic field are given by, respectively,

$$J_z = J_0\left(1 + \frac{1}{8}\Lambda e^{\psi_c} r^2\right)^{-2}, \tag{10.79}$$

$$B_\theta = \frac{2\pi}{c} J_0 r \left(1 + \frac{1}{8}\Lambda e^{\psi_c} r^2\right)^{-1}. \tag{10.80}$$

Given the same Λ, or equivalently FF', there are bifurcated solutions: a branch with flatter current profile and a branch with a more peaked current profile, as indicated in Eq. (10.79). Unequivocally, the flatter profile has a lower q_l/q_0 and better energy content, while the peaked profile has the opposite. The flatter profile would have higher q_0 and less likely to cuase the $m = 1$ internal kink instability, and consequently the sawtooth oscillation, for $q_0 > 1$ stabilizes the kink (cf. Sec. 10.2.4). The toroidal geometry appears to prohibit the magnetic energy and the pressure energy to exchange freely. As a result, they are not in thermal equilibration to have the same "temperature". While the relaxation process allows arbitray perturbations/instabilities that leave the total current intact, the minimum energy state are still prone to instabilities that destroy the constancy of the total plasma current. And, the surface instabilities playing no role during the relaxation process have to be checked also. The minimum energy state as governed by Eq. (10.75) has separated the relaxation process from serious MHD instabilities, and provides a unique current profile for given q_0 and q_l, the central and edge safety factors. Moreover, to achieve the consistent MHD profile, the plasma has to fine-tune the local energy content leading to the transport and confinement properties, presumably not the same as what the Fokker Planck model could predict.

10.4.3 *MHD Equilibrium and Plasma Transport*

Consider the scenario that while the MHD equilibrium is being maintained with the ohmic heating, the plasma energy is transported to keep the detailed balance. This readily provides us the information regarding the transport coefficient. The heat transport equation can be written as

$$\nabla \cdot (\chi \nabla T) + J_z E_z = 0. \tag{10.81}$$

Here we take from Eq. (9.8) by retaining only the energy source term $\vec{v} \cdot \vec{F}$ and a diffusion term. Replacing $\frac{4\pi}{c} J_z = (1/r)(d/dr) r B_\theta$, we find $\chi dT/dr = -E_z B_\theta$ by considering only the radial transport. The temperature profile is determined from the electrical conductivity with use of Eq. (10.79)

$$J_z = \sigma_0 E_z T^{3/2} = J_0 \left(1 + \frac{1}{8} \Lambda e^{\psi_c} r^2\right)^{-2}.$$

This yields the heat conduction

$$\chi \propto \left(1 + \frac{1}{8} \Lambda e^{\psi_c} r^2\right)^{3/4}.$$

We may further evaluate from

$$(\partial/\partial t) \frac{3}{2} n k_B T = \nabla \cdot (\chi \nabla T)$$

to arrive at the energy confinement time

$$\tau_E \approx \frac{3 n k_B a^2}{2\chi} \approx \frac{3 n k_B T}{2 E_z \langle J \rangle} \propto \frac{n T R^2}{V B_0} q_l, \tag{10.82}$$

where $V = 2\pi R E_z$ is the one-turn voltage.

10.4.4 *The Bifurcated Toroidal MHD Equilibrium*

Given the miminum energy state, there is still this question regarding which bifurcated solution will appear. The numerical program by the matrix inversion tends to find the H-branch of the flatter current profile much more easily. This however is a manifestation of the numerical algorithm. The shooting method would have no preference. Conversely, the edge plama condition in the experiment is crucial since it has to support the larger energy content at edge to support the flat profile. It is likely, however, that even one simple mechanism can be an obstacle to reach the bifurcated states. The relaxed MHD equilibrium state must survive all the disruptive effects.

We here further examine the toroidal MHD equilibrium. Solving for F^2 and p from Eq. (10.74), we find

$$\frac{1}{2}(F^2 - F_0^2) = \frac{c_F}{\lambda}(e^{\lambda\psi} - e^{\lambda\psi_0}),$$

and

$$p(\psi) = p_0 + \frac{4\Lambda\beta_J}{5a^2}\left(e^{\frac{5}{4}\lambda\psi} - e^{\frac{5}{4}\lambda\psi_0}\right) \tag{10.83}$$

Homework 10.11: MHD equilibrium — cylindrical model: Solve the magnetic flux function numerically as described by $d^2\psi/dr^2 + (1/r)d\psi/dr + \Lambda e^{\psi} = 0$ with use of matrix inversion.

where the subscript 0 refers to the corresponding values at the magnetic axis. If the ψ value vanishes at the plasma edge, the edge pressure is given by $p_1 = p_0 + \frac{4\Lambda}{5a^2}\beta_p(1 - e^{\frac{5}{4}\lambda\psi_0})$. Note that $p \sim (aR_0\lambda)^{-2}\Lambda\beta_J \sim \Lambda\beta_J B_p^2$. Therefore, $\beta_p \sim 8\pi\Lambda\beta_J$.

$$p(\psi) = p_0\frac{1 - e^{\frac{5}{4}\lambda\psi}}{1 - e^{\frac{5}{4}\lambda\psi_0}} + p_1\frac{e^{\frac{5}{4}\lambda\psi} - e^{\frac{5}{4}\lambda\psi_0}}{1 - e^{\frac{5}{4}\lambda\psi_0}}. \tag{10.83a}$$

Equation (10.75) has been shown to have bifurcated solutions in slab and cylinder models. We want to solve Eq. (10.75) in the toroidal geometry by taking $Z = r\sin\theta$, $R = R_0 + r\cos\theta$, and assuming the circular outermost flux surface $\psi(r = a) = 0$. Therefore, it can be cast into

$$\left(\frac{\partial^2}{\partial\bar{r}^2} + \frac{1}{\bar{r}}\frac{\partial}{\partial\bar{r}} + \frac{1}{\bar{r}^2}\frac{\partial^2}{\partial\theta^2} - \frac{\varepsilon\cos\theta}{(1+\varepsilon\bar{r}\cos\theta)}\frac{\partial}{\partial\bar{r}} - \frac{\varepsilon\sin\theta}{(1+\varepsilon\bar{r}\cos\theta)\bar{r}}\frac{\partial}{\partial\theta}\right)\Psi$$
$$= -\Lambda\left(e^{\Psi} + (1+\varepsilon\bar{r}\cos\theta)^2\beta_J e^{\frac{5}{4}\Psi}\right), \tag{10.84}$$

where $\bar{r} \equiv r/a$, $\varepsilon \equiv a/R_0$, $\Psi \equiv \lambda\psi$, and $\Lambda \equiv c_F a^2\lambda$.

The numerical algorithm to find the bifurcated toroidal MHD equilibria can be challenging, and rightfully so in most equations with bifurcated solutions. Many numerical algorithms often produce without hesitation one solution (the strong, the dominant), but not the other (the weak, the subdominant).

We may devise a method utilizing the **physical characteristics** to search for the bifurcated solutions. A successful implementation has been to identify the edge poloidal B field as the key physical variable to be evaluated from two independent functions: One is related to the LHS of Eq. (10.84), namely, $\langle B_p(r = a)\rangle = \langle|\nabla\psi/R|\rangle$ and the other the RHS, $\langle B_p(r = a)\rangle = \frac{4\pi}{c}\int d\vec{s}\cdot\vec{J}_\phi/2\pi a$, where $\langle\,\rangle \equiv \oint d\theta/2\pi$. At the start of iteration, these two values differ significantly but merge into one when the ψ solution is accurately solved and the iteration converges. The central ψ value is adjusted according to either the ratio of these two values or its inverse. Thus, the iterations may follow two different paths in searching for the bifurcated solutions. The concept can apply to any physical quantity that is derivable from the two parts of the equation. This methodology provides a powerful algorithm, not only for this particular problem but other bifurcation equations as well.

10.5 Relativistic MHD

For plasmas of extremely high energy particles, the relativistic effect is important. We derive the **relativistic fluid equations** in the following:

10.5.1 *Relativistic Fluid Equations*

The number density, mass density, particle flux, momentum density and the momentum flux in the lab frame are defined, respectively, as in the following:

$$n(\vec{r},t) \equiv \sum_i \delta(\vec{r} - \vec{r}_i(t)), \tag{10.85a}$$

$$\rho(\vec{r},t) = \sum_i m_i\delta(\vec{r} - \vec{r}_i(t)), \tag{10.85b}$$

$$\vec{\Gamma}(\vec{r},t) \equiv \sum_i \vec{w}_i\delta(\vec{r} - \vec{r}_i(t)), \tag{10.85c}$$

$$\vec{P}(\vec{r},t) \equiv \sum_i \vec{p}_i\delta(\vec{r} - \vec{r}_i(t)), \tag{10.85d}$$

$$\overleftrightarrow{K} \equiv \sum_i \frac{\vec{p}_i\vec{p}_i}{m_i}\delta(\vec{r} - \vec{r}_i(t)), \tag{10.85e}$$

where the notations have been definied in Chap. 4.

Consider the fluid moving at a constant flow velocity $\vec{U}$. In the flow frame, all the variables will be denoted with a prime. A particle, located at $\vec{r}$ position in the lab frame, is located in the flow frame at $\vec{r}'$ given by $\vec{r}' = \vec{r}_\perp + \gamma_U(\vec{r}_{||} - \vec{U}t)$. Here $\gamma_U \equiv 1/\sqrt{1-U^2/c^2}$ is the Lorentz factor due the flow velocity. Therefore, the particle density in the moving frame $n'(\vec{r},t) \equiv \sum_i \delta(\vec{r}' - \vec{r}_i'(t)) = \sum_i \delta(\vec{r}_\perp' - \vec{r}_{i\perp}'(t))\delta(\vec{r}_{||}' - \vec{r}_{i||}'(t)) = \sum_i \delta(\vec{r} - \vec{r}_i(t))/\gamma_U$ is related to the particle density in the lab frame by

$$n'(\vec{r},t) = \frac{1}{\gamma_U} n(\vec{r},t), \tag{10.86}$$

where $\vec{r}_\perp' - \vec{r}_{i\perp}'(t) = \vec{r}_\perp - \vec{r}_{i\perp}(t)$ and $\vec{r}_{||}' - \vec{r}_{i||}' = \gamma_U(\vec{r}_{||} - \vec{r}_{i||})$ have been applied. Equation (10.86) is the consequence of the relativistic **length contraction** since $n(\vec{r},t) = N/d\tau$ and $n'(\vec{r},t) = N/d\tau' = N/\gamma_U d\tau = n(\vec{r},t)/\gamma_U$, which is related to the volume element in the moving frame $d\tau'$ by $d\tau = d\tau'/\gamma_U$. Here N is the total number of particles in the volume element $d\tau$. It is straightforward to derive

$$\frac{\partial n(\vec{r},t)}{\partial t} = -\nabla \cdot \vec{\Gamma}, \tag{10.87}$$

where $\vec{\Gamma} \equiv \sum_i \vec{w}_i \delta(\vec{r} - \vec{r}_i(t))$ is the particle flux.

Homework 10.12: the continuity equation: Generalize Eq. (10.87) to the continuity equation of the mass density from the definition, namely, $\rho(\vec{r},t) = \sum_i m_i \delta(\vec{r} - \vec{r}_i(t))$. Assume the external force only arises from the electromagnetic fields.

Consider a particle with the momentum $\vec{p}$ and energy ε in the lab frame. Its momentum in the moving frame is given by $\vec{p}' = \vec{p}_\perp + \gamma_U(\vec{p}_{||} - \vec{U}\varepsilon/c^2)$, where the subscripts $||$ and $\perp$ refer to the components parallel and perpendicular to the flow, respectively. The fluid momentum defined by $\vec{P}(\vec{r},t) \equiv \sum_i \vec{p}_i \delta(\vec{r} - \vec{r}_i(t))$ in the lab frame is related to that in the moving frame by $\vec{P}' = \vec{P}_{||} - \rho\vec{U} + \vec{P}_\perp/\gamma_U$ since $\vec{P}'(\vec{r},t) \equiv \sum_i \vec{p}_i' \delta(\vec{r}' - \vec{r}_i'(t)) = \sum_i [\vec{p}_{i\perp} + \gamma_U(\vec{p}_{i||} - \vec{U}\varepsilon_i/c^2)]\delta(\vec{r} - \vec{r}_i(t))/\gamma_U$, where the energy of the ith particle $\varepsilon_i = m_i c^2$ has been applied and ρ is the mass density in the lab frame. By definition, the momentum perpendicular to the fluid flow vanishes, i.e., $\vec{P}_\perp = 0 = \vec{P}_\perp'$, so does the parallel momentum in the moving frame, i.e., $\vec{P}_{||}' = 0$. Therefore, $\vec{P}_{||} - \rho\vec{U} = 0$ defines the flow velocity

$$\vec{U} = \frac{\vec{P}}{\rho}. \tag{10.88}$$

It is important to recognize that *the total momentum in the flow frame has to vanish*, or the fluid element will move in the flow frame. This can be understood by an imaginary box that contains the fluid. The box is supposed to be stationary in the flow frame. Yet, if the total particle momentum is nonzero, the particles bombarding the box will impart momentum to it and cause the box to move, contradictory to the very definition of the flow frame. Eq. (10.87) defines the flow velocity which is simply the total momentum divided by the mass density, all in the lab frame.

> **Homework 10.13: the average mass in thermal plasma:** Given the relativistic thermal plasma with the Gibb's distribution function $f = N\exp(-\mu\gamma)$ where $N = \mu/4\pi K_2(\mu)$ is the normalization constant and $\gamma = \sqrt{p^2/m_0^2c^2+1}$ is the Lorentz factor, find the expected mass.

The mass density in the flow frame is given by $\rho'(\vec{r},t) \equiv \sum_i \gamma_i' m_0 \delta(\vec{r}' - \vec{r}_i'(t))$. Here, m_0 is the rest mass. Applying the energy relation $\gamma_i' m_0 = \gamma_U(\gamma_i m_0 - \vec{U} \cdot \vec{p}_i/c^2)$, we can relate the mass densities in the two frames as follows: $\rho' = \sum_i (\gamma_i m_0 - \vec{U}\cdot\vec{p}_i/c^2)\delta(\vec{r} - \vec{r}_i(t)) = \rho(1 - U^2/c^2) = \rho/\gamma_U^2$. The mass density relation

$$\rho' = \frac{1}{\gamma_U^2}\rho, \tag{10.89}$$

can be understood from the following: In the flow frame the mass density is given by $\rho'(\vec{r}',t) = N dM_0/d\tau'$, where the mass dM_0, which should not be confused with the true rest mass $n' m_0 d\tau'$ (cf. Homework 10.13). Nonetheless, it is not moving as a whole, and is thus related to the mass in the lab frame by $dM_0 = dM/\gamma_U$. Taking into account the volume element relation, we have $\rho'(\vec{r}',t) = N dM_0/d\tau' = N dM/d\tau/\gamma_U^2 = \rho(\vec{r},t)/\gamma_U^2$.

It is straightforward to derive the continuity equation of mass density (cf. Homework 10.12), which is due to the relativistic mass effect,

$$\begin{aligned}\frac{\partial\rho}{\partial t} + \nabla\cdot\rho\vec{U} &= \frac{q\vec{E}}{c^2}\cdot\vec{\Gamma} + \sum_i \vec{w}_i \cdot \vec{f}_i^{\,\text{int}}\delta(\vec{r} - \vec{r}_i(t)) \\ &= \frac{1}{c^2}\vec{J}\cdot\vec{E} + \frac{1}{c^2}\sum_i \vec{w}_i\cdot\vec{f}_i^{\,\text{int}}\delta(\vec{r} - \vec{r}_i(t)),\end{aligned} \tag{10.90}$$

where $\vec{J} \equiv q\vec{\Gamma}$ is the electrical current and $\vec{f}_i^{\,\text{int}}$ is the internal force on the ith particle. The ohmic heating or cooling by the external electric field $\vec{E}$ can alter the particle mass so to induce source or sink of mass density. Equation (10.90) can be regarded as the energy equation as well where the internal force will cause energy redistribution when the particles move along or against it. It can result in the damping of the flow velocity and the transfer of thermal energy due to the long range force in the plasma.

Homework 10.14: the mass density: Assume that in the flow frame the plasma is governed by the Gibb's distribution function $f = N\exp(-\mu\gamma)$. Show that the mass density ρ in the lab frame is given by $nm_0(K_3(\mu)/K_2(\mu) - 1/\mu)/\sqrt{1-U^2/c^2}$ where $\vec{U}$ is the flow velocity.

The temporal evolution of the particle flux is given by

$$\frac{\partial}{\partial t}\vec{\Gamma}(\vec{r},t) = -\nabla\cdot\sum_i \vec{w}_i\vec{w}_i\delta(\vec{r}-\vec{r}_i(t)) + \sum_i \vec{a}_i\delta(\vec{r}-\vec{r}_i(t)). \tag{10.91}$$

We may define the velocity flux tensor

$$\overleftrightarrow{\Xi}(\vec{r},t) \equiv \sum_i \vec{w}_i\vec{w}_i\delta(\vec{r}-\vec{r}_i(t)).$$

The particle flux in the flow frame is not necessarily nullified, and is given by

$$\vec{\Gamma}'(\vec{r},t) \equiv \sum_i \vec{w}'_i\delta(\vec{r}'-\vec{r}'_i(t)) = \frac{1}{\gamma_U^2}\sum_i \frac{\vec{w}_{i\perp}+\gamma_U(\vec{w}_{i\parallel}-\vec{U})}{(1-Uw_{i\parallel}/c^2)}\delta(\vec{r}-\vec{r}_i(t))$$

$$= \frac{1}{\gamma_U}\sum_i \frac{\vec{w}_{i\parallel}-\vec{U}}{1-Uw_{i\parallel}/c^2}\delta(\vec{r}-\vec{r}_i(t)).$$

Note that due to the symmetry in the perpendicular direction, that component is zero. The particle velocity in the lab frame given by $\vec{w}_i = \vec{p}_i/m_i$ is related to that in the flow frame by

$$\vec{w}'_i = \frac{\vec{p}'_i}{m'_i} = \frac{\vec{p}_{i\perp}+\gamma_U(\vec{p}_{i\parallel}-\vec{U}\varepsilon_i/c^2)}{\gamma_U(m_i-\vec{U}\cdot\vec{p}_i/c^2)} = \frac{\vec{w}_{i\perp}+\gamma_U(\vec{w}_{i\parallel}-\vec{U})}{\gamma_U(1-\vec{U}\cdot\vec{w}_i/c^2)}.$$

The momentum equation can be found by taking the time derivative on the fluid momentum to give

$$\frac{\partial}{\partial t}\vec{P}(\vec{r},t) = \frac{\partial}{\partial t}(\rho\vec{U}) = \sum_i \vec{f}_i\delta(\vec{r}-\vec{r}_i(t)) - \nabla\cdot\sum_i \vec{w}_i\vec{p}_i\delta(\vec{r}-\vec{r}_i(t))$$

$$= \vec{F} - \nabla\cdot\overleftrightarrow{K}. \tag{10.92}$$

Here $\overleftrightarrow{K} \equiv \sum_i \vec{p}_i\vec{p}_i\delta(\vec{r}-\vec{r}_i(t))/m_i$ is the momentum flux. We may define the pressure tensor to be the momentum flux in the flow frame, i.e.,

$$\overleftrightarrow{T}(\vec{r},t) \equiv \sum_i \frac{\vec{p}'_i\vec{p}'_i}{m'_i}\delta(\vec{r}'-\vec{r}'_i(t)). \tag{10.93}$$

The difference between the momentum fluxes in the lab frame and the flow frame defined by

$$\overleftrightarrow{W} \equiv \sum_i \vec{p}_i\vec{p}_i\delta(\vec{r}-\vec{r}_i(t))/m_i - \sum_i \vec{p}'_i\vec{p}'_i\delta(\vec{r}'-\vec{r}'_i(t))/m'_i$$

may be termed as the flow tensor. Therefore,

$$\frac{\partial}{\partial t}(\rho\vec{U}) = \vec{F} - \nabla \cdot \overleftrightarrow{W} - \nabla \cdot \overleftrightarrow{T}. \qquad (10.92a)$$

The pressure energy, same as the nonrelativistic counterpart, is given by $\varepsilon_p \equiv \frac{1}{2}trace\overleftrightarrow{T} = \frac{1}{2}\sum_i p_i'^2\delta(\vec{r}' - \vec{r}_i'(t))/m_i'$. Applying the relation, $p_i'^2 = (m_i'^2 - m_0^2)c^2$, we find (cf. Homework 10.15)

$$\begin{aligned}\varepsilon_p &= \frac{1}{2}\rho' c^2 - \frac{1}{2}m_0^2c^2\sum_i \frac{1}{m_i'}\delta(\vec{r}' - \vec{r}_i'(t))\\ &= \frac{1}{2}n' m_0 c^2\left(\frac{K_3(\mu)}{K_2(\mu)} - \frac{1}{\mu} - \frac{K_1(\mu)}{K_2(\mu)}\right) = \frac{3}{2}n' m_0 c^2\frac{1}{\mu}, \qquad (10.94)\end{aligned}$$

where the relativistic thermal plasma in the flow frame is governed by the Gibb's distribution. The pressure energy is $\varepsilon_p = \frac{3}{2}n'm_0c^2/\mu$. Taking $\mu = m_0c^2/k_BT$, we have $\varepsilon_p = \frac{3}{2}n'k_BT$. It has the identical form to the nonrelativistic case. However, in terms of variables in the lab frame, the pressure energy is $\varepsilon_p = \frac{3}{2}nm_0c^2/\mu/\gamma_U$ since $n'(\vec{r}, t) = n(\vec{r}, t)/\gamma_U$ as given in Eq. (10.85).

By utilizing the transformations of the momentum, $\vec{p}\,' = \vec{p}_\perp + \gamma_U(\vec{p}_{||} - \vec{U}\varepsilon/c^2)$, the volume element $\delta(\vec{r}' - \vec{r}_i') = \delta(\vec{r} - \vec{r}_i)/\gamma_U$, and the mass $m' = \gamma_U m(1 - Uw_{||}/c^2)$, the flow tensor $\overleftrightarrow{W}$ can be expressed as,

$$\begin{aligned}\overleftrightarrow{W} = \sum_i m_i \Bigg[&(\vec{w}_{i\perp} + \vec{w}_{i||})(\vec{w}_{i\perp} + \vec{w}_{i||}) - \frac{1}{1 - Uw_{i||}/c^2}\\ &\times\left(\frac{1}{\gamma_U}\vec{w}_{i\perp} + (\vec{w}_{i||} - \vec{U})\right)\left(\frac{1}{\gamma_U}\vec{w}_{i\perp} + (\vec{w}_{i||} - \vec{U})\right)\delta(\vec{r} - \vec{r}_i(t)).\end{aligned} \qquad (10.95)$$

> **Homework 10.15: the pressure energy** — Prove from the definition, namely, $\varepsilon_p = \frac{1}{2}trace\overleftrightarrow{T}$ that the pressure energy is given by $\frac{3}{2}n'k_BT$. Here in the flow frame, the relativistic thermal plasma is governed by the Gibb's distribution function $f = N\exp(-\mu\gamma)$ with the normalization constant $N = \mu/4\pi m_0^3c^3K_2(\mu)$.

By imposing the symmetry in the perpendicular velocity to make the expectation value of $\vec{w}_{i\perp}$ vanish thus eliminate the linear terms in $\vec{w}_{i\perp}$, Eq. (10.95) can be cast into (cf. Homework 10.16)

$$\begin{aligned}\overleftrightarrow{W} = \rho\vec{U}\vec{U} + \frac{U}{c^2}\sum_i m_i\left(\frac{U - w_{i||}}{1 - Uw_{i||}/c^2}\right)\\ \times(\vec{w}_{i\perp}\vec{w}_{i\perp} + \vec{w}_{i||}\vec{w}_{i||} - \vec{U}\vec{w}_{i||})\delta(\vec{r} - \vec{r}_i(t)).\end{aligned} \qquad (10.96)$$

Homework 10.16: the flow tensor — Prove Eq. (10.96).

The flow energy may be obtained from

$$\varepsilon_F \equiv \frac{1}{2} trace \overleftrightarrow{W} = \frac{1}{2}\rho U^2$$
$$+ \frac{1}{2}\frac{U}{c^2}\sum_i m_i \left(\frac{U - w_{i||}}{1 - Uw_{i||}/c^2}\right)(w_i^2 - Uw_{i||})\delta(\vec{r} - \vec{r}_i(t)).$$

Using $w_i^2 = c^2(1 - m_0^2/m_i^2)$ and $\sum_i m_i(U - w_{i||})\delta(\vec{r} - \vec{r}_i(t)) = 0$, we can eliminate the $c^2 - w_{i||}U$ term to yield the last term defined as ε_r,

$$\varepsilon_r \equiv \frac{1}{2}\frac{U}{c^2}\sum_i m_i \left(\frac{U - w_{i||}}{1 - Uw_{i||}/c^2}\right)(w_i^2 - Uw_{i||})\delta(\vec{r} - \vec{r}_i(t))$$
$$= -\frac{1}{2}m_0^2 U \sum_i \frac{1}{m_i}\frac{U - w_{i||}}{1 - Uw_{i||}/c^2}\delta(\vec{r} - \vec{r}_i(t)).$$

By adding and subtracting a c^2 term as follows,

$$\varepsilon_r = -\frac{1}{2}m_0^2 \sum_i \frac{1}{m_i}\frac{U^2 - c^2 + c^2 - Uw_{i||}}{1 - Uw_{i||}/c^2}\delta(\vec{r} - \vec{r}_i(t))$$
$$= \frac{1}{2}m_0^2 c^2 \sum_i \frac{1}{m_i}\delta(\vec{r} - \vec{r}_i(t))\left(\frac{1}{m_i\gamma_U^2}\frac{1}{1 - Uw_{i||}/c^2} - 1\right)$$
$$= \frac{1}{2}m_0^2 c^2 \left(\sum_i \frac{1}{m_i'}\delta(\vec{r}' - \vec{r}_i'(t)) - \sum_i \frac{1}{m_i}\delta(\vec{r} - \vec{r}_i(t))\right), \quad (10.97)$$

where we have applied the mass relation between the lab and flow frames, namely $m' = \gamma_U m(1 - Uw_{||}/c^2)$ and the length contraction $\delta(\vec{r}' - \vec{r}_i') = \delta(\vec{r} - \vec{r}_i)/\gamma_U$ to arrived at the last expression. It is clear that ε_r is equal to the difference of $nm_0^2c^2\langle 1/m\rangle$ in the flow and the lab frames, unambigously a relativistic effect.

The flow energy can be expressed as in the following (cf. Homework 10.17)

$$\varepsilon_F = \frac{1}{2}nm_0\gamma_U U^2\left(\frac{K_3(\mu)}{K_2(\mu)} - \frac{1}{\mu}\right)$$
$$+ \frac{1}{2}nm_0c^2\frac{K_1(\mu)}{K_2(\mu)}\left(\gamma_U - \frac{1}{\gamma_U}\right) \xrightarrow{\mu \gg 1} \frac{1}{2}\rho U^2 \xrightarrow{\mu \ll 1} \frac{3}{2}\rho U^2\frac{\gamma_U}{\mu}, \quad (10.98)$$

where we have applied $K_n(\mu) \approx \frac{1}{2}\Gamma(n)(\frac{1}{2}\mu)^{-n}$ for $n > 0$ as $\mu \to 0$, and $K_n(\mu) \approx \sqrt{\frac{1}{2}\pi/\mu}e^{-\mu}(1 + (4n^2 - 1)/8\mu + \cdots)$ for $n > 0$ as $\mu \gg 1$, where $\Gamma(n)$ is the gamma function.

We may define $\overleftrightarrow{\Delta} = \overleftrightarrow{W} - \rho\vec{U}\vec{U}$, which may be termed as the relativistic flow tensor, since it diminishes as the relativistic effect vanishes. The relativistic fluid equation of motion gets extra terms than its nonrelativistic counterpart of Eq. (4.19),

$$\rho\frac{\partial\vec{U}}{\partial t} + \rho\vec{U}\cdot\nabla\vec{U}$$
$$= \vec{F} - \nabla\cdot\overleftrightarrow{\Delta} - \nabla\cdot\overleftrightarrow{\mathrm{T}} - \vec{U}\left(\frac{1}{c^2}\vec{J}\cdot\vec{E} + \sum_i \vec{w}_i\cdot\vec{f}_i^{int}\delta\left(\vec{r} - \vec{r}_i(t)\right)\right). \tag{10.99}$$

Issues remain regarding the treatment of the pressure tensor, the thermal tensor, and the convection of the internal and external heat deposition.

Homework 10.17: the flow energy — Find the explicit expression for the relativistic term ε_r in the flow energy. Show that it vanishes in the nonrelativistic regime. Obtain the leading term in the ultra-relativistic case.

10.5.2 *Relativistic MHD Equations*

Neglecting the internal heat transfer, we may assume the continuity equation for the relativistic magnetohydrodynamics in the following,

$$\frac{\partial}{\partial t}\rho + \nabla\cdot(\rho\vec{U}) = \frac{1}{c^2}\vec{J}\cdot\vec{E} - \kappa U. \tag{10.100}$$

Here a damping coefficient κ due to the internal friction is taken into consideration that may be determined theoretically or experimentally.

In the low frequency and long wavelength regime, we will assume the charge neutrality holds true in the lab frame so that Eq. (10.86) summing up over the electron and the ion species to result in

$$\frac{\partial(n_e q_e + n_i q_i)}{\partial t} = -\nabla\cdot(q_e\vec{\Gamma}_e + q_i\vec{\Gamma}_i), \tag{10.101}$$

will give rise to

$$\nabla\cdot\vec{J} = 0, \tag{10.102}$$

where $\vec{J} = q_e\vec{\Gamma}_e + q_i\vec{\Gamma}_i$. Again, the low frequency regime gives the cross-field velocity as given by the E × B drift. Therefore,

$$\vec{E} + \frac{1}{c}\vec{U}\times\vec{B} = 0, \tag{10.103}$$

that makes the Faraday's law into

$$\frac{\partial}{\partial t}\vec{B} = \nabla \times (\vec{U} \times \vec{B}). \tag{10.104}$$

Once the magnetic field is found then the Ampere's law allows us to evaluate the current,

$$\vec{J} = \frac{c}{4\pi}\nabla \times \vec{B}, \tag{10.105}$$

and $\nabla \cdot \vec{J} = 0$ which is consistent with Eq. (10.96), and we will ignore the evolution equation of the particle flux.

To be consistent with Eq. (10.100), we neglect the last term of Eq. (10.99) and make an approximation by assuming isotropic pressure tensor and azimuthally symmetric thermal tensor. That leads to

$$\rho\left(\frac{\partial \vec{U}}{\partial t} + \vec{U} \cdot \nabla \vec{U}\right) = \vec{J} \times \vec{B} + \left(\frac{1}{2}\nabla_{\perp} - \nabla_{\parallel}\right)\varepsilon_u - 2\nabla_{\parallel}\varepsilon_r - \frac{2}{3}\nabla\varepsilon_p, \tag{10.106}$$

where $\varepsilon_u \equiv (U/c^2)\sum_i m_i' \vec{w}_{\perp}'^2 w_{\parallel}' \delta(\vec{r}' - \vec{r}_i')/(1+Uw_{\parallel}'/c^2)$ is of order $v_T^2 U^2/c^4$ and is therefore neglible in most practical cases. Note that the pressure $p = \frac{2}{3}\varepsilon_p = n' k_B T$ and ε_r and are functions of ρ, U and μ only, the relativistic MHD equations have a closure if the equation govering the evolution of μ is provided. In theory, the relativistic MHD energy principle can also be derived accordingly.

Homework 10.18: the thermal tensor — Assume the plasma is azimuthally symmetric in velocity space. Show that

$$\nabla \cdot \overleftrightarrow{\Delta} = 2\nabla_{\parallel}\varepsilon_r - \left(\frac{1}{2}\nabla_{\perp} - \nabla_{\parallel}\right)\left(\frac{U}{c^2}\sum_i \frac{m_i' \vec{w}_{\perp}'^2 w_{\parallel}'}{1 + Uw_{\parallel}'/c^2}\delta\left(\vec{r}' - \vec{r}_i'\right)\right).$$

The last term is of order $v_T^2 U^2/c^4$.

Further Reading

The original work on TAE can be found in Cheng, Chen and Chance (1985).

Balooning mode: Furth *et al.* (1966) Kulsrud (1966) and a kinetic treatment by Chu *et al.* (1978), Tearing Mode: Furth *et al.* (1963) and nonlinear dynamics, White *et al.* (1977).

The current profile in confined plasma was studied in terms of the maximum entropy state by Montgomery *et al.* (1979). Profile consistency was first proposed by Coppi (1981). Minimum energy state in tokamak confinement was first suggested by Kadamtsev (1987) and Hsu *et al.* (1987).

High poloidal beta experiment was carried out on TFTR and reported by Kesner *et al.* (1993).

Regarding the numerical methods to find bifurcated solutions, many numerical algorithms [Cliffe *et al.* (2011)], such as the analytical continuation, scaling iterative algorithm, the monotone iteration, the asymptotic numerical method [Cadou *et al.* (2001)], and the direct iteration algorithm [Demoulin and Chen 1974], often produce without hesitation one solution (the strong, the dominant), but not the other (the weak, the subdominant). The method following the physical characteristics is discussed in the next chapter.

Homework Hints

Homework 10.1: Prove the equilibrium properties as listed in Eq. (10.16). The following vector identities are useful:

$$\nabla(\vec{A}\cdot\vec{B}) = \vec{A}\times(\nabla\times\vec{B}) + \vec{B}\times(\nabla\times\vec{A}) + (\vec{A}\cdot\nabla)\vec{B} + (\vec{B}\cdot\nabla)\vec{A},$$

$$\nabla\times(\vec{A}\times\vec{B}) = \vec{A}(\nabla\cdot\vec{B}) - \vec{B}(\nabla\cdot\vec{A}) + (\vec{B}\cdot\nabla)\vec{A} - (\vec{A}\cdot\nabla)\vec{B}.$$

Multiplying the force balance equation, $(\nabla\times\vec{B})\times\vec{B}/4\pi = \nabla p$, with $\vec{B}$ gives

$$\vec{B}\cdot\nabla p = 0,$$

and similarly with $\vec{J}$ gives

$$\vec{J}\cdot\nabla p = 0.$$

Utilizing the $\nabla(\vec{A}\cdot\vec{B})$ expansion to replace $(\nabla\times\vec{B})\times\vec{B}$ in the force balance equation, we find

$$\frac{1}{4\pi}\vec{B}\cdot\nabla\vec{B} = \nabla\left(p + \frac{1}{8\pi}B^2\right).$$

Taking the curl of the force balance equation, we find

$$\nabla\times(\vec{J}\times\vec{B}) = 0 = \vec{J}(\nabla\cdot\vec{B}) - \vec{B}(\nabla\cdot\vec{J}) + (\vec{B}\cdot\nabla)\vec{J} - (\vec{J}\cdot\nabla)\vec{B}.$$

Since $\nabla\cdot\vec{B} = \nabla\cdot\vec{J} = 0$, we end with $\vec{B}\cdot\nabla\vec{J} = \vec{J}\cdot\nabla\vec{B}$. Multiplying the last equation with $\vec{B}$, we find

$$\vec{J}\cdot\nabla(B^2/2) = (\vec{B}\cdot\nabla\vec{J})\cdot\vec{B}.$$

From $\nabla(\vec{J}\cdot\vec{B}) = \vec{J}\times(\nabla\times\vec{B}) + \vec{B}\times(\nabla\times\vec{J}) + (\vec{J}\cdot\nabla)\vec{B} + (\vec{B}\cdot\nabla)\vec{J}$, we find $\vec{B}\cdot\nabla(\vec{J}\cdot\vec{B}) = \vec{B}\cdot((\vec{J}\cdot\nabla)\vec{B} + (\vec{B}\cdot\nabla)\vec{J}) = \vec{J}\cdot\nabla(B^2/2) + (\vec{B}\cdot\nabla\vec{J})\cdot\vec{B} = \vec{J}\cdot\nabla B^2$.

Therefore,

$$\vec{B}\cdot\nabla(\vec{B}\cdot\vec{J}) = \vec{J}\cdot\nabla(B^2).$$

Since $\vec{B}\times\nabla p = \vec{J}B^2/4\pi - \vec{B}\vec{J}\cdot\vec{B}/4\pi$, it gives $\nabla\cdot(\vec{B}\times\nabla p) = (\nabla\cdot\vec{J})B^2/4\pi - (\nabla\cdot\vec{B})\vec{J}\cdot\vec{B}/4\pi + \vec{J}\cdot\nabla(B^2/4\pi) - \vec{B}\cdot\nabla(\vec{B}\cdot\vec{J}/4\pi) = 0$. Thus,

$$\nabla\cdot(\vec{B}\times\nabla p) = 0$$

Homework 10.3: Derive Eq. (10.25) by adding the magnetic energy to Eq. (4.21).

Equation (4.21) has

$$\frac{\partial}{\partial t}\in(\vec{x},t) = \vec{F}^{ext}\cdot\vec{U} + \varsigma^{int} - \nabla\cdot(\vec{U}\in + \overline{\overline{P}}\cdot\vec{U} + \vec{Q}), \tag{4.21}$$

where $\in(\vec{x},t) \equiv \frac{1}{2}\sum_i m\vec{w}_i^2\delta(\vec{x}-\vec{x}_i) = \frac{1}{2}\rho U^2 + \frac{3}{2}p$, $\overline{\overline{P}} = p\overline{\overline{I}}$, $p = nk_BT$. Equation (10.23) gives

$$\begin{aligned}\frac{\partial}{\partial t}\left(\frac{1}{8\pi}\vec{B}^2\right) &= \frac{1}{4\pi}\vec{B}\cdot\nabla\times(\vec{U}\times\vec{B})\\ &= \frac{1}{c}\vec{U}\times\vec{B}\cdot\vec{J} + \frac{1}{4\pi}\nabla\cdot((\vec{U}\times\vec{B})\times\vec{B})\\ &= \vec{E}\cdot\vec{J} + \frac{c}{4\pi}\nabla\cdot(\vec{E}\times\vec{B})\end{aligned}$$

where $\vec{U} = c\vec{E}\times\vec{B}/B^2$ has been applied so that $\vec{U}\times\vec{B} = c\vec{E}$.

$$\begin{aligned}\therefore\ &\frac{\partial}{\partial t}\left(\frac{1}{2}\rho U^2 + \frac{1}{\gamma-1}p + \frac{1}{8\pi}\vec{B}^2\right)\\ &= \vec{E}\cdot\vec{J} - \nabla\cdot\left(\vec{U}\left(\frac{1}{2}\rho U^2 + \frac{\gamma}{\gamma-1}p\right) + \frac{c}{4\pi}\vec{E}\times\vec{B} + \vec{Q}\right).\end{aligned}$$

Homework 10.5: Examine the instability condition for the surface kinks of the current channel as described in Eq. (10.45) by including the plasma and vacuum δW.

We want to calculate the following for the uniform current profile:

$$\delta W_s = \frac{aL}{8}B_\theta^2(r=a)\left[\frac{f^2}{r}\left(1 - \frac{r\partial f}{f\partial r}\right)\right]_{r=a} = \frac{L}{8}B_\theta^2(r=a)\xi_0^2(1-\nu), \tag{10.5.1}$$

$$\delta W_p = \frac{L}{8} B_\theta^2(r=a) \int_0^a r dr \left(\frac{f}{a}\right)^2 \left(m^2 - 4 + \left(\frac{r\partial f}{f\partial r} - 1\right)^2\right), \qquad (10.5.2)$$

$$\delta W_v = \frac{L}{8} B_\theta^2(r=a) \int_a^\infty r dr \left(\frac{af}{r^2}\right)^2 \left[m^2 + \left(\frac{r\partial f}{f\partial r} - 1\right)^2\right]. \qquad (10.5.3)$$

```
function [Ws,Wp1,Wp2,Wv1,Wv2]=SurfaceKink
clear all; close all; clc
syms f0 nu lambda a m positive;
syms s;
nu=2; f0=0.5; b=0.25;
f=f0*(1+s)^nu*exp(-lambda*s^2);
inertia=int((s+1)*f^2,s,-1,b);
inertia=simplify(inertia),
df=diff(f,s);
F=(s+1)*df/f;
F=simplify(F),
I1=(s+1)*f^2;%*(m^2-4);
I2=(s+1)*f^2*(F-1)^2;
dWp1=int(I1,s,-1,0),
dWp2=int(I2,s,-1,0),
J1=f^2/(s+1)^3;%*m^2;
J2=f^2/(s+1)^3*(F-1)^2;
dWv1=int(J1,s,0,b),
dWv2=int(J2,s,0,b),
s=0; dWs=eval(f*(f-df));
Wp1=[]; Wv1=[]; Wp2=[]; Wv2=[]; Ws=[];
L=0.5:0.25:10;
for ic=1:length(L)
   lambda=L(ic);
   rho=eval(inertia);  Ws=[Ws,-eval(dWs)/rho];
   w=eval(dWp1);    Wp1=[Wp1,w/rho];
   w=eval(dWp2);    Wp2=[Wp2,w/rho];
   w=eval(dWv1);    Wv1=[Wv1,w/rho];
   w=eval(dWv2);    Wv2=[Wv2,w/rho];
end; plot(L,Wp1,'r-',L,Wv1,'g-',L,Wp2,'b-',L,Wv2,'k-',L,Ws,'c*');
xlabel('lambda'); ylabel('dWp dWv');
title('dW Components');
x=6; text(x,2.3,'Ws','color','c');
text(x+2,1.3,'Wv2','color','k'); text(x,1.3,'Wp2','color','b');
text(x,0.2,'Wp1','color','r'); text(x,0.4,'Wv1','color','g');
```

We normalize the length to the plasma radius a by defining $s = (r/a) - 1$. The virtual displacement is chosen as $f = \xi_0(1+s)^\nu \exp(-\lambda s^2)$ which is over weighted toward $s > 1$, so a cut off at $s = b$ is imposed in the vacuum region. The surface δW is negative, but its positive value is plotted for comparison. The plasma and vacuum δW's are split into two pieces. One piece is m independent and the other is not. The $m < 2$ modes have the additional free energy due to the internal kink. The high m modes on the other hand, benefiting from the plasma and vacuum δW for

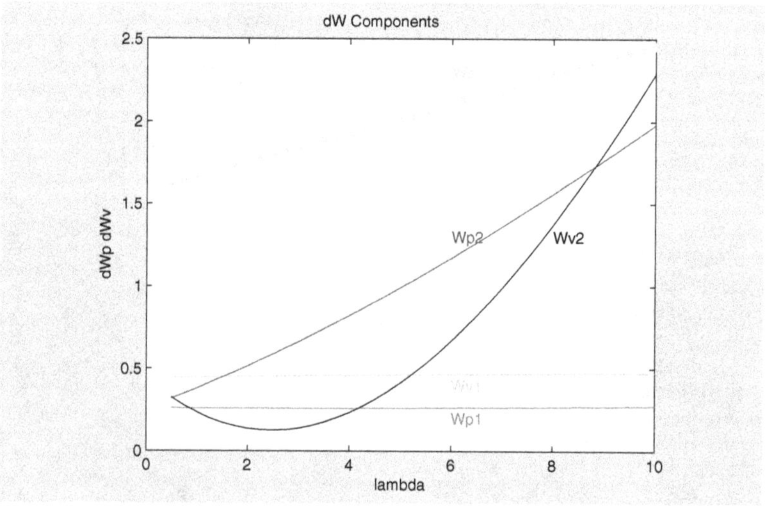

stability, would presumably be limited in the spatial extent, if not totally stabilized. The linear growth rate can easily be evaluated from the output of the code.

Homework 10.7: Show that the leading oder of the averaged magnetic well as described by Eq. (10.55) in the limit of large aspect ratio, $r/R_0 \ll 1$, is $1.5(r/R_0)^2$.

$$
\begin{aligned}
\frac{\langle \vec{B}^2 \rangle}{B_0^2} &= \oint \frac{d\theta}{2\pi} \left(1 + \frac{r}{R_0}\cos\theta\right)^{-2} \\
&= \oint \frac{d\theta}{2\pi} \left(1 - 2\frac{r}{R_0}\cos\theta + \frac{(-2)\cdot(-3)}{1\cdot 2}\frac{r^2}{R_0^2}\cos^2\theta + \cdots\right) \\
&\approx \left(1 + \frac{3}{2}\frac{r^2}{R_0^2} + \cdots\right)
\end{aligned}
\tag{10.7.1}
$$

Homework 10.9: Transform the differential operator Δ^* from(R, Φ, Z) coordinates to (r, θ, z) coordinates.

Define $R = R_0 + r\cos\theta$ and $Z = r\sin\theta$, and apply the chain rule to get

$$
\frac{\partial}{\partial r} = \frac{\partial}{\partial R}\cos\theta + \frac{\partial}{\partial Z}\sin\theta, \quad \text{and} \quad \frac{\partial}{r\partial\theta} = -\frac{\partial}{\partial R}\sin\theta + \frac{\partial}{\partial Z}\cos\theta.
$$

Repeat the process, we have

$$
\frac{\partial^2}{\partial r^2} = \frac{\partial^2}{\partial R^2}\cos^2\theta + \frac{\partial^2}{\partial Z^2}\sin^2\theta + \sin 2\theta \frac{\partial^2}{\partial R \partial Z},
$$

and

$$\frac{\partial^2}{\partial\theta^2} = \left(-r\sin\theta\frac{\partial}{\partial R} + r\cos\theta\frac{\partial}{\partial Z}\right)\left(-Z\frac{\partial}{\partial R} + (R-R_0)\frac{\partial}{\partial Z}\right)$$

$$= r^2\sin^2\theta\frac{\partial^2}{\partial R^2} + r^2\cos^2\theta\frac{\partial}{\partial Z^2} - r\sin\theta\frac{\partial}{\partial Z} - r\cos\theta\frac{\partial}{\partial R} - r^2\sin 2\theta\frac{\partial^2}{\partial R\partial Z}$$

so that

$$\frac{\partial^2}{r^2\partial\theta^2} = \sin^2\theta\frac{\partial^2}{\partial R^2} + \cos^2\theta\frac{\partial^2}{\partial Z^2} - \sin 2\theta\frac{\partial^2}{\partial R\partial Z} - \frac{\cos\theta}{r}\frac{\partial}{\partial R} - \frac{\sin\theta}{r}\frac{\partial}{\partial Z}.$$

Therefore, $\partial^2/\partial R^2 + \partial^2/\partial Z^2 = \partial^2/\partial r^2 + (1/r)\partial/\partial r + \partial^2/\partial z^2$. Utilizing, $\partial/\partial R = \cos\theta(\partial/\partial r) - \sin\theta(\partial/r\partial\theta)$, we then have

$$\Delta * \psi$$

$$= \left(\frac{\partial^2}{\partial r^2} + \frac{1}{r}\frac{\partial}{\partial r} + \frac{1}{r^2}\frac{\partial^2}{\partial\theta^2} - \frac{\cos\theta}{R_0\left(1+\frac{r}{R_0}\cos\theta\right)}\frac{\partial}{\partial r} + \frac{\sin\theta}{R_0\left(1+\frac{r}{R_0}\cos\theta\right)r}\frac{\partial}{\partial\theta}\right)\psi$$

> **Homework 10.11: MHD equilibrium — cylindrical model:** Solve the magnetic flux function numerically as described by $d^2\psi/dr^2 + (1/r)d\psi/dr + \Lambda e^{\psi} = 0$ with use of matrix inversion.

The differential equation is translated to the matrix form by the following:

$$\overleftrightarrow{M}\cdot\vec{\psi}$$

$$\equiv \frac{1}{\Delta r^2}\begin{pmatrix} 1 & 0 & & & & & \\ 1-\frac{1}{2}\Delta r & -2 & 1+\frac{1}{2}\Delta r & & & 0 & \\ & 1-\frac{1}{2}\Delta r & -2 & 1+\frac{1}{2}\Delta r & & & \\ & & \cdot & \cdot & \cdot & & \\ & & & \cdot & \cdot & 1 & \\ & 0 & & & 1-\frac{1}{2}\Delta r & -2 & 1+\frac{1}{2}\Delta r \\ & & & & & 0 & 1 \end{pmatrix}$$

$$\times\begin{bmatrix}\psi_1\\ \psi_2\\ \cdot\\ \cdot\\ \cdot\\ \psi_{N-1}\\ \psi_N\end{bmatrix} = \begin{bmatrix}\psi_c/\Delta r^2\\ -c_F e^{\lambda\psi_2}\\ \cdot\\ \cdot\\ \cdot\\ -c_F e^{\lambda\psi_{N-1}}\\ 0\end{bmatrix} \tag{10.11.1}$$

```
function MHDeqSlab(Lambda)
close all; clc;
global N x dx;
N=1000; dx=1/N;
x=0:dx:1; N=length(x); PSI=[x;x]*0;
[Minv,M]=setM;
error=1;
for ic=1:2
 psi0=ic^1.5; psi=psi0*(1-x.^2)';
for ip=1:20
   EQ1=M*psi; EQ1(1)=EQ1(1)*dx*dx;
   EQ2=Lambda*exp(psi);
   EQ=EQ1+EQ2;
   d=norm(EQ)/N;
   if(d<1.0e-3) break; end;
for it=1:30
   F=-Lambda*exp(psi); F(N)=0;          % For very small
F(1)=psi(1)/dx/dx;                      Λ = C_F a^2/λ,
   psi=Minv*F;                          the flat solution would pull the
   Ba=sum(Lambda*exp(psi))*dx;          solution over, and the algorithm
   Bp=-[0;diff(psi)]/dx; BpN=Bp(N);     would fail to find the bifurcated
   error=abs(Ba-BpN)/abs(Ba);           solution. This is in contrast with
   if(error<1.0e-6) break; end;         the experiment in that the flat
   if(ic==1) psi0=psi0*Ba/BpN;          solution tends to be more difficult
   else psi0=psi0*BpN/Ba; end;          to achieve.
   psi=psi/psi(1)*psi0;
   end;
   PSI(ic,:)=psi;
end;
end;
figure(1); hold on;
plot(x,real(PSI(1,:)),'g*',x,real(PSI(2,:)),'g*');
psi=getAnanlytical(Lambda,x);
plot(x,psi(1,:),'r-',x,psi(2,:),'r-');
text(0.8,1,'$\Lambda$=','interpreter','latex');
str=sprintf('%3.1f',Lambda); text(0.85,1,str);
title('The Bifurcated Solutions - Slab MHD');
xlabel('x'); ylabel('\psi'); hold off;
```

First we show that the distribution function is normalized. Note that $pdp = c^2m_0^2\gamma\, d\gamma$ and $p = m_0c\sqrt{\gamma^2-1}$. Therefore,

$$\begin{aligned}
\int_0^\infty 4\pi f p^2 dp &= 4\pi N m_0^3 c^3 \int_1^\infty e^{-\mu\gamma}\sqrt{\gamma^2-1}\,\gamma\, d\gamma \\
&= -4\pi N m_0^3 c^3 \frac{\partial}{\partial\mu}\int_1^\infty e^{-\mu\gamma}\sqrt{\gamma^2-1}\,d\gamma \\
&= -4\pi N m_0^3 c^3 \frac{\partial}{\partial\mu}\left(\frac{K_1(\mu)}{\mu}\right) \\
&= 4\pi N m_0^3 c^3 \frac{K_2(\mu)}{\mu} = 1,
\end{aligned} \tag{10.13.1}$$

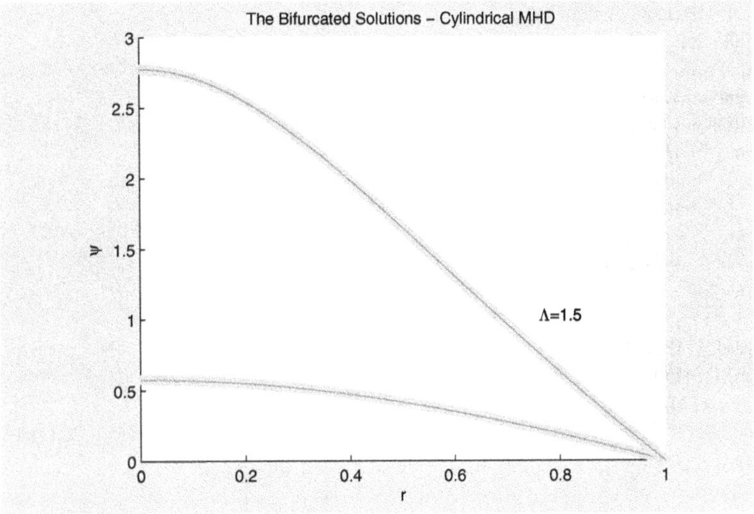

where we have utilized the integral representation of the modified Bessel function of the second kind,

$$K_\nu = \frac{\sqrt{\pi}\,(\frac{1}{2}\mu)^\nu}{\Gamma(\nu+\frac{1}{2})}\int_1^\infty e^{-\mu\gamma}\,(\gamma^2-1)^{\nu-\frac{1}{2}}d\gamma\,, \tag{10.13.2}$$

where $\Gamma(\frac{3}{2}) = \frac{1}{2}\sqrt{\pi}$.

The expected mass is given by

$$\begin{aligned}
\langle m\rangle &= 4\pi\, N m_0^4 c^3 \int_1^\infty e^{-\mu\gamma}\sqrt{\gamma^2-1}\,\gamma^2 d\gamma \\
&= 4\pi\, N m_0^4 c^3 \frac{\partial^2}{\partial\mu^2}\int_1^\infty e^{-\mu\gamma}\sqrt{\gamma^2-1}\,d\gamma \;= 4\pi\, N m_0^4 c^3 \frac{\partial^2}{\partial\mu^2}\left(\frac{K_1(\mu)}{\mu}\right). \\
&= -4\pi\, N m_0^4 c^3 \frac{\partial}{\partial\mu}\left(\frac{K_2(\mu)}{\mu}\right) \\
&= m_0\left(\frac{K_3(\mu)}{K_2(\mu)} - \frac{1}{\mu}\right) \xrightarrow{\mu\gg 1} m_0\left(1+\frac{3}{2}\frac{k_B T}{m_0 c^2}\right) \xrightarrow{\mu\ll 1} \frac{3m_0}{\mu} = 3\frac{k_B T}{c^2}.
\end{aligned} \tag{10.13.3}$$

Homework 10.13: the average mass in thermal plasma: Given the relativistic thermal plasma with the Gibb's distribution function $f = N\exp(-\mu\gamma)$ where $N = \mu/4\pi m_0^3 c^3 K_2(\mu)$ is the normalization constant and $\gamma = \sqrt{p^2/c^2 m_0^2 + 1}$ is the Lorentz factor, find the expected mass.

```
function MHDeqCylinder(cF)
close all; clc;
global N r dr;
a=1;lambda=1, R0=3;
LAMBDA=cF*a^2/lambda,
N=3000; dr=1/N; r=dr:dr:1;
figure(1); hold on;
[Minv,M]=setM;
for ic=1:2
  psi0=ic^1.5; psi=psi0*(1-r'.^2);
for ip=1:300
  EQ1=M*psi; EQ1(1)=EQ1(1)*dr*dr;
  EQ2=LAMBDA*exp(psi);
  EQ=EQ1+EQ2;
  d=norm(EQ)/N;
  if(d<1.0e-3) ip, break; end;
  psi0=psi(1);
for it=1:50
  F=-LAMBDA*exp(psi); F(N)=0;
F(1)=psi0/dr/dr;
  psi=Minv*(F);
  I=LAMBDA*getI(psi,r,dr)/R0;
  Ba=I/a/2/pi;
  Bp=-[0;diff(psi)]/dr/a/R0;
  BpN=Bp(N);
  error=abs((Ba-BpN)/(BpN+eps));
  if(ic==1) psi0=psi0*Ba/BpN;
  else psi0=psi0*BpN/Ba; end;
  psi=psi/psi(1)*psi0;
  if(error<1.0e-6) break; end;
  end;
end;
  ic, d, error, ip,
  psic=[1-sqrt(1-LAMBDA/2),1+sqrt(1-
LAMBDA/2)]/LAMBDA*4;
  psic=2*log(psic);
  PSI=psic(ic)-
2*log(1+LAMBDA/8*exp(psic(ic))*r'.^2);
  plot(r,real(psi),'g*',r,real(PSI),'r-');
end;
text(0.8,1,'$\Lambda$=','interpreter','latex');
str=sprintf('%3.1f',LAMBDA); text(0.85,1,str);
title('The Bifurcated Solutions - Cylindrical
MHD');
xlabel('r'); ylabel('\psi'); hold off;
```

% Shooting method would be able to find both solutions but it is limited to one dimensional problem (cf. Section 8.3.2 and Homework 8.5).

% The direct matrix inversion method listed here adopts the renomalization of the central ψ value by the ratio of the two poloidal magnetic fields evaluated at the plasma edge according to $\nabla\psi$ and $I/2\pi a$. By inverting the ratio, the bifurcated solutions are separated out.

```
function [Minv,M]=setM
global N r dr;
dr2=diag(ones(N-1,1),1)+diag(ones(N-1,1),-1)+diag(-2*ones(N,1));
dr1=diag(ones(N-1,1)./r(1:N-1)',1)-diag(ones(N-1,1)./r(2:N)',-1);
M=dr2/dr/dr+dr1/2/dr;
M(N,N)=1/dr/dr; M(N,N-1)=0; M(1,1)=1/dr/dr; M(1,2)=0;
Minv=inv(M);

function I=getI(psi,r,dr)
N=length(r);
I=exp(psi).*r';
I=sum(I)*dr*2*pi;
```

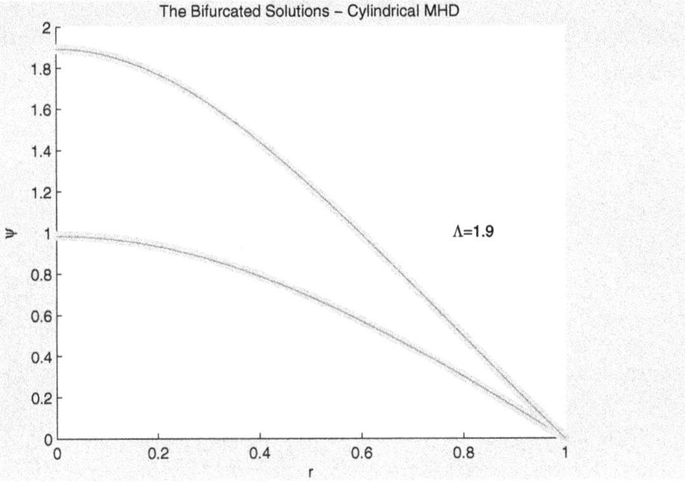

Note that we have applied the relation,

$$\left(\frac{1}{\mu}\frac{\partial}{\partial\mu}\right)^n\left(\frac{K_\nu(\mu)}{\mu^\nu}\right) = (-1)^n\frac{K_{\nu+n}(\mu)}{\mu^{\nu+n}}. \tag{10.13.4}$$

Homework 10.15: the pressure energy: Prove from the definition, namely, $\varepsilon_p = \frac{1}{2}trace\overleftrightarrow{T}$ that the pressure energy is given by $\frac{3}{2}n'k_BT$. Here in the flow frame, the relativistic thermal plasma is governed by the Gibb's distribution function $f = N\exp(-\mu\gamma)$ with the normalization constant $N = \mu/4\pi m_0^3c^3K_2(\mu)$.

We may express $\mu = m_0c^2/k_BT$, where T is the temperature. In the ultra relativistic regime, the mass is $3k_BT/c^2$. Note that $K_n(\mu) \approx \frac{1}{2}\Gamma(n)(\frac{1}{2}\mu)^{-n}$ for $n > 0$ as $\mu \to 0$, and $K_n(\mu) \approx \sqrt{\frac{1}{2}\pi/\mu}e^{-\mu}(1 + (4n^2 - 1)/8\mu + \cdots)$ for $n > 0$ as $\mu \gg 1$, where $\Gamma(n)$ is the gamma function.

The pressure energy is given by the trace of momentum flux tensor in the flow frame,

$$\varepsilon_p \equiv \frac{1}{2} trace\overleftrightarrow{T} = \frac{1}{2}\sum_i \frac{p_i'^2}{m_i'}\delta(\vec{r}' - \vec{r}_i'(t)) = \frac{1}{2}c^2 \sum_i \frac{m_i'^2 - m_0^2}{m_i'}\delta(\vec{r}' - \vec{r}_i'(t))$$

$$= \frac{1}{2}\rho' c^2 - \frac{1}{2}m_0 c^2 \sum_i \sqrt{1 - \frac{v'^2}{c^2}}\delta(\vec{r}' - \vec{r}_i'(t)). \qquad (10.15.1)$$

Here we have applied $\varepsilon^2 = p^2c^2 + m_0^2c^4 = m^2c^4$. Consider the relativistic thermal plasma with the Gibb's distribution function $f = N\exp(-\mu\gamma)$ in the flow frame, where $N = \mu/4\pi m_0^3 c^3 K_2(\mu)$is the normalization constant. We may then evaluate the last term,

$$\varepsilon_T \equiv \frac{1}{2}m_0 c^2 \sum_i \sqrt{1 - \frac{v'^2}{c^2}}\delta\left(\vec{r}' - \vec{r}_i'(t)\right)$$

$$= \frac{1}{2}\frac{\mu}{K_2(\mu)}n' m_0 c^2 \int_1^\infty e^{-\mu\gamma}\sqrt{\gamma^2 - 1}d\gamma$$

$$= \frac{1}{2}n' m_0 c^2 \frac{K_1(\mu)}{K_2(\mu)}. \qquad (10.15.2)$$

The integral is found by using Eq. (10.13.2). Thus, the pressure energy is given by

$$\varepsilon_p = \frac{1}{2}n' m_0 c^2 \left(\frac{K_3(\mu)}{K_2(\mu)} - \frac{1}{\mu} - \frac{K_1(\mu)}{K_2(\mu)}\right) = \frac{3}{2}n' m_0 c^2 \frac{1}{\mu}. \qquad (10.15.3)$$

Notice that $\mu = m_0 c^2/k_B T$ and $\varepsilon_p = \frac{3}{2}n' k_B T$, which has the identical form as the nonrelativistic case. However, in the relativistic case, the pressure energy is given by $\varepsilon_p = \frac{3}{2}n m_0 c^2/\mu/\gamma_U$ since $n'(\vec{r}, t) = n(\vec{r}, t)/\gamma_U$.

Homework 10.17: the flow energy — Find the explicit expression for the relativistic term ε_r in the flow energy. Show that it vanishes in the nonrelativistic regime. Obtain the leading term in the ultra-relativistic case.

The term in the flow frame is given by $\sum_i m_i'^{-1}\delta(\vec{r}' - \vec{r}_i'(t)) = (n'/m_0)(K_1(\mu)/K_2(\mu))$, while that in the lab frame is

$$\sum_i m_i^{-1}\delta(\vec{r} - \vec{r}_i(t)) = \frac{\gamma_U^2}{m_0}\sum_i \frac{1 + U w_{i\parallel}'/c^2}{\gamma_i'}\delta(\vec{r}' - \vec{r}_i'(t))$$

$$= \gamma_U^2 \frac{n'}{m_0}\frac{K_1(\mu)}{K_2(\mu)} + \frac{\gamma_U^2}{m_0}\sum_i \frac{U w_{i\parallel}'/c^2}{\gamma_i'}\delta(\vec{r}' - \vec{r}_i'(t)),$$

$$(10.17.1)$$

where we have made use of the transformation $\gamma_i m_0 = \gamma_U \gamma_i' m_0 (1 + \vec{U} \cdot \vec{w}_i' / c^2)$. Note that due to the symmetry of $w_{i\|}'$ in the flow frame, $\sum_i U w_{i\|}' (\gamma_i')^{-1} \delta(\vec{r}' - \vec{r}_i'(t)) = 0$. Therefore,

$$\varepsilon_r = \frac{1}{2} n m_0 c^2 \frac{K_1(\mu)}{K_2(\mu)} \left(\gamma_U - \frac{1}{\gamma_U} \right) \xrightarrow{\mu \gg 1} \frac{1}{2} n m_0 U^2 \left(1 - \frac{3}{2} \frac{1}{\mu} \right) \left(\gamma_U - \frac{1}{\gamma_U} \right)$$

$$\xrightarrow{\mu \ll 1} \frac{1}{4} n m_0 U^2 \mu \left(\gamma_U - \frac{1}{\gamma_U} \right) \tag{10.17.2}$$

In the nonrelativistic regime, the leading term is of order U^2/c^2 or $k_B T / m_0 c^2 = 1/\mu$. Nonetheless, it is vanishingly small when the relativistic effect is insignificant. The ultrarelativistic limit has the leading term given by $\frac{1}{4} n m_0 U^2 \mu \gamma_U$, which can be insignificant if the relativistic effect is only due to the thermal effect since $\mu = m_0 c^2 / k_B T$. Notice that we have applied $K_n(\mu) \approx \frac{1}{2} \Gamma(n) (\frac{1}{2}\mu)^{-n}$ for $n > 0$ as $\mu \to 0$, and $K_n(\mu) \approx \sqrt{\frac{1}{2}\pi/\mu} e^{-\mu} (1 + (4n^2 - 1)/8\mu + \cdots)$ for $n > 0$ as $\mu \gg 1$, where $\Gamma(n)$ is the gamma function.

Chapter Eleven

Nonlinear Topics

"It is vitally important that we supplement our specialized studies with serious attempts to take a crude look at the whole."

Murray Gell-Mann, Nobel Laureate 1969

While linear physics may provide a quick glimpse at the plasma topics of interest, more often than not, it is insufficient to explain away the natural phenomena. Nonlinear physics will ultimately reflect the truth since any deviation from the linear relationship would naturally be nonlinear. The complication arises as many competing terms might contribute. A consistent ordering is the rule of thumb to make sure the dominant effects are properly taken into consideration. Dimensionless variables will value themselves on an equal footing to their importance in the underlying principles even if they are of different physical entities, as discussed in the first Chapter. Beyond the inclusion of important terms, the well posed nonlinear equations can be tricky to solve. Arguments as supported by numerical, experimental, analytical and physical interpretations will be helpful to lead to the correct answers.

This chapter introduces a few nonlinear topics of interest. Besides examining the physics phenomena by the analytical means, we discuss the numerical algorithms that can often be challenging and are well worth the time to check them out.

11.1 Large Amplitude Plasma Wave

A large amplitude plasma wave (LAPW) can be described by the continuity equation,

$$\frac{\partial}{\partial t} n_e + \nabla \cdot n_e \vec{v}_e = D_e \nabla^2 n_e, \tag{11.1}$$

the momentum equation

$$\frac{\partial \vec{v}_e}{\partial t} + \vec{v}_e \cdot \nabla \vec{v}_e = \frac{e\vec{E}}{m_e} - \frac{k_B T_e}{n_e} \nabla n_e - \mu_e \nabla^2 \vec{v}_e, \tag{11.2}$$

and the Poisson's equation

$$\nabla \cdot \vec{E} = 4\pi q (n_0 - n_e). \tag{11.3}$$

Here, we assume isothermal electrons and uniform ion background. The diffusion and the viscosity, while weak, are included to help numerical stability, since the higher order derivatives can become extremely strong due to the steepening effect.

It is noteworthy that the analytical solution through the Lagrangian formulation can be found when the pressure and the damping effects are neglected.

Homework 11.1: Solve by the Lagrangian formulation the set of dimensionless equations for LAPW without the pressure and the damping effect.

```
function [n,v,e]=init
global diffusion viscosity N dz dt L z ni vi;
global w k G;
w=1.1; k=sqrt(w^2-1); L=4*pi/k; N=4000;
dz=L/(N+1); dt=dz/1.5;
z=0.5*dz:dz:(N)*dz;
v=0*z; e=v; ni=1+v; vi=v;
iterations=250;
diffusion=2e-3; viscosity=2e-3;
for i=1:iterations
  amplitude=1.2*i/iterations;
  v0=amplitude*cos(k*z); E0=amplitude*sin(k*z);
  x0=-E0; n0=1-amplitude*k*cos(k*z);
  n=n0./(1-k/w*amplitude*k*sin(k*z)*sin(w*i*dt)-amplitude*k/w^2*cos(k*z)*(1-cos(w*i*dt)));
  e=w*v0*sin(w*i*dt)-w^2*x0*cos(w*i*dt);
  v=v0*cos(w*i*dt)+w*x0*sin(w*i*dt);
  dn=D(n);
end;
g=sparse(N);
g=g+diag(-2+0*(1:N),0)+diag(1+0*(1:N-1),1)+ diag(1+0*(1:N-1),-1);
g(1,N)=g(1,2); g(N,1)=g(N,N-1);
G=inv(g);
```

Transforming the equations into the dimensionless variables: $\tau \equiv \omega_{pe}t$, $\varsigma \equiv x/\lambda_D$, $v \equiv v_e/V_{th,e}$, $n \equiv n_e/n_0$ and $\varepsilon \equiv qE\lambda_D/k_BT_e$, and treating the wave as a one-dimensional problem, we have:

$$\frac{\partial}{\partial\tau}n + \frac{\partial}{\partial\varsigma}nv = D\frac{\partial^2}{\partial\varsigma^2}n, \tag{11.1a}$$

$$\frac{\partial v}{\partial\tau} + v\frac{\partial v}{\partial\varsigma} = -\varepsilon - \frac{1}{n}\frac{\partial}{\partial\varsigma}n + \mu\frac{\partial^2}{\partial\varsigma^2}v, \tag{11.2a}$$

$$\frac{\partial}{\partial\varsigma}\varepsilon = 1 - n. \tag{11.3a}$$

For the small amplitude wave, there is no steepening effect. The LAPW, as shown in the movie clip generated by the computer program, by contrast, will develop strong steepening effect especially when the flow velocity approaches the electron thermal velocity (Mach number near unity). For the supersonic flow, the wave compresses the density and shock wave front is clearly visible.

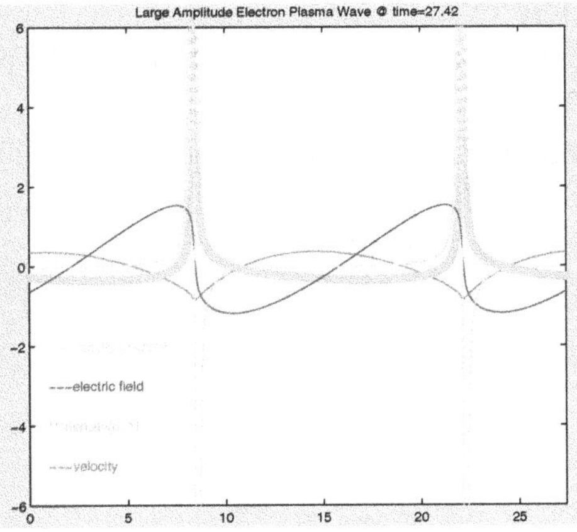


```
function LAPW
clear all; close all; clc;
global diffusion viscosity N dz dt L z ni vi;
global w k G;
[n,v,e]=init;
dt=dt/10;
iterations=30000;
[v,dn]=getv(n,v,e);
dt=dt*2; j=0;
NHistory=[];
for i=1:iterations
  n=getn(n,v);
  e=gete(n,e);
  [v,dn]=getv(n,v,e);
  if(mod(i,200)==0)
    figure(1);
    j=j+1;
    plot(z,v,'r-',z,n-1,'g*',z,e,'b-',z,dn,'c-');
    text(1,-5,'---velocity','color','r');
    text(1,-4,'***density(n-1)','color','g');
    text(1,-3,'---electric field','color','b');
    text(1,-2,'---density gradient','color','c');
s=sprintf('Large Amplitude  Plasma Wave time=%5.2f',dt*i);
    axis([0 L -6 6]);
    title(s);
    M(j)=getframe(gcf);
    NHistory=[NHistory;n];
    saveGIF(j,M);
    diffusion=diffusion/1.01; viscosity=viscosity/1.01;
   end;
end;
save LAPW NHistory N dz dt ;
```

```
function dn=D(n)
global diffusion viscosity N dz dt L z ni vi;
np=[n(2:N),n(1)];
nn=[n(N),n(1:N-1)];
dn=(np-nn)/2/dz;
```

The spatial structure of the LAPW is also analyzed by the **Hilbert-Huang Transform** (**HHT**) for its **Intrinsic Mode Function** (**IMF**) by the **Empirical Mode Decomposition**. HHT will be discussed in the next section. The dominant modes may further be analyzed by the Fast-Fourier Transform (FFT). Not surprisingly, an initial simple harmonic (top 2 figures) develops into many high harmonics (bottom 2 figures) to cause the **steepening effect**.

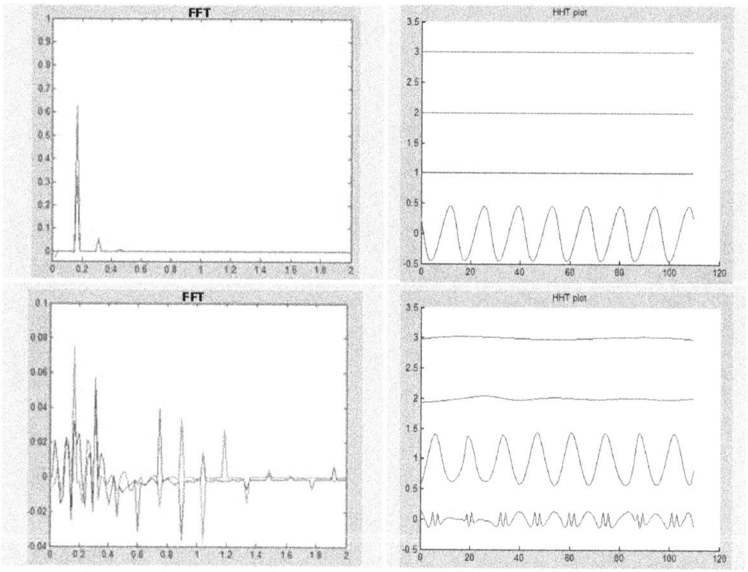


```
function n=getn(n,v)
global diffusion viscosity N dz dt L z ni vi;
dn=D(n);
dv=D(v);
d2n=D(dn);
n=n-(n.*dv+v. *dn)*dt+diffusion*d2n*dt;

function [v,dn]=getv(n,v,e)
global diffusion viscosity N dz dt L z ni vi;
dn=D(n);
dv=D(v);
d2v=D(dv);
v=v+dt*(-e-dn./n-v.*dv)+viscosity*d2v*dt;

function e=gete(n,e)
global diffusion viscosity N dz dt L z ni vi;
global w k G;
ro=ni-n;
phi=-(G*ro')';
e=-D(phi)*dz*dz;
```

11.2 Stochasticity

A classical dynamical system tends to be bounded by a **limit cycle** in space and a **Poincaré cycle** in time. The limit cycle is the spatial boundary that the dynamical system is to be confined within and the Poincaré cycle is the period the dynamical system will repeat itself in due time. In regular motion, the period of the Poincaré cycle is when the pattern of the trajectory repeats itself. As a trajectory becomes stochastic, that will increase until the process becomes irreversible in practice, and the limit cycle would expand until the impenetrable barriers disappear and every point in the relevant phase space becomes accessible.

Here we investigate a problem that is very much analytically and numerically tractable: the stochastic behavior of the particle motion in an electrostatic standing wave. It was predicted (Hsu *et al.* 1979) and verified subsequently in the plasma experiment (Doveil 1981). The governing equation can be cast into

$$\frac{d^2X}{dT^2} = p\sin X \sin T, \tag{11.4}$$

where $p = eE_0k_0/m\omega_0$ is the electric field strength, $X = k_0x$ and $T = \omega_0 t$ are the spatial and temporal coordinates in dimensionless units. The onset of chaos can be analytically identified from Mathieu's equation at $p = 0.456$ where the o-point in the phase space becomes unstable and a broadband spectrum of particle trajectory results. Stochasticity occurs in part of the phase space even when p is smaller than the critical value. But when the o-point, the most stable point, becomes unstable, it is expected that the stochasticity develops into the entire phase space the particle is capable to reach, as within the limit cycle.

The particle trajectory becomes stochastic when p exceeds a critical value. We may analyze the stability by the linearized equation of motion around the o-point,

$$\frac{d^2x}{dt^2} = px\sin t. \tag{11.5}$$

This is a particular form of Mathieu's equation with only one constant coefficient. It has an instability boundary at the critical value of $p_c = 0.456$. To understand the onset of this instability, let us Fourier transform Eq. (11.5). It gives

$$\omega^2X_\omega + \frac{p}{2i}(X_{\omega+1} - X_{\omega-1}) = 0. \tag{11.6}$$

Using Eq. (11.6) to have $X_{\omega+1} = \frac{1}{2}ip(X_\omega - X_{\omega+2})/(\omega+1)^2$ and $X_{\omega-1} = -\frac{1}{2}ip(X_\omega - X_{\omega-2})/(\omega-1)^2$ and neglecting terms higher than $O(p^2)$, we have the dispersion relation,

$$\left\{\omega^2 - \frac{p^2}{4}\left[\frac{1}{(\omega+1)^2} + \frac{1}{(\omega-1)^2}\right]\right\} X_\omega = -\frac{p^2}{4}\left[\frac{X_{\omega+2}}{(\omega+1)^2} + \frac{X_{\omega-2}}{(\omega-1)^2}\right]\right\}. \tag{11.7}$$

Note that $X_{\omega+n} \sim O(p^{1+n})$ and the RHS is $O(p^4)$, where $p < 1$ is assumed. Since the onset of the stochastic instability is expected to occur when $p < 1$, justified *a posteriori*, we may neglect the RHS term. The valid solution will have to lie within the said domain. Equation (11.7) has certain resemblance to the dispersion relation of the two stream instability, and has an instability threshold at $p_c = 0.475$, which has an error of $O(p^4)$ to the exact value of the Mathieu's equation.

A classical nondeterministic system may be quantified by the Lyapunov exponent for its chaotic nature of exponential sensitivity, and often exhibits the elliptical instability and broadband spectra. The exponential sensitivity to its initial infinitesimally small deviation characterizes the rapid departure of neighboring trajectories given however minute initial differences. The elliptical instability at o-point quantifies the transition to the full stochasticity and accounts well for the loss of deterministic trajectories. The broadband spectrum measures the degree of randomness due to the nonlinearity. These effects can be visualized by directly solving the particle trajectories given the initial phase space point.

A space of lower dimension is often selected to plot where the particle trajectory revisits this **Poincare section**. The **Poincare map** thus obtained shows the recurrence of the event. Its simplicity or complexity would reveal the degree of the chaos. Alternatively, one may choose the full space, like in this standing wave case since its phase space is only two dimensional, but record only one phase point per wave period. This is a stroboscopic plot, which discloses how the complexity sets in beyond the fundamental frequency and indicates the onset of subharmonics. The Fourier spectrum further quantifies the presence of these harmonics.

Three additional tools are presented here to analyze numerically the trajectories: the correlation dimension, the spectral entropy, and the Hilbert-Huang transform.

Homework 11.2: Find the critical value in Eq. (11.7) when the system has the complex roots, indicative of an instability.

11.2.1 *Correlation Dimension*

Dynamical systems often exhibit attractor behavior around a few fix points in phase space. A **strange attractor** tends to have zero measure in the embedding phase space and is well characterized by its fractal dimensionality f. This may be physically

```
function StandingWave(p)                %Solve the equation
dt=2*pi/2^6;  N=2^16;                   d2x/dt2=p*sin(t)*sin(x)
T=N*dt; t=0:dt:T;                       %The stochasticity
v=t*0;  x=t*0;                          boundary is at pc=0.456.
v(1)=0.5;                               %Run StandingWave(0.65)

X=(1:2^10)*0; V=X;
J=0;
for i=2:N
    v(i)=v(i-1)+dt*p*sin(i*dt)*sin(x(i-1));
    x(i)=x(i-1)+v(i)*dt;
   if(mod(i,2^6)==0) J=J+1; X(J)=x(i); V(J)=v(i); end;
end;
save SW X V;
figure;
X=mod(X,pi);
plot(X,V,'.g');
hold on;
n=fix(2*pi/0.1)+1;
X=zeros(n,6);  V=X;
for i=1:10
   E=0.1*i;
    vm=real(sqrt(2*E));
    V=vm*cos(t);                         %Solve d2x/dt2=p*sin(x)
    X=acos(E*sin(t)/p);
    plot(real(X),real(V),'m-');
end;                                     % Compare the regular and
y=abs(fft(v,N));                         the stochastic trajectories.
i=1:N;
M=fix(N/2*dt);
plot(i(10:M)*2*pi/dt/N,y(10:M),'g')
s=sprintf('fft of Standing Wave: p = %6.3f', p);
```

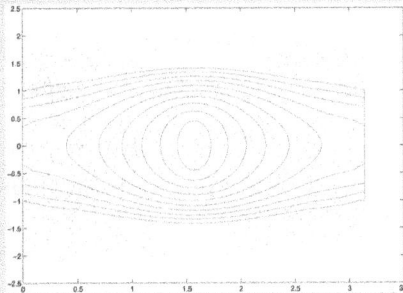

understood from the scattered trajectories in a 2d picture that seemingly span a surface more than a twisted, intertwined, or wiggled line of 1d nature. Therefore, understandably, the strange attractor in question would have $1 < f < 2$. On the other hand, a strange attractor exhibits for example, a seemingly 3d object would have $2 < f < 3$. Many fractals have the property of self-similarity which has parts as the reduced-size copy of the whole. A deterministic trajectory being limited to its

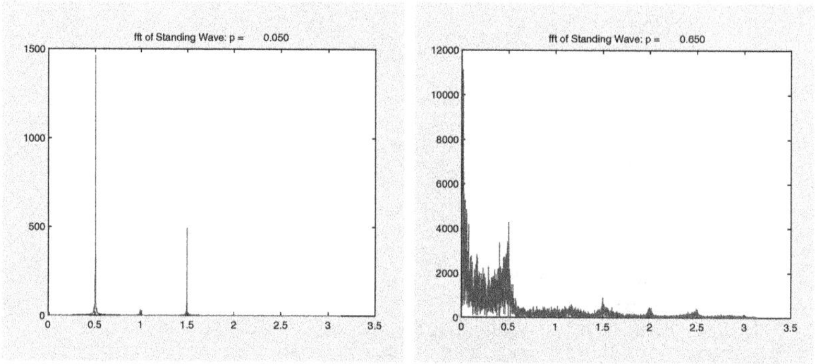

repeatable phase space trajectory will have a low correlation dimension. A simple curve, for example, has the correlation dimension of unity, while a trajectory expanding to the entire accessible space after the stochasticity onset, the correlation dimension would increase toward the dimension of the space.

By considering correlations between points of a long-time series on the attractor, Grassberger and Procaccia defined the correlation integral

$$C(r) \equiv \lim_{N\to\infty} \frac{1}{N^2} \sum_{i,j=1}^{N} \theta(r - |\vec{X}_i - \vec{X}_j|) \equiv \int_0^r d^n\vec{r}^{\,\prime} c(\vec{r}^{\,\prime}), \tag{11.8}$$

where N is the number of data points, n is the phase space dimension, θ is the Heaviside function, and $c(\vec{r})$ is the correlation function. $C(r)$ behaves as a power law of r for small r: $C(r) \propto r^\nu$, where ν is closely related to the fractal dimension. We take a Henon map:

$$x_{i+1} = y_i + 1 - ax_i^2, \quad y_{i+1} = bx_i, \tag{11.9}$$

to demonstrate the **correlation dimension**. The Henon map that generates a trajectory jumping around a few curved lines (see the figure above) has f around 1.15, not that far from a line of $f = 1$.

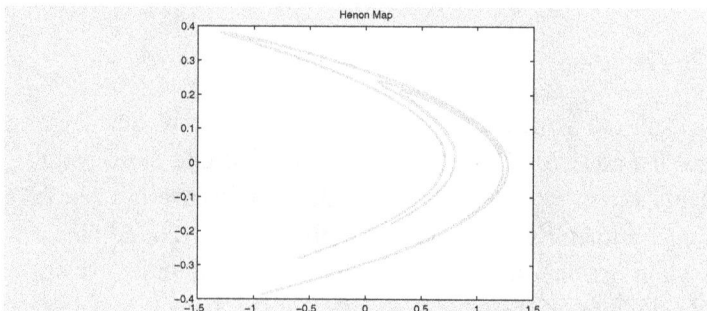

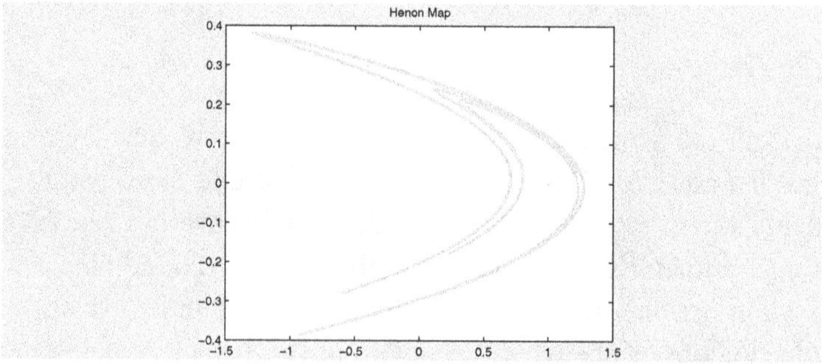

```
function HenonMap(Np)                % Np : number of points (20000 here)
x0=0; y0=0; a=1.4; b=0.3;
x=zeros(Np,1); y=x;
x(1)=y0+1-a*x0^2;
y(1)=b*x0;
for i=1:Npts-1
   x(i+1)=1-a*x(i)^2+y(i);
   y(i+1)=b*x(i);
end
plot(x,y,'g.');
title('Henon Map');
save CD Np x y;
```

```
function CorrelationDimension(N)
load CD Np x y;
D = sparse(Np);  X=repmat(x,1,Np);
D=(X-X').^2;     X=repmat(y,1,Np);
D=D+(X-X').^2;
D=sqrt(D);
rM = max(max(D,[],1),[],2);
rM = 2^ceil(log2(rM));
T=(1:N)';
rspc=rM*2.^(-(T-1));
Cr=[];
for jj=1:N
   r = rspc(jj);
   n = (D<r & D>0);
   s = sum(sum(n));
   Cr = [Cr; s/Npts^2];
end
dvs=4:N;
```

```
% Npts : number of points (10000 here)
% N: number of data in the graph (16)
% distance between any two points
% This shows that the Henon map has a fractal dimension equal to 1.1518. The number of data points was limited by the memory of the 32 bit MATLAB version.

% specified r
% correlation function
```

```
Rx=log2(rspc+eps);
Ry=log2(Cr+eps);
Mx=Rx(dvs); My=Ry(dvs);
 [coef,temp]=polyfit(Mx,My,1);
Dc=coef(1);
yfit=Dc*Rx+coef(2);
figure;
plot(Rx,Ry,'o',Rx,yfit,'r-');
axis tight, xlabel('log_2(r)');
ylabel('log_2(C(r))');
title(['D_c=',num2str(Dc)])
```

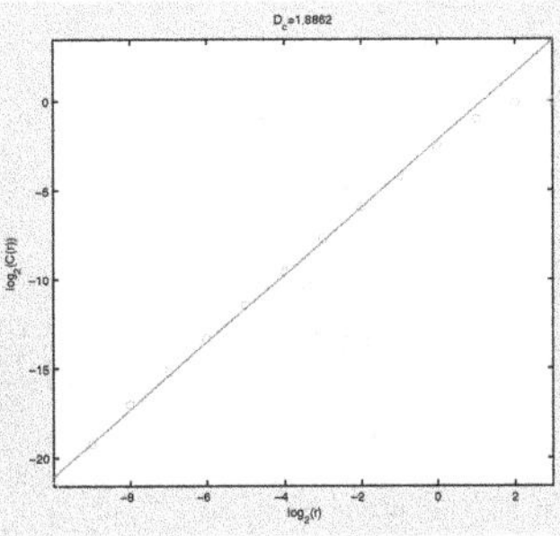

The velocity data from the standing wave is analyzed for the correlation dimension as function of p values. It is shown in the accompanying figure. The correlation dimension stays at around 1 at low electric field to 1.8 at high electric field. It rises sharply near the chaos onset. The data used for the correlation-dimension analysis was taken by the stroboscopic method that selects one point per pump wave period. Moreover, the velocity correlation shows the limit cycle that bounds the velocity range. Different sampling may yield substantial aberrations, but the general theme of a sudden jump in the correlation dimension after the threshold is evident.

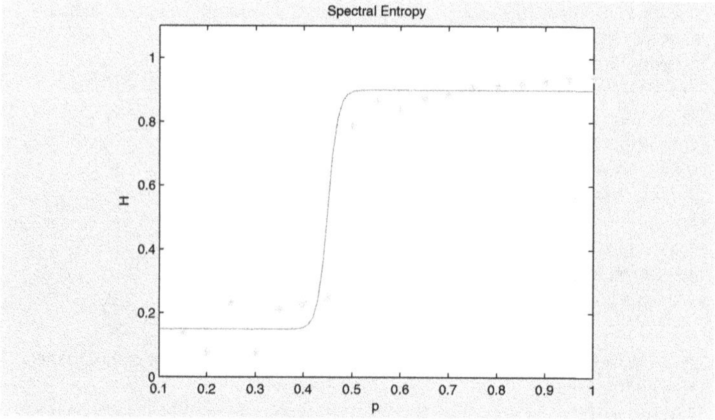

Homework 11.3: Instead of the phase space point at the wave period, choose two velocities at alternating periods. Plot the maps at $p < p_c$ and $p > p_c$ to show the sharp contrast.

11.2.2 *Spectral Entropy*

Drastic contrast of the Fourier spectrum exhibits between regular and stochastic trajectories. The regular trajectory is characterized by a few dominant frequency components, while the stochastic trajectory is characterized by the broadband spectrum. Further insight can be gained by the entropy measure of the Fourier spectrum as defined by (Powell and Percival 1979),

$$H(|S(\omega,t)|^2) = -\sum_{\omega=1}^{\Omega} P(|S(\omega,t)|^2)\log(P(|S(\omega,t)|^2)), \tag{11.9}$$

where $P(|S(\omega,t)|^2) = |S(\omega,t)|^2/\sum_{\omega=1}^{\Omega}|S(\omega,t)|^2$ is the probability of the frequency band, $S(\omega,t)$ is the magnitude of the corresponding spectrum at the frequency ω and at the time t, and Ω is the maximum number of discrete states. The spectral entropy measuring the states of configurations reflects the frequency spread of a given variable in a dynamical system. The entropy is a maximum when $S(\omega)$ follows the Gaussian white noise, and minimum when it has several discrete and individualistic pure tunes. The fast variation of the spectral entropy coincides well with the stochastic onset, as shown in the following computer program.

The spectral entropy and the bandwidth of the Fourier spectrum may serve as a measure of the Poincaré cycle. A full chaotic trajectory like a Gaussian white noise

```
function GetCorrelationDimension                    %Correlation dimension as
clear all;                                          function of p.
P=[]; C=[];
for p=0.1:0.05:1.0
[X, V]=STAND(p);
Ndtm=14; p,
N=length(X); x=V(1:2:N); y=V(2:2:N); N=N/2;
X=repmat(x,N,1); Y=repmat(y,N,1);
D=(X-X').^2+(Y-Y').^2;
D=sqrt(D);
rM = max(max(D,[],1),[],2),
rM = 2^ceil(log2(rM));
T=(1:Ndtm)';                                        % specified r
rspc=rM*2.^(-(T-1));
Cr=[];                                              % correlation function

for jj=1:Ndtm
   r = rspc(jj);
   n = (D<r & D>0);
   s = sum(sum(n));
   Cr = [Cr; s/N^2];
end
dvs=4:Ndtm;  clear X V x y D;
Rx=log2(rspc+eps); Ry=log2(Cr+eps);
Mx=Rx(dvs); My=Ry(dvs);
[coef,temp]=polyfit(Mx,My,1);
Dc=coef(1),
P=[P,p]; C=[C,Dc];
end;
yfit=Dc*Rx+coef(2);
figure(1);
plot(Rx,Ry,'go',Rx,yfit,'r-'); axis tight;
xlabel('log_2(r)'); ylabel('log_2(C(r))');
title(['D_c=',num2str(Dc)])
figure(2);
x=0.1:0.01:1;
F=1.4+0.4*tanh((x-0.45)*20);
plot(P,C,'g*',x,F,'r-'); axis tight;
title('Correlation Dimension as Function of p');
xlabel('p'); ylabel('D_c');
```

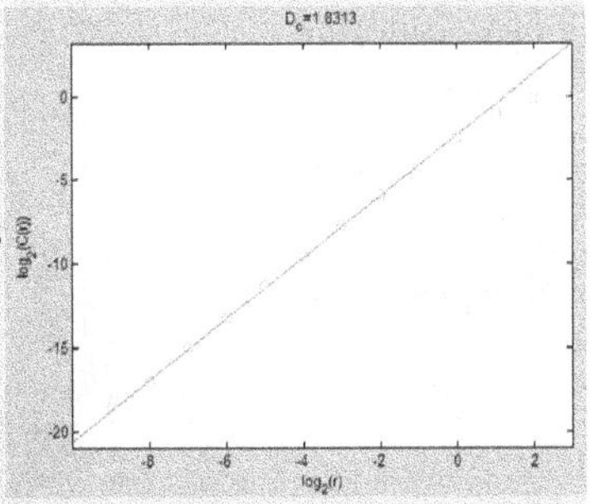


```
function [X,V]=STAND(p)
dt=0.01*pi; N=1.0e6;
x=0.05; v=0.05; dN=100;
V=[]; X=[];
x=x+v*dt/2;
for i=1:N
   t=dt*i;
   v=v+p*sin(x)*sin(t)*dt;
   x=x+v*dt;
   if(mod(i,dN)==0) X=[X,x];
V=[V,v]; end;
end;
```

has the highest spectral entropy, indicating the longest Poincaré cycle since

$$F(t) = \int_{-\infty}^{\infty} f(\omega)e^{-i\omega t}d\omega = \int_{-\infty}^{\infty} f(\omega)e^{-i\omega(t+T_p)}d\omega = F(t+T_p), \qquad (11.10)$$

where the Poincaré cycle T_p has to be the least common multiple of the inverse of the dominant frequency values assuming they are incommensurate. Therefore, the broader the spectrum, the longer the Poincaré cycle. This is consistent with the Boltzman's H-theorem in that the binary collisions eventually lead to the Maxwellian distribution of maximum entropy. It is also consistent with the central limit theorem which states that any probability distribution having the Markovian property of statistical independence follows Gaussian distribution.

```
function SpectralEntropy
P=[]; H=[];
for ip=1:10
   [X,V]=STAND(ip/10);
   Vk=fft(V);

Pv=Vk.*conj(Vk)/sum(V
k.*conj(Vk));
   Hv=-
sum(Pv.*log(Pv))/log(len
gth(Pv)),
   P=[P,ip/10];
   H=[H,Hv];
end
x=0.1:0.01:1;
z=5+3*tanh((x-0.45)*20);
plot(P,H,'g*',x,z,'r-');
axis([0.1 1 1 8]);
title('Spectral Entropy');
xlabel('p'); ylabel('H');


function
[X,V]=STAND(p)
dt=0.01*pi; N=1.0e6;
x=0.05; v=0.05; dN=100;
V=[]; X=[];
x=x+v*dt/2;
for i=1:N
   t=dt*i;

v=v+p*sin(x)*sin(t)*dt;
   x=x+v*dt;
   if(mod(i,dN)==0)
X=[X,x]; V=[V,v]; end;
end;
```

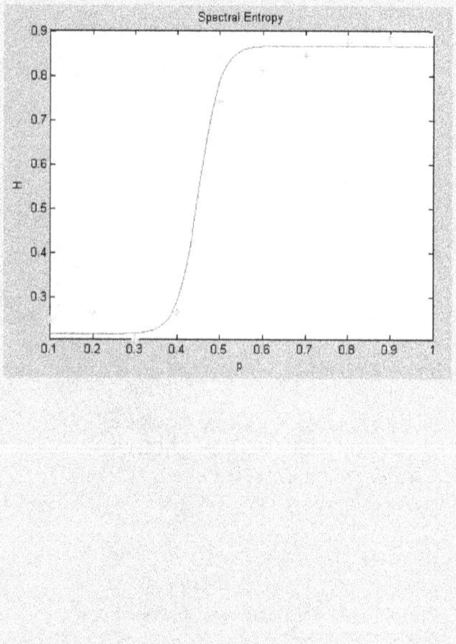

Homework 11.4: Show that the white noise in a normalized range of 0 to 1 has the spectral entropy equal to 0.27.

Homework 11.5: Show that the Gaussian white noise, a representation of the full chaos, has the spectral entropy equal to $(\log \pi + 1)/2 \sim 1$.

The entropy defined in Eq. (11.9), however, needs be divided by its phase volume $\log \Omega$ for the absolute measure.

11.2.3 *Hilbert-Huang Transform*

The **Hilbert-Huang transform** (**HHT**), proposed by Huang *et al.*, adopts the **empirical mode decomposition** (**EMD**) to break up a signal into the **intrinsic mode functions** (**IMF**) by the sifting process and the **Hilbert spectral analysis** (**HSA**) so to obtain the instantaneous frequency. HHT is adaptive and efficient, and can be applied to nonlinear and nonstationary processes. Moreover, good physical meanings have been shown by HHT. Almost all the studies reveal that the HHT gives results much sharper than any of the traditional analysis methods in time-frequency-energy representation. The MATLAB version, among others, of HHT analysis package can be downloaded from the HHT web.

The difference between the data and the mean of the upper and lower envelopes yields the first component. Ideally, it should satisfy the definition of an IMF, symmetric and having all maxima positive and all minima negative. Following the first round of sifting, the crest may become a local maximum. The first IMF contains the shortest timescale component of the signal. By taking it out of the data, the remainder contains the variations of the second shortest timescale. It is treated as the new data and subject to the same sifting process as described. This procedure can be repeated until finally when the residue becomes a monotonic function from which no more IMF can be extracted. The characteristic frequencies of IMF components recursively reduced to roughly half value in each sifting. EMD thus has the capability as a multiple timescale filter. EMD decomposes a signal into the IMF. By definition, an IMF is any function with the same number of ds and zero crossings, with its envelopes being symmetric with respect to zero. It guarantees a well-behaved Hilbert transform of the IMF.

For the original data $X(t)$, its Hilbert transform $Y(t)$ is,

$$Y(t) = \frac{1}{\pi} P \int_{-\infty}^{\infty} \frac{X(t')}{t - t'} dt', \tag{11.11}$$

where P indicates the Cauchy principal value. By taking $X(t)$ and $Y(t)$ as the complex conjugate pair, an analytical signal $Z(t) = X(t) + iY(t) = a(t)e^{i\theta(t)}$ is defined, where $a(t) = \sqrt{X(t)^2 + Y(t)^2}$ and $\theta(t) = \arctan(Y(t)/X(t))$. The

```
function getIMF(p)
FileName=sprintf('SWdata%d',fix(p/0.05));
load(FileName,'X', 'V', 'p', 'dt', 'N');
tic; time=datestr(now),
HH=eemd(V,0,1);
toc; time=datestr(now),
[m,n]=size(HH),
D=max(HH(:,n))-min(HH(:,n)),
sFile=sprintf('IMF%d.mat',fix(p/0.05)),
t=(1:N)*dt;
save(sFile,'HH', 't', 'p');

function PlotIMF(p)
sFile=sprintf('IMF%d.mat',fix(p/0.05)),
load(sFile,'HH', 't', 'p');
tic; time=datestr(now),
[m,n]=size(HH),
Figure(1);
s=sprintf('v plot @ p=%f',p);
title(s);
plot(t,HH(:,1),'b');
figure(2);
s=sprintf('IMF plot @ p=%f',p);
title(s);
for i=2:n
line(t,HH(:,i)-i*0.5);
end;

function GetOmega(p)
sFile=sprintf('IMF%d.mat',fix(p/0.05)),
load(sFile,'HH', 't', 'p');
tic; time=datestr(now),
[m,n]=size(HH),
figure;
title(s);
OMEGA=[];
for i=2:6
   i,
   y=HH(:,i);
   omega=ifndq(y,t/m);
   line(t,omega-2*p*i);
   OMEGA=[OMEGA,omega];
end;
toc; time=datestr(now);
```

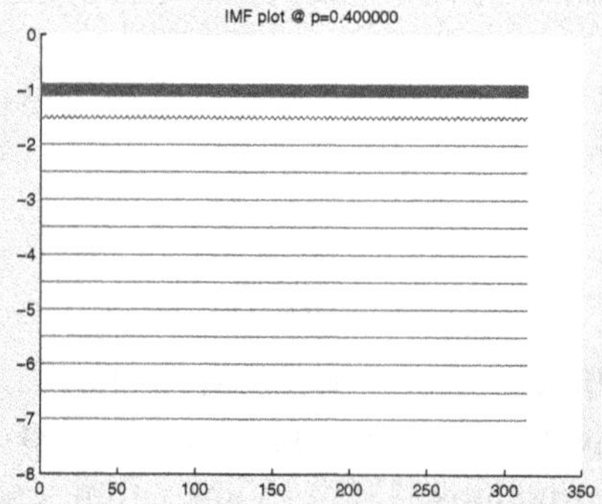

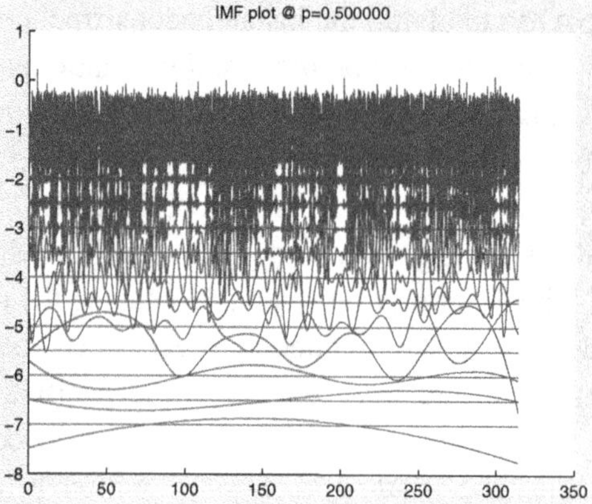

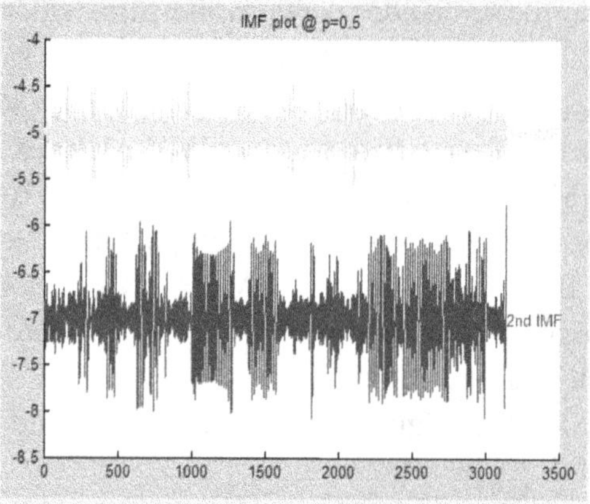

instantaneous frequency is given by

$$\omega(t) = \frac{d\theta(t)}{dt}. \tag{11.12}$$

The original data $X(t)$ can be expressed by its IMF components and the corresponding instantaneous frequencies as

$$X(t) = \sum_{j=1}^{n} a_j(t) e^{i \int \omega_j(t) dt}. \tag{11.13}$$

The first instantaneous frequency is the main characteristics of the dynamic system. For the regular trajectory of quasi-periodic nature, the frequency range is narrow and its corresponding first intrinsic mode function has smaller amplitudes, tantamount to smaller energy. The stochastic trajectory has broad frequency range and its first intrinsic mode function has larger amplitude. As p increases, the average instantaneous frequency in the first IMF increases from the pump wave frequency to twice that, and its rapid rise coincides with the stochasticity boundary. The empirical mode decomposition, capable as a **multiple-timescale filter**, gives in each IMF component the spectral entropy that can be compared with the Gaussian white noise.

11.3 Nonlinear Theory of Mode Conversion

Mode conversion occurs when two waves of different physical properties have the same frequency $\omega_1 = \omega_2$ and wave number $\vec{k}_1 = \vec{k}_2$ with the latter often compensated by the gradient of the equilibrium quantities, namely, $\vec{k}_1 = \vec{k}_2 + \vec{k}_L$, where $\vec{k}_L$ represents the effective wave number from the spatial gradient. When three waves satisfy the resonance conditions: $\omega_1 = \omega_2 + \omega_3$ and $k_1 = k_2 + k_3$, the large amplitude pump wave can transfer its energy to the other two waves, and the process can be reversible.

The three wave interaction is often termed the parametric instability. It is easy to recognize that at the quarter density where the pump wave frequency is twice the local plasma frequency, $\omega_0 \approx 2\omega_p$, the parametric decay into two plasmons can readily occur. This process is however not effective since the EM wave travels near light speed. On the other hand, the large amplitude plasma wave can also mode-mode couple to generate the second harmonic $\omega_2 = \omega_0 + \omega_0$, or the dc harmonic. The pump wave slows down at the layer of cutoff-resonance pair and the amplitude of wave can be greatly enhanced. Laser fusion and large amplitude wave heating also motivates us to examine the large amplitude wave in mode conversion. Due to the conservation of the canonical momentum $\vec{P} = \vec{p} + e\vec{A}/c$, where $\vec{\Omega} \equiv \nabla \times \delta\vec{v}$ is the vorticity, (cf. Sec. 9.4.3) it implies that the dc harmonic may result in both the **dc magnetic field** and **dc vortex flow**.

11.3.1 *Governing Equations*

To describe the electrostatic wave, we start with the continuity equation and the momentum equation for electrons,

$$\frac{\partial n}{\partial t} + \nabla \cdot (n\vec{v}) = 0, \tag{11.4}$$

$$\frac{\partial \vec{v}}{\partial t} + \vec{v} \cdot \nabla \vec{v} = -\frac{q}{m}\vec{E} - \frac{q}{mc}\vec{v} \times \vec{B} - \frac{\nabla p}{nm}, \tag{11.5}$$

where we take the electron charge $e = -q < 0$. Applying the time derivative to Eq. (11.4), and replacing $\partial \vec{v}/\partial t$ from Eq. (11.5), we find

$$\frac{\partial^2 n}{\partial t^2} - \nabla \cdot \left[\frac{\nabla p}{m} + \frac{nq}{m}\left(\vec{E} + \frac{1}{c}\vec{v} \times \vec{B}\right) + \nabla \cdot (n\vec{v}\vec{v})\right] = 0. \tag{11.6}$$

By taking $p = nk_BT \equiv nmv_{th}^2$, assuming constant and uniform temperature T, and utilizing the Poisson's equation, Eq. (11.6) becomes

$$\frac{\partial^2 n}{\partial t^2} - v_{th}^2 \nabla^2 n + \frac{4\pi q^2 n(n - n_0)}{m} - \frac{q\vec{E} \cdot \nabla n}{m}$$
$$-\nabla \cdot \left[\nabla \cdot (n\vec{v}\vec{v}) + \frac{nq}{mc}\vec{v} \times \vec{B}\right] = 0, \tag{11.7}$$

where n_0 is the equilibrium plasma density with a scale length L_0. The first three terms of Eq. (11.7) give the linear dispersion of the plasma wave. The fourth term is responsible for the mode conversion. The last term contains the nonlinear coupling effects due to the flow.

To describe the electromagnetic wave, we start from the Maxwell equations,

$$\nabla \times \vec{E} = -\frac{1}{c}\frac{\partial \vec{B}}{\partial t}, \tag{11.8}$$

$$\nabla \times \vec{B} = \frac{4\pi}{c}\vec{j} + \frac{1}{c}\frac{\partial \vec{E}}{\partial t}. \tag{11.9}$$

The current is given by $\vec{j} = -nq\vec{v}$. Taking the time derivative on Eq. (11.9) and eliminating the magnetic field by Eq. (11.8), we find,

$$\frac{\partial^2 \vec{E}}{\partial t^2} - 4\pi q\frac{\partial n\vec{v}}{\partial t} + c^2 \nabla \times (\nabla \times \vec{E}) = 0. \tag{11.10}$$

We assume the plasma density varies along $\hat{e}_z$ direction, the wave propagates on the $(\hat{e}_x,\hat{e}_z)$ plane, and the wave magnetic field is along $\hat{e}_y$ direction. Taking $\hat{e}_y \cdot \nabla\times$ operation on Eq. (11.10), replacing $\partial n\vec{v}/\partial t$ with use of Eq. (4.17) or the combination of Eqs. (11.4) and (11.5), and utilizing the identity $\nabla \times (\nabla \times \vec{E}) = \nabla(\nabla \cdot \vec{E}) - \nabla^2 \vec{E}$, we have

$$\left(\frac{\partial^2}{\partial t^2} - c^2\nabla^2 + \frac{4\pi n q^2}{m}\right)\widehat{e}_y \cdot \nabla \times \vec{E} + \widehat{e}_y \cdot \nabla\left(\frac{4\pi n q^2}{m}\right)$$

$$\times \vec{E} + 4\pi q \widehat{e}_y \cdot \nabla \times \left[\nabla \cdot (n\vec{v}\vec{v}) + \frac{nq}{mc}\vec{v} \times \vec{B}\right] = 0. \qquad (11.11)$$

The first three terms of Eq. (11.11) give the linear dispersion of the electromagnetic wave. The fourth term is responsible for the mode conversion. The last term contains the nonlinear coupling effects due to the flow.

We define two dimensionless variables:

$$\xi \equiv (n - n_0)/n_{00} = -\nabla \cdot \vec{E}/(4\pi e n_{00})$$

that measures the electrostatic charge density, and

$$\varsigma \equiv \widehat{e}_y \cdot \nabla \times \vec{E}/(4\pi e n_{00})$$

that represents the electromagnetic wave. Here, n_{00} is the critical density where the pump wave frequency matches the plasma frequency, $\omega_0 = \sqrt{4\pi e^2 n_{00}/m}$. We normalize the time and space variables to ω_0 and k_0, the frequency and wave number of the incident wave. We further define the dimensionless variables of velocity $\vec{u} = \vec{v}/c$, electric field $\vec{\varepsilon} = ek_0\vec{E}/(m\omega_0^2)$, magnetic field $\vec{b} = e\vec{B}/(mc\omega_0)$, thermal speed $\beta = v_{th}/c$ and equilibrium ion density $\bar{n} = n_0/n_{00}$. The dimensionless space and time variables are $\bar{x} = k_0 x$, $\bar{z} = k_0 z$ and $\tau = \omega_0 t$, and the corresponding dimensionless space and time derivatives are $\bar{\nabla} \equiv \nabla/k_0$ and $d/d\tau \equiv d/dt/\omega_0$. By taking $\bar{n} \sim O(1)$ and an expansion parameter $\delta \ll 1$, we make the following ordering $\bar{n} \gg \xi \sim \varsigma \sim |\vec{u}| \sim |\vec{b}| \sim |\vec{\varepsilon}| \sim 1/(k_0 L_0) \sim O(\delta)$, where L_0 is the scale length of equilibrium density n_0.

We want to solve Eqs. (11.7) and (11.11) to $O(\delta^2)$ accuracy. Inasmuch as they are, we only need the velocity and magnetic field to their first order. Equation (11.6) gives

$$\frac{\partial \vec{u}}{\partial \tau} = -\vec{\varepsilon} - \beta^2 \bar{\nabla}\xi, \qquad (11.12)$$

And Eq. (11.8) of the Faraday's law can be rewritten as

$$\frac{\partial b}{\partial \tau} = -\hat{e}_y \cdot \bar{\nabla} \times \vec{\varepsilon} = -\varsigma. \qquad (11.13)$$

Equations (11.7) and (11.11) therefore become

$$\left(\frac{\partial^2}{\partial\tau^2} - \beta^2\bar{\nabla}^2 + \bar{n} + \xi\right)\xi - \varepsilon_z\frac{\partial}{\partial\bar{z}}(\bar{n} + \xi) - \varepsilon_x\frac{\partial\xi}{\partial\bar{x}} - \bar{n}\bar{\nabla}\cdot\vec{\Gamma} = 0, \quad (11.14)$$

$$\left(\frac{\partial^2}{\partial\tau^2} - \bar{\nabla}^2 + \bar{n} + \xi\right)\varsigma + \varepsilon_x\frac{\partial}{\partial\bar{z}}(\bar{n} + \xi) - \varepsilon_z\frac{\partial\xi}{\partial\bar{x}} + \bar{n}\hat{e}_y\cdot\bar{\nabla}\times\vec{\Gamma} = 0, \quad (11.15)$$

where

$$\vec{\Gamma} \equiv \bar{\nabla}\cdot(\vec{u}\vec{u}) + \vec{u}\times\bar{b} = (\vec{u}\cdot\bar{\nabla})\vec{u} + \vec{u}\bar{\nabla}\cdot\vec{u} + \vec{u}\times\bar{b}. \quad (11.16)$$

The electric fields are found from $\xi = -\bar{\nabla}\cdot\vec{\varepsilon}$ and $\varsigma \equiv \hat{e}_y\cdot\bar{\nabla}\times\vec{\varepsilon}$, and solved by the Green's function method (cf. Eq. (7.21), Eq. (7.22) and Homework 7.10) in the numerical code. Due to the nonlinear mode coupling, we need to sum up more harmonics than what was done in the linear mode conversion case where only one harmonic was considered. Equations (11.12–16) are the basis for the nonlinear theory of mode conversion. Note that the equilibrium density gradient and the density perturbation are negligible in the $\vec{\Gamma}$ terms. The current approach reduces the problem to a manageable scope by investigating the relevant physics to $O(\delta^2)$.

Homework 11.6: Show that the canonical momentum $\vec{p} \equiv \vec{u} - \vec{a}$ is a constant of the motion in this nonlinear mode conversion process.

It can be shown (cf. Homework 11.6) that the canonical momentum $\vec{p} \equiv \vec{u} - \vec{a}$ is a constant of the motion, where $\vec{a}$ is the vector potential such that $\nabla\times\vec{a} = \vec{b}$. Thus, $\bar{\nabla}\times\vec{u} - \vec{b}$ is also a constant of the motion, null in the plasma before the wave arrives, and remains so through the nonlinear mode conversion process. As the dc harmonic is generated due to the nonlinear mode-mode coupling, the dc values of ξ and ς, and $\vec{\varepsilon}$ can readily occur. A dc magnetic field will result from Eq. (11.13). This leads to the spontaneous generation of both the vorticity and the magnetic field so to hold $\bar{\nabla}\times\vec{u} - \vec{b}$ unchanged. This mechanism indicates that the dc magnetic field will produce the vortex flow. On the other hand, the vortex flow in **solar flare** and the like will produce the dc magnetic field. It is a mechanism relevant to the generation of **magnetic loop** in the **solar prominence**.

From $\vec{\Gamma} \equiv \bar{\nabla}\cdot(\vec{u}\vec{u}) + \vec{u}\times\bar{b}$ of Eq. (11.16)., we can show (cf. Homework 11.7) that $\bar{\nabla}\times\vec{\Gamma} = \bar{\nabla}\times(\vec{u}\bar{\nabla}\cdot\vec{u})$. The magnetic field arises from ς, and the nonlinear coupling of ς can only result from the density perturbation ξ and $\bar{\nabla}\times\vec{\Gamma}$. The nonlinear effects of the transverse wave vanish if the compressibility of the electron flow is absent since $\bar{\nabla}\cdot\vec{u} = 0$ and the electron density follows $d\bar{n}/dt = 0 = \partial\bar{n}/\partial t + \vec{u}\cdot\nabla\bar{n} = 0$. *The compressibility* $\bar{\nabla}\cdot\vec{u} \neq 0$ *is therefore a necessary condition to have the dc magnetic field.* The magnetic field is synonymous with the vorticity as governed by Eq. (11.13). From Eqs. (11.14) and (11.15), the mode conversion process depending on the terms $\varepsilon_z \partial \bar{n}/\partial z$ and $\varepsilon_x \partial \bar{n}/\partial z$, which converts the EM wave energy to the ES wave. The latter is tantamount to the electron density perturbation, to produce the compressibility that couples with the electron flow to drive the nonlinear term $\bar{\nabla} \times \vec{\Gamma}$. Thus we have the dc component of ς in Eq. (11.13) that has the $\bar{\nabla} \times \vec{\varepsilon}$ component to generate the dc component of vorticity $\bar{\nabla} \times \vec{u}$ in Eq. (11.12).

Homework 11.7: Show that $\bar{\nabla} \times \vec{\Gamma} = \bar{\nabla} \times (\vec{u}\bar{\nabla} \cdot \vec{u})$, given $\vec{\Gamma} \equiv \bar{\nabla} \cdot (\vec{u}\vec{u}) + \vec{u} \times \bar{b}$.

A serious numerical instability in FDTD program can develop when the CFL condition is violated. As the wave propagates to the steepening density, the characteristic length of the system shortens since drastic spatial variation takes place in narrow region, especially when the short wavelength mode of electrostatic wave becomes significant. Thus, the effective Δx gets smaller. The effect can be more pronounced when the EM wave is launched at a larger oblique angle as the projection along the density gradient varies faster. But the most serious effect arises from the emerging short wavelength ES wave. The numerical stability may be improved by the following: reduce the wave amplitude which may sacrifice the strong nonlinear physics; improve the spatial derivatives which need higher order accuracy and slow down the computation; apply the implicit scheme that makes coding harder, add viscosity damping to trim the high harmonic spatial modes (cf. Homework 11.8), or cut down the time step that may increase the cumulative error due to the inaccuracy of spatial derivatives. The numerical code advances $\xi = -\bar{\nabla} \cdot \vec{\varepsilon}$ and $\varsigma \equiv \hat{e}_y \cdot \bar{\nabla} \times \vec{\varepsilon}$. It then integrates ς to find $\bar{b}$ accordingly. Error can incur when ς saturates but not vanishes. Since $\bar{b}$ is obtained by integrating Eq. (11.13), that can feed back to pump up ς as well as the vortex flow. The numerical runs must ensure that ς is vanishingly small and the dc magnetic field saturates when the EM wave moves away from the ES wave and no more energy to pump into the dc component. The numerical program listed below has selected the parameters near the stability boundary. The run at the marginal stability can go awry by increasing the amplitude of EM wave, or the launch angle.

The strength of the dc magnetic field varies with the incident angle, so does the mode conversion efficiency (cf. Sec. 7.4.2). It also depends on the wave amplitude. We take the NIF UV laser of angular frequency $\omega_0 \sim 5 * 10^{15}$ rad/sec to evaluate all effects. It is noteworthy that if b is of order unity, B is then around 0.3 gigagauss. To have the mode conversion layer residing inside the plasma, the plasma density needs be no less than $n_{00} \sim 10^{22}/\text{cm}^3$.

```
function NMC(RERUN)
close all; clc;
global A WavePacketWidth omega0 nt2 nt3 T Vth c L N Lx dx Tth;
global kx x dt t Lz dz z  nH nL zc Lc n0 dn0 ZZ;
init; datestr(now), tic;
Zt=zeros(N,N);Z=Zt; Ex=Zt; Ez=Zt; B=Zt; Vx=Zt; Vz=Zt;
X=Z;Xt=Zt; Tth=Vth^2; GAMMAx=0*X; GAMMAz=0*Z;
k0=sqrt(1-kx^2), theta=atan(kx/k0)*180/pi,
t=0; j=0; tf=nt3*dt; tend=nt2*dt; it=0;
fig=figure('position', [50 50 1000 650],'color','white');
if(RERUN)
    load('FDTD');
    it=fix(t/dt); t=t-dt; itstart=it; itf=6000; j=fix(t), ZS=0*x;
else
    itstart=1; itf=fix(50/dt); msg=sprintf('wave packet on @t=0'),
end;
for it=itstart:itf
    t=t+dt;
    ZS0=Z(:,1);
    if(it<fix(nt2))
        ZS=getZ(t); Z(:,1)=getZ(t-dt);
    else
        if(it<1.5*nt2) ZS=0*x;  Z(:,1)=ZS; end;
        if(it==fix(nt2)) X=X.*(repmat(n0,N,1)>0.26); msg=sprintf('wave packet off
@t=%4.1f',t), end;
    end
    if(it>=fix(2*nt2)) ZS=(0*ZS0+2*Z(:,1))/2;
        if(it==fix(2*nt2)) msg=sprintf('outgoing wave boundary condition on
@t=%4.1f',t), end;
    end;
[B,Vx,Vz,X,Z,Xt,Zt,Ex,Ez,GAMMAx,GAMMAz]=Propagator(Z,Zt,X,Xt,Ex,Ez,Vx,V
z,B,ZS,RERUN,GAMMAx,GAMMAz);
if mod(it,fix(1/dt))==0
    toc,tic,
    j=j+1; Smax=max(ZS);
    NameStr=sprintf('FDTD%d',j),
    save(NameStr,'X','Z','Ex','Ez','Vx','Vz','B','theta','t','N','Xt','Zt');
    PlotGraf(X,Z,Vx,Vz,B,theta,t,dt); % PlotGraf(X,Z,Ex,Ez,t)
    getframe;
    if(RERUN) CheckNonlinearity(Z,Zt,X,Xt,Ex,Ez,Vx,Vz,B); end;
    CheckConsistency(X,Z,Ex,Ez,N,dx,dz);
```

Homework 11.8: Add the viscosity in the equation of motion to rederive Eqs. (11.14) and (11.15).

Homework 11.9: Show that at $\delta \sim 1$, the electric field has a strength of $eE \sim k_0 mc^2 \sim 90\,\mathrm{GeV/cm}$, translated to a power density $P = c\|\vec{E}\|^2/(8\pi) \sim 1.2 * 10^{19}\,\mathrm{W/cm^2}$.

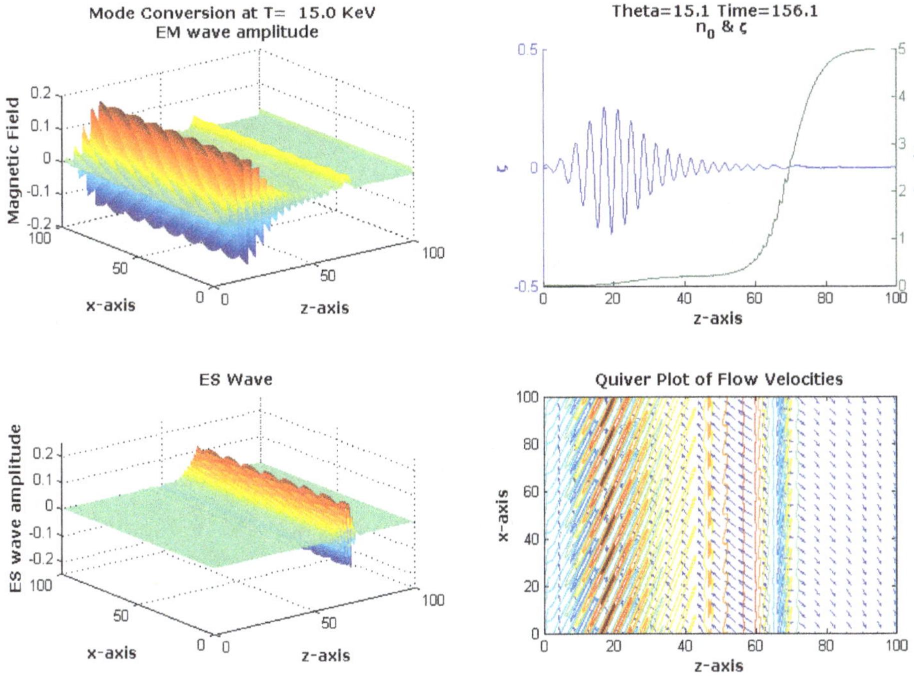


```
function W=getZ(time)
global A WavePacketWidth omega0 nt2 nt3 T Vth c L N Lx dx Tth;
global kx x dt t Lz dz z  nH nL zc Lc n0 dn0 ZZ;
kz=sqrt(1-kx^2);
w=WavePacketWidth -kz*time;
W=exp(-4*w^2/WavePacketWidth^2)*exp( -i*t+i*kx*x+i*kz*w);
W=real(W)*A;
```

11.4 Bifurcation Solutions via the Physical Characteristics

Finding numerically the bifurcated solutions can be rather challenging as discussed in the prior chapter. Taking the parameters of ITER machine: $R_0/a \sim 3.1$, $B_T \sim 6$ tesla, etc., the numerical algorithm utilizing the physical characteristics is demonstrated in the following to solve the zero β MHD equilibrium equation,

$$\Delta^* \psi = -c_F e^{\lambda\psi} = -\frac{4\pi}{c} R j_\phi. \tag{11.17}$$

The method adopts the physical characteristics of the edge poloidal B field as the key physical variable to be evaluated from two independent functions: One is related to the LHS of Eq. (11.17), i.e., $\langle B_p \rangle = \langle |\nabla\psi|/R \rangle$ by taking from Eq. (10.20), namely,

```
function CheckNonlinearity(Z,Zt,X,Xt,Ex,Ez,Vx,Vz,B)
function [B,Vx,Vz,X,Z,Xt,Zt,Ex,Ez,GAMMAx,GAMMAz]=
   Propagator(Z,Zt,X,Xt,Ex,Ez,Vx,Vz,B,ZS,RERUN,GAMMAx,GAMMAz)
global A WavePacketWidth omega0 nt2 nt3 T Vth c L N Lx dx Tth;
global kx x dt t Lz dz z nH nL zc Lc n0 dn0 ZZ;
global Zend Xend;
if(RERUN)
   divV=getDIV(Vx,Vz,N,dx,dz);
   curlG=getCURL(Vx.*divV,Vz.*divV,N,dx,dz);
   divG=getDIV(Vx.*divV,Vz.*divV,N, dx,dz);
   divG=divG+getDEL(Vx.^2+Vz.^2,N,dx,dz)/2;
end;
dt=0.05;
dt=dt/(1+4*(t>120));
Nc=4.95; NA=100;
Xend=X(:,N); Zend=Z(:,N);
Zt=Zt+dt*(getDEL(Z,N,dx,dz)-Z.*repmat(n0,N,1)-Ex.*repmat(dn0,N,1));
if(RERUN) nlZ=-X.*Z+Ez.*getDx(X,N,dx)-Ex.*getDz(X,N,dz)-curlG.*repmat(n0,N,1);
   Zt=Zt+dt*nlZ; clear nlZ; end;
Z=Z+dt*Zt; Z=Z-Z.*(repmat(n0,N,1)>Nc).*(1-exp(-NA*repmat(n0-Nc,N,1)));
Xt=Xt+dt*(Tth*getDEL(X,N,dx,dz)-X.*repmat(n0,N,1)+Ez.*repmat(dn0,N,1));
if(RERUN) nlX=-X.^2+Ez.*getDz(X,N,dz)+Ex.*getDx(X,N,dx)+divG.*repmat(n0,N,1);
   Xt=Xt+dt*nlX; clear nlX; end;
X=X+dt*Xt; X=X-X.*(repmat(n0,N,1)<0.25)*(t<100); X=X-X.*(repmat(n0,N,1)>Nc).*(1-
exp(-NA*repmat(n0-Nc,N,1)));
[Ex,Ez]=getE(X,Z,ZS);
Vx0=Vx; Vz0=Vz;
Vx=Vx-Tth*dt*getDx(X,N,dx)-dt*Ex;
Vz=Vz-Tth*dt*getDz(X,N,dx)-dt*Ez;
Vx=Vx-Vx.*(repmat(n0,N,1)<0.25)*(t<100);
Vz=Vz-Vz.*(repmat(n0,N,1)<0.25)*(t<100);
B=B-Z*dt;
B(:,1)=0; B(:,N)=0; X(:,N)=0; Z(:,N)=0; Xt(:,N)=0; Zt(:,N)=0;
Ex(:,N)=0; Ez(:,N)=0; Vx(:,N)=0; Vz(:,N)=0;
Vx=(Vx0+Vx)/2; Vz=(Vz+Vz0)/2;

function CheckConsistency(X,Z,Ex,Ez,N,dx,dz)
   divE=getDIV(Ex,Ez,N,dx,dz);
errX=X-divE;
    curlE=getCURL(Ex,Ez,N,dx,dz);
errZ=Z-curlE;
maxXerr=max(max(abs(errX),[],1),[],2);
maxX=max(max(abs(X),[],1),[],2),
maxZerr=max(max(abs(errZ),[],1),[],2);
maxZ=max(max(abs(Z),[],1),[],2),
```

$\vec{B} = \nabla\psi \times \nabla\phi + b(\psi)\nabla\phi$. The other is from the RHS with the relation $\langle B_p \rangle = \frac{4\pi}{c} \int d\vec{s} \cdot \vec{J}_\phi / 2\pi a$ by taking $\frac{4\pi}{c} J_\phi = -c_F e^{\lambda\psi}/R$. Here, the bracket represents the angular average at $r = a$, $\langle\rangle \equiv \oint d\theta/2\pi\,|_{r=a}$. At the start of iteration, these two values differ significantly but merge into one when the ψ solution is accurately solved and the iteration converges. The central ψ value is adjusted according to

```
function CheckNonlinearity(Z,Zt,X,Xt,Ex,Ez,Vx,Vz,B)
global A WavePacketWidth omega0 nt2 nt3 T Vth c L N Lx dx Tth;
global kx x dt t Lz dz z  nH nL zc Lc n0 dn0 ZZ;
V2=(Vx.^2+Vz.^2)/2;
maxRHO=max(max(X,[],1),[],2),
divV=getDIV(Vx,Vz,N,dx,dz);
curlV=getCURL(Vx,Vz,N,dx,dz);
FlowEnergy=sum(sum(X.*V2,1),2)/N/N/2,
Compressibility=sum(sum(divV,1),2)/N/N,
GAMMAx=B.*Vz+Vx.*divV+Vx.*getDx(Vx,N,dx)+Vz.*getDz(Vx,N,dx);
GAMMAz=-B.*Vx+Vz.*divV+Vx.*getDx(Vz,N,dx)+Vz.*getDz(Vz,N,dx);
curlG=getCURL(GAMMAx,GAMMAz,N,dx,dz);
maxCurlG=max(max(curlG,[],1),[],2),
divG=getDIV(GAMMAx,GAMMAz,N,dx,dz);
maxDivG=max(max(divG,[],1),[],2),
CanonicalMomentum=sum(sum(curlV-B,1),2)/N/N,

function curlP=getCURL(Px,Pz,N,dx,dz)
curlP=getDz(Px,N,dz)-getDx(Pz,N,dx);

function divP=getDIV(Px,Pz,N,dx,dz)
divP=getDx(Px,N,dx)+getDz(Pz,N,dz);

 function dxP=getDx(P,N,dx)
dxP=([P(2:N,:);P(1,:)]-[P(N,:);P(1:N-1,:)])/2/dx;

function dzP=getDz(P,N,dz)
dzP=([P(:,2:N),0*(1:N)']-[0*(1:N)',P(:,1:N-1)])/dz/2;

function d2P=getDEL(P,N,dx,dz)
d2P=(-2*P+[P(2:N,:);P(1,:)]+[P(N,:);P(1:N-1,:)])/dx/dx;
d2P=d2P+(-2*P+[P(:,2:N),0*(1:N)']+[0*(1:N)',P(:,1:N-1)])/dz/dz;

function init
global A WavePacketWidth omega0 nt2 nt3  T Vth c L N Lx dx Tth;
global kx x dt t Lz dz z  nH nL zc Lc n0 dn0 ZZ;
A=0.35; dOmega=0.25; dk=0.25; omega0=1;
dt=0.05;
WavePacketWidth=2*pi*2; %wave packet starts at t=0;
nt2=3*WavePacketWidth/dt;%wave packet ends
nt3=300/dt; %program ends
T=15; %keV
Vth=sqrt(T/500);
c=1;  N=1000; L=8*2*pi; %L=12*2*pi;
Lx=100; dx=Lx/N;  kx=L/Lx, x=(1:N)'*dx;
Lz=100; dz=Lz/N; z=(1:N)*dz;
ZZ=repmat(z,N,1);
[n0,dn0]=getDensity(Lz,z,dz,N);

function [n0,dn0]=getDensity(Lz,z,dz,N)
nH=4.8; nL=0.2;
zH=0.65*Lz; zL=0.2*Lz;
LH=8; LL=10;
n0=nH*(1+tanh((z-zH)/LH))/2+nL*(1+tanh((z-zL)/LL))/2;
dn0=diff(n0); dn0=[0,dn0];
```

```
function PlotGraf(X,Z,Vx,Vz,B,theta,t,dt)
global N Lx x Lz z n0 ZZ;
subplot(2,2,1);
H=surf(z,x,real(B));
set(H,'edgecolor','none');
str=sprintf('Nonlinear Mode Conversion at T=%6.1f \n magnetic field',t);
title(str,'fontsize',10,'fontname','verdana','fontweight','bold');
xlabel('z-axis','fontsize',10,'fontname','verdana','fontweight','bold')
ylabel('x-axis','fontsize',10,'fontname','verdana','fontweight','bold')
zlabel('EM wave amplitude','fontsize',10,'fontname','verdana','fontweight','bold');
view(20,30);

subplot(2,2,2)
[AX,H1,H2]=plotyy(z,real(Z(fix(36),:)),z,n0+real(X(fix(36),:)),'plot');
set(get(AX(1),'Ylabel'),'String','EM wave
amplitude','fontsize',10,'fontname','verdana','fontweight','bold')
set(get(AX(2),'Ylabel'),'String','density','fontsize',10,'fontname','verdana','fontweight','bold')
xlabel('z-axis','fontsize',10,'fontname','verdana','fontweight','bold')
str=sprintf('Theta=%4.1f dt=%5.3f Time=%5.1f \n n & ',theta,dt,t);
title(strcat(str,' \zeta'),'fontsize',10,'fontname','verdana','fontweight','bold');

subplot(2,2,3)
H=surf(z,x,real(X));
set(H,'edgecolor','none');
title('ES Wave','fontsize',10,'fontname','verdana','fontweig ht','bold');
xlabel('z-axis','fontsize',10,'fontname','verdana','fontweight','bold');
ylabel('x-axis','fontsize',10,'fontname','verdana','fontweight','bold');
zlabel('ES wave amplitude','fontsize',10,'fontname','verdana','fontweight','bold');
axis([0 100 0 100 -0.25 0.25]);
view(20,30);

subplot(2,2,4)
V=sqrt((real(Vx).^2)+(real(Vz).^2));
n=50;
zz=ZZ(1:n:N,1:n:N);
Vxx=Vx(1:n:N,1:n:N);Vzz=Vz(1:n:N,1:n:N);
quiver(zz,zz',real(Vzz),real(Vxx));
hold on
contour(ZZ,ZZ',V)
hold off
axis([0 Lz 0 Lx])
title('Flow Vel ocity','fontsize',10,'fontname','verdana','fontweight','bold');
xlabel('z-axis','fontsize',10,'fontname','verdana','fontweight','bold')
ylabel('x-axis','fontsize',10,'fontname','verdana','fontweight','bold')
```

either the ratio of these two values or its inverse. Thus, the iterations may follow two different paths in searching for the bifurcated solutions.

It turns out that the flatter profile is the dominant solution and will emerge so long as the iteration converges to a solution that satisfies the boundary condition. By normalizing the length to the minor radius, the following dimensionless variables are defined: $\Psi \equiv \lambda\psi$, $\Lambda \equiv \lambda c_F a^2$, $\bar{r} \equiv r/a$, $\bar{z} \equiv Z$, $\varepsilon \equiv a/R_0$. Equation (11.17) is cast into (cf. Eq. (10.57))

$$M\Psi \equiv \left(\frac{\partial^2}{\partial \bar{r}^2} + \frac{1}{\bar{r}}\frac{\partial}{\partial \bar{r}} + \frac{1}{\bar{r}^2}\frac{\partial^2}{\partial \theta^2} - \frac{\varepsilon \cos\theta}{(1+\varepsilon\bar{r}\cos\theta)}\frac{\partial}{\partial \bar{r}} + \frac{\varepsilon \sin\theta}{(1+\varepsilon\bar{r}\cos\theta)\bar{r}}\frac{\partial}{\partial \theta}\right)\Psi$$
$$= -\Lambda e^{\Psi}. \tag{11.18}$$

Since $\varepsilon \ll 1$, we take the cylindrical solution as the initial guess (cf. Eqs. (10.77–80)).

We will need to invert the matrix M no matter which bifurcated solution is to be solved. The implementation of the matrix M is not trivial. Considering the simpler case of circular cross section, we divide the plasma domain by $N_r \times N_t$ grids that are defined by $\bar{r} = dr/2 : dr : 1$ in the column direction and $\theta = d\theta : d\theta : 2\pi$ in the row direction. Therefore, the Ψ vector is defined as

$$\Psi = \Big[\underbrace{\psi(1,1), \psi(2,1), \ldots\psi(N_r,1)}_{\psi_1}, \underbrace{\psi(1,2), \psi(2,2), \ldots\psi(N_r,2)}_{\psi_2}, \ldots,$$
$$\underbrace{\psi(1,N_t), \psi(2,N_t), \ldots, \psi(N_r,N_t)}_{\psi_{N_t}}\Big]. \tag{11.19}$$

The central ψ_0 is evaluated by taking the average of the ψ values on the surrounding grids, namely, $\psi_0 = \frac{1}{N_t}\sum_{i=1}^{N_t} \psi(1,i)$. The derivatives along the θ direction follow the periodic boundary condition. The derivatives along $\bar{r}$ for the first element in the $\bar{r}$ vector are a bit tricky. It takes the central value ψ_0 to form the derivative. We have the matrix M as in the following:

$$M = \begin{pmatrix} m_{11} & m_{12} & 0 & . & . & . & 0 & m_{1,N_t} \\ m_{21} & m_{22} & m_{23} & 0 & . & . & 0 & 0 \\ 0 & m_{32} & m_{33} & m_{34} & 0 & . & . & . \\ . & 0 & . & . & . & 0 & . & . \\ . & . & 0 & . & . & . & . & . \\ . & . & . & 0 & . & . & m_{N_t-2,N_t-1} & 0 \\ 0 & . & . & . & . & m_{N_t-2,N_t-1} & m_{N_t-1,N_t-1} & m_{N_t-1,N_t} \\ m_{N_t,1} & 0 & . & . & . & 0 & m_{N_t,N_t-1} & m_{N_t,N_t} \end{pmatrix} \begin{pmatrix} \psi_1 \\ \psi_2 \\ . \\ . \\ . \\ . \\ . \\ \psi_{N_t} \end{pmatrix}. \tag{11.20}$$

The diagonal submatrixes have the following nonzero elements:

$$m_{ii}(j,j) = -\frac{2}{dr^2} - \frac{2}{r_j^2 d\theta^2}, \tag{11.21a}$$

$$m_{ii}(j, j+1) = \frac{1}{dr^2} + \frac{1}{(1+\varepsilon r_j \cos\theta_i)} \frac{1}{2r_j dr}, \tag{11.21b}$$

$$m_{ii}(j-1, j) = \frac{1}{dr^2} - \frac{1}{(1+\varepsilon r_j \cos\theta_i)} \frac{1}{2r_j dr}. \tag{11.21c}$$

Here $m_{ii}(j, j+1)$ is not applicable for $j = N_r$, but $\Psi(\bar{r} = 1 + dr, \theta) = 0$ makes that term irrelevant. Moreover, $m_{ii}(j, j-1) = 0$ is not applicable to $j = 1$, while the RHS of Eq. (11.18) has the first row vector modified into $-\alpha e^{\psi(1,i)} - \psi_0/dr^2 + \psi_0/(2r_1 dr(1 + \varepsilon r_1 \cos\theta_i))$. The off-diagonal submatrixes have the following nonzero elements to satisfy the periodic boundary condition in θ direction: For $1 < i < N_t$:

$$m_{i,i+1}(j, j) = \frac{1}{r_j^2 d\theta^2} + \frac{\varepsilon \sin\theta_i}{(1+\varepsilon r_j \cos\theta_i) r_j} \frac{1}{2d\theta}, \tag{11.21d}$$

$$m_{i-1,i}(j, j) = \frac{1}{r_j^2 d\theta^2} - \frac{\varepsilon \sin\theta_i}{(1+\varepsilon r_j \cos\theta_i) r_j} \frac{1}{2d\theta}, \tag{11.21e}$$

and

$$m_{1,N_t}(j, j) = \frac{1}{r_j^2 d\theta^2} - \frac{\varepsilon \sin\theta_1}{(1+\varepsilon r_j \cos\theta_1) r_j} \frac{1}{2d\theta}, \tag{11.21f}$$

$$m_{N_t,1}(j, j) = \frac{1}{r_j^2 d\theta^2} + \frac{\varepsilon \sin\theta_{N_t}}{(1+\varepsilon r_j \cos\theta_{N_t}) r_j} \frac{1}{2d\theta}. \tag{11.21g}$$

```
function [Minv,M]=setM  %cf. Eq(11.20) & Eq(11.21)
global Nr Nt dr dTheta r theta;
global alpha lambda epsilon PSI;
r=dr:dr:1;
m=diag(ones(Nr-1,1),1)+diag(ones(Nr-1,1),-1)+diag(-2*ones(Nr,1));
m=m/dr^2+diag(-2/dTheta^2./r.^2);
mp=diag(1/dTheta^2./r.^2); mn=mp;
M=sparse(Nr*Nt,Nr*Nt);
for i=1:Nt
        st=sin(theta(i)); ct=cos(theta(i));
        mii=m+diag(1/2/dr./r(1:Nr-1)./(1+epsilon.*r(1:Nr-1).*ct),1);
        mii=mii-diag(1/2/dr./r(2:Nr)./(1+epsilon.*r(2:Nr).*ct),-1);
        mip=mp+diag(epsilon*st/2./r./dTheta./(1+epsilon.*r.*ct));
        min=mn-diag(epsilon*st/2/dTheta./r./(1+epsilon.*r.*ct));
        i1=1+(i-1)*Nr; i2=Nr*i;
        M(i1:i2,i1:i2)= M(i1:i2,i1:i2)+mii;
        ip=i+1; if(ip>Nt) ip=1; end; in=i-1; if(in<=0) in=Nt; end;
        M(i1:i2,(in-1)*Nr+1:in*Nr)=M(i1:i2,(in-1)*Nr+1:in*Nr)+min;
        M((ip-1)*Nr+1:ip*Nr,i1:i2)=M((ip-1)*Nr+1:ip*Nr,i1:i2)+mip;
end;
Minv=inv(M);
```

```
function [dPr,dPt]=getGradPSI(psi0,P,r,dr,dt,Nr,Nt)
dPr=([P(2:Nr,:);zeros(1,Nt)]-[repmat(psi0,1,Nt);P(1:Nr-1,:)])/dr/2;
dPt=([P(:,2:Nt),P(:,1)]-[P(:,Nt),P(:,1:Nt-1)])./repmat(r',1,Nt)/2/dt;
```

```
function Bp=getBp %cf, Eq(11.22)
global Nr Nt dr dTheta r theta a R0;
global alpha beta epsilon PSI;
global X Y psi0 Jphi;
[dPr,dPt]=getGradPSI(psi0,PSI,r,dr,dTheta,Nr,Nt);
R=1+epsilon*r'*cos(theta);
Bpt=-dPr./R;
Bpr=dPt./R;
Bp=sqrt(Bpt.^2+Bpr.^2)*epsilon;
Bpedge=sum(Bp(Nr,:))/Nt,
clear Bpt Bpr dPr dPt R;
```

```
function [Jphi,I]=getCurrent % cf. Eq(11.23)
global Nr Nt dr dTheta r theta a R0;
global alpha epsilon PSI;
R=1+epsilon*r'*cos(theta);
Jphi=alpha*exp(PSI)*epsilon./R;
I=sum(sum(Jphi.*repmat(r',1,Nt),1),2)*dr*dTheta;
Jedge=sum(Jphi(Nr,:))/Nt;
J0=sum(Jphi(1,:))/Nt;
```

The matrix M is implemented as in the following program. The coding follows closely what described in the prior paragraph.

The MHDeq.m program solves Eq. (11.18) by the straightforward matrix inversion. It evaluates two different values for error checking: $D \equiv |1+M\Psi/\Lambda e^{\Psi}|$,

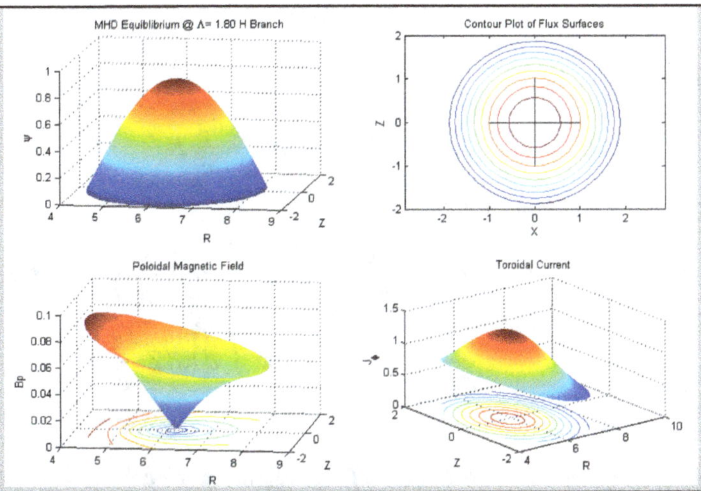

The MHD bifurcated equilibrium of flat profile.

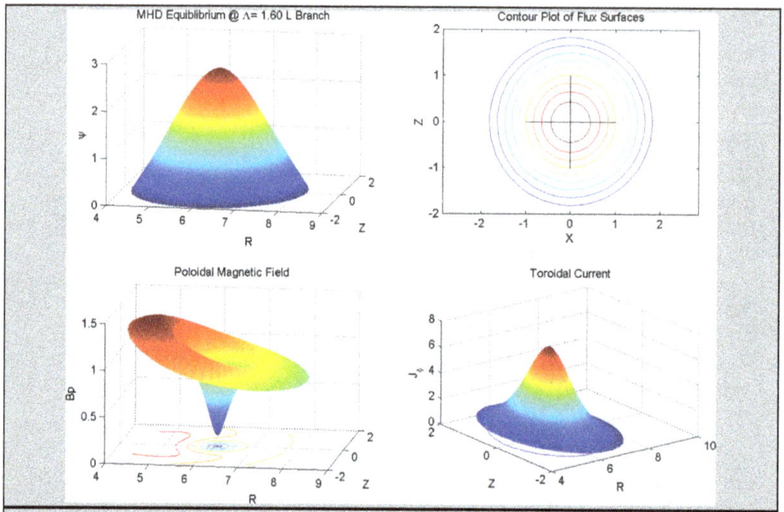

The MHD bifurcated equilibrium of flat profile.

and $d \equiv norm|\Psi - \Psi_0|/N_r/N_t$, where Ψ_0 is the prior solution of Ψ value. To ensure accuracy these two parameters are set to be smaller than 10^{-6}. The method however only yields the H-branch of flatter profile (cf. Homework 11.9) which has substantial current pedestal at the plasma edge. The current pedestal and the poloidal field are larger in the high field side than the low field side.

We have to evaluate the edge poloidal B field from the two sides of Eq. (11.17). The left hand side involves the differential operator,

$$\langle B_p \rangle = \langle |\nabla \psi|/R \rangle = \oint \frac{d\theta}{2\pi} \frac{1}{(R_0 + r\cos\theta)} \sqrt{\left|\frac{\partial \psi}{\partial r}\right|^2 + \left|\frac{\partial \psi}{r\partial \theta}\right|^2}\Bigg|_{r=a}, \tag{11.22}$$

From Eq. (10.17), we have

$$\Delta^* \psi = R^2 \nabla \cdot (R^{-2} \nabla \psi) = -\frac{4\pi}{c} R J_\phi.$$

Taking the volume integration gives,

$$\begin{aligned}\int dV \nabla \cdot (R^{-2} \nabla \psi) &= \oint d\vec{S} \cdot \frac{\nabla \psi}{R^2} = \oint dl \int R d\phi \frac{B_p}{R} = 2\pi \oint dl B_p \\ &= \frac{4\pi}{c} \int dV \frac{J_\phi}{R} = 2\pi \int dR dZ \frac{4\pi}{c} J_\phi,\end{aligned}$$

which simply yields

$$\oint dl B_p = \int r dr d\theta \frac{4\pi}{c} J_\phi = \int r dr d\theta \frac{4\pi}{c} \frac{c_F e^{\lambda \psi}}{R_0 + r\cos\theta}. \tag{11.23}$$

```
function MHDeq(Hmode)
% The algorithm iterates Eq(11.18) by evaluating the current  with use of the prior
Ψ
close all;  clc;
 % to find the new Ψ , and repeat the process.
global Nr Nt dr dTheta r theta a R0;
global alpha lambda epsilon PSI;
global X Y psi0 Jphi;
a=2; R0=6.2; B0=5.3; I0=15e6; V=850; ql=2;
cF=0.45, lambda=1, epsilon=a/R0;
alpha=cF*a^2/lambda,
Nr=500; Nt=60; dr=1/Nr; dTheta=2*pi/Nt;
r=dr:dr:1; theta=dTheta:dTheta:2*pi;
p=2*log(4/alpha*(1+[1;-1]*sqrt(1-alpha/2))),
if(Hmode) psi0=p(2), else psi0=p(1), end;
R=repmat(r',1,Nt); THETA=repmat(theta,Nr,1);
X=R.*cos(THETA); Y=R.*sin(THETA);
psi=psi0-2*log(1+alpha/8*exp(psi0)*r.^2);  %Eq(10.77)
PSI=repmat(psi',1,Nt);
% Initialize  by the cylindrical solution and make it into 2d structure
[Minv,M]=setM;
psi=psi0-0.1;

tic; datestr(now), factor=1;
for ip=1:200
          if(abs(psi-psi0)<1.0e-6) ip,  s='psi0 converged to spec', break; end
  psi0=psi;
  toc,tic,
  for it=1:10
          PSI0=PSI; % save a copy of the current solution
          F=-alpha*(exp(PSI));
          F(1,:)=F(1,:)-psi0/dr^2+psi0/2/dr./(1+epsilon*r(1)*cos(theta))/r(1);
          F=reshape(F,Nr*Nt,1);
          RATIO=norm(M*reshape(PSI,Nr*Nt,1))/norm(F);
          D=abs(RATIO-1);
          PSI=Minv*F;
          PSI=reshape(PSI,Nr,Nt);
        d=norm(PSI-PSI0)/Nr/Nt,
        psi=sum(PSI(1,:))/Nt,
        if(d<1.0e-6&&D<1.0e-6) s='error reduced to spec', break; end;
          end;
  end;
if(Hmode) filename='caseH'; else filename='caseL'; end;
save(filename,'PSI','M','Nr','Nt','r','theta','cF');
format long;
psi0, psi, alpha, epsilon, D, d,
plotGraf(Hmode);
toc, datestr(now),
```

```
function plotGraf(H)
global Nr Nt dr dTheta r theta a R0;
global alpha betap lambda epsilon PSI;
global X Y psi0 Jphi;
fig=figure('position', [10 20 950 650],'color','white');
[Jphi,I]=getCurrent;
Bp=getBp;
X=[X,X(:,1)]; Y=[Y,Y(:,1)]; PSI=[PSI,PSI(:,1)];
X=[zeros(1,Nt+1);X]; Y=[zeros(1,Nt+1);Y]; PSI=[psi0*ones(1,Nt+1);PSI];
subplot(2,2,1); %plot(R0+a*X,PSI); %
surfc(R0+a*X,a*Y,PSI);  view(-10,10);
xlabel('R'); ylabel('Z'); zlabel('\psi');
str=sprintf('%5.2f',alpha);
if(H==1) string=sprintf(' H Branch'); else string=sprintf('L Branch'); end;
title(strcat('MHD Equiblibrium @ \Lambda=',str,string));
view(10,15); shading flat;
subplot(2,2,2);
maxPSI=max(max(PSI,[],1),[],2); minPSI=min(min(PSI,[],1),[],2);
dV=(maxPSI-minPSI)/10;
contour(a*X,a*Y,PSI,minPSI*9/10:dV:maxPSI*9/10);
title('Contour Plot of Flux Surfaces'); xlabel('X'); ylabel('Z');
hold on;
x=-1:0.1:1; y=0*x; plot(x,y,'k-',y,x,'k-'); axis equal;
hold off;
subplot(2,2,3);
Bp0=sum(Bp(1,:)/Nt; Bp=[Bp,Bp(:,1)]; Bp=[Bp0*ones(1,Nt+1);Bp];
surfc(R0+a*X,a*Y,Bp); xlabel('R'); ylabel('Z'); zlabel('Bp');
str=sprintf('Poloidal Magnetic Field');
title(str); shading flat;
view(8,16);
subplot(2,2,4);
J0=sum(Jphi(1,:)/Nt; Jphi=[Jphi,Jphi(:,1)]; Jphi=[J0*ones(1,Nt+1);Jphi];
surfc(R0+a*X,a*Y,Jphi); shading flat;
title('Toroidal Current'); xlabel('R'); ylabel('Z'); zlabel('J_\phi');
```

At the start of iteration, these two values evaluated from Eqs. (11.22) and (11.23) differ significantly. The central ψ value is adjusted according to either the ratio of these two values or its inverse. Thus, the iterations may follow two different paths in searching for the bifurcated solutions. The poloidal magnetic fields are supposed to merge into one when the ψ solution is accurately solved and the iteration converges. The method does incur numerical error in solving the differential equation due to the finite grid size effect on evaluating the edge magnetic fields.

In principle, however, this concept is applicable to any type of bifurcation equation with a physical quantity that is derivable from the two parts of the equation. It provides a workable algorithm that remains to be explored to other bifurcation equations.

```
function MHDbifurcation(Hmode)
 % Utilizing the ratio as obtained from Eq(11.22) and Eq(11.23) to search for
close all;  clc;
% the bifurcated solutions.
global Nr Nt dr dTheta r theta a R0;
global alpha lambda epsilon PSI;
global X Y psi0 Jphi;
a=2; R0=6.2; B0=5.3; I0=15e6; V=850; ql=2;
cF=0.45, lambda=1, epsilon=a/R0;
alpha=cF*a^2/lambda,
Nr=600; Nt=50; dr=1/Nr; dTheta=2*pi/Nt;
r=dr:dr:1; theta=dTheta:dTheta:2*pi;
p=2*log(4/alpha*(1+[1;-1]*sqrt(1-alpha/2))),
if(Hmode) psi0=p(2), else psi0=p(1), end;
R=repmat(r',1,Nt); THETA=repmat(theta,Nr,1);
X=R.*cos(THETA); Y=R.*sin(THETA);
psi=psi0-2*log(1+alpha/8*exp(psi0)*r.^2);  %Eq(10.77)
PSI=repmat(psi',1,Nt);
% Initialize ψ by the cylindrical solution and make it into 2d structure
tic; datestr(now),
[Minv,M]=setM;
psi=psi0-0.1;
R=1+epsilon*(r'*theta);
factor=1; D=1; PSI0=PSI;
for ip=1:100
   if(abs(psi-psi0)<1.0e-6) ip,  s='psi0 converged to spec', break; end;
   psi0=psi;
   toc,tic,ip,
   for it=1:10
            PSI0=PSI; % save a copy of the current solution
            F=-alpha*(exp(PSI));
            F(1,:)=F(1,:)-psi0/dr^2+psi0/2/dr./(1+epsilon*r(1)*cos(theta))/r(1);
            F=reshape(F,Nr*Nt,1);
            RATIO=norm(M*reshape(PSI,Nr*Nt,1))/norm(F);
            D=abs(RATIO-1),
            PSI=Minv*F;
            PSI=reshape(PSI,Nr,Nt);
            d=norm(PSI-PSI0)/Nr/Nt,
          Bp=getBp;
            Bp1=Bp(Nr,fix(Nt/4));  %bp1=sum(Bp(Nr,:))/Nt,
            [Jphi,I]=getCurrent;
            bp2=I/2/pi,
            if(Hmode) ratio=bp1/(bp2+eps); else  ratio=bp2/(bp1+eps); end;
             PSI=PSI*(1-factor*(1-ratio));
            psi=sum(PSI(1,:))/Nt,
            if(d<1.0e-6&&D<1.0e-6) s='error reduced to spec', I, bp1, bp2,break; end;
            end;
```

```
for ip=1:100
   if(abs(psi-psi0)<1.0e-6) ip, s='psi0 converged to spec', break; end;
   psi0=psi;
   toc,tic,ip,
   for it=1:10
        PSI0=PSI; % save a copy of the current solution
        F=-alpha*(exp(PSI));
        F(1,:)=F(1,:)-psi0/dr^2+psi0/2/dr./(1+epsilon*r(1)*cos(theta))/r(1);
        F=reshape(F,Nr*Nt,1);
        RATIO=norm(M*reshape(PSI,Nr*Nt,1))/norm(F);
        D=abs(RATIO-1),
        PSI=Minv*F;
        PSI=reshape(PSI,Nr,Nt);
        d=norm(PSI-PSI0)/Nr/Nt,
      Bp=getBp;
        bp1=sum(Bp(Nr,:))/Nt,
        [Jphi,I]=getCurrent;
        bp2=I/2/pi,
        if(Hmode) ratio=bp1/(bp2+eps); else ratio=bp2/(bp1+eps); end;
         PSI=PSI*(1-factor*(1-ratio));
        psi=sum(PSI(1,:))/Nt,
        if(d<1.0e-6&&D<1.0e-6) s='error reduced to spec', I, bp1, bp2,break; end;
        end;
   end;
if(Hmode) filename='caseH'; else filename='caseL'; end;
save(filename,'PSI','M','Nr','Nt','r','theta','cF');
[Jphi,I]=getCurrent;
format long;
psi0, psi, alpha, epsilon, D, d,
plotGraf(Hmode);
```

Homework 11.10: Run MHDeq.mat by choosing the L mode, namely, type MHDeq(0). Observe that the solution converges to the H mode of flat profile.

Homework 11.11: Run MHDbifurcation.m by choosing some α value to find the solutions of both H and L modes.

Further Reading

A more detailed discussion on the analytical solution as also derived in Homework 11.1 to the large amplitude plasma wave can be found in Davidson (1972).

From regular motion to chaos, and vice versa as in the self organization (cf. Hasegawa and Mima 1977), the underlying physics can be very rich. It deserves the attention to bring out the most significant effect thereof. Plenty of tools are readily available. Further reading into the Hilbert-Huang Transform, check into the web site: http://rcada.ncu.edu.tw/research1.htm.

Mode conversion is a linear physics of resonance phenomena. The large amplitude effect can enhance both the linear and nonlinear effects and cause new effects that may otherwise not be triggered.

The original paper on "Nonlinear Theory of Mode Conversion" can be found in Wu and Hsu (2013).

One of the major driving forces behind the advancement of plasma physics is the fusion research, and magnetic fusion in particular. Many decades have passed and many more decades will come and go. The fusion community has over promised and under delivered the technology. Yet, the physics of fission reactors, accomplished just a few years of engineering work after the atomic bombs, can be understood from the individual neutron dynamics. Granted for its stochastic nature, simulation by the Monte Carlo would predict fairly accurately the chain reaction and the energy histogram. While there are different sets of challenges to construct, operate, and safe guard fission reactors, fusion reactors cannot be deemed any easier than fission reactors. The fundamental principles in fission reactor are not nonlinear many-body in nature. On the contrary, fusion plasma has the nonlinear many-body phenomena as its core physics, and the understanding of its principles can only be accomplished when the nonlinear physics is well studied.

Homework Hints

Homework 11.1: Solve by the Lagrangian formulation the set of dimensionless equations for LAPW without the pressure and the damping effect.

In terms of the dimensionless variables, the LAPW can be described by the following:

$$\frac{\partial}{\partial \tau} n + \frac{\partial}{\partial \varsigma} n v = 0, \tag{11.1.1}$$

$$\frac{\partial v}{\partial \tau} + v \frac{\partial v}{\partial \varsigma} = -\varepsilon, \tag{11.1.2}$$

$$\frac{\partial}{\partial \varsigma} \varepsilon = 1 - n. \tag{11.1.3}$$

Defining the Lagrangian variables (s, z) by $s = \tau$ and $z = \varsigma - \int_0^s ds' v(z, s')$, we transform the equations of the Eulerian variables (ς, τ) by applying the chain rule of derivative,

$$\frac{\partial}{\partial s} = \frac{\partial}{\partial \tau}\frac{\partial \tau}{\partial s} + \frac{\partial}{\partial \varsigma}\frac{\partial \varsigma}{\partial s} = \frac{\partial}{\partial \tau} + v \frac{\partial}{\partial \varsigma}, \tag{11.1.4}$$

$$\frac{\partial}{\partial z} = \frac{\partial}{\partial \tau}\frac{\partial \tau}{\partial z} + \frac{\partial}{\partial \varsigma}\frac{\partial \varsigma}{\partial z} = \left(1 + \int_0^s ds' \frac{\partial}{\partial z} v(z, s')\right) \frac{\partial}{\partial \varsigma}, \tag{11.1.5}$$

so that the convection along the flow is simply the time derivative of the Lagrangian variable s. We arrive at

$$\frac{\partial}{\partial \tau} n + \frac{\partial}{\partial \varsigma} n v = 0 = \frac{\partial n}{\partial s} + n \frac{\partial v}{\partial \varsigma} = \frac{\partial n}{\partial s} + \frac{n}{\left(1 + \int_0^s ds' \frac{\partial}{\partial z} v(z, s')\right)} \frac{\partial v}{\partial z}$$

that leads to

$$\left(1+\int_0^s ds'\frac{\partial}{\partial z}v(z,s')\right)\frac{\partial n}{\partial s}+n\frac{\partial v}{\partial z}=0=\frac{\partial}{\partial s}\left[n\left(1+\int_0^s ds'\frac{\partial}{\partial z}v(z,s')\right)\right] \tag{11.1.6}$$

Taking time derivative on Eq. (11.1.3), we find

$$\frac{\partial^2}{\partial\varsigma\partial\tau}\varepsilon=-\frac{\partial}{\partial\tau}n=\frac{\partial}{\partial\varsigma}nv\Rightarrow\frac{\partial}{\partial\tau}\varepsilon-nv=0=\frac{\partial\varepsilon}{\partial\tau}-v\left(1-\frac{\partial\varepsilon}{\partial\varsigma}\right).$$

Simplifying this equation with use of Eq. (11.1.14), we arrive at

$$\frac{\partial\varepsilon}{\partial s}=v. \tag{11.1.7}$$

Applying the same to Eq. (11.1.2), we have

$$\frac{\partial v}{\partial s}=-\varepsilon. \tag{11.1.8}$$

Combining these two equations, we have a simple harmonic oscillation in terms of the Lagrangian variables,

$$\frac{\partial^2 v}{\partial s^2}+v=0. \tag{11.1.9}$$

The general solution to Eq. (11.1.9) gives the flow velocity as $v(s,z)=v_0(z)\cos s+x_0(z)\sin s$, where $x_0(z)$ is the initial displacement, and $v_0(z)$ is the initial velocity. The electric field is then governed by $\varepsilon(s,z)=v_0(z)\sin s-x_0(z)\cos s$ as given by Eq. (11.1.8), while the electron density is from Eq. (11.1.6),

$$n(z,s)=\frac{n_0(z)}{\left(1+\frac{\partial v_0}{\partial z}\sin s+\frac{\partial x_0}{\partial z}(1-\cos s)\right)}. \tag{11.1.10}$$

The initial condition at $s=0$ requires $\varepsilon_0(z)=-x_0(z)$.

We will take the initial velocity to be zero, and choose a sinusoidal perturbation in the electric field, $\varepsilon_0(z)=-\Delta\sin z=-x_0(z)$. We expect $n_0(z)=1+\Delta\cos z$ by utilizing Eqs. (11.1.3) and (11.1.5). We find the solutions are

$$v(s,z)=\Delta\sin z\sin s, \tag{11.1.11}$$

$$\varepsilon(s,z)=-\Delta\sin z\cos s, \tag{11.1.12}$$

$$n(z,s)=\frac{1+\Delta\cos s}{1+\Delta\cos z(1-\cos s)}. \tag{11.1.13}$$

It is clear when $2\Delta \to 1$, there is a singularity that will cause the density to become highly steepened at the time when $s = (2n+1)\pi$.

When pressure is included, the physical interpretation of large Δ means the flow has to approach the sound speed. The Mach number needs to be $O(1)$ so to cause the steepening effect.

Homework 11.3: Instead of the phase space point at the wave period, choose two velocities at alternating periods. Plot the maps at $p < p_c$ and $p > p_c$ to show the sharp contrast.

```
function Homework11p3(p)
close all; clc;
rand('state',sum(100*clock));
dt=2*pi/100; N=200000;
x=0.05*(rand(1)-0.5); v=0.05*(rand(1)-0.5);
V=[]; X=[];
x=x+v*dt/2;
for i=1:N
   t=dt*i;
   v=v+p*sin(x)*sin(t)*dt;
   x=x+v*dt;
   if(mod(i,100)==0) X=[X,x]; V=[V,v]; end;
end;
v1=V(1:2:N/100); v2=V(2:2:N/100);
plot(v1,v2,'g*');
xlabel('V1'); ylabel('V2'); axis tight;
str=sprintf('Velocity Correlation @ p=%3.2f',p);
title(str);
end;
```

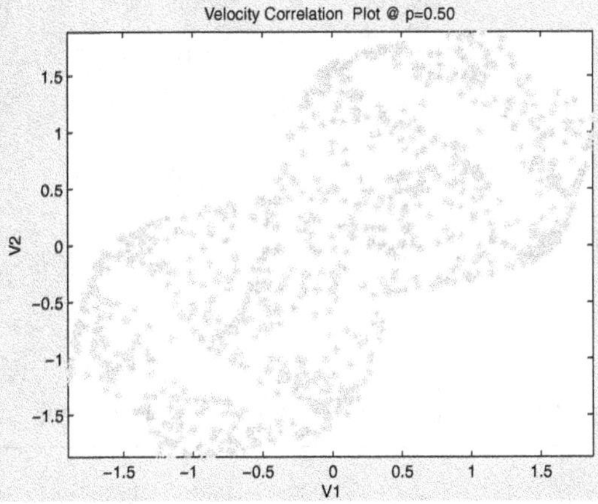

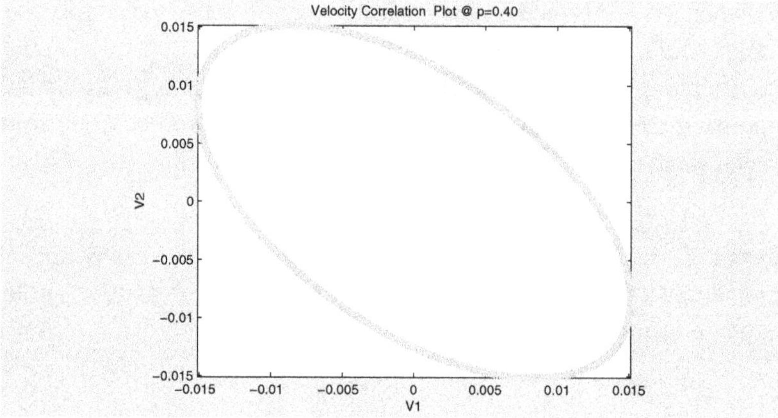

Homework 11.5: Show that the white noise in a normalized range of 0 to 1 has the spectral entropy equal to 0.27.

```
function Homework11p5
clear all; clc
syms a v k positive;
f=1;
variance=eval(int(v^2*f,v,0,
1)),
fk=int(f*exp(i*k*v),v,0,1);
Pk=fk*conj(fk);
N=int(Pk,k,0,1);
Pk=Pk/N;
S=-int(Pk*log(Pk),k,0,1),
S=eval(S),

rand('state',sum(100*clock));
sum(100*clock)));
tic,
N=10000000,
V=rand (1,N);
variance=sum(V.^2)/N,
Vk=fft(V);
Pk=Vk.*conj(Vk)/sum(Vk.*
conj(Vk));
Hv=-
sum(Pk.*log(Pk))/log(N),
toc,
```

```
variance =
   1

S =
log(pi)/2 + 1/2

S =
   1.072364942924700

N =
   60000000

variance =
   0.3333

Hv =
   0.2755
```

Homework 11.7: Show that $\bar{\nabla}\times\vec{\Gamma} = \bar{\nabla}\times(\vec{u}\bar{\nabla}\cdot\vec{u})$, given $\vec{\Gamma} \equiv \bar{\nabla}\cdot(\vec{u}\vec{u})+\vec{u}\times\bar{b}$.

$$\vec{\Gamma} \equiv \bar{\nabla}\cdot(\vec{u}\vec{u}) + \vec{u}\times\bar{b} = (\vec{u}\cdot\bar{\nabla})\vec{u} + \vec{u}\bar{\nabla}\cdot\vec{u} + \vec{u}\times\bar{b} = \bar{\nabla}u^2/2 - \vec{u}\times(\bar{\nabla}\times\vec{u}) + \vec{u}\bar{\nabla}\cdot\vec{u} + \vec{u}\times\bar{b} = \bar{\nabla}u^2/2 + \vec{u}\bar{\nabla}\cdot\vec{u} + \vec{u}\times\vec{p} = \bar{\nabla}u^2/2 + \vec{u}\bar{\nabla}\cdot\vec{u}.$$

Therefore, taking $\vec{p} = 0$, we have $\bar{\nabla} \times \vec{\Gamma} = \quad \bar{\nabla} \times (\vec{u}\bar{\nabla} \cdot \vec{u})$.

Homework 11.9: Run MHDeq.m by choosing the L mode, namely, type MHDeq(0). Observe that the solution converges to the H mode of flat profile.

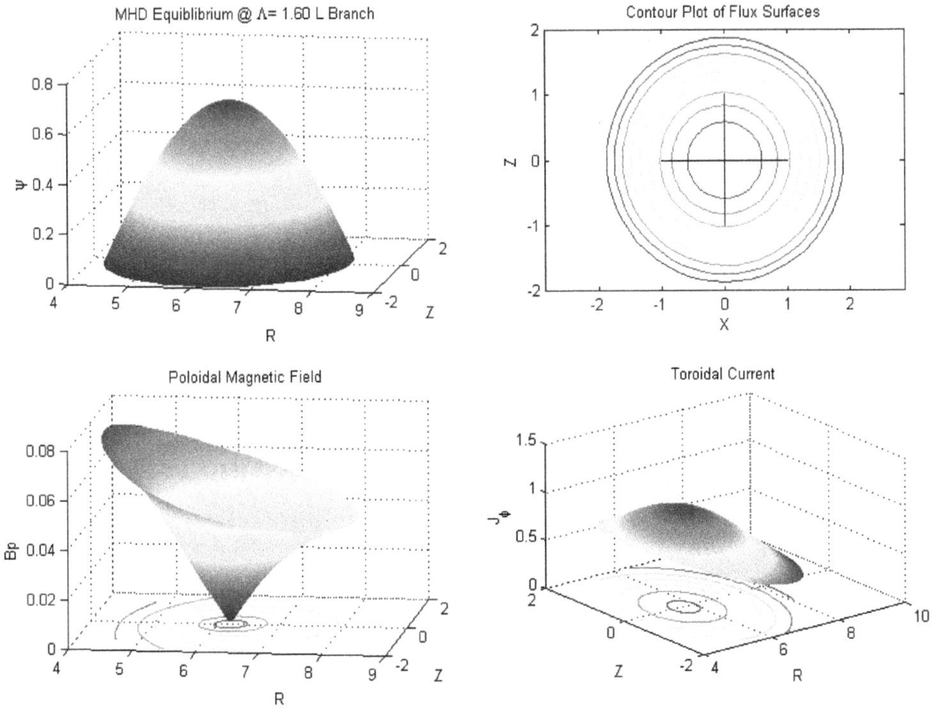

Bibliography

1. Alfven, H., Existence of electromagnetic-hydrodynamic waves, *Nature*, 150, 406 (1942).
2. Anderegg, F., X.-P. Huang, C. F. Driscoll, E. M. Hollmann, T. M. O'Neil and D. H. E. Dubin, Test particle transport due to long range interactions, *Phys. Rev. Lett.*, 78, 2128 (1997).
3. Bernstein, Ira B., Waves in a plasma in a magnetic field, *Phys. Rev.*, 109, 10 (1958).
4. Birdsall, Charles K. and A. Bruce Langdon, *Plasma Physics via Computer Simulations* (Adam Hilger, 1991).
5. Bohm, D., *The Characteristics of Electrical Discharges in Magnetic Fields*, edited by A. Guthrie and R. Wakerling, McGraw-Hill, New York, p. 201 (1949).
6. Boyd, T. J. M. and J. J. Sanderson, *Plasma Dynamics*, Barnes and Noble (1969).
7. Braginskii, S. I., Transport properties in a plasma, in *Reviews of Plasma Physics*, M. A. Leontovich (ed.) Consultants Bureau, New York, Vol. 1, 205 (1965).
8. Burrell, K. H., Effects of ExB velocity shear and magnetic shear on trubulence and transport in magnetic confinement devices, *Phys. Plasma*, 4, 1499 (1997).
9. Cadou, J. M., B. Cochelin, N. Damiland and M. Potier-Ferry, *Int. J. Numer. Meth. Engng.*, 50, 825 (2001).
10. Ceperley, D. M. and B. J. Alder, Ground state of the electron gas by a stochastic method, *Phys. Rev. Lett.*, 45, 566 (1980).
11. Cheng, C. Z., Liu Chen and M. S. Chance, High-n Ideal and resistive shear Alfven waves in tokamaks, *Annals of Physics*, 161, 21, (1985).
12. Chu, C., Magnetic spectra and electron transport of current-carrying plasmas, *Phys. Rev. Lett.*, 48, 246, (1982).
13. Chu, K. R., The electron cyclotron maser, *Rev. Mod. Phys.*, 76, 489, (2004).
14. Chu, K. R., H. Y. Chen, C. L. Hung, T. H. Chang, L. R. Barnett S. H. Chen and T. T. Yang, An ultra high gain gyrotron traveling wave amplifier, *Phys. Rev. Lett.*, 81, 4760 (1998).
15. Chu, M. S., C. Chu, G. Guest, J. Y. Hsu and T. Ohkawa, Kinetic analysis of the localized magnetohydrodynamic ballooning mode, *Phys. Rev. Lett.*, 41, 247, (1978).
16. Cliffe, K. A., A. Spence and S. J. Tavener, The numerical analysis of bifurcation problems with application to fluid mechanics, *Acta Numerica*, 39, 2008 (2011).
17. Coppi, B., Nonclassical transport and the "Principle of Profile Consistency", *Comments Plasma Phys. Control. Fusion*, 5, 261 (1980).
18. Cvitanović, P., R. Artuso, Mainieri, G. Tanner and G. Vattay, *Chaos: Classical and Quantum*, Niels Bohr Institute, Copenhagen (2005).
19. Davis, Harold T., *Introduction to Nonlinear Differential and Integral Equations*, Dover (1960).
20. Davison, Ronald C., *Methods in Nonlinear Plasma Theory*, Academic Press, (1972).
21. Dawson, J. M., H. Okuda and R. N. Carlile, *Phys. Rev. Lett.*, 27, 491 (1971).
22. De Brujin, N. G., *Asymptotic Methods in Analysis*, Dover Publications, Inc. (1961).
23. Demoulin, Yves-Marie J. and Y. M. Chen, An interation method for studying the bifurcation of solutions of the nonlinear equations, $L(\lambda)u+\varepsilon R(\lambda,u) = 0$, *Numer. Math.*, 23, 47 (1974).

24. Doveil, F., Stochastic Plasma heating by a large amplitude standing wave, *Phys. Rev. Lett.*, 46, 532 (1981).
25. Dupree, T. H., A perturbation theory for strong plasma turbulence, *Phys. Fluids*, 9, 1773, (1966).
26. Erckmann V. and U. Gasparino, Electron cyclotron resonance heating and current drive in toroidal fusion plasmas, *Plasma Phys. Control. Fusion*, 36, 1869 (1994).
27. Estrada, R. and R. P. Kanwal, *A Distributional Approach to Asymptotics — Theory and Applications*, Birkhäuser (2002).
28. Frieman, E. A. and Liu Chen, Nonlinear gyrokinetic equations for low-frequency electromagnetic waves in general plasma equilibria, *Phys. Fluids* 25, 502 (1982).
29. Furth, H. P., J. Killeen, M. N. Rosenbluth and B. Coppi, Stabilization by shear and negative V, *Plasma Phys. Control. Nuclear Fusion Res.*, 1,103 (1966).
30. Furth, Harold P. John Killeen, and Marshall N. Rosenbluth, Finite-resistivity instabilities of a sheet pinch, *Phys. Fluids*, 6, 459 (1963).
31. Galeev, R. Z. and A. A. Sagdeev, in *Nonlinear Plasma Theory,* revised and edited by T. M. O'Neil and D. L. Book, Benjamin, New York (1969).
32. Grassberger, P. and I. Procaccia, Characterization of strange attractors, *Phys. Rev. Lett.*, 50, 346 (1983).
33. Guevel, Y., H. Boutyour, J. M. Cadou, Automatic detection and branch switching methods for steady bifurcation in fluid mechanics, *J. Comp. Phys.* 230, 3614 (2011).
34. Harvey, R. W., M. G. McCoy, J. Y. Hsu and A. A. Mirin, Electron dynamics associated with stochastic magnetic and ambipolar electric fields, *Phys. Rev. Lett.*, 47, 102, (1981).
35. Hirshman, S. P. and D. J. Sigmar, Neoclassical transport of impurities in tokamak plasmas, *Nucl. Fusion*, 21, 1079 (1981).
36. Hsieh, C.-T., C.-M. Huang, C.-L. Chang, Y.-C. Ho, Y.-S. Chen, J.-Y. Lin, J. Wang and S.-Y. Chenal, Tomography of injection and acceleration of monoenergetic electrons in a laser-wakefield accelerator, *Phys. Rev. Lett.*, 96, 095001 (2006).
37. Hsu, J. Y., K. Matsuda, M. S. Chu, T. Jensen, Stochastic heating of a large amplitude wave, *Phys. Rev. Lett.*, 43, 203 (1979).
38. Hsu, J. Y., Frequency mismatch and stochastic heating in the cyclotron frequency range, *Phys. Fluids*, 25, 159 (1981).
39. Hsu, J. Y., R. W. Harvey and S. K. Wong, Anomalous impurity ion transport due to magnetic fluctuations, *Phys. Fluids*, 24, 2216 (1981).
40. Hsu, J. Y., V. S. Chan, R. W. Harvey, R. Prater and S. K. Wong, Resonance localization and poloidal electric field due to cyclotron wave heating in tokamak plasmas, *Phys. Rev. Lett.*, 53, 564 (1984).
41. Hsu, J. Y. and M. S. Chu, The tokamak equilibrium profile, *Phys. Fluids*, 30, 1221 (1987).
42. Hsu, J. Y., Relativistic theory of mode conversion at plasma frequency — the finite difference time domain simulation, *Comput. Phys. Commun.*, 182, 155 (2010).
43. Hsu, J. Y., K. Wu, S. K. Agarwal and C.-M. Ryu, The $B^{-3/2}$ diffusion in magnetized plasma, *Phys. Plasmas*, 17, 032104 (2013).
44. Huang, N. E., S. R. Long and Z. Shen, The mechanism for frequency downshift in nonlinear wave evolution, *Adv. Appl. Mech.*, 32, 59 (1996).
45. Jacques, S. A., Momentum and energy transport by waves in the solar atmospere and solar wind, *The Astrophys J.*, 215, 942, (1977).
46. Jardin, S., *Computational Methods in Plasma Physics*, CRC Press (2010).

47. Kadomtsev, B. B., *Sov. J. Plasma Physics*, 13, 443 (1987).
48. Kakurin, A. and I. Orlovsky, Hilbert-Huang transform in MHD plasma diagnostics, *Plasma Phys. Reports*, 31, 1054 (2005).
49. Kesner, J., J. Kesner, M. E. Mauel, G. A. Navratil, S. A. Sabbagh, M. Bell, R. Budny, C. Bush, E. Fredrickson, B. Grek, A. Janos, D. Johnson, D. Mansfield, D. McCune, K. McGuire, H. Park, A. Ramsey, E. Synakowski, G. Taylor, M. Zarnstorff, S. H. Batha and F. M. Levinton, High poloidal beta long pulse experiments in the tokamak fusion test reacto, *Phys. Fluids B*, 5, 2525 (1993).
50. Klimontovich, Y. L., *Kinetic Theory of Nonideal Gases and Nonideal Plasmas*, Pergamon Press, New York, (translated by R. Balescu) (1982).
51. Krall, N. A. and W. A. Trivelpiece, *Principles of Plasma Physics*, McGraw Hill (1973).
52. Kriesel J. M. and C. F. Driscoll, Measurements of viscosity in pure-electron plasmas, *Phys. Rev. Lett.*, 87, 135003 (2001).
53. Krommes, John A., Fundamental statistical descriptions of plasma turbulence in magnetic fields, *Phys. Report*, 360, 1, (2002).
54. Kruskal, M. D., *Asymptology*, Princeton University Plasma Physics Laboratory (1962).
55. Kulsrud, R. M., Plasma and Controlled Nuclear Fusion Research, 1, 127 (1966).
56. Kulsrud, R. M., Plasma Physics Controlled Nuclear Fusion Research, *Conf. Proceedings, Culham 1965*, 1, 127 (1966).
57. Kupiszezwki A., A High Frequency Microwave Amplifier by DSN Progress Report, May–June (1979).
58. Landau, L. D. and E. M. Lifshitz, *Fluid Mechanics*, (1969).
59. Landau, L. D., and E. M. Lifshitz, *Mechanics*, (1969).
60. Lee, P., B. J. Taylor, W. A. Peebles, H. Park, C. X. Yu, Y. Xu, N. C. Luhmann, Jr., and Jin, S. X., Observation of mode-converted ion Bernstein waves in the microtor tokamak, *Phys. Rev. Lett.*, 49, 205 (1982).
61. Lee, W. W., Gyrokinetic particle simulation model, *J. Comput. Phys.*, 72, 243 (1987).
62. Liu, C. S. and V. K. Tripathi, *Interaction of Electromagnetic Waves with Electron Beams and Plasmas*, World Scientific (1994).
63. Liu, J. Y., Chen, Y. I., Pulinets, S. A., Tsai, Y. B. and Chuo, Y. J., Seismo-ionospheric signatures prior to M >6.0 Taiwan earthquakes, *Geophys. Res. Lett.*, 27, 3113 (2000).
64. Lloyd, B., Overview of ECRH experimental results, *Plasma Phys. Control. Fusion*, 40, A119 (1998).
65. Madey, J., Stimulated emission of Bremsstrahlung in a periodic magnetic field, *J. Appl. Phys.*, 42, 1906, (1971).
66. Manz P., G. S. Xu, B. N. Wan, H. Q. Wang, H. Y. Guo, I. Cziegler, N. Fedorczak, C. Holland, S. H. Muller, S. C. Thakur, M. Xu, K. Miki, P. H. Diamond and G. R. Tynan, Zonal flow triggers the L-H transition in the experimental advanced superconducting tokamak, *Phys. Plasma*, 19, 072311 (2012).
67. Matsuda, K., and J.-Y. Hsu, Electron cyclotron damping in thermal and nonthermal plasma, *Phys. Fluids*, B3, 414 (1990).
68. Miyazawa, J., H. Yamada, S. Murakami, H. Funaba, S. Inagaki, N. Ohyabu, A. Komori, O. Motojima and LHD experimental group, Global confinement scaling for high-density plasmas in the Large Helical Device, *Plasma Phys. Control. Fusion* 48, 325 (2006).
69. Montgomery, D. C., *Theory of the Unmagnetized Plasma*, Gordon and Breach Science Publishers, New York, p.185 (1971).

70. Montgomery, D. and G. Joyce, Statistical mechanics of "negative temperature" states. *Phys. Fluids*, 17, 1139 (1974).
71. Montgomery, D. C., L. Turner and G. Vahala, Most probable states in magnetohydrodynamics, J. Plasma Phys., 21, 239 (1979).
72. Montogmery, D., W. T. Stribling, D. Martinez and S. Oughton, Relaxation in two dimensions and the Sinh-Poisson equation, *Phys. Fluids*, 4, 3, (1992).
73. Murray, J. D., *Asymptotic Analysis*, Clarendon Press, Oxford (1974).
74. O'Neil, T. M., New theory of transport due to like-particle collisions, *Phys. Rev. Lett.*, 55, 943 (1985).
75. O'Neil, T. M., Trapped plasmas with a single sign of charge, *Physics Today*, Feb p. 24 (1999).
76. Perkins, F. W., Heating tokamaks via the ion-cyclotron and ion-ion hybrid resonances, *Nucl. Fusion*, 17, 1197 (1977).
77. Press, W. H., S. A. Teukolsky, W. T. Vetterling and B. P. Flnnery, Numerical Recipes — the Art of Scientific Computing, Cambridge University Press, Third Edition in C++, (2007).
78. Powell, G. E. and I. C. Percival, A spectral entropy method for distinguishing regular and irregular motion of Hamiltonian systems, *J. Phys. A: Math. Gen.*, 12, 2053 (1979).
79. Pulinets, S. A., A. D. Legen'kab, T. V. Gaivoronskayab and V. Kh. Depuevb, Main phenomenological features of ionospheric precursors of strong earthquakes, *J. Atmospheric and Solar-Terrestrial Phys.*, 65, 1337 (2003).
80. Rosenbluth, M. N., Parametric instabilities in inhomogeneous media, *Phys. Rev. Lett.*, 29, 565 (1972).
81. Sauter, O. and Y. Martin, Considerations on energy confinement time scalings using present tokamak databases and prediction for ITER size experiments, *Nucl. Fusion*, 40, 955 (2000).
82. Schlichting, H., *Boundary-Layer Theory* (7th Edition), McGraw-Hill (1979).
83. Shaing, K. C. and E. C. Crume, Jr., Bifurcation theory of poloidal rotation in tokamaks: A model for L-H transition, *Phys. Rev. Lett.*, 63, 2369 (1989).
84. Shaing, K. C. and R. D. Hazeltine, Electron transport fluxes in potato plateau regime, *Phys. Plasma*, 4, 4331 (1997).
85. Shaing, K. C., M. S. Chu, C. T. Hsu, S. A. Sabbagh, J. C. Seol and Y. Sun, Theory for neoclassical toroidal plasma viscosity in tokamaks, *Plasma Phys. Control. Fusion*, 54, 124033 (2013).
86. Stamper, J. A., K. Papadopoulos, R. N. Sudan, S. O. Dean, E. A. McLean and J. M. Dawson, Spontaneous magnetic fields in laser-produced plasmas, *Phys. Rev. Lett.*, 26, 1012 (1971).
87. Stix, T., Radiation and absorption via mode conversion in an inhomogeneous collision-free plasma, *Phys. Rev. Lett.*, 15, 878 (1965).
88. Stix, T. H., *Waves in Plasmas*, American Institute of Physics (1992).
89. Sudan, R.N., Mechanism for the generation of 10^9 G magnetic fields in the interaction of ultraintense short laser pulse with an overdense plasma target, *Phys. Rev. Lett.*, 70, 3075 (1993).
90. Tachikawa M. and K. Fjuimoto, Emergence of multi-time scales in coupled oscillators with plastic frequencies, *Europhysics Lett.*, 78, 20004 (2007).
91. Tajima T. and J. M. Dawson, Laser electron accelerator. *Phys. Rev. Lett.*, 43, 267, (1979).
92. Tajima T., *Computational Plasma Physics*, Westview Press (2004).
93. Tatarakis, M., A. Gopal, I. Watts, F. N. Beg, A. E. Dangor, and K. Krushelnick, U. Wagner and P. A. Norreys, E. L. Clark, M. Zepf, R. G. Evans, Measurements of ultrastrong magnetic fields during relativistic laser–plasma interactions, *Phys. Plasma* 9, 2244 (2002).

94. Taylor, J. B. and B. McNamara, Plasma diffusion in two dimensions, *Phys. Fluids*, 14, 1492 (1971).
95. Terry, P. W. and P. H. Diamond, Theory of dissipative density-gradient-driven turbulence in the tokamak edge, *Phys. Fluids*, 28, 1419 (1985).
96. Thornton, S. T. and Jerry B. Marion, *Classical Dynamics of Particles and Systems*, Brooks/Cole, Cengage Learning, (2008).
97. Toi, K., J. Gernhardt, O. Kluber and M. Lornherr, Observation of precursor magnetic oscillations to the H-mode transition of the ASDEX tokamak, *Phys. Rev. Lett.*, 62, 430, 1989.
98. Vahala, G., Transport properties of the three-dimensional guiding-centre plasma, *J. Plasma Phys.*, 11, 159 (1974).
99. Weinstock, J., Formulation of a statistical theory of strong plasma turbulence, *Phys. Fluids*, 12, 1045 (1969).
100. White, R. B., D. A. Monticello, M. N. Rosenbluth and B. V. Waddell, Saturation of the tearing mode. *Phys, Fluids*, 20, 800 (1977).
101. Wilks, S. C., W. L. Kruer, M. Tabak, and A. B. Langdon, Absorption of ultra-intense laser pulses, *Phys. Rev. Lett.*, 69, 1383 (1992).
102. Wong, A. Y. , P. Y. Cheung, M. 3. McCarrick, J. Stanley, R. F. Wuerker, R. Close, B.
103. S. Bauer, E. Fremouw, W. Kruer, and B. Langdon, Large-scale resonant modification of the polar ionosphere by electromagnetic waves, *Phys. Rev. Lett.*, 63, 271 (1989).
104. Wu, Z., N. E. Huang, A Study of the Characteristics of White Noise Using the Empirical Mode Decomposition Method, *Proc. R. Lond. A* 460, 1597 (2004).
105. Yakhot, Victor, and Steven A. Orszag, *J. Scientific Computing*, 1, 3, (1986).
106. Yoshimura, Y., H. Igami, S. Kubo, T. Shimozuma, H. Takahashi, M. Nishiura, S. Ohdachi, K. Tanaka, K. Ida, M. Yoshinuma, Electron Bernstein wave heating by electron cyclotron wave injection from the high-field side in LHD, *Nucl. Fusion*, 53 063004 (2013).
107. Zagorodny A. and J. Weiland, Statistical theory of turbulent transport (non-Markovian effects), *Phys. Plasmas*, 6, 2359 (1999).

Program Index

Program Function	Description	Input	Page
FDTDmc	finite difference time domain simulation of mode conversion		226
FM	frequency mismatch resonance	pump amplitude	15
ForceFree	force free solution in toroidal geometry		338
gc2d	2d guiding center plasma diffusion		305
getA	example of asymptoloty		53
getB	homework 7.3 evaluate B field		240
getEIG	homework 2.5		57
getIMF	get intrinsic mode functions	pump amplitude	380
getK	homework 2.9		36, 58
gyration0	gyration in first order finite difference scheme		64
gyration1	gyration in Taylor fourth order finite difference scheme		65
gyration2	gyration by physical charateristics		66
gyration3	gyration in Runge Kutta fourth order	time step	81
gyroorbit	plot 3d gyro-orbit		31
h2p15	homework 2.15		60
HelicalB	plot flux funciton of a helical coil		33
HelmholtzCoil	B field in Helmholtz coil		82
HenonMap	Henon Map	number of data points	375
HH	homework 2.2 interaction potential of H atoms		56
HodgeDecomposition	Helmholtz Hodge decomposition	x,y,vx,vy,n,dx	104
Homework11p3	velcity space plot	pump amplitude	401
Homework11p5	Spectral Entropy		402
HW6p3	homework 6.3 composite solution		198
HW6p5	homework 6.5 composite solution		199
invGamma	inverse of Gamma function		18
KHIfft	Kelvin Helmholtz instability by FFT	rerun switch	109
Kink	internal kink	mode number	334
KxElectron	homework 5.15 perpendicular propagating electron waves		165
KxIon	perpendicular propagating ion waves		151
KzElectron	parallel propagating electron waves		148

Program Function	Description	Input	Page
ReleasedChain	homework 7.7 released chain		243
ResonanceLocalization	particle simulation of resonance localization	rerun switch	221
rk4	Runge Kutta fourth order algorithm	vector variable, time, time step	70
RTI	Raleigh Taylor instability	rerun switch	105
Separatrix	motion on the separatrix		185
ShootingMethod	shooting method		212
Simpson	Simpson integration	data, length of data, time interval	209
SpectralEntropy	Spectral Entropy		378
SPP	singular perturbative problem		182
Stairs	Fibonacci sequence		35
StandingWave	stochasticity in standing wave	pump amplitude	373
SurfaceKink	hmework 10.5		357
Taylor	Taylor state	lambda	274
Taylor4	Taylor expansion to the fourth order		64
TaylorGreenVortex	Taylor Green vortex solution	mode number	11
TestParticle	test particle pitch angle scattering	collision frequency	6
TPD	test particle diffusion	collision frequency, B field	13
TwoStream	two stream instability		140
VanderPol	Vander Pol equation		195
Vortex	vortex coalesce	rerun switch	112
WarmPlasmaVelocities	homework 5.17 velocities		166
WNSwhistler	wave normal surface of whistler wave		145

Index

www.ingramcontent.com/pod-product-compliance
Lightning Source LLC
Chambersburg PA
CBHW081501190326
41458CB00015B/5303